PHYLOGENETIC PERSPECTIVES ON THE VERTEBRATE IMMUNE SYSTEM

ADVANCES IN EXPERIMENTAL MEDICINE AND BIOLOGY

Recent Volumes in this Series

Volume 478
SHORT AND LONG TERM EFFECTS OF BREAST FEEDING ON CHILD HEALTH
Edited by Berthold Koletzko, Olle Hernell, and Kim Fleischer Michaelsen

Volume 479
THE BIOLOGY AND PATHOLOGY OF INNATE IMMUNITY MECHANISMS
Edited by Yona Keisari and Itzhak Ofek

Volume 480
BIOLOGY OF THE MAMMARY GLAND
Edited by Jan A. Mol and Roger A. Clegg

Volume 481
ELASTIC FILAMENTS OF THE CELL
Edited by Henk L. Granzier and Gerald H. Pollack

Volume 482
CHROMOGRANINS: Functional and Clinical Aspects
Edited by Karen B. Helle and Dominique Aunis

Volume 483
TAURINE 4: Taurine and Excitable Tissues
Edited by Laura Della Corte, Ryan J. Huxtable, Giampietro Sgaragli, and Keith F. Tipton

Volume 484
PHYLOGENETIC PERSPECTIVES ON THE VERTEBRATE IMMUNE SYSTEM
Edited by Gregory Beck, Manickam Sugumaran, and Edwin L. Cooper

Volume 485
GENES AND PROTEINS UNDERLYING MICROBIAL URINARY TRACT VIRULENCE: Basic Aspects and Applications
Edited by Levente Emődy, Tibor Pál, Jörg Hacker, and Gabriele Blum-Oehler

Volume 486
PURINE AND PYRIMIDINE METABOLISM IN MAN X
Edited by Esther Zoref-Shani and Oded Sperling

Volume 487
NEUROPATHOLOGY AND GENETICS OF DEMENTIA
Edited by Markus Tolnay and Alphonse Probst

Volume 488
HEADSPACE ANALYSIS OF FOODS AND FLAVORS: Theory and Practice
Edited by Russel L. Rous

PHYLOGENETIC PERSPECTIVES ON THE VERTEBRATE IMMUNE SYSTEM

Edited by

Gregory Beck
Manickam Sugumaran
University of Massachusetts Boston
Boston, Massachusetts

and

Edwin L. Cooper
University of California, Los Angeles
Los Angeles, California

Kluwer Academic / Plenum Publishers
New York, Boston, Dordrecht, London, Moscow

Library of Congress Cataloging-in-Publication Data

Phylogenetic perspectives on the vertebrate immune system/edited by Gregory Beck, Manickam Sugumaran, Edwin L. Cooper.
p. ; cm.—(Advances in experimental medicine and biology: v. 484)
Includes bibliographical references and index.
ISBN 0-306-46431-4
1. Immune system—Evolution—Congresses. 2. Invertebrates—Immunity—Congresses. I. Beck Gregory. Sugumaran, Manickam. III. Cooper, Edwin Lavern, 1919- .IV. Series.
QR182.2.E94 P494 2001
571.9′616—dc21

00-052732

Proceedings of the 1999 FASEB Summer Research Conference on Phylogenetic Perspectives on the Vertebrate Immune System, held July 11–16, 1999, at Copper Mountain, Colorado

ISBN 0-306-46431-4

233 Spring Street, New York, N.Y. 10013

http://www.wkap.nl/

10 9 8 7 6 5 4 3 2 1

A C.I.P. record for this book is available from the Library of Congress

Printed in the United States of America

PREFACE

This book contains the proceedings of the first meeting on invertebrate immunity ever sponsored as a summer research conference by the Federation of American Societies for Experimental Biology (FASEB). The conference was held in Copper Mountain, CO from July 11-16, 1999. It was a an extension of a New York Academy of Sciences meeting entitled "Primordial Immunity: Foundations for the Vertebrate Immune System" held on May 2-5, 1993 at the Marine Biological Laboratories in Woods Hole, MA. The proceedings of that meeting were published in *The Annals of the New York Academy of Sciences* (volume 712). At that meeting all the attendes agreed that this type of conference (a relatively small focused gathering) allowed for participation by investigators at all levels of their careers. We further agreed that we should search for a forum that would allow this meeting to continue. The FASEB Summer Research Conference was an excellent vehicle for this type of meeting. Furthermore, this year's participants decided to continue this meeting as a regularly scheduled FASEB sponsored event.

This was a unique conference in the sense that it focused upon mechanisms of development and defense in protostome and deuterostome invertebrates and lower vertebrates. There was a strong emphasis on evolutionary cell biology, phylogenetic inferences and the evolution of recognition and regulatory systems. Unlike typical immunology conferences which try to force the world of recognition within the dogmas of murine immunology, this conference appreciates the fact that organisms of distinct lineages, *e.g.*, insects as opposed to vertebrates, can have different induced protective mechanisms but may still share regulatory molecules such as neuropeptides and differentiation markers such as N-CAMs. The meeting was a relatively small one which brought together an international group of experts to discuss the origins of immunity in a setting that is conducive to intense exchange of ideas. There was no emphasis on current dogmas regarding the immune system, rather the focus of the conference was to consider the best information regarding molecular and cellular defenses and differentiation modes in lower animals and to attempt a synthesis of the general principles underlying recognition and defense in invertebrates and protochordates, the lineal precursors of vertebrates.

This conference was a vehicle for realizing the concept that invertebrates can contribute to the understanding and clarification of the immune/inflammatory response in higher animals. We expect this to lead to a more extensive development of invertebrate models, useful and unique in modern immunological and industrial laboratories. Speakers were drawn from diverse fields to afford interdisciplinary insights into evolution of antigen specific and nonspecific host defense systems. In this way we were able to stimulate further understanding of the origins of the molecules and cells of the immune system. Sessions included for example the following topics: MHC, transplantation-rejection, complement, coagulation, antibody-like molecules and the humoral response, cytokines, acute phase responses, the cells of the immune response and their activities, and lectins.

Survival depends upon an effective defense against foreign invaders. Host defense mechanisms have been evolving since the first organisms encountered one another hundreds of millions of years ago. All multi-cellular organisms face the survival necessity of distinguishing between self and nonself in response to parasitism or infection with microbial pathogens. Invertebrates are capable of recognition and defense as are the vertebrates, and, thus, lessons learned from invertebrates can provide a basis for strategies the solutions that vertebrate animals have adopted. In addition, invertebrates fall victim to neoplasms and attacks by viruses as well as bacteria. Investigations of the mechanisms of invertebrate strategies to thwart these affronts to homeostasis will surely shed light on the ways in which the vertebrate body deals with similar threats to survival. Data collected from studies of invertebrate defenses have extensive implications for understanding the evolution of immunity and problems of human health and disease.

Studies of invertebrate immunity are yielding new insights. First, new and diverse host defense functions characteristic of all living beings are being discovered. Second, we are analyzing the beginnings of immunity when we perform experiments using ancestral animals. Third, studying the relationship of parasite to host will contribute significantly to our understanding of diseases and processes that affect major portions of the earth's population. Fourth, new sources of therapeutic agents are be discovered in primitive species.

This is a time of great discoveries in innate immunity. Recent advances using the techniques of molecular biology and protein chemistry have led to rapid identification of products and characterization of mechanisms of immunoresponsiveness. Increasing public awareness is focused on not using controversial mammals for routine investigations. The high price of housing and feeding vertebrates renders the use of invertebrate model systems an attractive and perhaps necessary alternative. Invertebrate animals are noncontroversial, inexpensive and economical to care for, and to a certain degree, socially acceptable as experimental animals. In addition, invertebrate cell culture systems are as easily maintained and as practical and diverse as vertebrate systems.

We should not forget, as immunologists, that the first major breakthrough in basic immunology which heralded the cellular and humoral concepts to which we still adhere came from Metchnikoff's prescient experiments that revealed phagocytosis in simple invertebrates.

We would like to thank the Federation of American Societies for Experimental Biology, with special thanks to the Office of Scientific Meetings and Conference department members Geri Swindle, Jennifer LaFrankie, and especially Adele Hewitt. The conference was funded by generous donations from FASEB, the National Science Foundation, The International Society of Developmental and Comparative Immunology, and Merck & Co., Inc. I would also like to personally thank all the speakers who participated in the conference and for making it such a sucess. I would also like to thank my co-editors/organizors, Dr. Edwin Cooper and Dr. Manickam Sugumaran, without their help and input none of this would have been possible.

GREGORY BECK
University of Massachusetts Boston

TO REORDER YOUR UPS DIRECT THERMAL LABELS:

1. Access our supply ordering web site at **UPS.COM®** or contact UPS at 800-877-8652.
2. Please refer to label #0277400801 when ordering.

0277400801 RRDRF

CONTENTS

I. Evolution of Recognition Molecules

1. **New Major Histocompatibility Complex Class IIB Genes from Nurse Shark**
   Simona Bartl 1

2. **Evolutionary Factors in the Emergence of the Combinatorial Germline Antibody Repertoire**
   John J. Marchalonis, Miranda K. Adelman, Brian J. Zeitler, Paul M. Sarazin, P. Michael Jaqua, and Samuel F. Schluter 13

3. **Individual-Specific Repertoires of Immune Cells SRCR Receptors in the Purple Sea Urchin (*S. Purpuratus*)**
   Zeev Pancer 31

II. Cytokines and Antibacterial Peptides of Invertebrates

4. **Molecular Characterization of IL-1-like Molecules from Lower Vertebrates and Invertebrates**
   Gregory Beck and Nobel Truong 41

5. **The Detection and Isolation of a Novel Antimicrobial Peptide from the Echinoderm, *Cucumaria frondosa***
   Karen A. Beauregard, Nobel T. Truong, Haiyan Zhang, Wenyu Lin, and Gregory Beck 55

6. **Isolation and Characterization of an IL-1-like Protein from *Manduca sexta***
Nobel T. Truong, Karen A. Beauregard, Haiyan Zhang, Wenyu Lin, and Gregory Beck 63

7. **Clavanins and Styelins, α-Helical Antimicrobial Peptides from the Hemocytes of *Styela clava***
Robert I. Lehrer, In Hee Lee, Lorenzo Menzel, Alan Waring, and Chengquan Zhao 71

8. **The Discovery and Characterization of an Interleukin 6 Cytokine Family Antagonist Protein from a Marine Sponge, *Callyspongia* sp.**
Jane V. Peppard, Patricia Loo, Matthew A. Sills, Lawrence Wennogle, Amy Wright, Shirley Pomponi, and Maria Cueto 77

9. **Phylogeny of Vertebrate Cytokines**
C.J. Secombes, S. Bird, S. Hong, K.J. Laing, and J. Zou 89

10. **Antimicrobial Proteins in Crustaceans**
Valerie J. Smith and June R.S. Chisholm 95

11. **Form, Function and Phylogenetic Relationships of Mosquito Immune Peptides**
Carl A. Lowenberger 113

III. Cellular and Humoral Responses in Invertebrates

12. **Reactive Oxygen Species and Antimicrobial Defenses of Invertebrates: A Bivalve Model**
Robert S. Anderson 131

13. **A Role for Protease Inhibitors In Immunity of Long-Lived Animals**
Peter B. Armstrong and James P. Quigley 141

14. **Suppression of the *Drosophila* Cellular Immune Response by *Ganaspis xanthopoda***
Hsiling Chiu, Richard Paul Sorrentino, and Shubha Govind 161

15. **Annelid Humoral Immunity: Cell Lysis in Earthworms**
Edwin L. Cooper, Ellen Kauschke, and Andrea Cossarizza 169

16. **Horseshoe Crab Hemocyte-Derived Lectin Recognizing Specific O-Antigens of Lipopolysaccharides**
K. Inamori, T. Saito, D. Iwaki, T. Nagira, S. Iwanaga, F. Arisaka, and S. Kawabata 185

17. **A Family of C-type Lectins in *Manduca sexta***
Xiao-Qiang Yu and Michael R. Kanost 191

18. **Role of Tachylectins in Host Defense of the Japanese Horseshoe Crab *Tachypleus tridentatus***
S. Kawabata, H.-G. Beisel, R. Huber, W. Bode, S. Gokudan, T. Muta, R. Tsuda, K. Koori, T. Kawahara, N. Seki, Y. Mizunoe, S. N. Wai, and S. Iwanaga 195

19. **Hemolytic Capacity, Properties, and Activation Requirements of Cobra Venom Factor-Treated Chicken Serum**
T. L. Koppenheffer 203

20. **Molecular Studies of the Molluscan Response to *Digenean* Infection**
Eric S. Loker and Christopher J. Bayne 209

21. **Insect Lectins and Innate Immunity**
Shunji Natori 223

22. **Collagenous Lectins in Tunicates and the Proteolytic Activation of Complement**
David Raftos, Peter Green, Deepika Mahajan, Rebecca Newton, Sarina Pearce, Rodney Peters, Jennifer Robbins, and Sham Nair 229

23. **Immunological Properties of Heat Shock Proteins are Phylogenetically Conserved**
Jacques Robert, Antoine Ménoret, Pramod K. Srivastava, and Nicholas Cohen 237

24. ***Drosophila* Cellular Immune System: *Dorothy* Encodes a UDP Glycosyltransferase**
Z. Zhou, A. Rodriguez, C.-Y. Wu, and D. A. Kimbrell 251

25. **Eicosanoids in Insect Immune Signal Transduction**
D.W. Stanley and Ralph W. Howard 265

26. **Lectins from Tunicates: Structure-Function Relationships in Innate Immunity**
Gerardo R. Vasta, Michael S. Quesenberry, Hafiz Ahmed, and Nuala O'Leary 275

IV. Protease Cascades and Invertebrate Host Defenses

27. **Control Mechanisms of the Prophenoloxidase Cascade**
Manickam Sugumaran 289

28. **Activated Prophenoloxidase Engaged in the Cell Clump/Cell Adhesion of Coleopteran Insect, *Tenebrio molitor* Larvae**
Hyun Seong Lee, Mi Young Cho, and Bok Luel Lee 299

29. **Three Kinds of Encapsulation-Relating Factors in Coleopteran Insect, *Tenebrio molitor* Larvae**
Mi Young Cho, Hyun Seong Lee, and Bok Luel Lee 303

30. **A β-1,3-Glucan-Binding Protein from *Manduca sexta***
Congcong Ma and Michael R. Kanost 309

31. **Proteolytic Activation of Prophenoloxidase in an Insect *Manduca sexta***
Haobo Jiang, Yang Wang, and Michael R. Kanost 313

32. **Hemolymph Proteinases in Immune Responses of *Manduca sexta***
Michael R. Kanost, Haobo Jiang, Yang Wang, Xiao-Qiang Yu, Congcong Ma, and Yifei Zhu 319

33. **Cytotoxic Reactions Associated with Insect Immunity**
A. J. Nappi and E. Vass 329

34. **Genomic and cDNA Sequence of Prophenoloxidases from *Drosophila melanogaster***
Michael R. Chase and Manickam Sugumaran 349

35. **The Complement System in Sea Urchins**
L. Courtney Smith 363

36. **Paralytic and Plasmatocyte Spreading Peptide from *Manduca sexta***
Yang Wang, Haobo Jiang, and Michael R. Kanost 373

Contributors 377

Subject Index 379

NEW MAJOR HISTOCOMPATIBILITY COMPLEX CLASS IIB GENES FROM NURSE SHARK

Simona Bartl

Moss Landing Marine Laboratories, P.O. Box 450, Moss Landing, CA 95039-0450

INTRODUCTION

The Major Histocompatibility Complex (MHC) codes for proteins that process and present antigens to T cells. The actual presenting proteins are the classical MHC molecules and they are subdivided into two classes, I and II. Each class of MHC protein has an antigen binding cleft (ABC) that is formed by two membrane distal domains and that binds processed peptides. Classical MHC proteins tend to be highly polymorphic with most variable sites clustered around the ABC. This polymorphism has an effect on the repertoire of peptides that can bind and on the T cell receptors (TCRs) selected in the thymus by each MHC molecule (1, 2). The polymorphism appears to be generated by point mutations and recombination events (3-8). Analyses of mammalian genes suggests that blocks of residues can be transferred between MHC alleles or loci by recombination. These blocks have been identified as hypervariable

Phylogenetic Perspectives on the Vertebrate Immune System.
Edited by Beck *et al.*, Kluwer Academic/Plenum Publishers, 2001.

regions (HVRs) that are proposed to be preserved over time by selection for advantageous peptide binding (9). The ABC also contains conserved residues that serve as anchors by binding the main chain atoms of peptides and these appear conserved between mammals and cartilaginous fishes (10-12). The two membrane proximal domains form immunoglobulin (Ig) folds that display little inter-allelic polymorphism. Class II proteins are heterodimers of α and β chains, each chain contributing one domain for half of the ABC (called $\alpha 1$ and $\beta 1$) and one Ig-fold domain (called $\alpha 2$ and $\beta 2$). The genes for each chain, termed class IIA and IIB, are closely linked in the MHC of most vertebrates (13-15).

Additional proteins and genes that are structurally similar to the classical, antigen-presenting class I and II molecules have been found in a variety of vertebrates. These are termed non-classical MHC molecules since their functions vary. Although many non-classical class I genes have been found, only two sets of non-classical class II genes have been identified. In humans, they form two heterodimers, HLA-DM (encoded by DMA and DMB) and HLA-DO (encoded by DNA and DOB), that function together to edit and load peptides onto classical class II molecules (16-20). In contrast to classical MHC molecules, the non-classical molecules are less polymorphic and may lack the conserved peptide anchoring residues (21, 22).

MHC genes share many features in common across taxa (12). It appears that all jawed vertebrates have classical class I and class IIA and IIB genes. Multiple loci and highly polymorphic alleles appear to be common. Similarity in structural features, including conserved peptide binding residues, would argue that these genes code for proteins that function in the presentation of peptides to T cells. Additionally, it appears that all taxa with classical MHC molecules also possess some non-classical genes. Although non-classical class II molecules have only been demonstrated in mammals, phylogenetic analyses suggest an ancient lineage for DM genes indicating that homologs may be found in all the jawed vertebrates. Here, I report the isolation of two new class IIB genes from the nurse shark and analyze them with other class IIB genes, including two previously isolated shark clones and in the context of a possible evolutionary scheme.

STRUCTURAL FEATURES

Previously, two MHC class IIB genes were isolated from a spleen cDNA library of a single nurse shark (23). These clones, called 11 and 8, differed by 13% at the nucleotide level in the $\beta 1$ region with only 4 nucleotide differences in the rest of the protein-coding and the 3' untranslated (3' UT) regions. The clones reported here, named 1 and 2, were isolated from a second spleen cDNA library from a presumably unrelated individual using the $\beta 2$ region of clone 11 as a probe. Figure 1 shows an alignment of the nucleotide sequences of the new clones. Both sequences start within the signal peptide coding region. Nucleotide differences between the clones are spread other the entire sequence. The percent nucleotide divergence of the clones as compared to clone 8 is shown in Table 1. For primate MHC genes, intra-locus and inter-locus nucleotide

```
                                                                  Signal peptide <----||----> Beta 1
1     GCTGGGTGGCGTTGCTTCCCGCTCTCTGTGGATACGGATCCCTTTCCTTATCGCGGTCCTGGTTATCTTTAACGGCGAGAGGGACAGCGGAGCCGGTGCC   100
2                     ------C-------GC--------CC----C----G-ACAT-C-CCC--C----T--A-G---A---------A-T-AG--T

1     CATTGCTTCGATCAATTACATAGATGTGTGTTTAACAGCTCCAAGGACTGGGTCTTTCTAGAACAGTATGTTTATGACCAGGAGCTGATCGCGTATTATG   200
2     --C-C--AT-CAGGCA-CAC--CC----A------------GG----------A---------CT-------------------------------

1     ACTATAACCAGAGGAAGTACATCGCGGTGAAAGCCTGGATGAAGTCTAATGTGGACAGATGGAACAGTGAGGGCGCGGAAGAGCAGTACCAGGAAGCGAA   300
2     -----------------T---------------------------------------T---------------------G----G---------

                                                  Beta 1 <----||----> Beta 2
1     GGCTTACTGTGAACACAACATCCCGGTCATGTACGAGAGCGCCCTGGCCCGGCAAGTTGAACCAAAGGTGACAATCCGAACAAAAGAATCGACATATCCT   400
2     ---------C-GAGT---G--GG-A-T-C-G-GAGTGAG-TG---C-----AG----------CA--C--T------C-T----CC--TCA-C--T--

1     GGGCCCTCTGCCATACTCATTTGTTATGCTGTTGGATTTTACCCAGCAAAGATCAGTGTCACGTGGCTGAAAAATGGACAGAAGGTTTCTGATGCTGATG   500
2     ACA-AG-TG--TC-G---TCG---C---TGAC---------T--TC-TG----TGA-----T----T----------GCACCAA--C------G--TCA

1     TCACAGTTGAGTTGTTATCAAATGGTGACTGGACATATCAGGTTCGCCAGTATCTCCAGTATGAGCCAGTGTATGGTGATAAATACACCTGCCATGTGGA   600
2     ---ACAC--TCC--------G------------C--------GGAGG-TCTC---------C---------C---G--------T-----T-G------

                        Beta 2 <----||----> CP/TM/CY
1     ACACAGCAGCCTGACCTCACCAATGAGTGTGGATTGGGAAGTAGAGAGCACAAGTAAATCTGAAAAAACAAAGATCATTGTCGGAGTGCTGGGCTTTGGG   700
2     -------------GGAGT--T-G---------A-----G---C----T-TTCC-G----A-C-C-TC-C-----TG-A--T----CC---TT-----T-

1     TTTGGACTGATCATCTTGTTGGCTGGCGTGATAATGAGACTGAAGAATGCTAAAGCTATCCTGGATTCCA...ATCATGGCCCACGTCTGATGGGTCCAG   800
2     A-------A---------C-------A--T---------------------------T-----C----GCA-------------------A--A---

1     CTGTCAGTTAACTGATCACATCCACACAT.CAGCTGAATGATGGACAGAATTAAAACAGTTTGCAGACTCCTTGCAACTTATCCAGTGCAATTGCAGCTC   900
2     ---C-------TGG-TG-GAC-GG-GA---TG------C..--T-G--A---CTGC-C--..A-A-G---CT-----.----A--TG--CA---T--

1     ACTGGGCGTTAGAAGGCTTGGCCTTGGCTTTGTTTTGATTAACTTTACACTTGCAGTTATACACAGCTTTATATCCCACTTAAAACAAGAGGAGCATGTG  1000
2     --CA-ATA-CG-GCT----CA----------TGA-G-C--G-GG---ATG--ATGCAATGCT-TGA-AC-GC-CG-T-AC-TTT.T-TC-T-TTA-AACA

1     ACTTCTCAAACGTCCTGATCAGCGTGCCACCCTCCTCCCCCCTGTGCAGATGGGAAATAACTACATCTAACCACTGGACTTTGTTTCCAGTGCAGACTGT  1100
2     --C-G-A---ACAGA--C-T-AT--A-AG-TA-TG--A-AG-.-----.........-G-T--AC--CTC--T---C---CCCA-CA-CACC-TA-T--.

1     CTGGTAGGGTACATGCTTCTAGGCTCGTGCATCCAAACATGATTCTCCTGTGTTTTTAATGCTTTTATTTTCCTCTGTTCAGCCTTGTTGTAAATGTTTT  1200
2     ......AAA---TG-GA--AGT---T-G-TG-AG---....--CT-T-CTA---....-ACT----C---GT--G-TCA----TG--------------

1     AATCTCATTTGTAAATTCTTTGGGCTTTGTATCTAAATGAT.TTAATGCAGACTTTTTACCTCTTATGATCTCAAATGCAGCTGAAATGGTGGAACACTT  1300
2     G-AA--TACAT--TT-A---G-AAAA--T-----CTT-TG-A-CT-AA--A----AA-GAAA-AAT----T-T..........................

1     GATTAAAGAAATATTTGGC     1314
2     ...................     1224
```

Figure 1. Nucleotide sequences of nurse shark cDNA clones 1 and 2. Sequences were aligned using Clustal W. Dashes denote identity with the top sequence and dots are gaps in the alignment. Regions coding for the signal peptide, b1 domain, b2 domain, and the remaining domains [connecting peptide (CP), transmembrane (TM), and cytoplasmic (CY)] are defined above the sequences. Stop codons are underlined. GenBank accession numbers are AF194146 for clone 1 and AF194147 for clone 2.

differences are on the order of 0.1-6% and 5-11%, respectively, over all protein-coding regions (24). The percent divergence (7-9% overall) and the location of the divergent sites (mostly in the β1 domain) would indicate that clones 11, 8 and 1 are alleles. Clones 2 differs from clone 1 by 24% overall with differences spread over all the regions implying that clone 2 represents a new locus. However, for two distinct class IIB loci to be definitively defined in sharks awaits genetic mapping and segregation analysis. If clone 2 truly represents a new locus then cartilaginous fishes display a higher maximum proportion of polymorphic sites between alleles (6 *vs.* 9%) and between loci (11 *vs.* 24%) as compared to primates (24). This greater divergence may reflect longer time spans since the divergence of shark alleles and loci.

Table 1. Percent nucleotide variability by domain as compared to clone 8

	clone 8	clone 11	clone 1	clone 2
Signal Peptide	0	1	1	31
Beta 1	0	7	9	20
Beta 2	0	0	0	30
CP/TM/CY	0	1	0	21
3' UT	0	0	1	57

The deduced protein sequences for all known cartilaginous fish class IIB genes are aligned in Figure 2. Nucleotide differences in the β1 domain result in amino acid difference in the 3 alleles. Like the MHC alleles in other vertebrates the non-synonymous to synonymous substitution rate in this ABC-coding domain is greater than 1 and ranges from 5 to 7.7. Clone 2, the putative new locus, differs by many amino acids throughout the sequence. From comparisons of human and mice class IIB genes, Lundberg and McDevitt defined four hypervariable regions that varied within a species but were conserved across species (9). They performed statistical analyses that supported the direct descent of these regions and argued that the HVRs are maintained by selection for particular peptide binding capabilities. HVRI and HVRII were identified in both murine class IIB loci while HVRIII was identified in one locus. The last HVR (called either III or IV depending on the locus) covered a slightly different but overlapping region in each of the 2 loci (shown by the 2 underlines for HVRIV in Fig. 2). The first two HVRs (HVRI and HVRII) align well with regions of high amino acid variation in the 3 alleles. Two other regions of variability can be found shifted farther to the right of the mammalian HVR III and IV regions in the shark sequences (Fig. 2). The presence of HVRs implies similar processes for their generation and maintenance may be operating in lower vertebrates and that sequences at the HVRs may be conserved across species. However, a search of the current class IIB gene database did not find any similar sequences in bony fish or tetrapods genes. The shark HVRs may not be conserved over the large time scales that separate sharks from the other classes, or the selective agents that maintain the shark HVRs may not be operating in other taxa. Alternatively, as more fish class IIB genes are isolated, conserved HVR motifs may be found.

Structural features that are unique to the immunoglobulin family, in general, and to MHC genes, in particular, are found in the new class IIB clones as well. The two cysteines residues found approximately 55-60 amino acids apart form a disulfide bond and are found in both extracellular domains of classical class IIβ homologs (marked "*" in Fig. 2). In contrast, HLA-DMβ homologs contain 5 cysteines in the β1 domain (25-27). The presence of only 2 cysteines at positions that align with those in mammalian class IIβ molecules supports the idea that the shark clones are not DM homologs. Since both classical and non-classical (DOβ proteins) molecules are found that contain the two conserved cysteines, the shark clones could code for either of these homologs (28-31). The membrane proximal domains also contain conserved residues of the C1 set of immunoglobulin superfamily members, as do all MHC molecules (32). In addition to the two cysteines spaced 55-60 amino acids apart, Williams and Barclay (32)

```
BETA 1

           HVRI *  cho   HVRII                     HVRIII       HVRIV        *
11  DSGAGAHSEISLHRCVFNSTGDWVFLKQQVYDQELIAYYDYNQRKFIAVKAWMKSNVDRWNREGAEEQYESGKAYCEHNIPIVYESALARQV
8   -------CFDL-------------Y-E-Y------------------------------------------A-TF-------M-K-------
1   -------CFDQ--------SK-----E-Y----------------Y---------------S-------QEA---------VM---------
2   ---TE---YAGITT-E---S----Y-E-L------------------------------L-S-----R--EA----QS-VG-TESEV-P-R-
        |____sb_________________|                            p                   p

BETA 2

                        *                                                        *
11  EPKVTIRTKESTYPGPSAILICYAVGFYPAKISVTWLKNGQKVSDADVTVELLSNGDWTYQVRQYLQYEPVYGDKYTCHVEHSSLTSPMSVDW
8   ---------------------------------------------------------------------------------------------
1   ---------------------------------------------------------------------------------------------
2   --T----P-A-QHSTQL-L-S-HVT----PE-D-M-----APIP-GVINTV---D-------EDL---Q--S------R------R--R--E-
                            111  1          444444444444444                        1 1

CP/TM/CY

11  EVESTSKSEKTK    IIVGVLGFGFGLIILLAGVIM    RLKNAKAILDS.NHGPRLMGPAVS
8   ------------    ---------------------    -----------.------------
1   ------------    ---------------------    -----------.------------
2   --Q-IPE-AHP-    -V--A-F-VI-----------    -----------S-------ST-A-
```

Figure 2. Deduced amino acid sequences for new nurse shark MHC class IIB genes aligned with previously published genes. Each domain was aligned using Clustalw68k with minor adjustments performed by hand. Dashes denote identity with the top sequence and dots are gaps in the alignment. Potential glycosylation sites (cho), conserved cysteine residues (*) and regions that align with mammalian hypervariable regions (HVRI-IV) are marked above the top sequence. Residues that align with conserved sites in the C1 set of Ig family members (1), conserved peptide anchoring residues (p), and a major mammalian CD4-binding site (4) are labeled below the bottom sequence. A potential salt bridge (sb) is shown. Genbank accession numbers for clones 11 and 8 are L20274 and L20275, respectively.

identified an additional 8 amino acid positions that are characteristic of C1 set sequences. Residues at 6 of these positions are conserved in the class IIB clones (indicated with a "1" in Fig. 2).

Additional structural features are found in MHC proteins across taxa. The presence of glycosylation sites within certain domains seems to be conserved across taxa although the location can vary (labeled "cho" in Fig. 2). The shark sequences have a site at the same location as two bony fish class IIB sequences which is within one amino acid of the mammalian glycosylation site (12). Specific interaction sites between MHC class II proteins and the TCR co-receptor CD4 have been described. Based on alignments with mammalian sequences, a major CD4 binding site (marked "4" in Fig. 2) would be contained the β2 domain in the shark sequences (33, 34). This region is not conserved in bony fish sequences or in the shark sequences shown here (12). Specific residues that have been identified for binding peptides and anchoring them within the ABC (marked "p" in Fig. 2)(11). These anchoring residues are conserved for the classical MHC genes across taxa. The asparagine is conserved in all known classical class IIβ molecules. The tryptophan is conserved in the human, bass and in 3 shark class IIB sequences but not in zebrafish or clone 2 (12). If this site is examined in a variety of fish class IIB genes either aromatic or aliphatic amino acids are found (35). Neither site is conserved in DMβ molecules but both are found in DOβ molecules (25-31). Each of 4 salt bridges in the ABC forming domains is conserved within certain classes of mammalian MHC proteins. The bridges link the ends of two adjacent hairpins of the β sheet (36). Within the shark class II β1 domains a salt bridge can form between H7 and D32 (marked "sb" in Fig. 1) connecting two β-strands on the same side of the ABC. Corresponding salt links are conserved in many ABC-forming (membrane distal) domains. All vertebrate class I α1 and α2 domains and all bony and cartilaginous fish class II α1 domains have these salt links. In the class II β1 domains, they are found only in the shark and not the bony fish or tetrapod sequences (12). Since all ABC-forming domains in cartilaginous fish class I and II molecules contain this potential salt bridge, it may represent the primordial condition. Thus, throughout the jawed vertebrates class IIβ molecules share many common features. The only obivious exception is a conserved CD4-binding site.

EVIDENCE FOR RECOMBINATION

Both recombination and point mutations have been implicated in the creation of new mammalian MHC alleles (3-8). Recombination events seem to occur en bloc and leave a characteristic patchwork pattern (5-7, 37). As more shark MHC genes were analyzed, an obvious patchwork pattern of sequence motifs became apparent. This was most striking in class I sequences (38, 39). Within the shark class IIA genes, intraexonic recombination may be enhanced by a minisatellite core-like sequence in all shark class IIA sequences that is similar to a recombination promoting minisatellite sequence in mammalian class IIB genes (40). Preliminary evidence from the limited number of class IIB genes shown here supports the presence of sequence blocks that may be the result of recombination. These are seen in Figure 2 as the 2 regions that align with mammalian HVR I and II and the 2 additional regions of hypervaribility near the carboxy-terminal end of the β1 domain. Lundberg and McDevitt (9) provided evidence for ancient origins of HVRs and predicted that sequences similar to mammalian HVRs would be found in lower vertebrates. Finding similar motifs in more distantly related chondrichthyans such as the skates and rays may establish the evolutionary duration for MHC motifs. Thus, recombination as a mechanism for the generation of MHC allele diversity may operating in sharks as it does in mammals.

In addition to diversifying shark MHC genes, recombination also may be homogenizing these loci. The most striking examples are the three class IIB alleles shown here and two previously reported class I alleles with remarkably similar regions coding for membrane proximal domains and identical 3' untranslated regions (23, 38). Although clones 8 and 11 came from the same individual, clone 1 was isolated from a second, supposedly unrelated individual. This data along with previous Southern blot data using a class IIB 3' UT probe that detected identical bands in 2 other unrelated individuals indicates that this region is widely found within the species (23). Different genes with identical UT regions would argue for very recent recombination events with insufficient time for sequence divergence. On the other hand, evidence for shared motifs in the ABC of class I genes from widely divergent shark species (≥ 140 mya) would argue for the selected maintenance of advantageously recombined genes for very long time periods. The trans-species persistence of motifs as is seen in sharks has also previously been documented in teleosts (41).

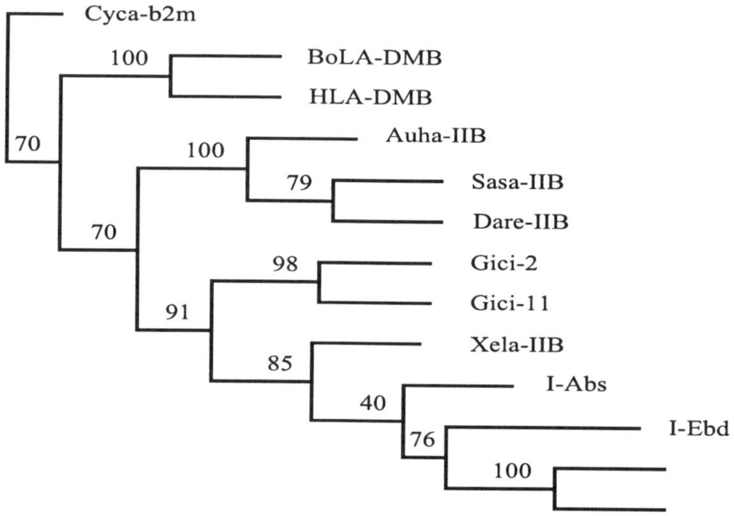

Figure 3. Phylogenetic relationships among selected MHC class IIB sequences. A comparison of fish and tetrapod class IIB genes is shown. The tree was constructed by the neighbor joining method (42) from deduced amino acid sequences of the β2 domains. Numbers at nodes indicate percent recovery of that node over 1000 bootstrap replications using Phylip 3-572. Sequences are for human DOB (HLA-DOB)(29), sheep DOB (Ovar-DOB)(30), mouse class IIB [I-Ebd(44) and I-Abs(45)], frog class IIB (Xela-IIB)(46), shark class IIB clone 11 (Gici-11)(23), zebrafish class IIB (Dare-IIB)(47), salmon class IIB (Sasa-IIB)(48), and cichlid class IIB (Auha-IIB) (49). Carp β2-microglobulin (Cyca-b2m)(50) was used as the outgroup.

PHYLOGENETIC ANALYSIS

Phylogenetic trees constructed by the neighbor joining method (42) produced three distinct clades of class IIβ molecules. One clade contains tetrapod (both classical and DO homologs) and shark genes. This grouping suggests that the shark and tetrapod genes share a direct ancestor and that the DO homologs represent more recently evolved non-classical genes. The second clade consists of bony fish genes which arose from an ancestor that was distinct from that of the shark/tetrapod lineage. The third cluster has the DM homologs implying an ancient origin for these non-classical genes and suggesting their presence in fish as well as mammals. Phylogenetic trees of class IIα molecules show the exact same clustering pattern as shown here (12).

MHC LOCI

Most vertebrates possess 1 to 3 functional loci for classical class IIB genes. It has been hypothesized that the number of loci is limited by functional considerations: large number of co-dominantly expressed proteins may severely limit the TCR repertoire selected in the thymus (3). In sharks the number of loci appears to be limited to no greater than 2. The houndsharks studied possess either 1 or 2 class I loci (38). The nurse sharks studied also have 1 or 2 class IIA loci (40). The number of bands seen on a Southern blot using a nurse shark class IIB gene probe is consistent with one or two loci (23). The presence of the highly divergent clone 2 also supports the presence of two class IIB loci. If clones 2 and 11 represent separate loci, they appear to have diverged along the chondrichthyan lineage (Fig. 3). Thus it appears the number of shark loci compares well with tetrapod loci. In contrast, some but not all teleosts studied contain much higher number of MHC loci. For example, at the population level, 17 different class IIB loci were identified in cichlids with each individual carrying 1-13 loci (43).

EVOLUTION OF MHC GENES

Since class I, class IIA and class IIB genes have been isolated from cartilaginous fishes it is clear that at least one ancestor for each was present prior to the evolution of jawed vertebrates. From analyses here and elsewhere (12) it appears that in fact at least 3 class IIB ancestors had emerged at this early time point. These ancestors and their descendents are shown schematically as lineages 1, 2 and 3 in Figure 4. Ancestor 1 gave rise to the most ancient clade in Figure 3 represented by DM homologs. Ancestors 2 and 3 split more recently and gave rise to the bony fish and the tetrapod/shark gene lineages, respectively. The orthologs of teleost class IIB genes may have been lost from the elasmobranchs and tetrapods or they await discovery. The same would be true for the descendants of ancestor 3 in the bony fishes. I would argue that the primordial ancestor of all MHC genes codes for the salt links that are observed in all present-day shark MHC genes. After ancestor 1 split off, the ABC of ancestor 2/3 evolved the peptide anchoring residues and polymorphic sites found in all classical

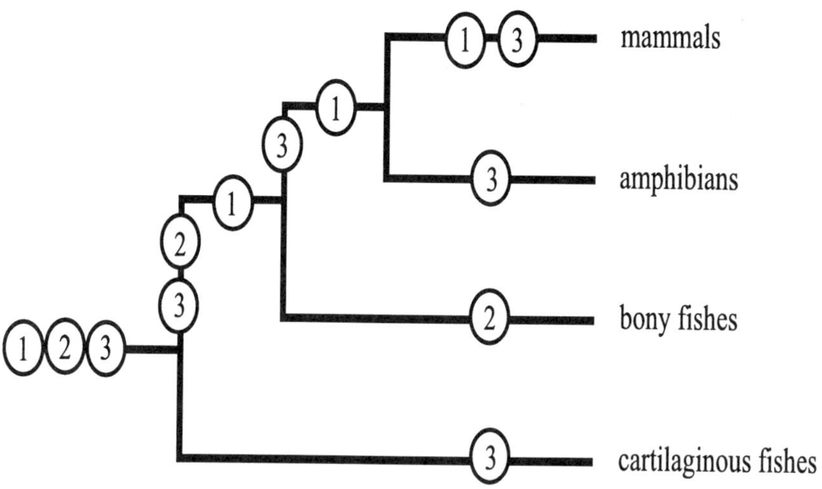

Figure 4. A schematic diagram of an evolutionary scenario for MHC class IIB molecules. From the phylogenetic tree (Fig. 3) three ancestors are inferred (numbered 1-3). The ancestors (shown on the left) and all of their descendents are shown along the vertebrate lineages by circles with the appropriate number.

MHC molecules. Alternatively, ancestor 1 may have lost these classical MHC features. The salt links seen in all shark MHC molecules would have been last in the descendants of ancestor 2 and along the tetrapod lineage of ancestor 3.

CONCLUSIONS

In many respects, the shark MHC class IIB genes are similar to classical tetrapod class IIB genes. First, shark class II genes cluster with the tetrapod classical and DO homologs. Like most tetrapods and unlike some bony fishes, sharks appear to have only a few (possibly 2) class IIB gene loci. Conversely, in some ways the shark genes reflect trends found in bony fish genes and are unlike tetrapod genes. For example, the shark genes do not exhibit an obvious CD4 binding site. Clearly, many features of shark MHC genes are unified by their persistence over long time periods. These include the salt bridge at the beginning of the membrane distal domains found in all shark MHC genes that may have been present in the original MHC ancestor. Also, distinct hypervariable regions and sequence motifs that appear the result of ancient recombination events in the ABC appear to be conserved over time. In contrast, almost identical 3' untranslated regions shared by 3 of 4 clones indicate that recent, homogenizing recombination events also have occurred.

ACKNOWLEDGEMENTS

I would like to thank Dr. Martin Flajnik for providing the nurse shark spleen cDNA library and Dr. Jonathan Geller for his enduring support and invaluable comments. I acknowledge the technical assistance of Amber Richard and the financial support of NSF (MCB-96-29300).

REFERENCES

1. Robey E, Fowles BJ. Selective events in T cell development. Annu Rev Immunol 1994;**12**:675-705.
2. Rothbard JB, Gefter ML. Interactions between immunogenic peptides and MHC proteins. Annu Rev Immunol 1991;**9**:527-565.
3. Lawlor DA, Zemmour J, Ennis PD, Parham P. Evolution of class-I MHC genes and proteins: from natural selection to thymic selection. Annu Rev Immunol 1990;**8**:23-63.
4. Klein J, Satta Y, O'hUigin C, Takahata N. The molecular descent of the major histocompatibility complex. Annu Rev Immunol 1993;**11**:269-295.
5. Gyllensten UB, Sundvall M, Erlich HA. Allelic diversity is generated by intraexon sequence exchange at the *DRB1* locus of primates. Proc Natl Acad Sci USA 1991;**88**:3686-3890.
6. Gaur LK, Nepom GT. Ancestral major histocompatibility complex *DRB* genes beget conserved patterns of localized polymorphisms. Proc Natl Acad Sci USA 1996;**93**:5380-5383.
7. Brunsberg U, Edfors-Lilja I, Andersson L, Gustafsson K. Structure and organization of pig *MHC* class II *DRB* genes: evidence for genetic exchange between loci. Immunogenetics 1996;**44**:1-8.
8. Ohta T. On the pattern of polymorphisms at major histocompatibility complex loci. J Mol Evol 1998;**46**:633-638.
9. Lundberg AS, McDevitt HO. Evolution of major histocompatibility complex class II allelic diversity: Direct descent in mice and humans. Proc Natl Acad Sci USA 1992;**89**:6545-6549.
10. Madden DR, Gorga JC, Strominger JL, Wiley DC. The three-dimensional structure of HLA-B27 at 2-1Å resolution suggests a general mechanism for tight peptide binding to MHC. Cell 1992;**70**:1035-1048.
11. Brown JH, et al. Three-dimensional structure of the human class II histocompatibility antigen HLA-DR1. Nature 1993;**364**:33-39.
12. Bartl S. What sharks can tell us about the evolution of MHC genes. Immunol Rev 1998;**166**:317-332.
13. Klein J. Natural History of the Major Histocompatibility Complex. New York:Wiley and Son; 1986.
14. Trowsdale J. "Both man & bird & beast": comparative organization of MHC genes. Immunogenetics 1995; **41**:1-17.
15. Jones EY. MHC class I and class II structures. Curr Opin Immunol 1997;**9**:75-79.
16. Kroppshofer H, Hammerling GJ, Vogt AB. How HLA-DM edits the MHC class II peptide repertoire: survival of the fittest? Immunol Today 1997;**18**:77-82.
17. Kropshofer H, Vogt AB, Moldenhauer G, Hammer J, Blum JS, Hammerling GJ. Editing of the HLA-DR-peptide repertoire by HLA-DM. EMBO J 1996;**15**:6144-6154 .
18. Vogt AB, Moldenhauer G, Hammerling GJ, Kropshofer H. HLA-DM stabilizes empty HLA-DR molecules in a chaperone-like fashion. Immunol Lett 1997;**57**:209-211.
19. Kropshofer H, Vogt AB, Thery C, Armandola EA, Li BC, Moldenhauer G, Amigorena S, Hammerling GJ. A role for HLA-DO as a co-chaperone of HLA-DM in peptide loading of MHC class II molecules. EMBO J 1998;**17**:2971-2981.
20. Liljedahl M, Kuwana T, Fung-Leung WP, Jackson MR, Peterson PA, Karlsson L. HLA-DO is a lysosomal resident which requires association with HLA-DM for efficient intracellular transport. EMBO J 1996;**15**:4817-4824.
21. Parham P. The rise and fall of great class I genes. Semin Immunol 1994;**6**:373-382.
22. Kasahara M, Flajnik MF, Ishibashi T, Natori T. Evolution of the major histocompatibility complex: a current overview. Transplant Immunology 1995;**3**:1-20.
23. Bartl S, Weissman IL. Isolation and characterization of major histocompatibility complex class IIB genes from the nurse shark. Proc Natl Acad Sci USA 1994;**91**:262-266.
24. Lawlor DA, Ward FE, Ennis PD, Jackson AP, Parham P. HLA-A and B polymorphisms predate the divergence of humans and chimpanzees. Nature 1988;**335**:268-271.

25. Kelley AP, Monaco JJ, Cho S, Trowsdale J. A new human HLA class II-related locus, *DM*. Nature 1991;**353**:571-573.
26. Cho S, Attaya M, Monaco JJ. New class II-like genes in the murine MHC. Nature 1991;**353**:573-576.
27. Niimi M, Nakai Y, Aida Y. Nucleotide sequences and the molecular evolution of the DMA and DMB genes of the bovine major histocompatibility complex. Biochem Biophys Res Commun 1995;**217**:522-528.
28. Tonnelle C, DeMars R, Long EO. DO beta: a new beta chain gene in HLA-D with a distinct regulation of expression. EMBO J 1985;**4**:2839-2847.
29. Servenius B, Rask L, Peterson PA. Class II genes of the human major histocompatibility complex. The DO beta gene is a divergent member of the class II beta gene family. J Biol Chem 1987;**262**:8759-8766.
30. Wright H, Redmond J, Ballingall KT. The sheep orthologue of the HLA-DOB gene. Immunogenetics 1996;**43**:76-79.
31. Larhammar D, Hammerling U, Rask L, Peterson PA. Sequence of gene and cDNA encoding murine major histocompatibility complex class II gene A beta 2. J Biol Chem 1985;**260**:14111-14119.
32. Williams AF, Barclay AN. The immunoglobulin superfamily-domains for cell surface recognition. Ann Rev Immunol 1988;**6**:381-405.
33. König R, Huang L-Y, Germain RN. MHC class II interaction with CD4 mediated by a region analogous to the MHC class I binding site for CD8. Nature 1992;**356**:796-798.
34. Cammarota G, et al. Identification of a CD4 binding site on the β2 domain of HLA-DR molecules. Nature 1992;**356**:799-801.
35. Graser R, O'hUigin C, Vincek V, Meyer A, Klein J. Trans-species polymorphism of class II *Mhc* loci in danio fishes. Immunogenetics 1996;**44**:36-48.
36. Brown JH, Jardetzky T, Saper MA, Samraoui B, Bjorkman PJ, Wiley DC. A hypothetical model of the foreign antigen binding site of class II histocompatibility molecules. Nature 1988;**332**:845-850.
37. Ohta T. On the evolution of multigene families. Theor Popul Biol 1983;**23**:216-240.
38. Okamura K, Ototake M, Nakanishi T, Kurosawa Y, Hashimoto K. The most primitive vertebrates with jaws possess highly polymorphic MHC class I genes comparable to those of humans. Immunity 1997;**7**:777-790.
39. Bartl S, Baish MA, Flajnik MF, Ohta Y. Identification of class I genes in cartilaginous fish, the most ancient group of vertebrates displaying an adaptive immune response. J Immunol 1997;**159**:6097-6104.
40. Kasahara M, McKinney EC, Flajnik MF, Ishibashi T. The evolutionary origin of the major histocompatibility complex: polymorphism of class II α chain genes in the cartilaginous fish. Eur J Immunol 1993;**23**:2160-2165.
41. Hansen JD, Strassburger P, Du Pasquier L. Conservation of an alpha 2 domain within the teleostean world, MHC class I from the rainbow trout *Oncorhynchus mykiss*. Dev Comp Immunol 1996;**20**:417-425.
42. Saitou N, Nei M. The neighbor-joining method: A new method for constructing phylogenetic trees. Mol Biol Evol 1987;**4**:406-425.
43. Málaga-Trillo E, et al. Linkage relationships and haplotype polymorphism among cichlid *Mhc* class II *B* loci. Genetics 1998;**149**:1527-1537.
44. Saito H, Maki RA, Clayton LK, Tonegawa S. Complete primary structures of the E β chain and gene of the mouse major histocompatibility complex. Proc Natl Acad Sci USA 1983;**80**:5520-5524.
45. Estess P, Begovich AB, Koo M, Jones PP, McDevitt HO. Sequence analysis and structure-function correlations of murine q, k, u, s, and f haplotype I-A beta cDNA clones. Proc Natl Acad Sci USA 1986;**83**:3594-3598.
46. Sato K, Flajnik MF, Du Pasquier L, Katagiri M, Kasahara M. Evolution of the MHC: isolation of class II β-chain cDNA clones from the amphibian Xenopus laevis. J Immunol 1993;**150**:2831-2843.
47. Ono H, et al. Major histocompatibility complex class II genes of zebrafish. Proc Natl Acad Sci USA 1992;**89**:11886-11890.
48. Hordvik I, Grimholt U, Fosse VM, Lie Ø, Endresen C. Cloning and sequence analysis of cDNAs encoding the MHC class II β chain in Atlantic salmon (*Salmo salar*). Immunogenetics 1993;**37**:437-441.
49. Klein D, Ono H, O'hUigin C, Vincek V, Goldschmidt T, Klein J. Extensive MHC variability in cichlid fishes of Lake Malawi. Nature 1993;**364**:330-334.
50. Dixon B, Stet PJM, van Erp SHM, Pohajdak B. Characterization of β2-microglobulin gene in the zebrafish. Immunogenetics 1993;**38**:27-34.

EVOLUTIONARY FACTORS IN THE EMERGENCE OF THE COMBINATORIAL GERMLINE ANTIBODY REPERTOIRE

John J. Marchalonis, Miranda K. Adelman, Brian J. Zeitler, Paul M. Sarazin, P. Michael Jaqua, and Samuel F. Schluter

Microbiology and Immunology College of Medicine, University of Arizona, Tucson AZ 85724

INTRODUCTION

Although life began on earth approximately 3.5 billion years ago, the combinatorial immune response apparently arose in a "big bang" approximately 450 million years ago, [1-4] coincident with the emergence of jawed vertebrates. Preceding this event was the so-called Cambrian explosion occurring approximately 545 million years ago that resulted in the seemingly rapid appearance of virtually all living forms as represented by the fossil record [5, 6]. However, molecular investigations seeking to calibrate evolutionary clocks and analyze phylogenetic relationships indicate that the explosive phases of evolution implied by the fossil record may have been preceded by extended periods of inconspicuous innovation [5, 6] in possible living organisms that

Phylogenetic Perspectives on the Vertebrate Immune System.
Edited by Beck *et al.*, Kluwer Academic/Plenum Publishers, 2001.

did not become part of the currently available fossil record. The necessary elements of the combinatorial immune system, immunoglobulins (Igs), T-cell receptors (TCR), MHC products and recombinase activator genes (RAG) are clearly present in even the most primitive jawed vertebrates, the chondrichthian fishes [7-10] which appeared in evolution approximately 450 million years ago. Definitive evidence for these elements is thus far lacking in agnathan vertebrates and in lower deuterostomes. Nevertheless, many primordial elements upon which the combinatorial system is built may well have preceded the split in evolution between protostomes and deuterostomes and their origins may even extrapolate back to ancient times corresponding to the origin and evolution of bacteria.

Here, we will consider four aspects of the "early" evolutionary emergence (Figure 1) of the combinatorial immune system. **Genesis**: This period lasted approximately 3 billion years and includes the emergence of archaeobacteria, eubacteria, and eucaryotes, horizontal transfer of organelles such as mitochondria, and widespread horizontal transfer of genes among these emerging and evolving organisms [11]. **Big Bang:** This is the emergence of the combinatorial immune system following from duplication of primordial immunoglobulin genes and the horizontal transfer of site -specific recombination mechanisms [2-4, 12, 13]. This complex event lasted an order of magnitude of ten million years. **Equilibration under Stringent Selection:** Once *bona fide* immunoglobulin genes had developed, they were under stringent evolutionary selective pressure based upon their capacity to recognize antigen and carry out effector functions. A time estimate for this is approximately ten million years duration. **Evolution:** The T-cell receptor and immunoglobulin domains present in the ancestral sharks and their kin underwent evolution under relatively modest selective pressure consistent with the evolution within that group and also the emergence of the other vertebrate classes. This process has continued for approximately 450 million years.

GENESIS

The Ig "domain" defined by the Ig fold was first thought to be restricted to extracellular spaces [14] where it was involved in antigen recognition either by lymphocytes (TCR or Igs), or by antigen presenting cells (MHC). Based upon current evidence, *bona fide* immunoglobulin V and C (C1) domains are restricted to jawed vertebrates and to specific immune type recognition functions. Ig domains of C2 type and strongly related molecules such as fibronectin type III (Fn3), however, are widely dispersed intracellularly, as well as extracellularly, and occur in species as primitive as bacteria [15, 16]. Overall, the Ig superfamily includes not only the *bona fide* immunoglobulins but hundreds of proteins involved in animal cell-cell attachment of proteins and in communication among cells. The intracellular domains are found in large structures including skeletal muscle C-protein, smooth muscle light chain kinase and long connection proteins such as titin, a vertebrate muscle, twitchin of the nematode *C. elegans*, as well as projectin of *Drosophila*. These large intracellular muscle type molecules consist of regular arrays of Fn3 and Ig C2 domains. Recognition structures involved with protein kinases or phosphatases likewise can consist of runs of these domains associated in some cases with lectin domains as well as transmembrane and

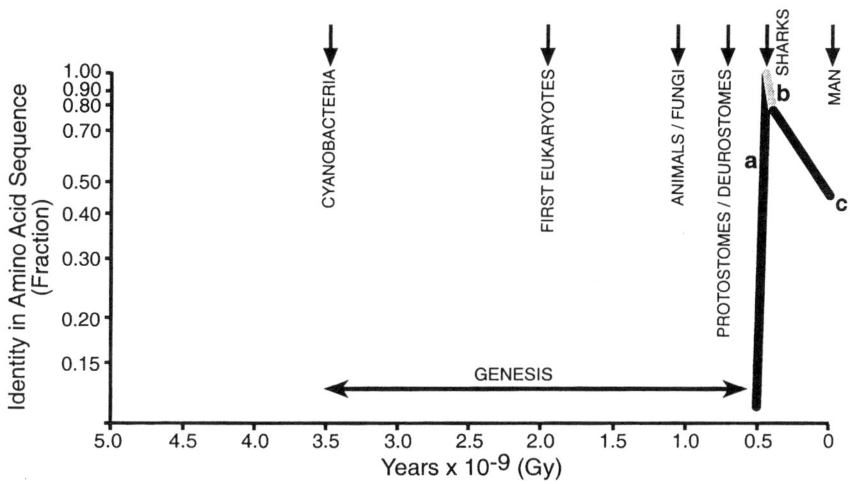

Figure 1. Diagram to scale of the evolutionary appearance of living organisms, divergences of major groups and the rapid emergence (a), and the rapid equilibration under stringent selection (b) and evolution (c) of the combinatorial immune system of jawed vertebrates Based upon Marchalonis and Schluter [2].

intracellular kinases. Immunoglobulins and T-cell receptors, by contrast, have never been observed to form such long run-on-arrays and do not form covalent associations with other domains. It has been suggested that the Ig-C2 and Fn domains, most probably arose from an ancestral form that resembled bacterial PapD domains [16], based upon 3-dimensional structures which erode less rapidly in evolution than do amino acid sequences [15]. The peptide binding domains of MHC products are not related to immunoglobulin domains although the membrane proximal Ig-C-like domains meet the statistical criteria for inclusion with the combinatorial immunoglobulins. A bacterial chaperone has a structure remarkably similar to the immunoglobulin fold, but has no detectable sequence homology [17]. Thus, molecules having immunoglobulin superfamily domains are widely spread, if not ubiquitous among living organisms, but the *bona fide* recognition and effector Ig domains are apparently restricted to jawed vertebrates.

Table 1 illustrates both ancient and relatively universal features that preceded the origin of jawed vertebrates in evolution and may have been retained partially or completely in the latter system. Klein [18] and Hughes [19] question the actual homology of recognition in effector systems in protostomes and acoelomate invertebrates to the molecules of the immune system. We [1, 2] agree with the overall distribution of elements as stated by Klein [18, 20], but have proposed that the combinatorial system coopted existing mechanisms for recognition and cell differentiation. Certainly, phagocytic cells are universal and lectin type recognition is probably a wide-spread feature of living multicellular organisms. Acute phase proteins such as pentraxins occur in arthropod species such as the horseshoe crab (an ancient arachnoid), the ancestral divergence of which certainly preceded the deuterostome/ protostome split. Complement components and their related molecules including $\alpha 2$

macroglobulin are of a relatively ancient origin which clearly preceded the emergence of jawed vertebrates [21-23]. These molecules provide an effector bridge not only to antibodies but to pentraxins including C-reactive protein because both fix complement by the classical pathway. As *caveats*, however, there are inducible anti-bacterial systems that clearly are not related to the combinatorial immune system although they are considered parts of "innate immunity" [24-26]. A clear example of an inducible defense system that is not related to the immune system is that of the cytochrome oxidases involved in detoxification in the liver; e.g. cytochrome P-452E1 normally occurs at constitutive levels but is inducible by high levels of ethanol, which it oxidizes as a specific substrate [27]. Furthermore, the induction of Nf-κB does not establish a conclusive tie-in with "immunological" mechanisms because this transcription factor is generated in response to various kinds of stress such as oxidative stress and is manifested by hepatic tumors and colon cancers in response to bile salts and oxidative stresses [28].

Table 1. Comparison of Defense Reactions

	Insect/Arthropod	Lower Chordate Cyclostome	Sharks
Innate System			
Phagocytic cells	Yes	Yes	Yes
Inducible inflammatory system (cytokines, Toll, NfκB)	Yes	Yes	Yes
Acute Phase Proteins	pentraxins	Yes	Yes
	Ig superfamily	Yes	Probably
Complement/ α2 macroglobulin	Yes	Yes	Yes
Combinatorial Immune System ("adaptive")			
Lymphocytes/ lymphoid system	No	Yes	Yes
Antibodies (bona fide V and C1 domains)	No	?	Yes
T-cell Receptors (11)	No	?	Yes
Major histocompatibility complex (MHC)[Ig-C1]	No	?	Yes
RAG genes	No	No	Yes

While the lamprey, in particular, has gut associated lymphoid tissue and cells with the morphological appearance of plasma cells [29], the extant cyclostomes, hagfish and lampreys, present an enigma because, while they are clearly vertebrates, they appear not to possess the vertebrate type immune system. Although there have been a number of reports of inducible antigen specific molecules in these species [30-32], the levels are low there have been no definitive characterization of either the expressed molecules or gene specifying immunoglobulins, TCRs, MHCs, or RAGs. In contrast to these results, there are numerous reports of molecules found in cyclostomes which are highly conserved throughout the vertebrates. In a collaborative study, Whitfield and Dang (K. Whitfield and H. Dang, unpublished observations) have obtained a lamprey cDNA clone which is clearly a homolog of the human Vitamin D receptor (VDR). The overall homology between the lamprey homolog and the human VDR was 59%, with the functional domains for DNA binding (two zinc finger motifs) the 1,25 $(OH)_2$ Vitamin D_3 hormone ligand binding exhibiting 84% homology. Between these two domains, however, was a region of poor conservation (less than 20% homology), including a significant difference in the number of residues with the lamprey clone having 24 fewer amino acids in this region. Preliminary functional testing of a full-length lamprey clone in a mammalian transfection system indicates that the lamprey VDR homolog binds 1-25$(OH)_2$ D_3 with high affinity, justifying its designation as a *bona fide* receptor for Vitamin D.

Similarly, a cDNA fragment corresponding to a highly evolutionarily conserved region of the major anion transport protein AE1 (band 3) was cloned from lamprey mRNA using PCR homology probing [33]. The lamprey fragment consisted of 481 bases in the coding region with the fragment sharing 74% identity to the human homolog at the nucleic acid level and 73% at the amino acid level (Figure 2). The translated fragment corresponded to the membrane-associated segment and contained an external loop, two transmembrane helices and two internal structures. In comparisons of the homologous regions of band 3 molecules of lamprey, human, mouse, chicken and trout, 76 residues or 47% were absolutely conserved indicating conserved function of the segment in all the species. The lamprey AE1 fragment overall showed 63% identity to the human homolog. In addition, other workers have recently isolated a lamprey cDNA clone from leukocyte mRNA that is homologous to IL-8 of chicken (40% identity) and human (32-33% identity) [34].

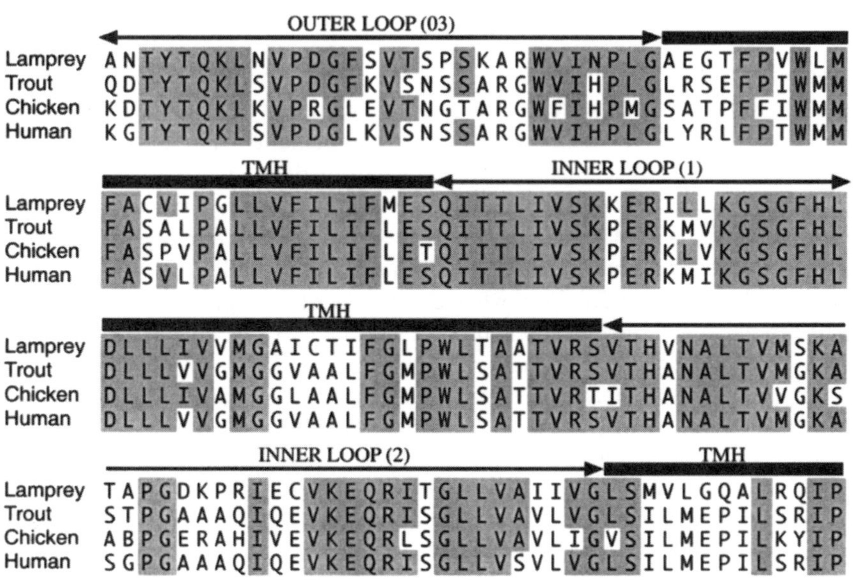

Figure 2. Alignment of the translated cDNA fragments corresponding to the membrane associated segment of the major anion transport protein AE-1 from lamprey and jawed vertebrate species. (Data taken from [33]).

BIG BANG

As outlined in Table 1 above, the essential genetic elements of the combinatorial immune system are present as a complete ensemble in the most primitive extant jawed vertebrates, the sharks and their kin, but apparently are not present in the even more primitive living vertebrates the ostracoderm-derived agnathans. Based upon the evolutionary distribution of *bona fide* Ig domains and RAG genes [1, 2] and functional studies of these [12, 13], it was suggested that the combinatorial immune response emerged in evolution as the result of an accidental event in which genes capable of catalyzing site-directed recombination and duplication were transferred from microbes to ancestral vertebrates. It is not clear whether retroviral vectors or other means of transferring DNA were involved, although truncated RAG1/RAG2 clusters can act as transposons [12, 13]. Molecular genetic analyses have now established that horizontal gene transfer was a general phenomenon around the time of the emergence of the first vertebrates [35, 36] and was fairly frequent among microbes, plants and animals [11]. Britten [37] has hypothesized that genes transferred in this way may have played a major role in facilitating duplication, recombination and functional activation or suppression of genes within the host. It is documented that retroviral insertions have occurred within the MHC cluster of humans [38] and in the Ig heavy chain cluster of the catfish [39]. The accidental, horizontal transfer of elements defining the combinatorial systems renders many of the former teleological questions that were frequently asked meaningless. For example, the response to the question of why do long lived invertebrates that are subject to the same variety of pathogens that vertebrates living in their environment encounter lack a vertebrate type or combinatorial immune response becomes trivial. The genes allowing the combinatorial scrambling were not present in the appropriate ancestors of the invertebrate, whereas they were present in the ancestors of the vertebrate.

A segmental analysis of the shark RAG1 gene product that contains 1,116 residues also indicates that this molecule, which is apparently restricted to jawed vertebrates, has homologs in microbes [39], and has functional domains with various degrees of conservation. The initial segments of RAG1 in comparison of shark with human are only about 25% identical whereas the C-terminal halves of the molecules are better than 80% identical. These segmental differences in fractional identity between large proteins emphasizes a concern in calculation of phylogenetic trees because some domains can show strong relatedness whereas others can be marginal. Therefore, overall comparisons of entire sequences might prove misleading because of "jumping domains" [40] as a consequence of "exon shuffling" [16].

Even if one accepts the present recombinant DNA evidence and the lack of molecules homologous to RAG1 except for microbial integrases below the phylogenetic level of jawed vertebrates as conclusive evidence for the "big bang", one can still enquire whether Ig domains of the C1 or *bona fide* V type also emerged only within the same ten million year period. The time of appearance of definitive Ig molecules can be estimated by calculation of the rates of evolution of Ig and TCR domains [2, 7, 41]. For these calculations, we have focused upon the constant domains and have averaged these for light chains, heavy chains and TCR α, β, γ and δ sequences for available species ranging from sharks to mammals [7]. Using Kimura's [42] δ or distance the average rate for TCR C-domains computed to 2.8±0.1x10^{-9} substitutions per site per year and that of Ig constant domains of λ and κ light chains, and Cμ1, Cμ2, Cμ3 and Cμ4 computes to 1.1±0.1x10^{-9} substitutions per site per year. The rate of evolutionary substitution of the TCR C domains is significantly faster than that of the Ig constant domains which is comparable to that of the superoxide dismutase. Coincidentally, superoxide dismutase contains the immunoglobulin fold but has no obviously discernible homology to immunoglobulins. This molecule arose early in evolution (approximately 3 billion years ago), and underwent strong selective change with the appearance of oxygen in the atmosphere approximately two billion years ago. However, the rate of evolution appears constant for a billion years [43] and is comparable to that of Ig-C domains. In Figure 3, we used these rates to predict when in evolution Ig domains emerged. Klein [44] raised the issue that once protein sequences diverged by more than 75%, it is very difficult if not impossible to make the call whether or not they are homologous. As shown in the previous segmental comparisons for the Vitamin D receptor and RAG1 sequences, we hesitate to accept this assumption because clearly related domains can occur within two proteins that differ markedly in other stretches of sequence. Taking these strictures into account, the tentative conclusion is that *bona fide* TCRs emerged at approximately the time of origin of the jawed vertebrates (450-500 million years ago) whereas *bona fide* Ig-C domains might have emerged approximately one billion years ago, roughly at the time of divergence between ancestral animals and fungi. The conclusion is congruent with the analyses of Cooper and Fortey [5] and Wray *et al.*, [6]. Thus, it is possible that *bona fide* members of the Ig family, as opposed to distally related members of the superfamily may have emerged in evolution prior to the appearance of jawed vertebrates. This raises the possibility that such molecules may be found in agnathan vertebrates or lower deuterostomes and may even occur in protostomes. The conclusion follows that the transferred enzymes facilitating site-specific recombination and duplication found pre-existing genes upon which they could operate during the insertion to an ancestral jawed vertebrate. Although the combinatorial system appears to have emerged in one complex series of operations, there may well have been a series of steps involved with the catalyst for duplication and rearrangement provided by the horizontal transfer of the recombinase genes.

SHAPING THE PRIMORDIAL/GERMLINE (?) REPERTOIRE

Although sharks can mount strong immune responses to a variety of antigens, repeated immunizations do not cause substantial incremental increases in titers and no evidence has been obtained for affinity maturation comparable to that of the secondary response of mammals [45-47]. Evidence for somatic mutation in shark V_H [48] and NAR molecules [49] has been reported, but its impact upon selection of combining sites has not yet been established. The fact that shark Ig light chain and heavy chain gene segments occur in individual clusters [7-9, 50] suggests that the V_H or V_L expression would be largely if not completely based upon germline sequence. In the case of the λ-like light chains where the Vλ and Jλ are fused in register in the germline [51, 52], the germline configuration would be expected except for somatic hypermutation. The μ and ω heavy chain clusters contain V segments separated from Ds and Js with appropriate recombination signal sequences, so recombination would be required for activation of the particular cluster. A variety of natural antibodies have been studied in mammals [53, 54] and lower vertebrates including teleosts [55] and sharks [8, 56-58]. Such molecules that occur in the serum in the absence of purposeful immunization may provide a clue to the common or conserved primordial antigen recognition repertoire of all jawed vertebrates. At this juncture, we would emphasize that recent studies of monoclonal natural antibodies of humans and mice have brought about a reexamination of concepts of antibody specificity. Landsteiner's concept of "exquisite specificity" [59] should be replaced by a more general concept of "epitope promiscuity" [60-63].

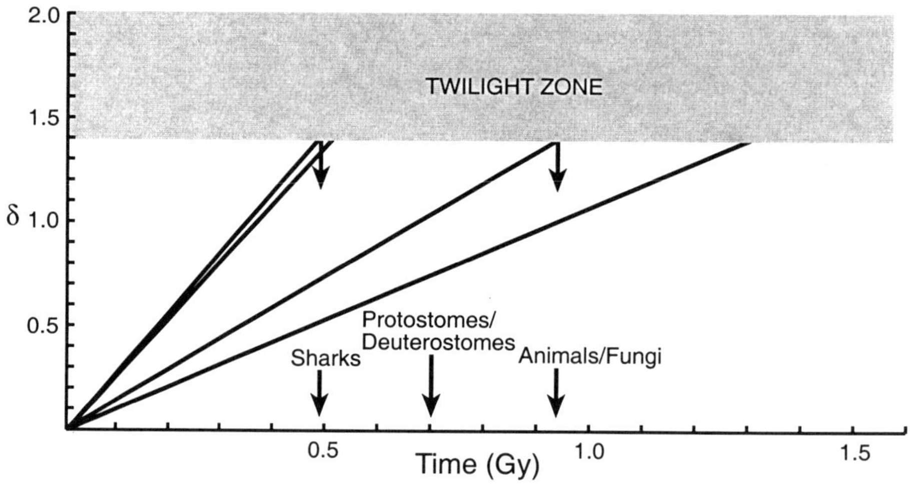

Figure 3. Calculated time of emergence of constant domains of immunoglobulins {Cκ, Cλ Cμ1, Cμ2, Cμ3, Cμ4; average} and T-cell receptors {average of Cα, Cβ, Cγ, and Cδ}. The two lines for each illustrate the range of error for each set. The Twilight Zone represents the area corresponding to identity of <25% where the exact relationship becomes uncertain. Arrows associated with respective rate line indicate the earliest time of computed evolutionary emergence. Arrows beneath the organisms indicate emergence time of the organism or divergence of the groups.

We carried out studies designed to determine whether serum of two carcharhine sharks, the tiger shark (*Galeoderdo cuverii)* and the sandbar shark (*Carcharhinus plumbeus*), had detectable levels of naturally occurring antibodies with specificities that had been observed and characterized in mammals. Our approach was predicated upon two hypotheses. Firstly, that IgM antibodies of sharks were comparable to those of mammals in "quality", *i.e.* binding affinity, avidity and specificity for antigens. Secondly, that functional strictures of the *internal milieu* as well as contact through infections with ubiquitous organisms such as viruses and bacteria shaped the antigen-recognition repertoires of all vertebrates that express IgM antibodies. Specificity is an operational property that is contingent upon the set of antigens used in the determination. On the other hand, affinity of the individual combining site for monovalent hapten is a physical property that provides a clear index of the strength of binding. Interactions between polymeric IgM immunoglobulins, or even "monomers" having two combining sites, with complex antigens reflect avidity rather than individual combining site activity. We have carried out binding studies of affinity purified naturally occurring shark and human antibodies to the dinitrophenyl hapten (DNP) using the fluorescence quenching method of Eisen [64] and found that these natural Ab molecules have comparable affinities in a range appropriate for study, but considerably less than that of MOPC 315. Since no evidence for affinity maturation has been shown for sharks [10, 46] or teleosts [10, 65], these results indicate that the affinity purified molecules have binding affinities in the range typical of IgM antibodies in a primary response. A second consideration is obtaining enough material for further studies of binding and partial characterization. The molecules we have studied range in percentage composition of shark IgM immunoglobulin from 0.01 to 0.4% which provides adequate material for partial biochemical characterization.

Table II compares natural antibody systems of sharks and humans. One set of test specificities falls under the category of "housekeeping" and includes antibodies against self antigens such as DNA, thyroglobulin and the major anion transporter band 3 (AE1). In mammals, the anti-band 3 antibodies are IgG molecules directed against the senescent cell antigen which is a neo-antigen formed on aged red cells, most probably as a result of oxidative damage [66]. This natural scavenger antibody detects defective red cells which are then phagocytosed and removed. Free DNA would be toxic and natural antibodies exist for its removal. Autoantibodies to thyroglobulins occur frequently at low levels, but molecules of distinct specificities can predispose to autoimmune diseases as can anti-DNA autoantibodies of particular specificities. A second major set of natural antibodies can be termed regulatory or immunomodulatory. This group includes antibodies against Ig-C determinants *e.g.* the constant domains of IgG, where the antibodies are termed rheumatoid factors and are elevated in rheumatoid arthritis. These autoantibodies can also carry out a normal immunoregulatory role shutting down IgG production following immunization [67]. In addition, naturally occurring autoantibodies with anti-idiotype specificity for V-region epitopes occur in sera of humans and mice [68, 69]. We have found that autoantibodies against TCR variable region epitopes occur in humans [70] and mice [71], where they most probably carry out an immunoregulatory role. These two broad subsets of naturally occurring antibodies are directed against autologous components and play either a regulatory or housekeeping role. Exogenous antigens in the form of microbes, both viruses and bacteria, are frequent pathogens of all vertebrates. Our studies also considered antibodies against phosphorylcholine (PC) which is a major structural component of gram positive bacteria such as *Streptococcus pneumoniae* where it is associated with a polysaccharide component of the cell wall teichoic acid [72]. This molecule has been found in a large number of gram positive organisms, as well as gram-negative bacteria such as *Hemophilus influenzae*, and eucaryotic disease-causing organisms including *Leishmania, Trypanosomas,* fungi, and parasitic trematodes, tapeworms and trematodes. In humans, two types of circulating molecules bind specifically to PC. The first is the ancient C-reactive protein which requires calcium for binding, and the second is antibody. In an evolutionary context, it is noteworthy that CRP activates the classical complement pathway via binding to C1q. In mice, the antibodies that bind PC often express the T15 idiotype first associated with the murine κα myeloma protein TEPC15 [73]. Some monoclonal antibodies recognizing this idiotope also react with C-reactive protein [73] and with *Limulus* hemagglutinin/CRP of similar specificity [74]. Murine naturally occurring IgM antibodies to PC are effective in preventing infection with *S. pneumoniae* [75]. Humans have natural IgM antibodies directed against bacterial components and retroviral glycoproteins, notably those of HIV-1 [53]. This is, of course, not an exhaustive list but one which is directed towards obtaining information on the overall repertoire of unimmunized sharks and facilitating comparison with human molecules.

Our studies included affinity purification of shark antibody on insolubilized antigens (bound covalently to Sepharose) with separate studies using IgM-enriched human Cohn Fraction III as a parallel control. Affinity purified molecules were studied for their capacity to bind the cognate selecting antigen as well as other antigens to assess cross-reaction, were analyzed by polyacrylamide gel electrophoresis and isoelectricfocusing, and were transferred to immobilon blots for N-terminal sequence analysis. Since we have available approximately 30 Vλ, and Vμ sequences in our own databases [7, 8, 76] as well as studies of other investigators with other elasmobranchs [9, 48, 77-79], this approach allowed us to make identifications of dominant sequences of the affinity purified oligoclonal antibodies with available gene sequences.

The overall spectrum of natural activities of the shark, at least to this set of antigens, was comparable to that of the natural human antibody population. With respect to naturally occurring molecules that bound to PC, we obtained the interesting result that our available tiger shark serum contained both IgM antibodies and a CRP component clearly homologous to CRP of the smooth dogfish *Mustelus canis* [80] and to human neuronal pentraxin 2 and apexin of the guinea pig, rather than to classical serum CRP.

Table 2. Examples of Natural Antibody Systems Found in Sharks and Humans

	Shark	Human
Housekeeping*		
Anti-DNA	polyreactive	polyreactive
Anti-thyroglobulin	protein specific	protein specific
Anti-band 3 (AE1)	peptide specific	SCA, peptide, protein specific
Regulatory		
Anti-Ig	anti-human BJ	anti-Cγ (RF); anti-idiotype
Anti-TCR (V)	anti-Vα/Vβ	anti-Vα/Vα Fv; anti-peptide
Infections		
Anti-bacterial components	anti-PC	anti-PC
Anti-viral envelope glycoproteins	anti-gp160; anti-peptide	anti- HIV gps

* Variants of anti-DNA and anti-thyroglobulin autoantibodies have been found to cause autoimmune disease in mammals. BJ, Bence-Jones proteins (monoclonal light chains) from multiple myeloma; SCA, senescent cell antigen; RF, rheumatoid factor.

Affinity purified shark NAbs to ssDNA show strong cross-reactive binding to protein antigens including ovalbumin (62%) and thyroglobulin (45%). IgM human NAbs to ssDNA likewise cross-react strongly with thyroglobulin (105%). By contrast, shark NAbs to thyroglobulin show diminished capacity to bind ssDNA (19%) and bind slightly to fetuin (30%), but show the expected cross-reactivity between bovine and porcine thyroglobulins (87%). Furthermore, the shark antibody to thyroglobulin binds to the same protease-derived fragment of thyroglobulin as does the human antibody. Although it is not known why, all vertebrate species tend to have a relatively large percentage of natural antibodies to the dinitrophenyl hapten. Our preliminary studies indicate that affinity purified shark anti-DNP is highly specific. This material was prepared by first passing serum or purified Ig preparations through a column of bovine serum albumin coupled to Sepharose and taking the material which did not bind and reacting it with a column of DNP-derivatized bovine serum albumin. Our affinity-purified NAb differs from older studies reporting that natural anti-DNP molecules were highly polyreactive [57, 58]. However, the purification procedures were quite distinct and prior attempts to measure affinity were made using equilibrium dialysis instead of fluorescence quenching.

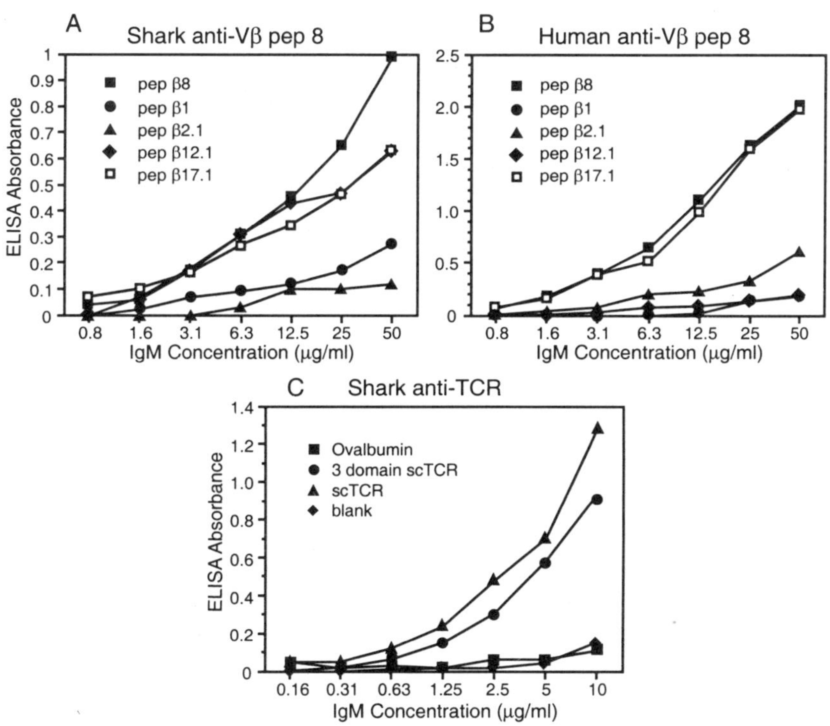

Figure 4. ELISA profiles of binding of affinity-purified sandbar shark (A, C) or human (B) natural antibodies to either synthetic 16-mer peptides corresponding to the third frameworks of Vβ gene products (A, B) or a recombinant VαVβ construct (C) containing the intact Vα and Vβ sequences of the TCR of Jurkat. In C, two recombinant TCRs were used: scTCR = Vα/Vβ, 3 domain scTCR = Vα/VβCβ.

Figure 4 illustrates profiles of shark natural antibody activity against either a recombinant single chain TCR Vα/Vβ construct (C), or human (B) and shark antibodies (A) directed against a 16-mer peptide corresponding to the Fr3 segment of TCR Vβ. The affinity purified shark NAb to the human scTCR reacts with both the Vα/Vβ construct and to a recombinant construct containing Vα and Vβ and Cβ domains. It does not bind to other protein antigens as indicated here by ovalbumin. Both the human and shark affinity purified NAbs to a particular Vβ Fr3 peptide show some cross-reaction with homologous members of the set. Thus, the recognition capacity of the shark and human NAbs to TCR Vβ epitopes parallel one another, although the exact specificities are not identical. Litman and his colleagues have shown strong homology between corresponding segments of shark [81] and human TCR V region sequences. Our result gives preliminary credence to the hypothesis that natural regulatory systems involving recognition of "public idiotopes" (defined by segments of the Vβ gene product) occur in the most divergent jawed vertebrate species to have the combinatorial immune system.

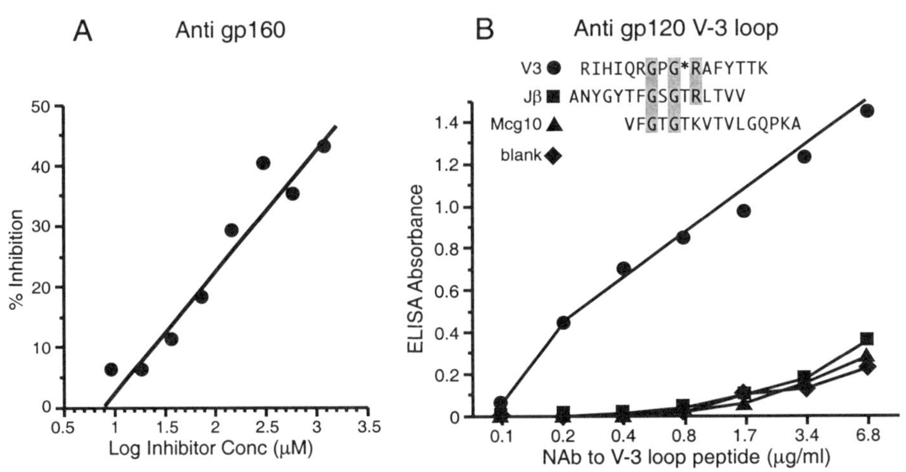

Figure 5. ELISA binding of sandbar shark IgM natural antibodies to antigens of human retrovirus HIV-1. (A) Inhibition of binding to recombinant HIV-1 glycoprotein 160 by soluble gp160. (B) Affinity purified sandbar shark natural antibody to the conserved portion of the hypervariable V3-loop tested against partially related peptide sequences of human TCR Jβ and Mcg Jλ.

A subset of sandbar shark IgM reacts with either HIV recombinant glycoproteins or synthetic peptides defining particular epitopes. The NAb shows relatively strong binding to recombinant gp160 that is much stronger than binding to ovalbumin, fetuin, gp120 or a peptide derived from sv40 when tested under comparable conditions. Figure 5 illustrates inhibition of the ELISA binding by recombinant gp160 (A) and ELISA binding of affinity-purified anti-V3 loop peptide (B) to the cognate peptide and marginally related peptides. The curve in Figure 5A has not yet reached saturation, but can be fit with the least squares line with r = 0.97 and the 50% binding level extrapolated to 2.5 μM. The right figure (B) illustrates the specificity of shark IgM antibody isolated on an insolubilized peptide corresponding to the V3-loop of HIV-1 gp120. Studies with sera of HIV infected humans showed that affinity purified antibodies to this conserved portion of the V3 loop cross-reacted with peptides corresponding to the joining segment of human TCRβ chains [82]. By contrast, the affinity-purified shark NAb to the retroviral variable loop reacted only with the viral epitope peptide. Thus, shark natural antibodies are capable of showing high levels of specificity for precisely defined peptide epitopes.

Figure 6 illustrates an SDS-polyacrylamide gel pattern under reducing conditions of the affinity purified sandbar shark NAb to the hapten PC. Only μ chains and light chains are apparent in the preparation which was affinity purified from whole shark serum. Isoelectricfocusing patterns involving the light chains of many affinity purified preparations both of humans and mice, indicate that the preparations are oligoclonal by comparison with the polyclonality shown by the entire set of shark or human serum light chains. Furthermore, the degree of polydispersity can be diminished, but not eliminated, by absorption with Sepharose alone and other antigens coupled covalently to the Sepharose matrix. N-terminal sequence analysis of the shark anti-PC NAbs and other samples is shown in Figure 7. Parallel studies were carried out with affinity purified human IgM NAbs to ssDNA and thyroglobulin. The dominant N-terminal sequences of the human NAb subpopulations corresponded to Vκ3 and V_H 3. The V_H sequences of the shark NAbs to thyroglobulin and PC corresponded closely to those of our V_H clone 12 [76]. The ten residues obtained for the anti-thyroglobulin gave an exact match, and the anti-PC differed by only one residue with an arginine (R) instead of a histidine (H) at position #13. The sandbar shark NAb V_L sequences for DNA and thyroglobulin corresponded exactly to our gene sequences V_L C_L U11, and V_L C_L 5.1 respectively. Since sandbar shark Vλ and Jλ gene segments are fused in register in the germline, with the exception of somatic mutation in the CDRs, the complete antibody VJ should resemble closely the sequence of the molecule in our database.

The situation for the sandbar shark NAb to PC was noteworthy because a dominant sequence comprising about 80% of the sample corresponded to Vκ sequences very similar to that of induced nurse shark antibody to *Streptococcal A* polysaccharide as determined by Sledge *et. al.* [83] and to the VκI sequences of PC-specific murine IgM monoclonal antibodies raised against heat killed *S. pneumoniae* [84]. The minor V_L sequence was a λ sequence which differed in two out of eleven residues from the consensus Vλ pooled sequence. The Vκ result was unexpected, because the major light chain families in carcharhine sharks correspond to λ-like light chains [7, 8, 85]. However, we have isolated serum molecules containing Cκ peptide sequences, and have cloned a Cκ gene fragment. κ-like molecules predominate in the nurse shark. It is, thus, possible that a strong selection has occurred for a particular light chain, or restricted set of light chains for their reactivity against the phosphorylcholine determinant which is a component of ubiquitous pathogens. The anti-PC Vκ sequences of mice that resemble the shark molecule differ from the Vk sequence used by the anti-PC prototype TEPC 15.

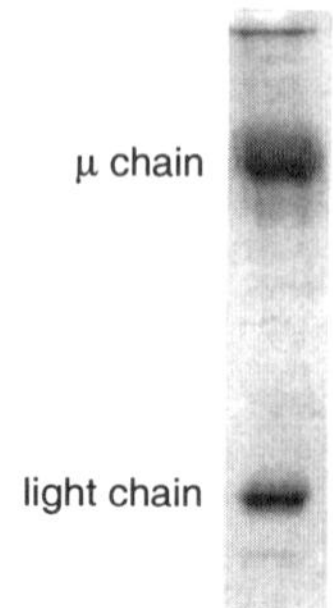

Figure 6. SDS Polyacrylamide gel electrophoresis under reducing conditions of sandbar shark natural antibody to phosphorylcholine isolated from whole serum by affinity chromatography.

V_H sequences	1 5 10 15
NAb anti-thyro	D I V L T Q P E A E
NAb anti-PC	D I V L T Q P E A E T G R P G G S
SbS V_HC12	D I V L T Q P E A E T G H P G G S
V_L sequences	1 5 10 15
NAb anti-PC (maj)	E I T M T E T P P A L S V S L G E
Nurse anti-strepA	D I T M T E S P P V L S V G L G Q
Mu anti-PC (CBA/J6 F9)	D I V M T Q S P S S L S V S A G E
NAb anti-PC (min)	D P V L L E P G S I S
SbS V_λ concensus	D P V L T G P G S I S
NAb anti-ssDNA	D I V L T Q T P
SbS V_L Cl U11	D I V L T Q T P
NAb anti-thyro	D P V L T Q P G
SbS V_L Cl 5.1	D P V L T Q P G

Figure 7. Alignments of N-terminal light chain and heavy chain sequences of affinity purified shark natural antibodies with best matches from the shark database. Residue positions that are not identical in the pairwise comparisons as shaded. The Vκ of a mouse IgMκ antibody [84] to PC also shows strong identity to the sandbar shark anti-PC Vκ.

CONCLUSION

Overall, in these studies and in comparisons of sequences of carcharhine shark V_H and V_L structure with those of humans, it appears that these widely divergent species in evolution have conserved functional recognition repertoires most probably generated in the period immediately following the "big bang". Pending definitive comparisons of combining site structures, we tentatively conclude that the IgM repertoires of vertebrates are comparable in recognition capacity and in configurations of the V_H/V_L domains. It is of considerable interest to obtain monoclonal shark antibodies specific for the same epitopes and intact antigenic molecules against which human and murine monoclonal NAbs have been generated in order to compare the arrangement of relevant residues in the combining sites. The shark IgM system, like the mammalian, gives an indication of high degree of specificity in some cases, but epitope cross-reactivity and polyreactivity in others. Detailed structural and genetic analysis of shark antibodies should provide novel and comprehensive insight into the fundamental nature of the antigen-combining site and further define the basis of the epitope-recognition spectra shown by antibodies of all jawed vertebrate species.

ACKNOWLEDGEMENTS

This work was supported in part by Grant #MCB 9631846 from the National Science Foundation. We thank Ms. Andrea Garza for expert technical assistance and Ms. Diana Humphreys for valuable assistance in preparation of the manuscript. Data for PMJ were generated in partial fulfillment of the Master of Science Degree in Microbiology and Immunology. Those of PMS and BJZ were obtained in forfilling the requirements for Honors Degrees in Biochemistry and Ecology and Evolutionary Biology, respectively.

REFERENCES

1. Marchalonis, J.J., and S.F. Schluter. 1990. On the relevance of invertebrate recognition and defense mechanisms to the emergence of the immune response of vertebrates. *Scand. J. Immunol.* 32:13.
2. Marchalonis, J.J., and S.F. Schluter. 1998. A stochastic model for the rapid emergence of specific vertebrate immunity incorporating horizontal transfer of systems enabling duplication and combinatorial diversification. *J. Theo. Biol.* 193:429.
3. Schluter, S.F., R.M. Bernstein, H. Bernstein, and J.J. Marchalonis. 1999. 'Big Bang' emergence of the combinatorial immune system. *Dev. & Comp. Immunol.* 23:107.
4. Thompson, C.B. 1995. New Insights into V(D)J recombination and its role in the evolution of the immune system. *Immunity* 3:531.
5. Cooper, A., and R. Fortey. 1998. Evolutionary explosions and the phylogenetic fuse. *Tree* 13:151.
6. Wray, G.A., J.S. Levinton, and L.H. Shapiro. 1996. Molecular evidence for deep precambrian divergencies among metazoan phyla. *Science* 274:568.
7. Marchalonis, J.J., S.F. Schluter, R.M. Bernstein, and A.B. Edmundson. 1998. Phylogenetic emergence and molecular evolution of the immunoglobulin family. *Adv. in Immunol.* 70:417.
8. Marchalonis, J.J., S.F. Schluter, R.M. Bernstein, and V.S. Hohman. 1998. Antibodies of Sharks: revolution and evolution. *Immunol. Rev.* 166:103.
9. Litman, G.W., M.K. Anderson, and J.P. Rast. 1999. Evolution of antigen binding receptors. *Annu. Rev. Immunol.* 17:109.
10. Du Pasquier, L., and M. Flajnik. 1999. Origin and evolution of the vertebrate immune system. *In* Fundamental Immunoloty, vol. 4th Edition. Raven Press, New York. 199.
11. Doolittle, W.F. 1999. Phylogenetic Classification and the Universal tree. *Science* 284:2124.
12. Agrawal, A., Q.M. Eastman, and D.G. Schatz. 1998. Transposition mediated by RAG1 and RAG2 and its implications for the evolution of the immune system. *Nature* 394:744.
13. Hiom, K., M. Melek, and M. Gellert. 1998. DNA transposition by the RAG1 and RAG2 proteins: a possible source of oncogenic translocations. *Cell* 94:463.

14. Williams, A.F., and A.N. Barclay. 1988. The immunoglobulin superfamily - domains for cell surface recognition. *Ann.Rev. Immunol.* 6:381.
15. Bork, P., L. Holm, and C. Sander. 1994. The Immunoglobulin Fold. Structural classification, sequence patterns and common core. *J. Mol. Biol.* 242:309.
16. Doolittle, R.F. 1995. The multiplicity of domains in proteins. *Annu. Rev. Biochem.* 64:287.
17. Holmgren, A., M.J. Kuehn, C.-I. Branden, and S.J. Hultgren. 1992. Conserved immunoglobulin-like features in a family of periplasmic pilus chaperones in bacteria. *EMBO J.*:1617.
18. Klein, J. 1997. Homology between immune responses in vertebrates and invertebrates: does it exist ? *Scand. J. Immunol.* 46:558.
19. Hughes, A.L. 1998. Protein phylogenics provide evidence of a radical discontinuity between arthropod and vertebrate immune systems. *Immunogenet* 47:283.
20. Klein, J. 1989. Are invertebrates capable of anticipatory immune responses? *Scand. J. Immunol.* 29:499.
21. Lambris, J.D., K.B.M. Reid, and J.E. Volanakis. 1999. The Evolution, structure, biology and pathophysiology of complement. *Immunol. Today* 20:207.
22. Smith, L.C., C.-S. Shih, and S.G. Dachenhausen. 1998. Coelomocytes express SpBf, a homologue of factor B, the second component in the sea urchin complement system. *J. Immunol.* 161:6784.
23. Armstrong, P.B., W.F. Mangel, J.S. Wall, J.F. Hainfield, K.E. Van Holde, A. Ikai, and J.P. Quigley. 1991. Structure of a2-macroglobulin from the arthropod *Limulus polyphemus*. *J. Biol. Chem.* 266:2526.
24. Hoffman, J.A., F.C. Kafatos, C.A.J. Janeway, and R.A.B. Ezekowitz. 1999. Phylogenetic perspectives in innate immunity. *Science* 284:1313.
25. Fearon, D.T., and R.M. Locksley. 1996. The instructive role of innate immunity in the acquired immune response. *Science* 272:50.
26. Medzhitov, R., and C.A. Janeway. 1997. Innate Immunity: The virtues of a nonclonal system of recognition. *Cell* 91:295.
27. Lieber, C.S. 1997. Cytochrome P-4502E1: Its physiological and pathological role. *Physiol. Revs* 77:517.
28. Payne, C.M., C. Crowley, D. Washo-Stutltz, M. Briehl, H. Bernstein, C. Bernstein, S. Beard, H. Holubec, and J. Warneke. 1998. The stress-response protein poly(ADP-ribose) polymerase and NK-kB oritect against bile salt-induced apoptosis. *Cell Death & Different.* 5:623.
29. Zapata, A., C.F. Ardavin, R.P. Gomariz, and J. Leceta. 1981. Plasma cells of the amniocoete of *Petromyzon marinus*. *Cell Tiss. Res.* 221:203.
30. Marchalonis, J.J., and G.M. Edelman. 1968. Phylogenetic origins of antibody structure. III. Immunoglobulins from the sea lamprey, *Petromyzon marinus*. *J. Exp. Med.* 127:891.
31. Raison, R.L., and W.H. Hildemann. 1984. Immunoglobulin-bearing blood leucocytes in the Pacific hagfish. *Dev. Comp. Immunol.* 8:99.
32. Varner, J., P. Neame, and G.W. Litman. 1991. A serum heterodimer from hagfish (*Eptatretus stoutii*) exhibits structural similarity and partial sequence homology with immunoglobulin. *Proc. Natl. Acad. Sci. USA* 88:1746.
33. Kay, M.M.B., C. Cover, S.F. Schluter, R.M. Bernstein, and J.J. Marchalonis. 1995. Band 3, the anion transporter, is conserved during evolution: implications fro aging and vertebrate evolution. *Cell. & Mol. Biol.* 41:833.
34. Najakshin, A.M., L.V. Mechetina, B.Y. Alabyev, and A.V. Taranin. 1999. Identification of an IL-8 homolog in lamprey (*Lampetra fluviatilis*) : early evolutionary divergence of chemokines. *Eur. J. Immunol.* 29:375.
35. Goodier, J.L., and W.S. Davidson. 1994. Tc1 Transposon-like sequences are widely distributed in Salmonids. *J Mol Biol* 241:26.
36. Ivics, Z., P.B. Hackett, R.H. Plasterk, and Z. Izsvak. 1997. Molecular reconstruction of *Sleeping Beauty* a *Tc1*-like transposon from fish, and its transposition in human cells. *Cell* 91:501.
37. Britten, R.J. 1997. Mobile elements inserted in the distant past have taken on important functions. *Gene* 205:177.38.
38. Gaudieri, S., C. Leelayuwat, D.C. Townend, J.K. Kulski, and R.L. Dawkins. 1997. Genomic characterization of the region between HLA-B and TNF: implications for the evolution of multicopy gene families. *J. Mol. Evol.* 44:S147.
39. Ghaffari, S.H., and C.J. Lobb. 1999. Structure and genomic organization of a second cluster of immunoglobulin heavy chain gene segments in the channel catfish. *J. Immunol.* 162:1519.

40. Marchalonis, J.J., G.R. Vasta, G.W. Warr, and W.C. Barker. 1984. Probing the boundaries of the extended immunoglobulin family of recognition moelcules: Jumping domains, convergences and minigenes. *Immunol. Today* 5:133.
41. Andersson, E., and T. Matsunaga. 1993. Complete cDNA sequence of a rainbow trout IgM gene and evolution of vertebrate IgM constant domains. *Immunogenet* 38:243.
42. Kimura, M. 1969. The rate of molecular evolution considered from the standpoint of population genetics. *Proc. Natl. Acad. Sci. USA* 63:1181.
43. Ayala, F.J. 1997. Vagaries of the molecular clock. *Proc Natl Acad Sci USA* 94:7776.
44. Klein, J. 1998. In an immunological twilight zone. *Proc. Natl. Acad. Sci. USA* 95:11504.
45. Papermaster, B.W. 1966. Genetic considerations of immunoglobulin evolution in vertebrates. *In* Phylogeny of Immunity. R.T. Smith, P.A. Miescher and R.A. Good, editors. University of Florida Press, Gainesville, FL. 118.
46. Makela, O., and G.W. Litman. 1980. Lack of heterogeneity in anti-hapten antibodies of a phylogenetically primitive shark. *Nature* 287:639.
47. Clem, L.W., W.E. McLearn, and V. Shankey. 1975. Quantitative and habitative aspects of the antibody library of sharks. *Adv. Exp. Med. Biol.* 64:231.
48. Hinds-Frey, K.R., H. Nishikata, R.T. Litman, and G.W. Litman. 1993. Somatic variation precedes extensive diversification of germline sequences combinatorial joining in the evolution of immunoglobulin heavy chain diversity. *J Exp Med* 178:815.
49. Diaz, M., J. Velez, M. Singh, J. Cerny, and M.F. Flajnik. 1999. Mutational pattern of the nurse shark antigen receptor gene (NAR) is similar to that of mammalian Ig genes and to spontaneous mutations in evolution: the translesion synthesis model of somatic hypermutation. *Intl. Immunol.* 11:825.
50. Hinds, K.R., and G.W. Litman. 1986. Major reorganization of immunoglobulin VH segmental elements during vertebrate evolution. *Nature (Ldn)* 320:546.
51. Hohman, V.S., D.B. Schuchman, S.F. Schluter, and J.J. Marchalonis. 1993. Genomic clone for sandbar shark l light chain: generation of diversity in the absence of gene rearrangement. *Proc Natl Acad Sci USA* 90:9882.
52. Rast, J.P., M.K. Anderson, T. Ota, R.T. Litman, M. Margittal, M.J. Shamblott, and G.W. Litman. 1994. Immunoglobulin light chain class multiplicity and alternative organizational forms in early vertebrate phylogeny. *Immunogenet.* 40:83.
53. Harindranath, N., H. Ikematsu, A.L. Notkins, and P. Casali. 1993. Structure of the V_H and V_L segments of polyreactive and monoreactive human natural antibodies to HIV-1 and *escherchia coli* b-galactosidase. *Int Immunol* 5:1523.
54. Coutinho, A., M.D. Kazatchkine, and S. Avrameas. 1995. Natural autoantibodies. *Curr. Opin. in Immunol* 7:812.
55. Gonzalez, R., J. Charlemagne, W. Mahana, and S. Avrameas. 1988. Specificity of natural serum antibodies present in phylogenetically distinct fish species. *Immunology* 63:31.
56. Marchalonis, J.J., V.S. Hohman, C. Thomas, and S.F. Schluter. 1993. Antibody production in sharks and humans: a role for natural antibodies. *Dev Comp Immunol* 17:41.
57. Rudikoff, S., E.W. Voss, and S. M.M. 1970. Biological and Chemical properties of natural antibodies in the nurse shark. *J. Immunol.* 105, no. 6:1344.
58. Leslie, G.A., and L.W. Clem. 1970. Reactivity of normal shark immunoglobulins with nitrophenyl ligands. *J. Immunol.* 105:1547.
59. Landsteiner, K. 1962. The specificity of Serological Reactions. Dover Publications, Inc., New York.
60. Marchalonis, J.J., A. Garza, W.J. Landsperger, S.F. Schluter, and A.-C. Wang. 1997. Binding of human IgG myeloma proteins to autologous T-cell receptor determinants. *Crit. Rev. in Immunol.* 17:497.
61. Marchalonis, J.J., I. Robey, S.F. Schluter, and D.E. Yocum. 1999. Epitope promiscuity of human monoclonal autoantibodies to T-cell receptor combining site determinants. *App. Biochem. & Biotech.* submitted.
62. Kramer, A., T. Keitel, K. Winkler, W. Stocklein, W. Hohne, and J. Schneider-Mergener. 1997. Molecular basis for the binding promiscuity of an anti-p24 (HIV-1) monoclonal antibody. *Cell* 91:799.
63. Keitel, T., A. Kramer, H. Wessner, C. Scholz, J. Schneider-Mergener, and W. Hohne. 1997. Crystallographic analysis of Anti-p24 (HIV-1) monoclonal antibody cross-reactivity and polyspecificity. *Cell* 91:811.
64. Velick, S.F., C.W. Parker, and H.N. Eisen. 1960. Excitation energy transfer and the quantitative study of the antibody hapten reaction. *Proc. Natl. Acad. Sci. USA* 46:1470.

65. Arkoosh, M.R., and S.L. Kaattari. 1991. Development of immunological memory in rainbow trout (*Oncorhynhus mykiss*) I an immunochemical and cellular analysis of the B cell response. *Dev. & Comp. Immunol.* 15:279.
66. Kay, M.M.B. 1984. Localization of senescent cell antigen on band 3. *Proc. Natl. Acad. Sci. USA* 81:5753.
67. Carson, D.A., P.P. Chen, R.I. Fox, T.J. Kipps, F. Jirik, R.D. Goldfien, G. Silverman, V. Radoux, and S. Fong. 1987. Rheumatoid factor and immune networks. *Annu. Rev. Immunol* 5:109.
68. Yan, X., S.V. Evans, M.J. Kaminki, S.D. Gillies, R.A. Reisfeld, A.N. Houghton, and P.B. Chapman. 1996. Characterization of an Ig VH idiotope that results in specific homophilic binding and increased avidity for antigen. *J. Immunol.* 157:1582.
69. Bergenbrant, S., A. Osterborg, G. Holm, H. Mellstedt, and A.K. Lefvert. 1991. Anti-idiotypic antibodies in patients with monoclonal gammopathies: relation to the tumour load. *Brit. J. Haematology* 78:66.
70. Marchalonis, J.J., S.F. Schluter, E. Wang, K. Dehghanpisheh, D. Lake, D.E. Yocum, A.B. Edmundson, and J.B. Winfield. 1994. Synthetic autoantigens of immunoglobulins and T-cell receptors: their recognition in aging, infection and autoimmunity. *Proc. Soc. Expt. Biol.* 207:129.
71. Dehghanpisheh, K., and J.J. Marchalonis. 1997. Retrovirally induced mouse anti-TCR monoclonals can synergize the *in vitro* proliferative T cell response to bacterial superantigens. *Scand. J. Immunol.* 45:645.
72. Harnett, W., and M.M. Harnett. 1999. Phosphorylcholine: friend or foe of the immune system? *Immunol. Today* 28:125.
73. Volanakis, J.E., Y. Xu, and K.J. Macon. 1990. Human C-reactive protein and host defense. *In* Defense Molecules. J.J.M.a.C.L. Reinisch, editor. Wiley-Liss, Inc., New York. 161.
74. Vasta, G.R., J.J. Marchalonis, and H. Kohler. 1984. Invertebrate recognition protein cross-reacts with an immunoglobulin idiotype. *J Exp Med* 159:1270.
75. Briles, D.E., J.L. Claflin, K. Schroer, and C. Formann. 1981. Mouse Igg3 antibodies are highly protective against infection with Streptococcus pneumoniae. *Nature* 294:88.
76. Shen, S.Y., R.M. Bernstein, S.F. Schluter, and J.J. Marchalonis. 1996. Heavy chain variable regions in carcharhine sharks: development of a comprehensive model for the evolution of VH domains among the gnathanstomes. *Immunol. & Cell. Biol.* 74:357.
77. Rast, J.P., C.T. Anmemiya, R.T. Litman, S.J. Strong, and G.W. Litman. 1998. Distinct patterns of IgH structure and organization in a divergent lineage of chondrichthyan fishes. *Immunogenet.* 47:234.
78. Greenberg, A.S., L. Steiner, M. Kasahara, and M.F. Flajnik. 1993. Isolation of a shark immunoglobulin light chain cDNA clone encoding a protein resembling mammalian k light chains: implications for the evolution of light chains. *Immunology* 90:10603.
79. Greenberg, A.S., D. Avila, M. Hughes, A. Hughes, E.C. McKinney, and Flajnik. 1995. A new antigen receptor gene family that undergoes rearrangement and extensive somatic diversification in sharks. *Nature* 374:168.
80. Robey, F.A., T. Tanaka, and T.Y. Liu. 1983. Isolation and characterization of two major serum proteins from the dogfish, *Mustelus canis*, C-reactive protein and amyloid P component. *J. Biol. Chem.* 258:3889.
81. Hawke, N.S., J.P. Rast, and G.W. Litman. 1996. Extensive diversity of transcribed TCR-beta in a phylogenetically primitive vertebrate. *J Immunol* 156:2458.
82. Lake, D.F., S.F. Schluter, E. Wang, R.M. Bernstein, A.B. Edmundson, and J.J. Marchalonis. 1994. Autoantibodies to the ab T-cell receptors in human immunodeficiency virus (HIV) infection: dysregulation and mimcry. *Proc. Natl. Acad. Sci. USA* 91:10849.
83. Sledge, C., L.W. Clem, and L.E. Hood. 1974. Antibody structure: amino terminal sequence of nurse shark light and heavy chains. *J. Immunol.* 112:941.
84. Clarke, S.H., J.L. Chaflin, M. Potter, and S. Rudikoff. 1983. Polymorphism in anti-phosphorylcholine antibodies reflecting evolution of immunoglobulin families. *J. Exp. Med.* 157:98.
85. Hohman, V.S., S.F. Schluter, and J.J. Marchalonis. 1992. Complete sequence of a cDNA clone specifying sandbar shark immunoglobulin light chain: gene organization and implications for the evolution of light chains. *Proc Natl Acad Sci U S A* 89:276.

INDIVIDUAL-SPECIFIC REPERTOIRES OF IMMUNE CELLS SRCR RECEPTORS IN THE PURPLE SEA URCHIN (*S. PURPURATUS*)

Zeev Pancer

Department of Biology, California Institute of Technology, Pasadena, CA 98765

INTRODUCTION AND BACKGROUND

Invertebrates comprise the vast majority of contemporary animal species, yet their immune system been poorly characterized, especially at the molecular level (Klein, 1995). Nevertheless, invertebrates possess elaborate mechanisms to identify and eliminate life-threatening pathogens (reviews by: Smith and Davidson, 1992; Sima, 1993; Smith and Davidson, 1994; Hoffmann et al., 1996; Medzhitov and Janeway, 1998). While the evolutionary elaboration of anti-pathogen responses in a hostile environment seems inevitable, the evolution of invertebrate and vertebrate individuality determining systems presents an evolutionary enigma. A basic function of the vertebrate immune system is to maintain a representative "image" of the cells and molecules of which individuals consist, and to mediate in the conflict between cooperation and competition among individual cells. The development and elaboration of such molecular and cellular surveillance systems might have accompanied the emergence of multicellular animals, along with the evolution of higher levels of molecular and cellular complexity.

The most studied early deuterostome innate immune system is that of the Purple Sea Urchin *Strongylocentrotus purpuratus* (Smith et al., 1996; Al-Sharif et al., 1998; Pancer et al., 1999; Smith et al., 1998; reviews by: Smith and Davidson, 1992; Smith and Davidson, 1994). The immune effector cells of sea urchins are the coelomocytes, free wandering cells that populate the coelomic cavity. In *S. purpuratus* there are 1-5x10^6 coelomocytes per ml of coelomic fluid, about two-thirds of which are phagocytic (Smith and Davidson, 1992; Smith et al., 1995). Other types of coelomocytes are vibratile cells, and colorless and red spherule cells. Experimental observations demonstrate that coelomocytes carry out many different functions: these include formation of cellular clots, chemotactic accumulation at sites of injury, and allograft rejection; coelomocytes also clear bacteria and other foreign substances from the coelomic cavity with great effectiveness (reviewed by Smith and Davidson, 1994).

Recently two sea urchin proteins of the complement system were reported, one homologous to Factor B (Smith et al., 1998) and the other a homologue of the vertebrate C3/C4/C5 complement proteins (Smith et al., 1996; Al-Sharif et al., 1998). Coelomocytes respond sharply to injection of LPS, as well as of bacteria (Smith et al., 1995). Furthermore, we have shown that coelomocytes mount an extensive transcription-level response to immune challenge. Introduction of live sea urchin surface bacteria sharply affects the levels of three different mRNAs encoding the immune related transcription factors SpNFκB, SpRunt-1 and SpGATAc. Obviously this implies the existence of batteries of facultatively expressed genes, that must be activated downstream of these transcription factors. Sea urchins require an effective immune system for they are long lived animals that inhabit environments which sometimes bear heavy concentrations of fungal and bacterial pathogens (Pancer et al., 1999).

For the first time we have revealed individual-specific heterogeneity in the nature of the immune capabilities of an early deuterostome invertebrate. Sea urchin coelomocytes express unique repertoires from amongst their multigene family of heterogeneous SRCR-containing receptors (Pancer et al., 1999). These receptors are members of the metazoan superfamily of cell-surface and secreted proteins featuring one or more Scavenger Receptor Cysteine-rich Repeats [SRCR], whose mammalian members include several macrophage receptors as well as receptors of B and T cells such as CD5 and CD6.

METAZOAN SRCR RECEPTORS

SRCR domains define a metazoan protein superfamily. In fact, protein members featuring one or more SRCRs were reported from phyla encompassing several branches of the animal kingdom, such as marine sponges (*Geodia cydonium*; Pancer et al., 1997; Blumbach et al., 1998), nematodes (*Caenorhabditis elegans*; Accession

E330348), echinoderms (Dangott et al., 1989; Pancer et al., 1999), sea lampreys (*Petromyzon marinus*; Mayer and Tichy, 1995), as well as various mammalian members (Reviewed by: Aruffo et al., 1997; O'Keeffe et al., 1999). The SRCR domain consists of a 110 aa-residue motif with conserved spacing of six to eight cysteines (Fig.1), that form intradomain disulfide bonds (Resnick et al., 1994; Hohenester et al., 1999).

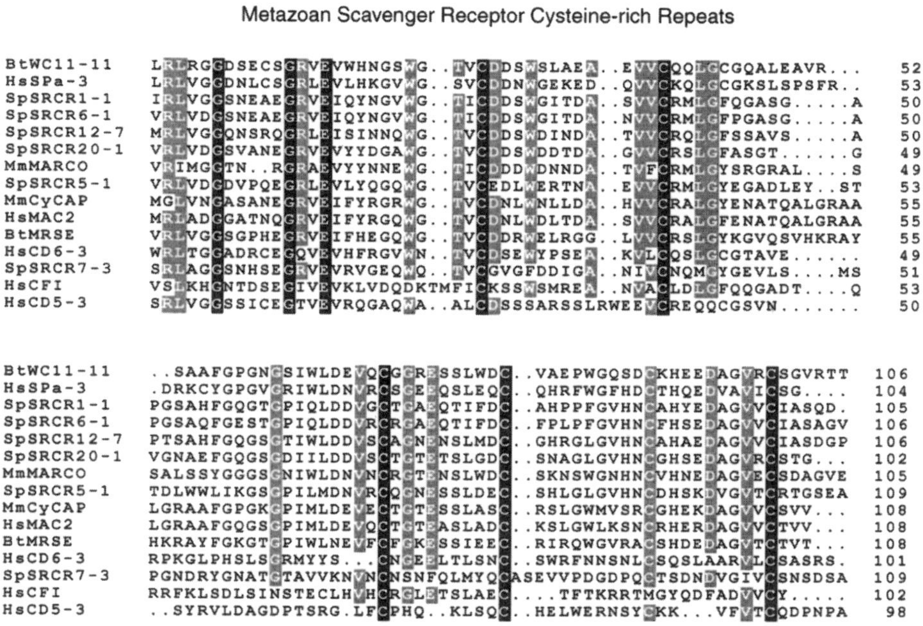

Figure. 1. Metazoan SRCR receptors. Multiple alignment of representative SRCR domains from *S. purpuratus* coelomocyte SRCR proteins, as well as mammalian SRCR containing proteins of macrophages and B and T cells. The SRCR domains in the alignment correspond to the following proteins: BtWC11 (bovine T cell WC1 or T19 antigen homologous, 11th of 11 SRCRs, aa 1155-1260, P30205); HsSPa (human lymphoid Spa, 3d of 3 SRCRs, aa 245-347, U82812); SpSRCR1 (sea urchin SRCR type 1, 1st of 4 or 5 repeats, aa 620-724); SpSRCR6 (sea urchin SRCR type 6, 1st of 3 repeats, aa 136-241); SpSRCR12 (sea urchin SRCR type 12, 7th of 20 repeats, aa 684-789); SpSRCR20 (sea urchin SRCR type 20, 1st of 7 repeats, aa 24-125); MmMARCO (murine macrophage bacteria-binding receptor, single SRCR, aa 423-528; X99321); SpSRCR5 (sea urchin SRCR type 5, 1st of 2 repeats, aa 302-410); MmCyCAP (murine macrophage cyclophilin-C-binding protein, single SRCR, aa 24-125; A48231); HsMAC2 (human macrophage MAC2-binding protein, single SRCR, aa 24-124; A47161); BtMRSE (bovine macrophage type I scavenger receptor, single SRCR, aa 387-488; P21758); HsCD6 (human B and T cell CD6, 3d of 3 SRCRs, aa 265-365, P30203); SpSRCR7 (sea urchin SRCR type 7, 3d of 7 repeats, aa 268-376); HsCFI (human complement factor I, single SRCR, aa 114-215; P05156); HsCD5 (human B and T cell CD5, 3d of 3 SRCRs, aa 277-373, P06127). Residues conserved among all sequences are shown in inverted type, while residues conserved in 80% or more of the sequences are light gray shaded.

Most of the known SRCR proteins function as recognition and immune system molecules. Vertebrate SRCR proteins are implicated in the development of the immune system and regulation of immune responses (reviews in: Aruffo et al., 1997; O'Keeffe et al., 1999). Mammalian macrophages express several SRCR-containing receptors, which are involved in binding bacteria (MARCO, Elomaa et al., 1995), binding lectins (cyclophilin-C-binding protein, Friedman et al., 1993; MAC2-binding protein, Koths et al., 1993), in regulation of the complement cascade (Complement Factor I, Goldberger et al., 19987), and in endocytosis (MRSE scavenger receptor, Freeman et al., 1990). Among the cell surface SRCR proteins of lymphoid cells are CD5 (Yu et al., 1990) and CD6 (Aruffo et al., 1991), both involved in T-cell activation. One ligand of CD5 is CD72, a B cell antigen homologous to C-type lectins (Van de Velde et al., 1991), while CD6 binds to Activated Leucocyte Cell Adhesion Molecule [ALCAM], a cell surface member of the immunoglobulin superfamily (Whitney et al., 1995). Spa, a secreted SRCR protein related to CD5 and CD6, is expressed in lymphoid tissues and binds to peripheral blood monocytes (Gebe et al., 1997).

Of special interest is a T-cells multigene family of SRCR receptors of cattle, pig and sheep, collectively named T19 or WC1 antigen homologous (Wijngaard et al., 1992; O'Keeffe et al., 1999). Multiple SRCR receptors are selectively expressed by discrete subsets of ungulate T cells populations, distinguished by amino acid polymorphism and the presence or absence of complete protein domains, as well as by alternatively spliced transcripts (Kanan et al., 1997; O'Keeffe et al., 1999). This multigene family is characterized by many variable genomic versions of the genes (more than 50 genes in sheep; Walker et al., 1994). Although no ligand for T19 SRCR proteins has yet been identified, evidence suggests a role for T19 receptors in signaling T cells growth arrest via downregulation of ERK2/MAP kinases cascades (Kirkham et al., 1997; Takamatsu et al., 1997). Thus far the only documented multigene families within the SRCR superfamily are T19 and the sea urchin coelomocyte SRCR molecules (O'Keeffe et al., 1999; Pancer et al., 1999).

Invertebrate members of the SRCR protein superfamily that function as recognition molecules include the marine sponge SRCR proteins, that appear in several membrane-bound and secreted forms, one of which is the putative sponge aggregation receptor (Blumbach et al., 1998). Another SRCR member is the sea urchin sperm egg-peptide receptor (Dangott et al., 1989).

PURPLE SEA URCHIN COELOMOCYTE SRCR RECEPTORS

Six types of sea urchin SRCR-containing proteins have been completely sequenced (Fig. 2: types 1, 5, 6, 7, 12, 20; 457-2153 aa long), all are expressed only in coelomocytes (Plate 1, Fig. 1), as demonstrated by hybridization to gel blots of RNA from embryonic and larval stages, adult tissues (gut, testes, ovary and lantern muscle) and coelomocytes. This is of special importance since coelomocytes are the immune effector cells of sea urchins, capable of mounting an extensive response to immune challenges such as pathogenic bacteria (Pancer et al., 1999). As shown in Plate 1, Fig. 2, coelomocyte SRCR gene types 1, 5, and 7 are expressed as individual-specific heterogeneous patterns of transcripts. The 15 sea urchins included in this experiment differ in the number and length of bands detected by each probe, and the intensity of hybridization. For example, coelomocytes of animals 12 and 15 barely express SRCR type 1, although expression of type 7 is at high levels in both, as is the expression of type 5 in animal 12. Likewise, animal 10 transcribes high levels of types 1 and 7 but barely detectable messages of type 5. Animal 4 expresses low levels of all three SRCR gene types, although its Thymosin levels, included as a loading control, are similar to other animals (SpThymosin is a general coelomocyte marker; Pancer et al., 1999).

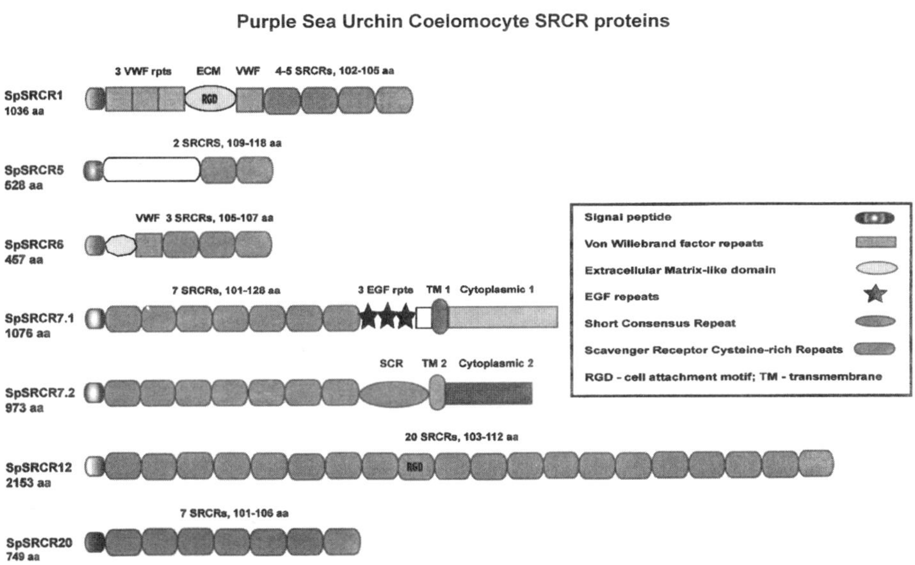

Fig. 2. Structural domains of six Purple Sea Urchin coelomocyte specific SRCR-containing proteins. Five of these proteins appear to be secreted (SpSRCR1, 5, 6, 12 and 20) and one cell bound (SpSRCR7), with two completely different C-terminal variants immediately following the 7th SRCR (SpSRCR7.1 and SpSRCR7.2). Four of the SRCR-containing proteins are mosaic, additionally featuring domains known from other proteins. These include Von Willebrand factor [VWF] repeats in SpSRCR1 and SpSRCR6 (61-62 aa) and another domain in SpSRCR1 (ECM, 347 aa), which is similar to a gastrula-specific Extra Cellular Matrix of the green sea urchin (*Lytechinus variegatus*). SpSRCR7.1 features three Epidermal Growth Factor-like [EGF] repeats (34-38 aa), while in SpSRCR7.2 there is one Short Consensus Repeat [SCR] of the complement control superfamily (68 aa). Other domains are not similar to any published sequences, like the N-terminal domains in SpSRCR5 (285 aa) and SpSRCR6 (53 aa), and the domain the separates the EGF-like repeats from the membrane retention domain in SpSRCR7.1 (52 aa). The genes SpSRCR1 and 6 are closely related; the leader peptides, VWF repeats and SRCRs are nearly identical. In contrast, the SRCRs of other types are highly divergent; SpSRCR12 consists of twenty repeats of 103-112 aa, each with unique sequence: 33-83% pairwise aa identities; the seven SRCRs in SpSRCR20 are of 101-106 aa, 51-91% pairwise aa identities.

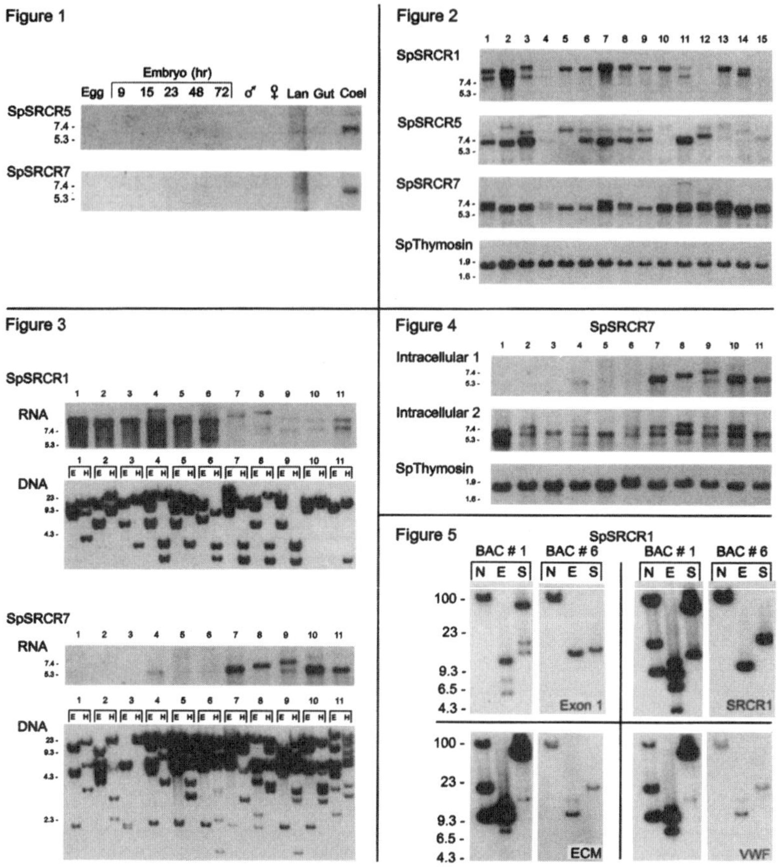

Plate 1, Fig. 1. Coelomocyte specific transcription of *S. purpuratus* SRCR messages SpSRCR5 (Accession AF076514) and SpSRCR7. Hybridization of probes representing these cDNAs is shown to gel blot of total RNA (10 µg per lane) extracted from coelomocytes, gut, lantern muscle, ovary, testes, eggs, embryos and larvae (0, 9, 15 23 48 and 72 hr). Probes were: SpSRCR5, a 459 bp *Cla*I/*Pst*I fragment encompassing the C-terminus of the N-terminal domain and the 1st SRCR (nt 927-1386 in SpSRCR5); SpSRCR7, a 615 bp *Cla*I/*Sac*I fragment encompassing the C-terminal half of the 6th and the complete 7th SRCRs (nt 1972-2587 in SpSRCR7.1).

Plate 1, Fig. 2. Individual specific patterns of expression of SpSRCR1 (AF076513), SpSRCR5 and SpSRCR7. Northern blot of coelomocyte total RNA from 15 individual sea urchins (10 µg per lane) was repeatedly probed with SpSRCR1, SpSRCR5 and SpSRCR7. The bottom panel shows the same blot probed with SpThymosin (AF076515), as an RNA loading control (SpThymosin is a general coelomocyte marker; Pancer et al., 1999). Probes other than those described in the legend of Plate 1, Fig. 1, were: SpSRCR1 (SRCR1), a 237 bp product of PCR amplification across one SRCR repeat (nt 2009-2246 in SpSRCR1); SpThymosin, 408 bp PCR product from the 5' end (nt 1-408 in SpThymosin).

Plate 1, Fig. 3. Phenotypic variation in expression of sea urchin SRCR genes types 1 and 7. Blots, either of coelomocyte total RNA (10 µg per lane), or *EcoR*I (E) and *Hind*III (H) digested genomic DNA (5 µg per lane) from 11 sea urchins, were repeatedly probed with SpSRCR1 and SpSRCR7.1. Probes were: SpSRCR1 (ECM), 533 bp product of PCR amplification across the ECM domain (nt 807-1340 in SpSRCR1);

SpSRCR7.1 (Intracellular 1), 312 bp product of PCR amplification across the transmembrane and intracellular domains of the first C-terminal variant (nt 3088-3400 in SpSRCR7.1).

Plate 1, Fig. 4. Alternative expression of two intracellular variants of SpSRCR7. Coelomocyte total RNA blot from 11 animals (10 µg per lane) was repeatedly probed with the first and second C-terminal variants of SpSRCR7, and with SpThymosin as a loading control. Probes other than those described in the legend of Plate 1, Figs. 1 and 3, were: Intracellular 2, 185 bp product of PCR amplification across the membrane retention segment and membrane proximal half of the intracellular domain in the second variant SpSRCR7.2.

Plate 1, Fig. 5. Clustered gene arrays encoded in two type 1 SRCR loci. Clones of recombinant sea urchin genomic DNA in Bacterial Artificial Chromosome [BAC] plasmid vector, were detected with a genomic SRCR type 1 5' flanking probe. BACs DNA (BAC #1 133 kb; BAC #6 140 kb) was digested with *Not*I (N), *EcoR*I (E) or *Sal*I (S) and the blots were probed with specific gene segments of SpSRCR1. A cluster consisting of at least three closely related type 1 SRCR genes was detected by all four probes in BAC #1 and two in BAC #6, perhaps more distantly related genes. Probes other than those described in the legend of Plate 1, Figs. 2 and 3, were: Exon 1, 223 bp product of PCR amplification (nt 521-744 in genomic clone gSRCR1.59, corresponds to 5' UTR and leader peptide in SpSRCR1); VWF, 311 bp product of PCR amplification (nt 151-462 in SpSRCR1); SRCR type 1 5' flanking probe, 1078 bp (gSRCR1, 110 nt overlap with the 5' end of SpSRCR1).

Though transcription of given SRCR messages in coelomocytes may range from high to undetectable levels among individuals, all possess genomic loci encoding these genes. This is evident when comparing the extent of hybridization in RNA gel blots from 11 animals and in corresponding genome blots (Plate 1, Fig. 3). For example, in animals 7-11 transcripts of SpSRCR1 are barely detectable, and animals 1-3 do not express detectable levels of the first C-terminal variant of SpSRCR7 (see below), yet all reveal the presence of the corresponding gene segments in their genome. Hence, variability in the expression of these SRCR genes is phenotypic, reflecting animal to animal differences in the levels of transcription of the same or highly related SRCR genes.

Coelomocyte SRCRs heterogeneity is generated in part due to alternative usage of gene segments. For example, the sequence of two cDNA clones of SpSRCR1 differed in the number of SRCRs, four or five repeats. Likewise, two completely different forms of the C-terminus were documented in SpSRCR7 transcripts, encompassing the membrane-proximal segment, membrane retention and intracellular domains. Animals may express either one only or both of these intracellular variants of SpSRCR7, as shown in Plate 1, Fig. 4. The gene products of the other five types of coelomocyte SRCRs seem to be secreted proteins, as none features membrane anchorage domains (Fig. 2; types 1, 5, 6, 12 and 20). However, it is possible that gene segments encoding alternatively spliced cell surface forms of these SRCR genes do exists, or that heterodimeric associations result in cell bound forms also of these SRCR genes.

Four of the sea urchin SRCR-containing proteins are mosaic (Fig. 2; types 1, 5, 6 and 7), including additional domains. These include Von Willebrand factor [VWF] repeats in SpSRCR1 and SpSRCR6, each 61-62 aa long, which in mammals participate in platelet-vessel wall interactions at sites of vascular injury, and a domain in SpSRCR1 (ECM; 347 aa) that is similar to a gastrula-specific extracellular matrix domain of another sea urchin species (*Lytechinus variegatus*). The two cell bound forms of SpSRCR7 differ in the C-terminus, immediately following the 7th SRCR: SpSRCR7.1 features three Epidermal Growth Factor-like [EGF] repeats (34-38 aa, unknown function), while SpSRCR7.2 features one Short Consensus Repeat [SCR] of the complement control superfamily (68 aa, unknown function). Other domains are not similar to any published sequences, like the N-terminal domains in SpSRCR5 (285 aa) and SpSRCR6 (53 aa), and the domain that separates the EGF-like repeats from the membrane retention domain in SpSRCR7.1 (52 aa). The SRCRs of gene types 1 and 6 are nearly identical, with respect to repeat domains *per se*, and the same is true for the VWF repeats and leader peptides. In contrast, the SRCRs of other types are highly divergent; SpSRCR12 consists of 20 repeats of 103-112 aa, each with unique sequence (33-83% pairwise aa identities) and the 7 SRCRs in SpSRCR20 are of 101-106 aa (51-91% pairwise aa identities).

The sea urchin genome harbors numerous heterogeneous SRCR gene-loci, which in some cases consist of clustered gene arrays. The number SRCR gene-loci encoding types 1 and 7 was estimated by hybridization of specific gene-segments probes to an arrayed genomic library, constructed in Bacterial Artificial Chromosome vector [BAC]. This BAC library, with inserts of an average length of 130 kb, represents about 13 haploid copies of the *S. purpuratus* genome. For SRCR type 1, 26 clones were identified by hybridization to a genomic SRCR1 5' flanking probe. Detailed analysis was preformed on two of these BAC clones (Plate 1, Fig. 5). Blots of BACs plasmid DNA digested with *Not*I, *EcoR*I or *Sal*I were hybridized with probes detecting Exon 1 of SpSRCR1; the SRCR repeat element; the ECM and the VWF domains. A cluster of at least three closely related SRCR type 1 genes is detected by all four probes in BAC #1 and perhaps two in BAC #6. For SpSRCR7, 13 clones hybridized with an SRCR7 probe and with a probe for the intracellular C-terminal region of variant SpSRCR7.2 (Intracellular 2 probe). Seven clones hybridized with the probe for the first C-terminal variant of SpSRCR7.1, but only two of these were the same clones identified with the Intracellular 2 probe. The locus encoding the first intracellular domain must be located at least tens of kilobases away from the locus encoding the second intracellular domain of SpSRCR7.

CONCLUSION

Sea urchin coelomocytes express multiple SRCR genes from amongst numerous SRCR encoding gene-loci, many of which are transcribed as individual-specific repertoires.Our data shows that SRCR gene types 1, 5 and 7 are present in the genomes of all animals that were tested, whether or not transcripts of these are detectable in coelomocytes of the same animals. It is possible that the genetic apparatus of individual sea urchins responds to immune stimuli by the selection of a specific repertoire of SRCR genes to be expressed. However, what determines the individual-specific transcript levels of given SRCR genes is unknown, nor is it known whether levels of expression vary over time or are fixed. Coelomocytes of individual sea urchins could transcribe similar SRCR genes in all descendents from the same cell lineage, or cells of common origin may exhibit clonal variation in the transcribed repertoire of SRCR genes. Hence, animals might differ in: 1) the regulatory mechanisms that control expression of given coelomocyte SRCR genes; 2) the numbers of circulating coelomocytes of given classes; or 3) in their history of exposure to the stimuli that induce specific SRCR gene expression. Experiments are underway to determine genotypic and phenotypic determinants of the sea urchin SRCR repertoire, and its function.

The most unexpected discovery in regard to sea urchin immune system function is the differential heterogeneity in expression of coelomocyte-specific genes in individual sea urchins. At an abstract level the outcome from individual-specific expression of coelomocyte SRCR molecules could be amazingly reminiscent of that produced by vertebrate adaptive immune system, even though the effector molecules are different. For the first time we now possess molecular tools to probe into the function and meaning of non-vertebrate heterogeneous and polymorphic immune cells receptors, as putative participants in an evolutionary ancient innate host defense system. We may be seeing the tip of the iceberg with respect to the molecular definition of immune individuality in our distant deuterostome cousins.

ACKNOWLEDGMENTS

Supported by HFSP Grant RG-333/96 and by NIH Training Grant HD-07257.

REFERENCES

Al-Sharif, W.Z., Sunyer, J.O., Lambris, J.D., and Smith, L.C. Sea urchin coelomocytes specifically express a homologue of the complement component C3. J. Immunol., **160**, 2983-2997, 1998.

Aruffo, A., G.S., Bowen, M.A., Patel, D.D., Hynes, B.F., Starling, G.C., Gebe, J.A. and Bajorath, J. CD6-ligand interactions: a paradigm for SRCR domain function? Immunol. Today, **18**, 498-504, 1997.

Aruffo, A., Melnick, M.B., Linsley, P.S. and Seed, B. The lymphocyte glycoprotein CD6 contains a repeated domain structure characteristic of a new family of cell surface and secreted proteins. J. Exp. Med., **174**, 949-952, 1991.

Blumbach, B., Pancer, Z., Diehl-Seifert, B., Steffen, R., Münkner, J., Müller, I. and Müller, W.E.G. The putative sponge aggregation receptor: Isolation and characterization of a molecule composed of scavenger receptor cysteine-rich domains and short consensus repeats. J. Cell Sci., **111**, 2635-2644, 1998.

Cheng, H., Bjerknes, M. and Chen, H. CRP-ductin: gene expressed in intestinal crypts and in pancreatic and hepatic ducts. Anat. Rec., **244**, 327-343, 1996.

Dangott, L.J., Jordan, J.E., Bellet, R.A. and Garbers, D.L. Cloning of the mRNA for the protein that crosslinks to the egg peptide speract. Proc. Natl. Acad. Sci. USA, **86**, 2128-2132, 1989.

Elomaa, O., Kangas, M., Sahlberg, C., Tuukkanen, J., Sormunen, R., Liakka, A., Thesleff, I., Kraal, G. and Tryggvason, K. Cloning of a novel bacteria-binding receptor structurally related to scavenger receptors and expressed in a subset of macrophages. Cell, **80**, 603-609, 1995.

Freeman, M., Ashkenas, J., Rees, D.J., Kingsley, D.M., Copeland, N.G., Jenkins, N.A. and Krieger, M. An ancient, highly conserved family of cysteine-rich protein domains revealed by cloning type I and type II murine macrophage scavenger receptors. Proc. Natl. Acad. Sci. USA, **87**, 8810-8814, 1990.

Friedman, J., Trahey, M. and Weissman, I. Cloning and characterization of cyclophilin C-associated protein: a candidate natural cellular ligand for cyclophilin C. Proc. Natl. Acad. Sci. USA, **90**, 6815-6819, 1993.

Gebe, J.A., Kiener, P.A., Ring, H.Z., Li, X., Francke, U. and Aruffo, A. Molecular cloning, mapping to human chromosome 1 q21-q23, and cell binding characteristics of SPa, a new member of the scavenger receptor cysteine-rich (SRCR) family of proteins. J. Biol. Chem., **272**, 6151-58, 1997.

Goldberger, G., Bruns, G.A., Rits, M., Edge, M.D. and Kwiatkowski, D.J. Human complement factor I: analysis of cDNA-derived primary structure and assignment of its gene to chromosome 4. J. Biol. Chem., **262**, 10065-10071, 1987.

Hoffmann, J.A., Reichhart, J-M. and Hertru, C. Innate immunity in higher insects. Cur. Opinion Immunol., **8**, 8-13, 1996.

Hohenester, E., Sasaki, T. and Timpl, R. Crystal structure of a scavenger receptor cysteine-rich domain sheds light on an ancient superfamily. Nature Struct. Biol., **6**, 228-232, 1999

Kanan, J.H.C., Nayeem, N., Binns, R.M. and Chain, B.M. Mechanisms for variability in a member of the scavenger-receptor cysteine-rich superfamily. Immunogenetics **46,** 276-82, 1997.

Kirkham, P.A., Takamatsu, H-H. and Parkhouse R.M.E. Growth arrest of gd T cells induced by monoclonal antibody against WC1 correlates with activation of multiple tyrosine phosphatases and dephosphorylation of MAP kinases erk2. Eur. J. Immunol., **27**, 717-725, 1997.

Klein, J. Ehrlich and Darwin: homochauvinism in immunology. Immunol. Cell Biol., **73,** 103-108, 1995.

Koths, K., Taylor, E., Halenbeck, R., Casipit, C. and Wang, A. Cloning and characterization of a human Mac-2-binding protein, a new member of the superfamily defined by the macrophage scavenger receptor cysteine-rich domain. J. Biol. Chem., **268**, 14225-14249, 1993.

Krieger, M. Molecular flypaper and arteriosclerosis: structure of the macrophage scavenger receptor. Trends Biochem. Sci., **17**, 141-146, 1992.

Law, S.K., Micklem, K.J., Shaw, J.M., Zhang, X.P., Dong, Y., Willis, A.C. and Mason, D.Y. A new macrophage differentiation antigen which is a member of the scavenger receptor superfamily. Eur. J. Immunol., **23**, 2320-2325, 1993.

Mayer, W.E. and Tichy, H. A cDNA clone from the sea lamprey *Petromyzon marinus* coding for a scavenger receptor Cys-rich (SRCR) domain protein. Gene, **164**, 267-271, 1995.

O'Keeffe, M.A., Metcalfe, S.A., Cunningham, C.P. and Walker, I.D. Sheep $CD4^+$ ab T cells express novel members of the T19 multigene family. Immunogenetics, **49**, 45-55, 1999.

Pancer, Z., Münkner, J., Müller, I. and Müller, W.E.G. A novel member of an ancient superfamily: sponge (*Geodia cydonium*, Porifera) putative protein that features scavenger receptor cysteine-rich repeats. Gene, **193,** 211-218, 1997.

Pancer, Z., Rast, J.P. and Davidson E.H. Origins of immunity: transcription factors and homologues of effector genes of the vertebrate immune system expressed in sea urchin coelomocytes. Immunogenetics, **49**, 773-786, 1999.

Resnick, D., Pearson, A. and Krieger, M. The SRCR superfamily: a family reminiscent of the Ig superfamily. Trends Biochem. Sci., **19**, 5-8, 1994.

Sima, P. Evolution of immune reactions. Crit. Rev. Immunol., **13,** 83-114, 1993.

Smith, L.C., and Davidson, E.H. The echinoid immune system and the phylogenetic occurrence of immune mechanisms in deuterostomes. Immunol. Today, **13**, 356-361, 1992.

Smith, L.C., and Davidson, E.H. () The echinoderm immune system. Ann. NY Acad. Sci., **712**, 213-226, 1994.

Smith, L.C., Britten, R.J. and Davidson, E.H. SpCoel1: a sea urchin profilin gene expressed specifically in coelomocytes in response to injury. Mol. Biol. Cell, **3**, 403-414, 1992.

Smith, L.C., Britten, R.J. and Davidson, E.H. Lipopolysaccharide activates the sea urchin immune system. Dev. Comp. Immunol., **19,** 217-224, 1995.

Smith, L.C., Chang, L., Britten, R.J. and Davidson, E.H. Sea urchin genes expressed in activated coelomocytes are identified by expressed sequence tags. Immunology, **156,** 593-602, 1996.

Smith, L.C., Shih, C-S. and Dachenhausen, S.G. Coelomocytes express SpBf, a homologue of factor B, the second component in the sea urchin complement system. J. Immunol., **161**,6784-6793. 1998.

Takamatsu, H-H., Kirkham, P.A. and Parkhouse R.M.E. A gd T cell specific surface receptor (WC1) signaling G0/G1 cell cycle arrest. Eur. J. Immunol., **27**, 105-100, 1997.

Van de Velde, H., von Hoegen, I., Luo, W., Parnes, J.R. and Thielemans, K. The B-cell surface protein CD72/Lyb-2 is the ligand for CD5. Nature, **351**, 662-65, 1991.

Walker, I.D., Glew, M.D., O'Keeffe, M.A., Metclafe, S.A., Clevers, H.C., Wijngaard, P.L.J., Adams, T.E. and Hein, W.R. A novel multi-gene family of sheep T cells. Immunol., **83**, 517-523, 1994.

Whitney, G.S., Bowen, M.A., Neubauer, M. and Aruffo, A. Cloning and characterization of murine CD6. Mol. Immunol., **32**, 89-92, 1994.

Whitney, G.S., Starling, G.C., Bowen, M.A., Modrell, B., Siadak, A.W. and Aruffo, A. The membrane-proximal scavenger receptor cysteine-rich domain of CD6 contains the activated leukocyte cell adhesion molecule binding site. J. Biol. Chem., **270,** 18187-18190, 1995.

Wijngaard, P.L., Metzelaar, M.J., MacHugh, N.D., Morrison, W.I. and Clevers, H.C. Molecular characterization of the WC1 antigen expressed specifically on bovine CD4-CD8-gamma delta T lymphocytes. J. Immunol., **149**, 3273-3277, 1992.

Yu, Q., Reichert, M., Brousseau, T., Cleuter, Y., Burny, A. and Kettmann, R. Sequence of bovine CD5. Nucleic Acids Res., **18**, 5296, 1990.

MOLECULAR CHARACTERIZATION OF IL-1-LIKE MOLECULES FROM LOWER VERTEBRATES AND INVERTEBRATES

Gregory Beck and Nobel Truong

University of Massachusetts - Boston, Department of Biology, 100 Morrissey Blvd., Boston, MA 02125

INTRODUCTION

The immune system plays a crucial role in protecting the host from attack by foreign pathogens. The evolution of defense mechanisms can be traced back through the invertebrates (Abbas *et al.*, 1997). Vertebrate host defense mechanisms have evolved from their invertebrate ancestors (Sima and Vetvicka, 1993). In vertebrates there are two specific arms of the immune response, humoral and cell-mediated immunity. The central mediators of the humoral response are antibodies, defense molecules present in body fluid such as blood, mucus, and milk. The cell-mediated response is mediated by the lymphocytes, cells that recognize and respond to foreign antigens.

Invertebrates have immune systems that are also composed of humoral and cell-mediated responses. Humoral responses include the synthesis and release of antibacterial peptides such as lysozyme, cecropins, and attacins (Pathak, 1993). Cell-mediated responses involve phagocytosis, nodule formation, and encapsulation of foreign antigens. Phagocytosis is an ancient form of host defense, a process in which a cell ingests a foreign invader. This process is common to both vertebrate and

Phylogenetic Perspectives on the Vertebrate Immune System.
Edited by Beck *et al.*, Kluwer Academic/Plenum Publishers, 2001.

invertebrate immune systems. However, invertebrates do not have antigen-specific lymphocytes and do not produce antibodies (Abbas *et al.*, 1997).

Some invertebrate species contain phagocytes capable of producing cytokine-like molecules similar to those of the vertebrates (Beck and Habicht, 1986; Ellis, 1996). Cytokines are low molecular weight soluble proteins that act as chemical messengers of the immune system. Cytokines are produced by leukocytes (and in some instances other cell types), and play an essential role in regulating animal host defenses. Their role in the vertebrate immune response includes control of cellular proliferation and differentiation, and regulation of hematopoiesis (Abbas *et al.*, 1997). Examples of cytokines include, tumor necrosis factor (TNF), the interferons (IFN), and the members of the interleukin (IL) family to name a few.

INTERLEUKIN-1

Interleukin-1 (IL-1) is a family of three related proteins, IL-1α, IL-1β, and IL-1ra (receptor antagonist) and was one of the first cytokines to be discovered. Before discussing the characteristics of IL-1, its role as a mediator of responses to diseases and effect on acute phase response, it is important to mention its history. As early as the 1940s, researchers hypothesized that a factor synthesized and released by activated mononuclear phagocytic cells might mediate fever (Dinarello, 1984). Some investigators identified such a factor and named it endogenous pyrogen (EP), and others, lymphocyte activating factor (LAF). In the late 1960s, it was found that EP/LAF was released from stimulated monocytes and acted on the hypothalamus to induce fever (Gery and Schmidt, 1985; Sission, 1989). Researchers in the early 1970s reported that lipopolysaccharide (LPs), an endotoxin produced by gram-negative bacteria, stimulated human and murine adherent cells (mainly macrophages) and induced the production of EP/LAF (Gery and Waksman, 1972). It wasn't until the late 1970s that these activities were found to be identical and the name interleukin-1 was given to the soluble factor synthesized and released by monocytes. The monocyte is not the only type of cell that is capable of synthesis and release of IL-1. Diverse cell types such as epithelial and endothelial cells are also capable of producing IL-1 (Dinarello, 1984). Additional studies have revealed that the role of IL-1 in the immune system is more than a fever activating agent (Dinarello, 1984; Luger, 1983). IL-1 acts as a signal affecting immunocompetent cells that results in the amplification of specific immune responses (Mizel, 1979).

Early biological analyses showed EP/LAF is present in human and mouse macrophages (cell line $P388D_1$). It possesses a molecular weight in the range of 12 to 17 kilodaltons with isoelectric points ranging from 4.5 to 5.5 and 6.8 to 7.4 (Dinarello *et al.*, 1974; Krakauer *et al.*, 1982; Mizel, 1979; Mizel and Rosenstreich, 1979). Mizel and colleagues also found that EP/LAF has an inactivated precursor form with higher molecular weights, 26 to 39 kilodaltons and isoelectric points 5.2 to 5.9 (Mizel, 1979; Mizel and Rosenstreich, 1979). They found that the precursor form is synthesized by activated macrophages and upon exposure to a second stimulus, the high molecular weight inactive protein is converted to the active low molecular weight protein and secreted. However, they had not ruled out that a single stimulation to the macrophages may trigger the same mechanism. Further analysis of this mechanism shows that the process in which the inactive high molecular weight form converts to the active low molecular weight form could be due to protein complex dissociation or enzymatic action (Dinarello, 1984; Mizel and Rosenstreich, 1979).

In 1984, two cDNAs clones were isolated: IL-1α was cloned from murine macrophage cell line $P388D_1$, and IL-1β was cloned from human blood monocytes (Auron *et al.*, 1984; Lomedico *et al.*, 1984). IL-1α and IL-1β have isoelectric points of 5.3 and 7.2, respectively, sharing similar pIs with EP/LAF (Dinarello, 1992). Subsequently, IL-1α was cloned in human, rabbit, and rat; IL-1β was cloned in rat, cow, rabbit, and mouse. Human IL-1α and IL-1β are coded by genes on chromosome 2 and possess all the biological properties related to EP/LAF (Sission, 1989). The genomic sequence for the human IL-1α gene is 10.5 kilobases while the IL-1β gene is 7.8 kilobases (Clark *et al.*, 1986; Furutani *et al.*, 1986). Each gene contains seven exons which are translated as a 31 kilodalton precursor IL-1 polypeptide. The polypeptide is then cleaved by specific proteases (ICE) in a posttranslational process resulting in a 17 kilodalton final product. The precursor and the mature IL-1α are both active, whereas the IL-1β precursor form is inactive and requires posttranslational cleavage to be active. The two precursor forms of IL-1 polypeptide share only 26 percent amino acid homology. In the mature forms of IL-1, the amino acid sequences are moderately conserved, 60 to 70 percent between species, for IL-1α, and 75 to 78 percent for IL-1β. The amino acid sequence homology between IL-1α and β in humans is only 20 percent and in mice 23 percent. Despite the low amino acid homology, both proteins are recognized by the same receptors and share similar functions. The region of highest similarity is found on the carboxyl ends of the mature IL-1 forms. This could be the sites recognized by the IL-1 receptors (Dinarello, 1989). These regions also share some amino acid sequence homology to other cytokines such as interferon-α2 and interferon-β 1. Blocking these sites using antibodies do not inhibit the biological activities of IL-1. However, antibodies against the C- terminal and N-terminal inhibit IL-1 from binding to its receptors thus reducing its biological functions (Gabay *et al.*, 1997). Therefore, it appears that both terminals are involved in receptor binding activities (Gabay *et al.*, 1997).

A third member of the IL-1 family is IL-1ra, an anti-inflammatory protein that inhibits the biological effects of IL-1α and IL-1β by binding to the IL-1 receptors (Gabay *et al.*, 1997). The IL-1ra gene is also located in chromosome 2, contains four exons which code a 17 kilodalton polypeptide. The protein shares 19 percent amino acid sequence homology with IL-1α and 30 percent with IL-1β (Dripps *et al.*, 1991; Granowitz *et al.*, 1991).

MOLECULAR EVOLUTION OF VERTEBRATE CYTOKINES

Although having very low sequence homology it appears that IL-1α, β and the ra evolved from a common ancestral molecule (Eisenberg *et al.*, 1991). They all share similar exon-intron junction number and position. IL-1α and β both consist of 7 exons coding for similar regions of the molecules (Eisenberg *et al.*, 1991). IL-1ra's second, third and fourth exons align with α and β's fifth, sixth and seventh exons (Eisenberg *et al.*, 1991; Graves *et al.*, 1990; Schreuder *et al.*, 1995). The crystal structure for human IL-1α (Graves *et al.*, 1990), β (Priestle *et al.*, 1988), and ra (Schreuder *et al.*, 1995) have been determined. In general, the overall structure is composed of 12 β- strands. Six of these form a barrel which is closed at one end by the other 6 strands. The structure can best be thought of as a tetrahedron with 3 antiparellel β- strands forming each triangular face (Priestle *et al.*, 1988). Although they are structurally similar, only α and β are biologically active. Indeed, despite the similarity in size, biological characteristics, and receptor binding IL-1α and β possess only 23% amino acid sequence identity (Auron *et al.*, 1984, March *et al.*, 1985).

The molecular data of Eisenberg *et al.* clearly indicate that the IL-1ra and IL-1β are more homologous to each other than either of them is to IL-1α (Eisenberg *et al.*, 1991). They also reported that based on their comparison of the human and mouse genes, IL-1α is evolving over twice as fast as IL-1β and the IL-1ra. They attribute this difference to a difference in the mutation rate. They gave two hypotheses for the origin of the IL-1 family: the IL-1ra gene evolved from an ancestral IL-1α/β gene, or the ancestral IL-1α/β genes evolved from an IL-1ra-like gene. Other studies have shown that the IL-1α family appears to be evolving faster than the IL-1β family, which implies that IL-1β is closer to the common ancestral protein and perhaps retains more of its properties (Young and Sylvester, 1989). It has also been suggested that IL-1ra evolved from a duplicated IL-1β gene (Hughes, 1994).

Several groups have mapped IL-1 to determine functional areas. To localize the active regions of IL-1, Mosley *et al.* (Mosley *et al.*, 1987) constructed a series of truncated forms of human IL-1 by removing amino acids from both the amino-terminal and carboxy-terminal ends. They demonstrated that core sequences of 147 amino acids for IL-1β (120-266) and 140 amino acids for IL-1α (128-267) must be left intact to retain full biological activity and that the biological activities of the IL-1 polypeptides parallel their receptor binding capabilities. Auron *et al.* (Auron *et al.*, 1985) showed regional homology between human IL-1β and murine IL-1α. This homology is strongest within the carboxy halves of the molecules [in regions they labeled as C (amino acids 150-162), D (165-185), and E (219-240)]. Using surface plasmon resonance mapping, D'Ettorree *et al.* (D'Ettorree *et al.*, 1997), characterized the IL-1β molecular domains relevant to receptor binding and biological activity. They found several discrete areas of the molecule that were necessary for full biological activity. Two areas appear to be most important: The amino acid stretches 177-186 and 218-273 are critical for binding to the IL-1 receptor. These regions correspond to β- strands 5 and 6 and β- strands 8 and 9, respectively.

More recently, studies of lower vertebrate cytokines have resulted in some unexpected findings. Secombes' group (Secombes *et al.*, 1997 and this volume) has found that the trout IL-1β molecule is missing the first β- strand. This implies that the molecule could be ra-like and not an agonist, but this can not be since the molecule is biologically active (Secombes *et al.*, 1997). What it does suggest is that PCR studies using primers based on nucleotides located in the N-terminal region may be ineffective in amplifying any IL-1 sequences from lower vertebrates. Other odd findings concerning lower vertebrate cytokines have been reported. Several groups have attempted to clone chicken IL-2 for over 7 years. Sundick and Gill-Dixon (Sundick and Gill-Dixon, 1997) finally had to rely on an expression vector system and a complex biological activity assay to succeed. When they finally cloned the molecule the reason for the difficulty was revealed. The chicken IL-2 protein is very similar to both mammalian IL-2 and IL-15. Similarly, Dixon *et al.* (Dixon *et al.*, 1997) were forced to pick random sequences from a trout cDNA library in order to find a β- chemokine-like cDNA. These studies, as well as our own, attest to the time consuming and difficult nature of these vertebrate-invertebrate cloning strategies.

CYTOKINE-LIKE MOLECULES IN LOWER ANIMAL SPECIES

Cytokines (or molecules with cytokine-like properties) have been isolated from a variety of invertebrate species (reviewed in Beck *et al.*, 1989 and Beck and Habicht, 1991a). In general they are released by phagocytes and act on various aspects of the inflammatory response. Chain and Anderson have isolated plasmatocyte depletion factor from the waxmoth, *Galleria mellonella.* This protein, released by hemocytes, induces the disappearance of plasmatocytes from circulation (Beck and Habicht, 1991a). Cherbas has described haemokinin, a protein isolated from several moths that activates hemocytes (Beck and Habicht, 1991a). Phagocytosis-stimulating mediator has been isolated from the hemolymph of *G. mellonella* by Mohrig and Schittek. As its name implies this protein stimulates phagocytosis within minutes of exposure (Beck and Habicht, 1991a). A snail IL-1-like activity has been reported (Granath *et al.*, 1994). Recently, the same group (Connors *et al.*, 1995) reported that human IL-1 stimulates phagocytosis and superoxide production by *B. glabrata* hemocytes. Burke and Watkins (Burke and Watkins 1991) showed that an IL-1-like factor could be detected in *Asterias* coelomic fluids. Prendergast and colleagues isolated a protein from starfish (sea star factor) that inhibited vertebrate immune responses to T-dependent antigens and inhibited macrophage migration among other effects (Prendergast *et al.*, 1970). Müller and colleagues (Muller *et al.*, 1992) have shown that xenografts between two sponge species contain a TNF-like molecule. Dissous and colleagues (Dissous *et al.*, 1994) have characterized an immunoreactive TNF-like molecule in *B. glabrata.* Stefano and colleagues have detected, by immunoreactivity, several cytokines (IL-1, IL-6, and TNF) in a mollusk (*M. edulis*) (Hughes *et al.*, 1990 and 1991). Studies by Hoffmann and colleagues (Hoffmann *et al.*, 1995) have shown a sequence homologous to an interferon consensus response element in the diptericin promoter of *Drosophila.* This suggests that interferon-related molecules may be present in insects and are part of an insect host defense response. They also showed that the gene for the *Drosophila* defense peptide, diptericin, has sequences with putative transcription regulatory sequences similar to vertebrate cytokine consensus motifs.

EXPERIMENT OBJECTIVES

Previous work from our laboratory showed that invertebrates can synthesize and release cytokine-like proteins (IL-1α– and β-like and IL-6-like proteins) in response to LPS stimulation. Beck and Habicht have successfully isolated and purified a primitive IL-1-like proteins from the starfish, *Asterias forbesi*, using a vertebrate assay system (the murine thymocyte proliferation assay) (Beck and Habicht, 1986; Beck and Habicht, 1991). The primitive IL-1-like proteins isolated from starfish has a molecular weight of approximately 20,000 daltons (Beck and Habicht, 1986). Recently, IL-1-like and IL-6-like proteins were detected in the hemolymph of the tobacco hormworm, *Manduca sexta* (Beck *et al.*, 1996; Ellis, 1996; Savaz, 1997; Beck *et al.*, 1998; Truong *et al*, this volume), with molecular weights of approximately 20,000 and 32,000 daltons, respectively (Beck *et al.*, 1996; Savaz, 1997; Truong *et al*, this volume). The isolation and characterization of these cytokine-like proteins suggests that vertebrates and invertebrates have similar immune responses and that cytokines are ancient and functionally conserved proteins (Beck and Habicht, 1986).

MATERIALS AND METHODS

Bioinformatics

DNA sequences were analyzed using MacVector software (Oxford Molecular Group, UK) and BLAST search through GenBank (National Center for Biotechnology Information, Bethesda, MD). Interleukin-1 nucleic acid and amino acid sequences were retrieved using search programs in Genbank, EMBL (European Molecular Biology Laboratory, Heildenberg, Germany), and MacVector software. Interleukin-1 sequences were analyzed using search algorithms such as BLAST and PSI-BLAST to find percentage identities in nucleic acid and amino acid sequences. An ENTREZ program was used to search for sequence literature. Amino acid sequences were used to run a TBLASTN search against the public database of ESTs (expressed sequence tag) and the MOTIF pattern search database. Sequence alignments were performed using CLUSTAL X (National Center for Biotechnology Information, Bethesda, MD).

mRNA extraction

Frozen activated *Manduca* cells were thawed in preparation for mRNA extraction using Dynabeads- oligo(dT)25 (Oslo, Norway). To the cells, 1 ml of the lysis-binding buffer was added. The lysate was drawn through a 22 gauge needle 3 times to disrupt the cells. The lysate was added to a tube containing Dynabeads and annealed to the oligo(dT)25 at 22°C for 5 minutes. The Dynabeads containing the mRNA were washed with washing buffer with SDS and then again with washing buffer. The mRNA was eluted with EDTA (pH 8.0) at 65°C and stored at -80°C until use.

Genomic DNA extraction

Activated *Manduca* cells were homogenized in 4 ml buffer A. The composition of buffer A is shown in table 4. The supernatant was transferred to a centrifuge tube and incubated at 65°C for 30 minutes. To the supernatant, 960 µl of 5 M potassium acetate were added and incubated on ice for 20 minutes and then centrifuged at 14,000 *x* g for 5 minutes to remove debris. The supernatant was transferred to a new tube and centrifuged at 14,000 *x g* for 5 minutes to ensure all insoluble debris were removed. The supernatant was precipitated with an equal volume of ethyl alcohol, the pellet (genomic DNA) was collected by centrifugation and resuspended in 2.4 ml 10 mM Tris-HCl (pH 8.0) and 1.0 mM EDTA. The genomic DNA (gDNA) was further purified by adding 240 ml of a 2 M sodium acetate solution followed by 2 volumes of ethyl alcohol and precipitated at 22°C for 5 minutes. Genomic DNA was collected by centrifugation, washed with 70% ethyl alcohol, and resuspended in 10 mM Tris-HCl (pH 8.0) and 1.0 mM EDTA and stored at -20°C. Ten to 20 μg of gDNA was used in PCR reactions.

Reverse transcriptase

The mRNA (0.5 μg) was denatured by heating at 90°C for 5 minutes then quickly frozen on dry ice and mixed with 1 ml of 4.3 pmol/ml clamping oligo(dT) primer (3' – V(A/G/C)TTTTT...- 5'), 2.5 μl of 10X *Taq* buffer (15 mM $MgCl_2$, 500 mM KCl, 100 mM Tris-HCl (pH 9.0), 1% Triton X-100) and adjusted to a total volume of 25 μl with nuclease-free water. The sample was incubated on ice until used. The cDNA was synthesized by adding 2.5 μl of 10X *Tag* buffer, 5 μl of dNTP (10 mM), 16.5 μl nuclease-free water, and 1 μl (200 units) of MMLV reverse transcriptase (Promega Co.) to the sample and incubated at 37°C for 45 minutes. The reaction was stored at -20°C until use. Five μl of cDNA was used in a PCR reaction.

PCR amplification of cDNA and gDNA

Primers were selected from the conserved regions of human, mouse and rabbit IL-1 sequences. PCR reactions (50 μl) containing 50 mM KCl, 0.1% Triton X-100, 2.5 mM $MgCl_2$, 10 mM Tris-HCl (pH 9.0), a specified amount of cDNA or gDNA, 20 to 30 pmol of each primer. Also added to the PCR tube was 0.25 mM dNTP, and 2.5 units *Taq* polymerase (Sigma). Each tube was overlaid with mineral oil and amplified using a HyBaid Thermocycler: 1 minute denature at 94°C, followed by 30 cycles, 92°C for 1 minute, 40°C for 1 minute, 72°C for 1 minutes, 5 minutes extension at 72°C. To ensure that the PCR-amplified DNAs were not due to contamination, every experiment included a negative control in which no nucleic acid was added to the reaction. Gene specific primers to human β- actin were added to every experiment as a positive control to ensure cDNA or gDNA is present in the sample. The PCR products were detected by electrophoresis on a 0.8% ethidium bromide (EtBr) agarose gel and viewed under ultraviolet light.

Cleaning, cloning and sequencing of cDNA

A single cDNA band was excised and purified using a Geneclean kit (Bio 101 Inc.). The purified DNA fragment was ligated into pGEM-T Vector system (Promega Co.) and incubated overnight at 12°C according to Promega's directions.

The vector was transformed (1800 Volts and 150 Ω) into DH5α competent cells by electroporation and grew on LB plates with ampicillin/IPTG/X-Gal overnight (The Electroporator II Manual, Invitrogen, San Diego, CA). White bacteria colonies were randomly isolated and grew in LB broth with ampicillin. The selected colonies were used to run a second PCR reaction and the products were run on a 0.8% EtBr agarose gel to determine if the isolated colonies contain the inserted DNA fragment. Plasmid was isolated from colony that contained the inserted DNA fragment using Wizard™ Minipreps (Promega Co.), and sequenced using an Perkin Elmer automatic DNA Sequencer (Foster City, CA).

RESULTS AND DISCUSSION

Human IL-1 amino acid sequences were used to search against a non-redundant GenBank BLASTP database designated nr. The BLASTP program uses a local alignment algorithm to align the most similar sequences. Several amino acid sequences that show significant homology to human IL-1 were identified. Interestingly, IL-1 has been found and sequenced in other organisms besides mammals such as chicken and fish (Weining *et al.*, 1998; Fujiki *et al.*, 1998; Zou *et al.*, 1999). Amino acid sequences for IL-1α share higher percentage identities compared to IL-1β. This suggests that IL-1α is a more conserved protein compared to IL-1β. However, the amino acid sequences in the IL-1ra family have the highest percentage identities (>72%) when compared with human amino acid sequence. The IL-1ra is the most conserved protein family compared to both IL-1α and IL-1β. Within the IL-1α protein family, bovine, dog, goat, horse, and sheep have the highest percentage (70% to 72%) amino acid identities when compared to human amino acid. Rabbit protein sequence share the highest similarity (71%), whereas carp shares the lowest (27%) amino acid identities with human in the IL-1β protein family. Amino acid sequences for both IL-1α and IL-1β were compared with each other using BLAST between two sequence local alignments. Both IL-1 protein forms show a low amino acid homology (range 12-28%) with the highest is seen in rats (28%). Surprisingly, rabbit IL-1α and IL-1β proteins have no significant similarity between each other. The same analysis was performed with IL-1β and IL-1ra resulting in a 30% to 36% amino acid sequence identity. The IL-1β proteins have higher amino acid identities with IL-1ra as compared to IL-1α. There was no significant similarity between IL-1α and IL-1ra protein sequences when both were compared. Nucleotide sequences for IL-1 were used to perform a BLASTN search on GenBank non-redundant database. The result of this search was that IL-1 nucleotide sequences exhibit significant homology and are highly conserved.

All three forms of the IL-1 protein in each organism were subjected to homology domains search using the MOTIF database. The MOTIF database contains the PROSITE pattern library, the PROSITE profile library, ProDom database, BLOCKS database, and PRINTS database. The PROSITE pattern library contains descriptors for functionally important protein motifs. IL-1 proteins in all organisms contain the interleukin-1 signature site, cAMP- and cGMP-dependent protein kinase phosphorylation site, Casein kinase II phosphorylation site, *N*-glycosylation site, *N*-myristoylation site, protein kinase c phosphorylation sites and tyrosine phosphorylation sites which are located in the middle of the protein. We did not find any motif in the PROSITE profile. The BLOCKS, ProDom, and PRINTS are motif databases similar to PROSITE; however, PRINTS is an advanced database. All three databases show IL-1 proteins share a similar region called the interleukin-1 signature domain (99) with 21 amino acids in length located near the carboxyl terminal (FESAAFPNWYISTSQAEHKPV).

Multiple sequence alignments were performed with IL-1α, IL-1β, and IL-1ra amino acid sequences using ClustalX. The alignment shows IL-1α NH2-terminal is highly conserved. It contains stretch of 36 amino acids starting with the translational initiation peptide methionine and ending with a tyrosine. The multiple alignment of IL-1α, IL-1β and IL-1ra show that all three proteins are conserved at the IL-1 signature region.

The IL-1 signature region of all three proteins was realigned as shown in Figure 1. The conserved amino acids are phenylalanine, serine, proline, lysine, isoleucine, threonine, and valine. A phylogenetic tree of IL-1α, generated with NJPlot, reveals mouse and rat proteins are the more ancient proteins compared to that of human, rabbit, sheep, goat, bovine, pig, horse, cat, and dog. The IL-1β phylogenetic tree shows mammal IL-1β proteins were derived from chicken and that they share a common ancestor protein with carp and trout.

IL-1α	
Goat	FKSVAHPKLFIATKQEKLVHM
Sheep	FKSVAHPKLFIATKQEKLVHM
Bovine	FKSVAHPKLFIATKQEKLVHM
Horse	FKSVAHPKLFIATKQGKLVHM
Pig	FKSAAHPKLFIATRQEKLVHM
Cat	FKSVAHPKLFIATQEEQLVHM
Dog	FKSVAQPKLFIATQERKLVHM
Rabbit	FKSAANPQLFIATKPEHLVHM
Human	FTSVAHPNLFIATKQDYWVCL
Mouse	FTSAAYPELFIATKEQSRVHL
Rat	FTSAAFPELLIATKEQSQVHL
IL-1β	
Pig	FESVLYPNWYISTSQIEEKPV
Goat	FESVLYPNWYISTSQIEEKPV
Sheep	FESVLYPNWYISTSQIEEKPV
Bovine	FESVLYPNWYISTSQIEERPV
Human	FESAQFPNWYISTSQAENMPV
Rabbit	FESAQFPNWYISTSQTEYMPV
Cat	FESSQFPNWYISTSQAEEMPV
Rat	FESAQFPNWYISTSQAEHRPV
Mouse	FESAEFPNWYISTSQAEHKPV
Horse	FESAMYPNWYISTSQAEKKPV
Chicken	FESAAFPGWFICTSLQPRQPV
Carp	FESVKYPGWFISTAFDDWEKV
IL-1ra	
Human	FESAACPGWFLCTAMEADQPV
Mouse	FESAACPGWFLCTTLEADRPV
Rat	FESLACPGWFLCTTLEADHPV
Bovine	FESAACPGWFLCTSLEADQPV
Rabbit	FESASCPGWFLCTALEADQPV
Horse	FESAACPGWFLCTAQEADRPV
IL-1 Consensus	FESAACPGWFICTSLQADVPV

Figure 1. Multiple alignment of IL-1 signature domains. The alignment was performed with the human, mouse, rat, bovine, rabbit, and horse sequences of three IL-1 proteins.

The consensus interleukin-1 signature domain amino acid sequence was used to search the GenBank EST database using a TBLASTN algorithm. The result of this search was the identification of several lower vertebrate and invertebrate ESTs that exhibit significant homology to IL-1β. An isolated zebra fish EST, a 5' cDNA clone isolated from fin regeneration (accession # AI330756), has a 43% sequence identity when compared to the human IL-1β sequence (Figure 2). *Drosophila* ESTs are 5' cDNA clones isolated from embryo (accession #'s AA246678 and AA978741). Both share a 28% sequence identity when compared to human IL-1β sequence. These ESTs were retrieved and translated into proteins using MacVector software. The translated products were then searched against a nonredundant database using BLASTP algorithm. The result of this search was that the retrieved ESTs shows significant homology to both lower vertebrate and mammal IL-1β sequences. The translated zebra fish product revealed a 48% and 25% amino acid identity when compared with the carp and horse protein sequences, respectively. The translated *Drosophila* product revealed a 27% amino acid identity when compared with the carp protein sequence.

MMACGLYEVTIAPKNLWETDSAVYSDSDEMDCSDPLAMSYRCDMHEGIRLEMWTSQ
HKMKQLVNVIIALNRMKHIKPQSTEFGEKEVLDMLMANVIQEREVNVVDSVPSYTK
TKNVLQ**CTICDQYKKSLVRSGGSPHLQAVTL**RAGSSDLKVRF

Figure 2. Partial translation of a zebra fish EST. Translated EST (AI330756) is a 152 amino acid sequence coding for IL-1β and it contains an IL-1 domain (highlighted) found on the PRINTS database of the MOTIF search.

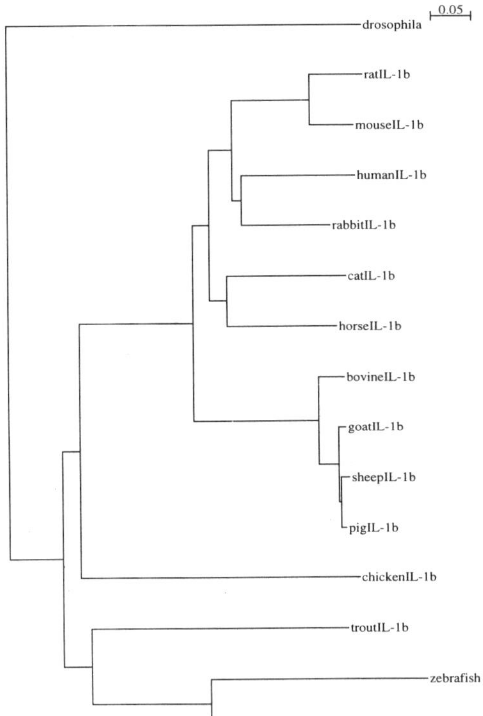

Figure 3. Phylogenetic tree (NJ Plot) of IL-1β with the zebra fish and *Drosophila* translated products included.

RT-PCR of Manduca mRNA and PCR of gDNA

The mRNA extracted from LPS activated *Manduca* hemocytes was used for RT-PCR. Primers were selected to amplify the 3'- end of the IL-1 DNA sequence. PCR was repeated several times with decreasing of the annealing temperature from 60°C to 40°C in increments of 5°C. No PCR product was observed using ethidium bromide staining. However, a 700 bp band was detected in the positive control lane assuring the RT step worked and cDNA is present in the samples (data not shown). Amplification of gDNA using primers with an annealing temperature of 50°C reveals PCR products of approximately 950 bp. The DNA fragment was isolated from the gel and ligated into Promega pGEM-T Vector and electroporated into *E. coli* competent cells. Several colonies were spotted on the plates yielding a blue or white colony.

White colonies were isolated and checked using PCR for inserts. Plasmid was extracted from the colony and the sequence was determined. The isolated sequence was used to search the GenBank nr database using BLASTN algorithm. The results of this search reveal the isolated DNA does not code for IL-1. The search revealed the isolated DNA has a small region of 19 nucleotides exhibiting sequence homology to rat fibroblast growth factor receptor 1-like protein (sequence identity of 100%), *Saccharomyces cerevisiae* mitochondrion COX/OXI3 gene encoding for cytochrome oxidase subunit 1 (sequence identity of 100%), *Drosophila* Val-tRNA-3b gene from site 84D (sequence identity of 100%), and *Saccharomyces cerevisiae* mitochondrion OXI3 gene (sequence identity of 100%).

CONCLUSION

In silico experiments with vertebrate IL-1 amino acid and nucleotide sequence have shed light on our approach to isolating invertebrate cDNA's encoding IL-1-like molecules. Searches using vertebrate IL-1 nucleic acid and amino acid sequences were performed against the public GenBank molecular database. From the searches, it could be concluded that the IL-1 protein family is highly conserved within the vertebrates. Interestingly, studies performed within the mammalian IL-1 sequences revealed they have evolved rapidly, indicating a high mutation rate (Young and Sylvester, 1989; Hughes, 1994). Since trout and carp are considered lower vertebrates, it is predicted that they share low amino acid homology compared to the higher vertebrate organisms. From this, it can be assumed that the invertebrate IL-1 amino acid sequence will be poorly conserved compared to the vertebrates. This will present a problem in isolating invertebrate cDNA encoding IL-1 using molecular techniques, as we have been observing.

The amino acid sequence of insect IL-1-like molecules have not been determined. However, experiments are under way to sequence isolated insect IL-1-like proteins. Data from the sequence will aid us in developing primers to isolate the cDNA encoding other invertebrate IL-1-like proteins using molecular techniques. The existence of IL-1 in insects (Beck *et al.*, 1998; Truong *et al,* this volume) suggests that IL-1 is an ancient and functionally conserved protein. To better support this theory it is necessary to determine the invertebrate IL-1 sequences. Success in this effort would further allow us to study the evolution of the immune system of invertebrates and vertebrates.

ACKNOWLEDGEMENTS

This research was supported by a grant to GB from the National Institutes of Health (AI 39685-01). We would like to thank Karen Beauregard, Linda Mario, Haiyan Zhang, and Wenyu Lin for laboratory help and discussions.

REFERENCES

Abbas A. K., Lichtman A. H. and Pober J. S. (1997) *Cellular and molecular immunology*. Saunders, Philadelphia.

Auron P. E., Webb A. C., Rosenwasser L. J., Mucci S. F., Rich A., Wolff S. M. and Dinarello C. A. (1984) Nucleotide sequence of human monocyte interleukin 1 precursor cDNA. *Proc Natl Acad Sci U S A* **81**, 7907-11.

Auron, P., Rosenwasser, L., Matsushima, K., Copeland, T., Dinarello, C., Oppenheim, J. and A. Webb, (1985). Human and murine interleukin 1 possess sequence and structural similarities. J. *Mol. Cell. Immunol.* 2:169-177.

Beck G., Cardinale S., Wang L., Reiner M. and Sugumaran M. (1996) Characterization of a defense complex consisting of interleukin 1 and phenol oxidase from the hemolymph of the tobacco hornworm, *Manduca sexta*. *J Biol Chem* **271**, 11035-8.

Beck, G., O'Brien, R. and G. Habicht, (1989). Invertebrate cytokines: the phylogenetic emergence of interleukin-1. *BioEssays* 11:62-67.

Beck G. and Habicht G. S. (1986) Isolation and characterization of a primitive interleukin-1-like protein from an invertebrate, *Asterias forbesi*. *Proc Natl Acad Sci U S A* **83**, 7429-33.

Beck G. and Habicht G. S. (1991) Purification and biochemical characterization of an invertebrate interleukin 1. *Mol Immunol* **28**, 577-84.

Beck, G. and Habicht, G. (1991a) Primitive cytokines: Harbingers of vertebrate defense. *Immunol. Today* 12, 180-183.

Binkley S. W., Dickerson W. A., and West W. R. (1975) The Tobacco Hornworm. *Carolina Tips* **38**.

Boyer, O., Porchet, E., Capron, A. and C. Dissous, (1994). Characterization of immunoreactive TNFα molecules in the gastropod *Biomphalaria glabtata*. *Dev. Comp. Immunol.* 18:211-218. 1-4.

Burke, R.D. and Watkins, R.F. (1991) Stimulation of starfish coelomocytes by interleukin-1. *Biochem. Biophys. Res. Comm.* 180, 579-584.

Clark B. D., Collins K. L., Gandy M. S., Webb A. C. and Auron P. E. (1986) Genomic sequence for human prointerleukin 1 beta: possible evolution from a reverse transcribed prointerleukin 1 alpha gene [published erratum appears in Nucleic Acids Res 1987 Jan 26;15(2):868]. *Nucleic Acids Res* **14**, 7897-914.

Connors, V., DeBuron, I. and Granath, W, (1995). *Schistosoma mansoni*: Interleukin-1 increases phagocytosis and superoxide production by hemocytes and decreases output of cercariae in schistosome-susceptible *Biomphalaria glabrata*. *Exp. Parasitol.* 80:139-148.

D'Ettorree, C., De Chiara, G., Casadei, R., Boraschi, D. and Tagliabue, A. (1997). Functional epitope mapping of human interleukin-1β by surface plasmon resonance. *Eur. Cytokine Netw.* 8, 161-171.

Dinarello C. A. (1984) Interleukin-1. *Rev Infect Dis* **6**, 51-95.

Dinarello C. A. (1989) Interleukin-1 and its biologically related cytokines. *Adv Immunol* **44**, 153-205.

Dinarello C. A. (1992) The biology of interleukin-1. *Chem Immunol* **51**, 1-32.

Dinarello C. A. (1994) The interleukin-1 family: 10 years of discovery. *FASEB J* **8**, 1314-25.

Dinarello C. A. (1996) Biologic basis for interleukin-1 in disease. *Blood* **87**, 2095-147.

Dinarello C. A., Goldin N. P. and Wolff S. M. (1974) Demonstration and characterization of two distinct human leukocytic pyrogens. *J Exp Med* **139**, 1369-81.

Dixon, B., Shum, B., Adams, E., Magor, K. and Parham, P. (1997). Isolation of a beta-chemokine like cDNA from rainbow trout (*Oncorhynchus mykiss*). *Dev. Comp. Immunol.* 21, 187.

Dower S. K., Kronheim S. R., March C. J., Conlon P. J., Hopp T. P., Gillis S. and Urdal D. L. (1985) Detection and characterization of high affinity plasma membrane receptors for human interleukin 1. *J Exp Med* **162**, 501-15.

Dripps D. J., Brandhuber B. J., Thompson R. C. and Eisenberg S. P. (1991) Interleukin-1 (IL-1) receptor antagonist binds to the 80-kDa IL-1 receptor but does not initiate IL-1 signal transduction. *J Biol Chem* **266**, 10331-6.

Ebberink R. H. M., Smit A. B. and Minnen J. V. (1989) The insulin family: Evolution of structure and function in vertebrates and invertebrates. *Biol. Bull.* **177**, 1776-182.

Einsenberg S. P., Brewer M. T., Verderber E., Heimdal P., Brandhuber B. J., and Thompson R. C. (1991) Interleukin-1 receptor antagonist is a member of the interleukin-1 gene family: Evolution of a cytokine control mechanism. *Proc Natl Acad Sci U S A* **88**, 5232-36.

Ellis T. (1996) Identification of interleukin-1 secreting hemocytes in *Manduca sexta*. University of Massachusetts, Boston.

Evans R. J., Bray J., Childs J. D., Vigers G. P., Brandhuber B. J., Skalicky J. J., Thompson R. C. and Eisenberg S. P. (1995) Mapping receptor binding sites in interleukin (IL)-1 receptor antagonist and IL-1 beta by site-directed mutagenesis. Identification of a single site in IL-1ra and two sites in IL-1 beta. *J Biol Chem* **270**, 11477-83.

Fujiki K., Nakao M., Shin D., and Yano,T. (1998) cDNA cloning of a carp homologue of mammalian interleukin-1 betas. *In press*.

Furutani Y., Notake M., Fukui T., Ohue M., Nomura H., Yamada M. and Nakamura S. (1986) Complete nucleotide sequence of the gene for human interleukin 1 alpha *Nucleic Acids Res* **14**, 3167-79.

Gabay C., Smith M. F., Eidlen D. and Arend W. P. (1997) Interleukin 1 receptor antagonist (IL-1Ra) is an acute-phase protein [see comments]. *J Clin Invest* **99**, 2930-40.

Georgel, P., Kappler, C., Langley, E., Gross, I., Nicolas, E., Reichhart, J. and J. Hoffmann, (1995). *Drosophila* immunity. A sequence homologous to mammalian interferon consensus response element enhances the activity of the diptericin promoter. *Nuc. Acid Res.* 23:1140-1145.

Gery I. and Schmidt J. A. (1985) Human interleukin 1. *Methods In Enzymology* **116**, 456-67.

Gery I. and Waksman B. H. (1972) Potentiation of the T-lymphocyte response to mitogens. II. The cellular source of potentiating mediator(s). *J Exp Med* **136**, 143-55.

Granath, W., Connors, V. and Tarlet, R. (1994) Interleukin-1 activity in hemolymph from strains of the snail *Biomphalaria glaba* varying in susceptibility to the human blood fluke *Schistosoma mansoni:* Presence, differential expression, and biological function. *Cytokine* 6, 21-30.

Granowitz E. V., Clark B. D., Mancilla J. and Dinarello C. A. (1991) Interleukin-1 receptor antagonist competitively inhibits the binding of interleukin-1 to the type II interleukin-1 receptor. *J Biol Chem* **266**, 14147-50.

Graves, B., Hatada, M., Hendrickson, W., Miller, J., Madison, V. and Satow, Y. (1990). Structure of interleukin-1α at 2.7 Å resolution. *Biochemistry* 29, 2679-2684.

Ha S. H., Park T. H. and Kim S.-E. (1996) Silkworm hemolymph as a substitute for fetal bovine serum in insect cell culture. *Biotechnology Techniques* **10**, 401-406.

Hughes A. L. (1994) Evolution of the interleukin-1 gene family in mammals. *J Molec Evol* **39**, 6-12.

Hughes, T., Smith, E., Chin, R., Cadet, P., Sinisterra, J., Leung, M., Shipp, M., Scharrer, B. and G. Stefano, (1990). Interaction of immunoactive monokines (interleukin 1 and tumor necrosis factor) in the bivalve mollusc *Mytilus edulis. Proc. Natl. Acad. Sci. USA.* 87:4426-4429.

Hughes, T., Smith, E. and G. Stefano, (1991). Detection of immunoreactive interleukin 6 in invertebrate hemolymph and nervous tissue. *Prog. NeuroEndocrinImmunology* 4:234-239.

Krakauer T., Mizel D. and Oppenheim J. J. (1982) Independent and synergistic thymocyte proliferative activities of PMA and IL 1. *J Immunol* **129**, 939-41.

Kuby J. (1997) *Immunology*. W.H. Freeman and Company, New York.

Lomedico P. T., Gubler U., Hellmann C. P., Dukovich M., Giri J. G., Pan Y. C., Collier K., Semionow R., Chua A. O. and Mizel S. B. (1984) Cloning and expression of murine interleukin-1 cDNA in *Escherichia coli. Nature* **312**, 458-62.

Luger T. a. O., J. J. (1983) *Charateristics of Interleukin-1 Epidermal-Cell-Derived Thymocytes Activating Factor*. Raven Press, New York.

March C., Mosley B., Larsen A., Cerretti D., Braedt G., Price V., Gillis S., Henney C., Kronheim S., Grabstein K., Colon P., Hopp T. and D. Cosman, (1985). Cloning, sequence, and expression of two distinct human interleukin 1 complementary DNAs. *Nature* 315:641-652.

McGee D. W., Vitkus S. J. and Lee P. (1996) The effect of cytokine stimulation on IL-1 receptor mRNA expression by intestinal epithelial cells. *Cell Immunol* **168**, 276-80.

McMahan C. J., Slack J. L., Mosley B., Cosman D., Lupton S. D., Brunton L. L., Grubin C. E., Wignall J. M., Jenkins N. A., Brannan C. I. and et al. (1991) A novel IL-1 receptor, cloned from B cells by mammalian expression, is expressed in many cell types. *Embo J* **10**, 2821-32.

Mizel S. B. (1979) Physicochemical characterization of lymphocyte-activating factor (LAF). *J Immunol* **122**, 2167-72.

Mizel S. B. and Rosenstreich D. L. (1979) Regulation of lymphocyte-activating factor (LAF) production and secretion in P388D1 cells: identification of high molecular weight precursors of LAF. *J Immunol* **122**, 2173-9.

Mosley, B., Dower, S., Gillis, S. and Cosman, D. (1987). Determination of the mininumpolypeptide lengths of the functionally active sites of human interleukins 1α and 1β. *Proc. Natl. Acad. Sci. USA* 84, 4572-4576.

Pathak J. P. N. (1993) *Insect immunity*. Kluwer Academic Publishers, Dordrecht; Boston.

Pfeifer, K., Schroder, H., Rinkevich, B., Uhlenbruck, G., Hanisch, F., Kurelec, B., Scholz, P. and Muller, W. (1992) Immunological and biological identification of tumor necrosis-like factor in sponges: Endotoxin that mediates necrosis formation in xenografts. *Cytokine* 4, 161-169.

Prendergast, R. and M. Suzuki, (1970). Invertebrate protein simulating mediators of delayed hypersensitivity. *Nature*. 227:277-279.

Priestle J. P., Schar H. P. and Grutter M. G. (1989) Crystallographic refinement of interleukin 1 beta at 2.0 A resolution. *Proc Natl Acad Sci U S A* **86**, 9667-71.

Savaz S. (1997) Characterization and partial isolation of an IL-6-like- molecule from *Manduca sexta*. University of Massachusetts, Boston.

Schreuder, H., Tardif, C., Trump-Kallmeyer, S., Soffientini, A., Sarubbi, E., Akeson, A., Bowlin, T., Yanofsky, S. and Barrett, R. (1995). Refined crystal structure of the interleukin-1 receptor antagonist. Presence of a disulfide link and a *cis* proline. *Eur. J. Biochem.* 227, 838-847.

Sima P. and Vetvicka V. (1993) Evolution of immune reactions. *Critical Reviews in Immunology* **13**, 83-114.

Sims J. E. (1990) Molecular Chracterization of Cytokines and Their Receptors. In *Molecular Aspects of Immune Response and Infection Diseases* (Edited by H. Kiyono E. J., and C. DeSimone), p. 23. Raven Press, New York.

Sission S. and Dinarello C. (1989) *Interleukin-1*. Academic Press, London.

Sundick, R. and Gill-Dixon, C. (1997). A cloned chicken lymphokine homologous to both mammalian IL-2 and IL-15. *J. Immunol.* 159, 721-7725.

Weining K.C., Sick C., Kaspers B., and Staeheli P. (1998) A chicken homolog of mammalian interleukin-1 beta: cDNA cloning and purification of active recombinant protein. *Eur J Biochem* **258**, 994-1000.

Wesche H., Neumann D., Resch K. and Martin M. U. (1996) Co-expression of mRNA for type I and type II interleukin-1 receptors and the IL-1 receptor accessory protein correlates to IL-1 responsiveness. *FEBS Lett* **391**, 104-8.

Wielgus J. J., Caldwell G. A., Nichols R. L. and White C. F. (1990) Purification, properties, and titer of a hemolymph trofic factor in larvae and pupae of *Manduca sexta*. *Insect Biochem* **20**, 65-72.

Young P. R., and Sylvester D. (1989) Cloning of rabbit interleukin-1β: differential evolution of IL-1α. *Protein Eng* **2**, 545-51.

Zou J., Cunningham C., and Secombes C.J. (1999) The rainbow trout *Oncorhynchus mykiss* interleukin-1β gene has a different organisation to mammals and undergoes incomplete splicing. *Eur. J. Biochem.* **259**, 901-8.

THE DETECTION AND ISOLATION OF A NOVEL ANTIMICROBIAL PEPTIDE FROM THE ECHINODERM, *Cucumaria frondosa*

Karen A. Beauregard, Nobel T. Truong, Haiyan Zhang, Wenyu Lin and Gregory Beck

Department of Biology, University of Massachusetts, Boston MA 02125-3393

INTRODUCTION

Microorganisms and their toxins are the cause of many diseases that harm multicellular organisms and the invertebrates are as susceptible to infection by bacteria, fungi and viruses as vertebrates (Marchalonis and Schluter, 1990). The invertebrates have developed highly effective immune systems for detecting and eliminating foreign invaders, consisting of both cellular and humoral/innate components (Boman *et al.*, 1991).

The echinoderm cellular immune response involves phagocyte-mediated effector functions, including the phagocytosis of foreign material, primarily by amoebocytes circulating in the coelom, which may be regarded as the self/nonself discriminating cells (Sima and Vetvicka, 1993). It also includes the encapsulation of foreign substances, the formation of "brown bodies" and the subsequent degradation of the microorganisms by hydrolytic enzymes found within the cell (Canicatti, 1990). Echinoderms have also been shown to exhibit other immunologic responses that are very similar to those of higher vertebrates. The sea urchin has been shown to effectively reject allografts in a manner similar to that of vertebrates through the means of specialized phagocytic coelomocytes (Smith and Davidson, 1994). An invertebrate cytokine, an interleukin (IL)-1-like protein, has been detected in the coelomic fluid of a starfish (Beck and Habicht, 1986). The starfish coelomocytes have been experimentally shown to produce an IL-1-like molecule similar enough to mammalian IL-1 to be active in assays on mammalian cells (Beck and Habicht, 1991).

The humoral immune response of invertebrates consists of the production of proteins that are sequestered within the cells, having hemolytic and hemagglutinating properties (Canicatti, 1989) and also includes the synthesis of antimicrobial peptides that are secreted from the cell (Kaaya, 1993). The invertebrates have been shown to have both naturally-occurring and inducible antimicrobial peptides that exhibit a broad spectrum of activity against fungi and Gram positive and/or negative bacteria as part of their humoral/innate immune defense (Kaaya, 1993).

Antimicrobial peptides are proving to be of vital importance for both vertebrate and invertebrate innate immunity and are thought to play a significant role in the survival against invading bacteria and fungi for many different species (Boman and Hultmark, 1987; Kaaya, 1993). These peptides are defined as cationic, low molecular weight proteins, typically less than 10 kDa and the primary target of these peptides is the bacterial cell membrane. Antimicrobial peptides have been classified into three major groups, based on their amino acid sequence, secondary structure and functional similarities. Group one includes the linear peptides that are devoid of the amino acid cysteine and form amphipathic alpha helices. Group two includes the antiparallel ß-sheet peptides with 2 - 4 intramolecular disulfide bridges and the third group includes the proline/arginine and proline/glycine rich, linear peptides that are devoid of the amino acid cysteine.

Peptides exhibiting antimicrobial activity have been identified in a diverse range of animal groups. For example, they have been found in insects (Boman and Hultmark, 1987; Kaaya, 1993) mammals (Ganz and Lehrer, 1998), amphibians (Bevins and Zasloff, 1990), tunicates (Lee *et al.*, 1997b) and crustaceans (Schnapp *et al.*, 1996) but none had yet been isolated from an echinoderm. Recently, we have detected antibacterial activity in the coelomic fluid obtained from echinoderms, which were not bacterially challenged, the orange-footed Sea cucumber, *Cucumaria frondosa*. The antibacterial activity has been detected against several species of both gram-positive and gram-negative bacteria. The sea cucumber provides an ideal experimental model from the echinoderm family for isolating antibacterial peptides. They are relatively abundant in nature and are of a size that yields workable quantities of cell rich coelomocytes. From an evolutionary perspective, the isolation and study of endogenous antimicrobial peptides from echinoderms could have extensive implications for understanding the evolution of immunity and also aid in the development of new drugs to help in the fight against disease in humans.

MATERIAL & METHODS

Animals

The orange-footed sea cucumber (*Cucumaria frondosa*) was collected by dredging in the Gulf of Maine by our collaborators at Coastside Research in Stonington, ME. The animals were bled and the pooled coelomic fluids were collected in large glass containers.

Isolation and Purification of Peptides

Proteins in the coelomic fluids were fractionated utilizing ammonium sulfate precipitation. The fractionated proteins were dialyzed against dH_2O and phosphate buffered saline using MWCO: 500 dialysis tubing (Spectrum Medical Ind., Inc.), centrifuged and filter-sterilized using a 0.45 μm sterile Acrodisc (Gelman Sciences). The fractionated protein samples were then concentrated by ultrafiltration with an Amicon YC-05 filter (MWCO: 500) and the protein concentration was determined by the Bradford method.

The antimicrobial activity was monitored throughout the purification process using the radial diffusion and bacterial lysis assays (Lehrer, 1991; Steinberg, 1997).

Size Exclusion Chromatography

A 2 ml sample of the fractionated coelomic fluids were applied to a Sephadex G-50 (Sigma) column (50 x 1.5 cm) and eluted with phosphate buffered saline (PBS) pH 7.4 or 20 mM HCl pH 6.0 at a flow rate of 0.70 ml/min. The protein fractions of 3 ml were collected using a Bio-Rad fraction collector at 4°C and the absorbance of the fractions was measured using a Milton-Roy 601 spectrophotometer at 280 nm. The column fractions were assayed in a bacterial lysis assay and the column fractions exhibiting antimicrobial activity were pooled and either lyophilized or concentrated by ultrafiltration with a YC-05.and used for further purification.

The column was calibrated with Blue dextran (approximately 2 million Da), carbonic anhydrase (29,000 Da), cytochrome c (12,000 Da) and aprotinin (6,500 Da) as molecular weight markers. The elution volume (Ve) of each molecular weight marker was determined spectrophotometrically by measuring the effluent volume from the point of sample application to the center of effluent peak. The approximate molecular weight of the Sea cucumber column fractions exhibiting antibacterial activity was estimated from the standard curve.

Gel Electrophoresis

Tris-Tricine Gel electrophoresis of the AS fractionated Sea cucumber fluids was performed following the methods of Schagger and Von Jagow, (1987) using 4% stacking gels, 10% spacer gels and 16.5% separating gels with the following modification. The sample buffer was prepared following the protocol of Biorad, (0.2M Tris-HCl, 2% SDS, 40% Glycerol, 0.04% Coomassie G-250, pH 6.8). As per the original Schagger and Von Jagow protocol the gel buffer (3.0M Tris, 0.3% SDS), cathode buffer (0.1M Tris-HCl, 0.1M Tricine, 0.1% SDS, pH 8.25), anode buffer (0.1M Tris-HCl, pH 8.9) and the stock solution of acrylamide and bisacrylamide (49.5%T, 3%C) were prepared and stored at RT.

The sea cucumber column fraction samples were prepared in sample buffer at a ratio of 2:1 and incubated for ≈20 min at 37°C after which 15 ul samples were loaded into the gel. The gel was subjected to electrophoresis in a Hoefer Mighty Small II (SE250/SE260) apparatus at 4°C and 90V for ~ 16 hrs. After electrophoresis, the protein bands were stained by silver staining (Bio-Rad) or with Brilliant Blue G. The gel was calibrated with Sigma's Ultralow color molecular weight markers, consisting of Triosephosphate Isomerase (26,600 Da), Myoglobin (17,000 Da), ß-Lactalbumin (14,200 Da), Aprotinin (6,500 Da), Insulin Chain B (3,496 Da) and Bradykinin (1,060 Da).

Bacterial Strains

The following bacterial strains were obtained from Dr. M. P. Shiaris and T. Durant, (Dep't of Microbiology, U. Mass. - Boston); Gram-negative strains used *are Klebsiella pneumoniae, Proteus mirabillis, Enterobactor aerogenes, Alcaligenes faecalis* and *Pseudomonas aeruginosa.* The gram-positive strains used are *Staphylococcus aureus, Enterococcus faecalis and Bacillus subtillus.* Additional strains, *Escherichia coli* (-) strain K-12 and *Micrococcus luteus* (+) (ATCC 4698) were purchased from Sigma chemicals. All bacterial strains were cultured and maintained in Bacto Nutrient Broth, (Difco Laboratories).

Radial Diffusion Assay

The radial diffusion assay utilizes an underlayer agarose gel containing 10mM sodium phosphate buffer, 0.3 mg/ml of TSB and 1% (w/v) low-electroendosmosis (EEO) type agarose (Fisher Biotech) and an overlayer agarose consisting of a 2x nutrient broth solution with 1% w/v low EEO agarose. The underlayer is inoculated with approximately 1.2 x 10^7 CFU's mid-log phase cultures of bacteria.

In the radial diffusion assay, approximately 2mm diameter wells are punched into the underlayer gel and 5 to 10 ul of test or control material is added to each well. After incubating for 3 h at 37°C ≈7 ml of the overlayer agarose was added to the plates and then they were incubated for an additional 18-24 h at 37°C. The amount of antibacterial activity can be determined by measuring the zone of clearing surrounding the wells.

Bacterial Lysis Assay

The bacterial lysis assay utilizes mid-log phase bacterial cultures that are centrifuged, washed and resuspended in a 10 mM Sodium phosphate buffer (NaPB) pH 7.4 with 100 mM NaCl. The bacterial cultures are diluted in a liquid testing medium (LTM) consisting of 100 mM NaPB pH 6.5 with 1% *v/v* Nutrient broth, 2% *v/v* 5 M NaCl. The bacterial lysis assay employs the use of spectrophotometric turbidity measurements in a 96 well plate assay to determine antimicrobial activity. The activity was measured by reading the absorbance at OD_{600} of the bacterial cultures in LTM after the addition of sea cucumber fluid samples or G-50 column samples. Readings were taken initially and then hourly for the first 3 – 4 hours and the plates were incubated at 37°C between readings. Additional readings are taken at 18, 22, and 24 hrs, with the plates being incubated between readings. No growth detected or death of the bacterial cultures indicates antibacterial activity.

RESULTS & DISCUSSION

In this study, we isolated an antimicrobial peptide from the coelomic fluid of the Orange-footed sea cucumber, *Cucumaria frondosa*. In contrast to the antimicrobial peptides isolated from insects, it has been shown that marine invertebrates do not require stimulation by wounding or injection of bacteria in order to induce antimicrobial peptide synthesis. Constitutively produced peptides have been isolated from the hemolymph of the mollusc, *Mytilus edulis* (Charlet *et al.*, 1996), the scorpian, *Androctonus australis* (Ehret-Sabatier *et al.*, 1996), the shrimp, *Penaeus vannamei* (Destoumieux *et al.*, 1997) and the shore crab *Carcinus maenas*,(Schnapp *et al.*, 1996).

The sea cucumber coelomic fluids obtained from unchallenged animals were fractionated by ammonium sulfate (AS) precipitation and then assayed for antimicrobial activity. The results of bacterial lysis assays performed on the fractionated coelomic fluids are shown in Table 1. Antimicrobial activity was observed in all fractions, with the 40% fraction exhibiting the highest level of activity. The data also demonstrates that there is a broad range of antimicrobial activity against both Gram positive and Gram negative species.

Table 1. Bacterial Lysis Assay of Ammonium Sulfate Fractions

	% Bacterial Cell Death* (± SE)			
		Ammonium Sulfate Fractions (%)		
Bacteria (Gram +/-)	**Starting material**	**20%**	**40%**	**60%**
Klebsiella pneumoniae (-)	2.38±1.9	12.97±5.5	17.41±1.8	12.43±2.2
Escherichia coli (-)	1.12±0.0	0	1.23±0.0	0
Pseudamonas aeruginosa (-)	9.17±1.6	10.83±0.8	19.17±1.6	8.33±1.6
Alcaligenes faecalis (-)	16.88±2.0	22.76±1.7	29.80±2.0	28.90±2.8
Proteus mirabillis (-)	0	0	2.95±0.0	0
Enterobactor aerogenes (-)	0	11.01±0.9	16.87±1.2	12.54±0.9
Staphylococcus aureus (+)	10.10±0.9	13.20±0.0	18.73±0.04	6.00±1.0
Enterococcus faecalis (+)	34.00±1.6	32.57±0.3	48.08±1.52	38.57±5.0
Micrococcus luteus (+)	0	0	13.77±1.2	5.71±0.8

*Microtiter plates were innoculated with bacteria in rows A (1-12) and E (1-12) 1:1 bacteria to LTM and serial diluted down columns 1-12 into rows B, C, D and F,G,H. Sea cucumber AS fractioned samples were added to the plate in quadruplicate. Positive Control (bacteria and LTM) = (A-D, 1-4), Starting material = (A-D, 5-8), 20% = (A-D, 9-12), 40% = (E-H, 1-4), 60% = (E-H, 5-8) and Negative control (LTM alone) = (E-H, 9-12).

The fractionated AS samples were electrophoresed using a Tris-Tricine gel (Figure 1) which is good for resolving low M_r proteins. Silver staining of the gel showed the detection of bands in the 40% and 60% AS fractions with a M_r of ~6,500.

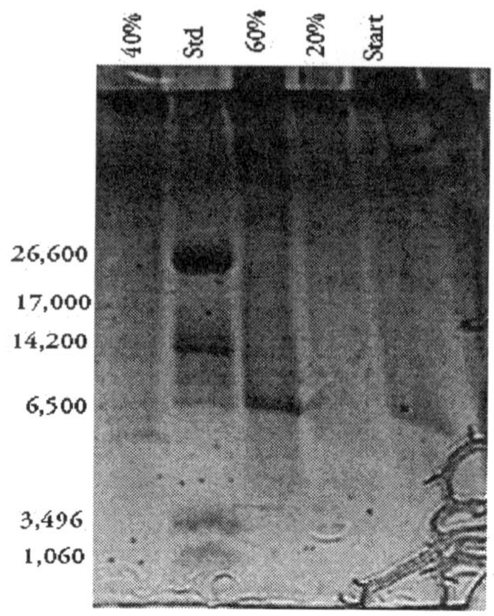

Figure 1. Tris-Tricine PAGE. Gel electrophoresis of the AS fractionated Sea cucumber fluids was performed following the methods of Schagger and Von Jagow, (1987) using 4% stacking gels, 10% spacer gels and 16.5% separating gels. The sea cucumber column fraction samples were prepared in sample buffer at a ratio of 2:1, incubated for ≈20 min at 37°C and 15 ul samples were loaded into the gel. After electrophoresis, the protein bands were visualized by silver staining.

The sea cucumber AS fractions were further purified by molecular sieve chromatography. A typical elution profile of the 40% AS fraction on Sephadex G-50 is shown in Figure 2.

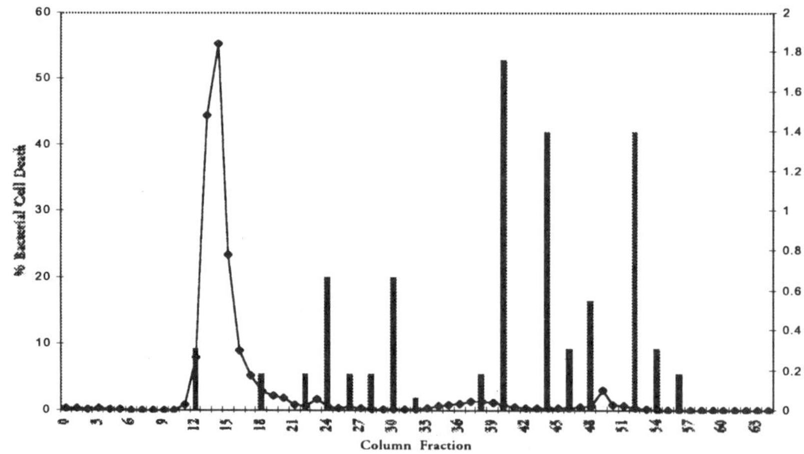

Figure 2. Size exclusion chromatography and a bacterial lysis assay of the 40% AS fraction from the C. frondosa coelomic fluids. Two ml of the 40% AS fraction was applied to a Sephadex G-50 column (50 x 1.5 cm) equilibrated in 20 mM HCl pH 6.0 and eluted with the same buffer at a flow rate of 0.70 ml/min. Absorbance was monitored at 280nm (line graph). Three ml fractions were collected and 100 μl samples of the fractions were tested in triplicate in a bacterial lysis assay against *S. aureus.* (Bars indicate % cell death).

A bacterial lysis assay was used to test the column fractions for antimicrobial activity and the percent bacterial cell death is represented by the bar graph shown in Figure 2. The graph shown indicates antimicrobial activity occurring in several fractions. The activity in fractions 39 – 45 at a M_r of ~6,000 we propose to be the peptide. The activity observed in fractions 24 – 30 is evidence of the presence of lysozyme. Additional activity was observed in fractions 51 – 57, indicating there may be additional low molecular weight peptides circulating in the coelomic fluid.

The column fractions exhibiting antimicrobial activity were pooled, concentrated and tested for activity in a radial diffusion assay. The results of the radial diffusion assay indicate antimicrobial activity against both Gram negative bacteria, shown in Figure 3a and Gram positive bacteria, shown in Figure 3b. The antibacterial activity observed in this assay appears to be pH dependent. The highest amount of activity was observed at pH 5.5 and 6.5, with lower levels being detected at pH 7.4. Higher activity was also observed from the G-50 column fractions eluted with 20 mM HCl pH 6.0. These results are similar to those reported for the clavanins, which are peptides isolated from the solitary tunicate *Styela clava* (this volume and Lee *et al.*, 1997a).

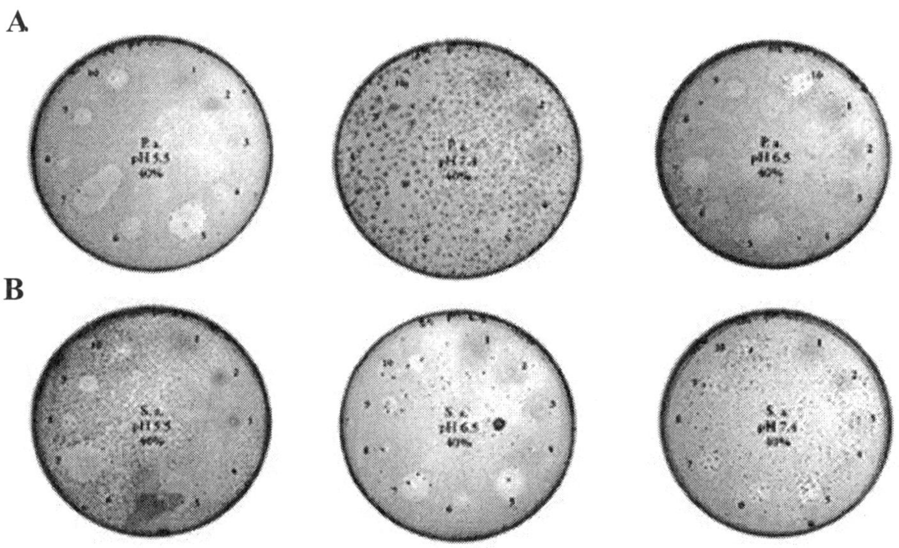

Figure 3. Radial diffusion assay's of fractionated coelomic fluid fractions. (A) Radial diffusion assay of 40% AS Sea cucumber G-50 column fractions using the Gram negative bacterium *Pseudomonas aruginosa.* 10 µl samples were loaded into each well and plates were incubated for 3 hours at 37°C before overlayer agarose was added. (B) Radial diffusion assay of the 40% AS Sea cucumber G-50 column fractions using the Gram positive bacterium, *Staphylococcus aureus.* (Well 1, Starting material; Well 2, 40% AS Fraction; Well 3, 60% AS Fraction; Well 4, Negative Control; Well 5, Fractions 17 – 31, pH 6; Well 6, Fractions 32 - 41, pH 6; Well 7 Fractions 42 - 64, pH 6; Well 8, Fractions 17 – 31, pH 9; Well 9, Fractions 32 - 41, pH 9; Well 10, Fractions 42 - 64, pH 9).

CONCLUSION

We have detected the presence of antimicrobial peptides in the coelomic fluid of the sea cucumber, *Cucumaria frondosa.* One peptide has a M_r of ~6000 Da, but there was also activity observed at lower M_r's. The antimicrobial activity appears to be pH dependent with maximal activity observed at pH 5.

ACKNOWLEDGEMENTS

This research was supported by a grant to GB from the National Institutes of Health (AI 39685-01).

REFERENCES

Beck G. and Habicht G. S. (1986) Isolation and characterization of a primitive interleukin-1-like protein from an invertebrate, *Asterias forbesi*. *Proc Natl Acad Sci U S A* **83**, 7429-7433.

Beck G. and Habicht G. S. (1991) Purification and biochemical characterization of an invertebrate interleukin 1. *Mol Immunol* **28**, 577-584.

Bevins C. L. and Zasloff M. (1990) Peptides from frog skin. *Annu Rev Biochem* **59**, 395-414.

Boman H. G., Faye I., Gudmundsson G. H., Lee J. Y. and Lidholm D. A. (1991) Cell-free immunity in Cecropia. A model system for antibacterial proteins. *Eur J Biochem* **201**, 23-31.

Boman H. G. and Hultmark D. (1987) Cell-free immunity in insects. *Annu Rev Microbiol* **41**, 103-126.

Canicatti C. (1990) Lysosomal Enzyme Patterns in *Holothuria polii* Coelomocytes. *J. Invertebr Pathol* **56**, 70-74.

Canicatti C. a. D. A. G. (1989) Cellular Aspects of *Holothuria polii* Immune Response. *J. Invertebr Pathol* **53**, 152-158.

Ganz T. and Lehrer R. I. (1998) Antimicrobial peptides of vertebrates. *Curr Opin Immunol* **10**, 41-44.

Kaaya G. P. (1993) Inducible Humoral Antibacterial Immunity in Insects. In *Insect Immunity* (Edited by Pathak J. P. N.), p. 69-89. Oxford & IBH Publishing Co. PVT. LTD., New Delhi, India.

Lee I. H., Cho Y. and Lehrer R. I. (1997) Styelins, broad-spectrum antimicrobial peptides from the solitary tunicate, Styela clava. *Comp Biochem Physiol B Biochem Mol Biol* **118**, 515-521.

Lehrer R. I., Michael Rosenman, Sylvia S.S.L. Harwig, Richard Jackson, Patricia Eisenhauer. (1991) Ultrasensitive assays for endogenous antimicrobial polypeptides. *Journal of immunological methods* **137**, 167-173.

Marchalonis J. J. and Schluter S. F. (1990) On the relevance of invertebrate recognition and defence mechanisms to the emergence of the immune response of vertebrates. *Scand J Immunol* **32**, 13-20.

Schagger H. and von Jagow G. (1987) Tricine-sodium dodecyl sulfate-polyacrylamide gel electrophoresis for the separation of proteins in the range from 1 to 100 kDa. *Anal Biochem* **166**, 368-379.

Schnapp D., Kemp G. D. and Smith V. J. (1996) Purification and characterization of a proline-rich antibacterial peptide, with sequence similarity to bactenecin-7, from the haemocytes of the shore crab, Carcinus maenas. *Eur J Biochem* **240**, 532-539.

Sima P. and Vetvicka V. (1993) Evolution of immune reactions. *Crit Rev Immunol* **13**, 83-114.

Smith L. C. and Davidson E. H. (1994) The echinoderm immune system. Characters shared with vertebrate immune systems and characters arising later in deuterostome phylogeny. *Annals Of The New York Academy Of Sciences* **712**, 213-226.

Steinberg D. A. a. L., Robert I. (1997) *Designer Assays for Antimicrobial Peptides*. Humana Press, Inc., Totowa, New Jersey.

ISOLATION AND CHARACTERIZATION OF AN IL-1-LIKE PROTEIN FROM *Manduca sexta*

Nobel T. Truong, Karen A. Beauregard, Haiyan Zhang, Wenyu Lin, and Gregory Beck

Department of Biology, University of Massachusetts Boston, Boston, MA 02125-3393

INTRODUCTION

Invertebrates such as insects have similar immune systems when compared to vertebrates, which are composed of humoral and cell-mediated responses (Pathak, 1993). Humoral response involves the synthesis and release of antibacterial peptides such as lysozyme, cecropins, and attacins. Cell-mediated response involves phagocytosis, nodule formation, and encapsulation of foreign antigens. Phagocytosis is an ancient form of host defense, a process in which a cell, usually macrophage, ingests a foreign antigen. This process can be observed in both vertebrates and invertebrate immune systems. In insects, plasmatocytes and granular hemocytes are the main cells with phagocytic activity. However, invertebrates do not have antigen-specific lymphocytes and do not produce specific immune recognition molecules (immunoglobins) (Abbas *et al.*, 1997). Some invertebrate species contain macrophage-like cells capable of producing cytokine-like molecules similar to that of vertebrate (Beck and Habicht, 1986; Ellis, 1996). Cytokines are low molecular weight proteins, some examples are the members of the interleukin (IL) family, tumor necrosis factor

Phylogenetic Perspectives on the Vertebrate Immune System.
Edited by Beck *et al.*, Kluwer Academic/Plenum Publishers, 2001.

(TNF), and interferon (INF) to name a few. Their role in the vertebrate immune response includes control of cellular proliferation and differentiation, and regulation of hematopoiesis (Abbas *et al.*, 1997).

Interleukin (IL) -1 is a family of three related proteins, IL-1α, IL-1β, and IL-1ra (receptor antagonist) and was one of the first cytokines discovered. IL-1α and IL-1β have two isoelectric points 5.3 and 7.2, respectively. IL-1 mRNAs are translated as 31 kDa precursor polypeptides. The polypeptide is then cleaved by proteases in a posttranslational process result in a 17 kDa final product. The precursor and the mature IL-1a are both active, whereas the IL-1β precursor form is inactive and requires posttranslational cleavage to be active. IL-1ra is an anti-inflammatory protein with a molecular weight of 17 kDa that inhibits the biological effects of IL-1α and IL-1β by binding to the IL-1 receptors (Gabay *et al.*, 1997).

Previous work from our laboratory shows invertebrates can synthesize and release cytokine-like proteins (IL-1-like and IL-6-like proteins) in response to lipopolysaccharide (LPS) stimulation. Beck and Habicht have successfully isolated and purified a primitive IL-1-like protein from the starfish, *Asterias forbesi*, using a vertebrate assay system (the murine thymocyte proliferation assay) (Beck and Habicht, 1986; Beck and Habicht, 1991). The primitive IL-1-like protein isolated from starfish has a molecular weight of approximately 20,000 Da (Beck and Habicht, 1986). Recently, IL-1-like and IL-6-like proteins were detected in the hemolymph of the tobacco hormworm, *Manduca sexta* (Beck *et al.*, 1996; Ellis, 1996; Savaz, 1997), with molecular weights of approximately 20,000 and 32,000 Da, respectively (Beck *et al.*, 1996; Savaz, 1997). The isolation of these cytokine-like proteins suggests that vertebrates and invertebrates have similar immune responses and that cytokines are ancient and functionally conserved proteins (Beck and Habicht, 1986).

MATERIALS AND METHODS

Animals

Manduca sexta larvae were reared on a synthetic medium and kept at 25°C with a 16 hr light – 8 hr dark photoperiod. Fifth instar larvae were used in all experiments.

Activation of IL-1 and Collection of Hemolymph

The animals were injected with 2.5 mg LPS using 27 gauge needles. The next day, hemolymph was collected from each animal. Hemolymph was extracted by making a small incision on the second leg and it was dripped into a sterile 50 ml centrifuge tube containing ice cold 20 ml of the anticoagulant solution (100 mM glucose, 15mM sodium chloride, 30 mM trisodium citrate, and 26 mM citric acid, pH 4.6). The sample was centrifuged at 110 *x* g for 10 minutes at 15°C to remove the hemocytes. The hemolymph was concentrated by ultrafiltration in a stirred cell using 10,000 nmwl filters (PM 10 and stirred cell; Amicon, Lexington, MA).

Sephacryl S-200-HR Chromatography of the Stimulated Hemolymph

The stimulated hemolymph was loaded on a Sephacryl S-200-HR column (50 cm *x* 1.5 cm) pre-equilibrated with 50 mM sodium phosphate buffer (PBS), pH 7.5. The column was eluted with PBS at a flow rate of 0.70 ml/min and protein fractions of 3 ml were collected using a Bio-Rad fraction collector at 4°C. The absorbance of the fractions was measured using a Milton-Roy 601 spectrophotometer at 280 nm. The protein fractions containing IL-1 were pooled and used for further purification.

Blue dextran (approximately 2 million Da), albumin (66,000 Da), carbonic anhydrase (29,000 Da), and cytochrome c (12,000 Da) were used as molecular weight markers. Blue dextran was used to determine the void volume (Vo) of the column. The elution volume (Ve) of each molecular weight marker was determined spectrophotometrically by measuring the effluent volume from the point of sample application to the center of effluent peak. The log of their molecular weight was plotted using Excel Graph (Microsoft Corp., Redmond, WA) against their elution volume to obtain the standard graph. IL-1 from hemolymph was chromatographed on the same column and the elution volume was determined. The elution volume of IL-1 was plotted on the standard graph and the approximate molecular weight was determined.

Preparative Isoelectric Focusing

Fractions that contained IL-1 were concentrated by ultrafiltration in a stirred cell using 10,000 nmwl filters (PM 10 and stirred cell; Amicon, Lexington, MA), and then dialyzed against distilled water. To the concentrated column eluate, 2.0 ml of ampholyte solution (pH range 5.08.0, 40% w/v) was added, and the sample was diluted to 40 ml with distilled water. Focusing was done at 12 W constant power at 4°C until V*H 3,400. Twenty fractions were collected and the pH and absorbance of each were recorded. Fractions with pH values ranging from 5.0 to 5.7 and 7.0 to 7.7 were pooled separately and used for further purification.

Gel Electrophoresis

Each fraction was subjected to electrophoretic analysis. Bio-Rad broad range SDS-PAGE molecular weight standards were used to calibrate the gel. The standard is consisted of myosin (200,000 Da), b-galactosidase (116,250 Da), phosphorylate b (97,400 Da), serum albumin (66,200 Da), ovalbumin (45,000 Da), carbonic anhydrase (31,000 Da), trypsin inhibitor (21,500 Da), lysozyme (14,400 Da), and aprotinin (6,500 Da). The gel was subjected to electrophoresis in a Bio-Rad mini PROTEIN II apparatus with a constant current 25 milliamps per gel at room temperature. After electrophoresis, detection of protein bands was obtained by silver staining (Bio-Rad).

A375 Cell Growth Inhibition Bioassay

A375 cells are human melanoma cell line and are used to detect the presence of IL-1 (Beck and Habicht, 1986; Beck and Habicht, 1991). Recombinant human IL-1a was used to construct standard curves to determine the number of units of IL-1 in each experiment. Fractions from the isoelectric focusing procedure were diluted 1:4, 1:8, 1:16 and 1:32 in RPMI with 0.01 M HEPES. Each fraction was added to a 96 well tissue culture plate to a volume of 100 ml in triplicate. 100 ml of the A375 cell suspension was then added to the wells. The plates were incubated for 72 hours at 37°C with 5% CO_2. After 72 hours, the media is discarded from the wells and replaced with 25 ml/ well of crystal violet (0.5% in 20% methanol). The plates were incubate at 37°C for 10 minutes and then washed with distilled water and dried to remove excess crystal violet and unattached (dead) cells. 100 ml of 1% SDS was added to each well and incubated at 37°C for 30 minutes for extraction of dye. The plates were read at 590nm. Data are expressed in absorbance units at 590nm.

Sf9 Cell Proliferation Bioassay

The Sf9 insect cell line is a clonal isolate derived from the parental cell line IPLB-Sf-21-AE isolated from immature ovaries of *Spondoptra frugiperda* pupae (Vaughn *et al.*, 1971). Fractions from the isoelectric focusing procedure were diluted 1:4, 1:8, 1:16 and 1:32 in TC-100. Each fraction was added to a 96 well tissue culture plate to a volume of 100 ml in triplicate. 100 ml of the Sf9 cell suspension was then added to the wells. The plates were incubated for 72 hours at 22°C. After 72 hours, 20 ml/well of Alamar Blue was added aseptically and incubation continued for approximately 1 hour. Alamar Blue is a colorimetric growth indicator and is blue when in the oxidized state. If the cells are proliferating, Alamar Blue is taken up by the cells and reduced which causes a color change to pink. Plates were read at 1 hour and 24 hours at 600 nm. Data are expressed in absorbance units at 600 nm.

RESULTS AND DISCUSSION

In this study we attempted to characterize a low M_r secreted form of the IL-1-like molecule from the hemolymph of *Manduca sexta*. Samples were fractionated by gel sieve column chromatography. A typical separation of *Manduca* hemolymph is presented in Figure 1. IL-1 activities were detected at fractions 25 – 38 at approximate M_r ranges 8,000-20,000. The IL-1 peak has an approximate molecular weight of 10,000 Da. Preparative isoelectric focusing of the *Manduca* low M_r peak resulted in the separation of three peaks with IL-1-like activity (Figure 2). One fraction had a pI of 5.0, and another had a pI of 7.2. These physicochemical values are similar to those reported for vertebrate IL-1 and echinoderm IL-1-like molecules (Beck and Habicht, 1991; Dinarello, 1989). All IL-1 and IL-1-like molecules studied to date have revealed two separate protein species, an acidic form (pI 5.0) and a neutral form (pI 7.0).

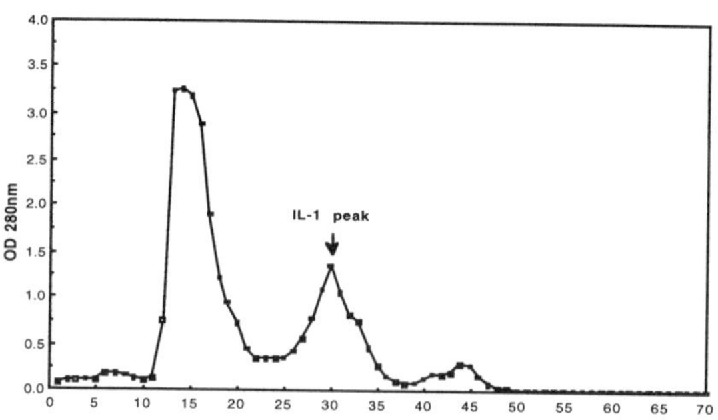

Figure 1. A typical Sephacryl S-200-HR elution profile of *Manduca* hemolymph. IL-1like activities were detected at fractions 25 - 38 at distinct ranges of 8,000 - 20,000 daltons. IL-1-like protein peak appears at column fraction 30. The IL-1 peak is at an approximate value of 10,000 daltons.

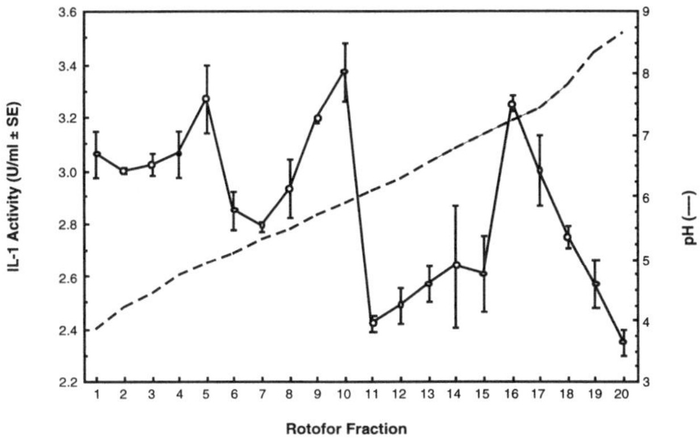

Figure 2. Isoelectric focusing of IL-1-like activity from *M. sexta* hemolymph. Preparative isoelectric focusing resulted in the separation of three peaks with pI values of 5.0, 6.0 and 7.2 with IL-1-like activity.

The identity of the pI 6.0 fraction is unknown and under investigation at this time. Fractions with pH values ranging from 5.0 to 5.7 and 7.0 to 7.7 were pooled separately and fractionated to achieve better pH gradients from 4.5 to 5.7 and 7.2 to 7.5, respectively.

The fractions collected from the columns were assayed in the A375 assay. A375 growth inhibition were detected in fractions with pH values 5.0–5.2. Gel electrophoresis analysis shows these fractions contain a single protein band with an approximate molecular weight of 18,000 Da. Fractions with pH values 6.5 – 6.9 also inhibited A375 growth. Gel electrophoresis analysis shows these fractions contain a single protein band with an approximate molecular weight of 18,000 Da. The acidic form (pI 5.1) of the IL-1-like protein was found to be capable of stimulating Sf9 cells proliferation (Figure 3 and 4). However, the neutral species of the IL-1-like protein has no effect on the Sf9 cells.

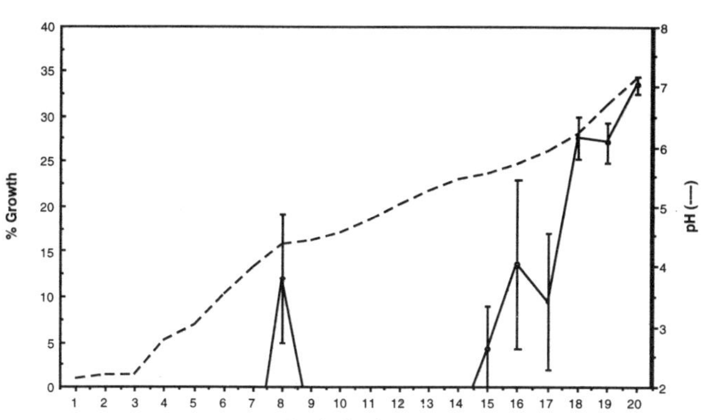

Figure 3. Sf9 proliferation assay (RotoLyte pH 4.5-5.7). Rotofor fraction 8, with pH 5.1 was found to be capable of stimulating Sf9 cells proliferation.

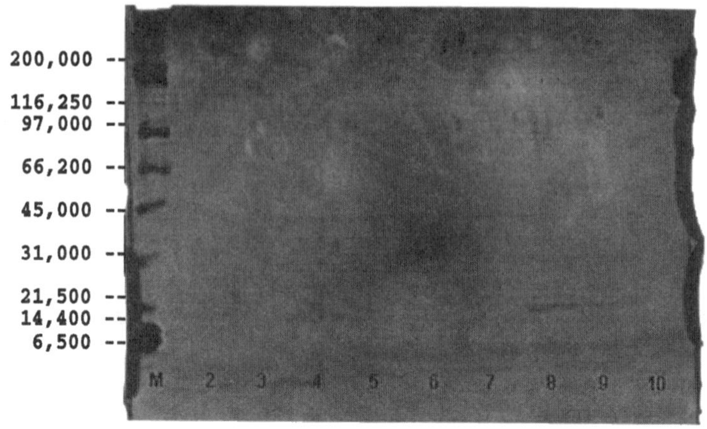

Figure 4. SDS-PAGE of RotoLyte pH 4.5-5.7 fractions. Lanes 2 to 10 are fraction samples from the isoelectric focusing procedure. A single silver stain protein band was detected in fraction sample 8 with M_r approximately 18,000 daltons.

CONCLUSION

We have shown that the insect *Manduca sexta* hemolymph contains a primitive IL-1-like protein. Evidence points to similarities between vertebrate IL-1 and this protein (1) The observed M_r of approximately 18,000 of the insect IL-1 is in good agreement with M_r that other investigators have found for vertebrates and invertebrates (Dinarello, 1989; Beck and Habicht, 1986; Beck and Habicht, 1991); 2) Isoelectricfocusing of insect IL-1-like proteins reveals two species with pI values 5.1 and 7.0; 3) The Manduca IL-1-like protein inhibits the growth of the human tumor cell line A375; and 4) The acidic form of the insect IL-1-like protein is capable of stimulating the growth of the *Spondoptra frugiperda* ovarian cell line.

ACKNOWLEDGEMENTS

This research was supported by a grant to GB from the National Institutes of Health (AI 39685-01).

REFERENCES

1. Abbas A. K., Lichtman A. H. and Pober J. S. (1997) *Cellular and Molecular Immunology*. Saunders, Philadlephia.
2. Beck G., Cardinale S., Wang L., Reiner M. and Sugumaran M. (1996) Characterization of a defense complex consisting of interleukin 1 and phenoloxidase from the hemolymph of the tobacco hornworm, *Manduca sexta*. *J Biol Chem* **271**, 11035-11038.
3. Beck G. and Habicht G. S. (1986) Isolation and characterization of a primitive interleukin-1-like protein from an invertebrate, *Asterias forbesi*. *Proc Natl Acad Sci U S A* **83**, 7429-7433.
4. Beck G. and Habicht G. S. (1991) Purification and biochemical characterization of an invertebrate IL-1. *Mol Immunol* **28**, 577-584.
5. Dinarello C. A. (1989) Interleukin-1 and its biologically related cytokines. *Adv Immunol* **44**, 153-205.
6. Ellis T. (1996) Identification of interleukin-1 secreting hemocytes in *Manduca sexta*. University of Massachusetts Boston.

7. Gabay C., Smith M. F., Eidlen D. and Arend W. P. (1997) Interleukin 1 receptor antagonist (IL-1Ra) is an acute-phase protein [see comments]. *J Clin Invest* **99**, 2930-2940.
8. Pathak J. P. N. (1993) *Insect immunity*. Kluwer Academic Publishers, Dordrecht ; Boston.
9. Savaz S. (1997) Characterization and partial isolation of an IL-6-like- molecule from *Manduca sexta*. University of Massachusetts Boston.
10. Vaughn J. L., Goodwin R. H., Tompkins G. J., and McCawley P. (1971) The establishment of two cell lines from the insect *Spodoptera frugiperda* (Lepidoptera; Noctuidae). *In Vitro* **13**, 213-17.

CLAVANINS AND STYELINS, α-HELICAL ANTIMICROBIAL PEPTIDES FROM THE HEMOCYTES OF *Styela clava*

Robert I. Lehrer[1, 2], In Hee Lee, Lorenzo Menzel[1], Alan Waring[1,3] and Chengquan Zhao[1]

Molecular Host Defense Section, Dep't of Medicine[1], Molecular Biology Institute [2] and Dep't of Pediatrics[3], UCLA School of Medicine, Los Angeles CA 90095

INTRODUCTION

Many different antimicrobial peptides are expressed by mammalian and avian leukocytes [reviewed in 1]. These peptides, which serve as endogenous antibiotics, are amphipathic and cationic molecules comprised of 12-40 amino acid residues. Most, if not all, target and perturb microbial membranes, often by forming voltage-gated channels. Many antimicrobial peptides have α-helical secondary structures, while others contain cystine-stabilized β-sheets. A few antimicrobial peptides are histidine or proline-rich.

Tunicates are marine protochordates, whose hemocytes correspond to the leukocytes found in higher vertebrates. This report describes two families of antimicrobial α-helical antimicrobial peptides that are expressed by the hemocytes of *Styela clava*, a noncolonial tunicate [2-6].

Phylogenetic Perspectives on the Vertebrate Immune System.
Edited by Beck *et al.*, Kluwer Academic/Plenum Publishers, 2001.

MATERIALS AND METHODS

Hemocytes

As described elsewhere in detail [2,3], hemolymph was obtained from recently collected *Styela clava* and anticoagulated with EDTA. Examination of these hemocytes by phase, light and electron microscopy revealed the presence of at least six different hemocyte morphotypes. The mixed hemocytes were washed and then sonicated in cold 5% acetic acid. After overnight stirring at 4° C, the sonicates were clarified by centrifugation. When prepared from the hemocytes of 50 tunicates (approximately 2 x 10^8 cells) such supernatants contained ~15 mg protein, and provided the material used for peptide isolation.

Purification of antimicrobial peptides

Clavanins and styelins were purified from the supernatants by a three step procedure, that included: 1) ultrafiltration through a 10 kDa cutoff membrane; 2) either gel permeation chromatography or preparative acid-urea PAGE; 3) several cycles of reverse phase HPLC [2,3]. Throughout the purification, fractions were tested for antimicrobial activity by a two stage radial diffusion assay that has been described elsewhere in detail [7,8].

Peptide characterization and cDNA Cloning

Amino acid analysis and N-terminal sequencing were performed by standard techniques. Mass spectrometric analyses were performed at the UCLA Center for Molecular and Medical Mass Spectrometry. Circular dichroism spectra were obtained on an AVIV model 62DS spectropolarimeter (AVIV Associates, Lakewood, NJ). Total RNA was isolated from tunicate pharyngeal tissues and purified using a total RNA Separator Kit (Clontech, Palo Alto, CA). Additional details of the cDNA cloningcan be found elsewhere [5, 6].

Bacteriologic studies

Peptides were mixed with mid-logarithmic phase bacteria in 10 mM sodium phosphate buffer supplemented with various concentrations of NaCl (0-0.4 M) and dilute nutrients [2-4]. The mixtures, typically in a final volume of 50-100 µl, were incubated in a shaking water bath and 10 µl aliquots were removed at intervals, appropriately diluted and plated on nutrient agar. Colonies were counted after an overnight incubation. Two stage radial diffusion assays were performed as previously described [7,8].

Synthetic clavanins

Synthetic clavanins were synthesized by one of us (A.W.) or by a company (SynPep, Dublin, CA) using F-moc (9-fluoroenylmethyloxycarbonyl) chemistry and purified to virtual homogeneity by RP-HPLC.

Immunoelectron microscopy

Polyclonal rabbit antibody to synthetic clavanin A was generated conventionally, and was found to recognize all five clavanins. Light level immunostaining was done with a 1:800 dilution of primary antibody, using a secondary antibody (goat anti-rabbit

IgG-HRP conjugate) purchased from Dako. Immunostains were developed with DAB reagents purchased from Sigma. Ultrastructural immunocytochemistry was done by an immunogold procedure.

RESULTS AND DISCUSSION

The primary sequences of clavanins A, B, C, D and E are shown in Figure 1. Eighteen of the 23 clavanin residues (78.3%) were identical in all of the peptides. The circular dichroism (CD) spectrum of synthetic clavanin A showed a well-defined double minimum at ~208 and 222 nm – the characteristic CD signature of an α-helix [2]. Similar spectra were obtained in trifluoroethanol and in a dispersion of liposomes whose composition simulated that present in *S. aureus* membranes [2].

Clavanin A	V	**F**	Q	F	**LG**	K	**IIHHVGNFV**	H	**GFSHVF***
Clavanin B	V	**F**	Q	F	**LG**	R	**IIHHVGNFV**	H	**GFSHVF***
Clavanin C	V	**F**	H	L	**LG**	K	**IIHHVGNFV**	Y′	**GFSHVF***
Clavanin D	A	**F**	K	L	**LG**	R	**IIHHVGNFV**	Y′	**GFSHVF***
Clavanin E	L	**F**	K	L	**LG**	K	**IIHHVGNFV**	H	**GFSHVF***

Figure 1. Primary sequences of Clavanins A-E. The sequences are shown in standard single letter code. The asterisk (*) signifies C-terminal amidation; and Y′ denotes methyl-tyrosine. Invariant residues are shown in bold type and are boxed.

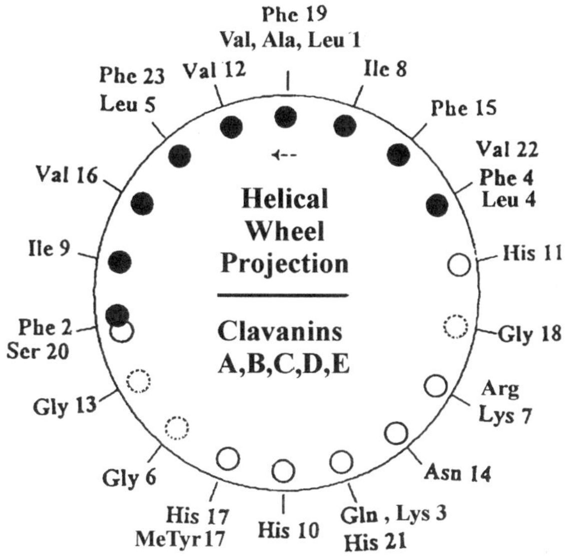

Figure 2. Helical wheel projection of Clavanins A-E: Apolar residues are denoted by solid circles, polar residues by open circles, and glycines by dotted circles.

Figure 2 shows a helical wheel projection of clavanins A-E. Note that all of the apolar residues (phenylalanine, isoleucine, leucine, valine and alanine) are clustered on one face of the helix, while polar residues (histidine, methyl tyrosine, lysine, asparagine, arginine, glutamine, and serine) appear only on the other face. This spatial segregation of apolar and polar residues establishes that the clavanin α-helix is amphipathic.

The N-terminal sequences of Styelins A and B were obtained by peptide level analyses [3], and the complete sequences of styelins C-E were ascertained by cDNA cloning [5]. These are shown in Figure 3. Only 8-9 of the first 20 N-terminal residues were identical in all five styelins. No fewer than 17 of 20 N-terminal residues were identical in styelins A and B, and 31 of 33 residues were identical in styelins D and E. If mature styelins are amidated, then the C-terminal glycine$_{33}$ of the precursor would be absent from the mature peptide. The primary structure of styelin C was intermediate between those of the A/ B and D/ E styelin pairs.

```
Styelin A.  G X F G K A F X S V S N F A K K H K T A ......
Styelin B.  G X F G P A F H S V S N F A K K H K T A .......
Styelin C.  G W F G K A F R S V S N F Y K K H K T Y I H A - G L S A A T L L G
Styelin D.  G W L R K A A K S V G K F Y Y K H K Y Y I K A A W Q I G K H A L G
Styelin E.  G W L R H A A K S V G K F Y Y K H K Y Y I K A A W K I G R H A L G
```

Figure 3. Primary sequences of Styelins A-E. The N-terminal sequences of styelins A and B were determined by peptide-level sequencing, and those of styelins C, D, and E were determined by cDNA cloning. Each residue shown as a lysine (K) in styelins A and B was actually found to be a hydroxylysine. Consequently, some or all of the lysines shown in cDNA-derived sequences of styelins C, D, and E are likely to undergo similar post-translational modification.

The precursors of styelins and clavanins contain one or two anionic domains, respectively. Prostyelins D and E have an identical anionic domain with a net charge of –7 (DMTDEEFQDFMKEVEQAREEELQSRQ). Since the styelin D domain contains 7 lysines or hydroxylysines, 2 histidines and 1 arginine, its net charge would range from +8 and + 10, depending on the ambient pH (or from +9 to +11 if the peptide is amidated). By reducing the net positive charge of the propeptide, the anionic domain could render the styelin domain inactive and noncytotoxic until the mature styelin peptide gains its release from its precursor by limited proteolysis.

Similar considerations apply to clavanin precursors, since the mature clavanin domain is surrounded by two anionic domains. The N-terminal anionic domain (LEERKSEEEK) has a net charge of –2 and the C-terminal anionic domain has a net charge of –4. Together, these would effectively balance the net charge of clavanins A-E, which will range from +2 to +7, depending on the ambient pH. This marked dependence of clavanin's charge to pH results from its histidine-rich nature and accounts for the enhanced potency of clavanins at acidic pH levels [4].

The activity of clavanins A and E against four human bacterial pathogens is compared in Figure 4. These radial diffusion assays were performed at pH 5.5, with an underlay gel that contained 0.2 M NaCl, 9 mM sodium phosphate + 1 mM citrate, and dilute trypticase soy broth. In such assays, the minimal inhibitory concentrations are defined by the X intercepts of the regression lines [7, 8]. It can be seen that clavanin E was more potent than clavanin A against all of these organisms. The effects of salinity and pH on the antimicrobial properties of clavanins [4] and styelins [3] have been described elsewhere.

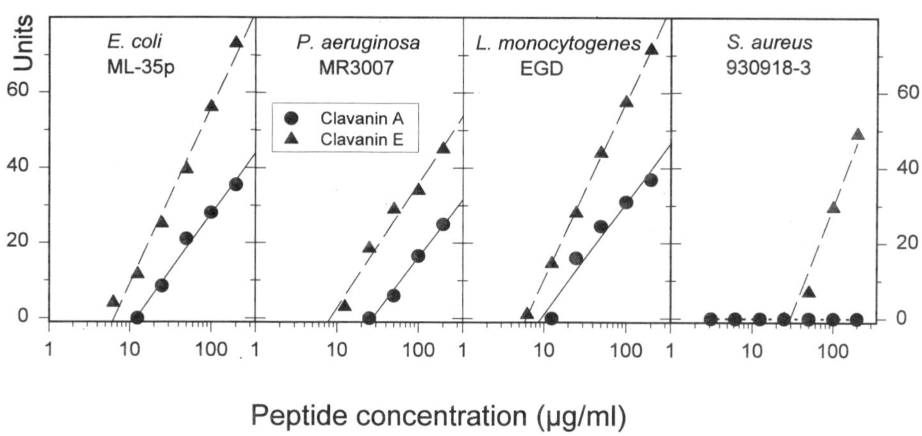

Figure 4. Antimicrobial activity of Clavanin A. Serial dilutions of clavanin A were tested against *E. coli* ML-35p, S. aureus 930918-3, *L. monocytogenes* EGD and *C. albicans* in radial diffusion assays. The assays were performed at pH 5.5. Low salt underlays contained 10 mM sodium phosphate buffer, and high salt underlays contained 10 mM sodium phosphate buffer + 0.2 M NaCl. Zone diameters are expressed in units (10 units = 1 mm). The lines are least mean square regressions, and the minimal inhibitory concentrations correspond to the X-intercepts of these lines.

We used a polyclonal antibody to clavanins to perform immunolocalization studies by light and electron microscopy. Clavanins were present in at least four different hemocyte types, generally within their distinctive cytoplasmic granules. These studies will be published elsewhere.

In addition to the ten styelins and clavanins described in this report, we have preliminary evidence that many other bioactive peptides are present in *Styela clava*. One of these is an a-helical peptide that we have named "clavaspirin" and will describe elsewhere. Unlike clavanins A-E, clavaspirin is potently hemolytic and may therefore be a mediator of hemocyte-mediated cytotoxicity. Although the mature domain of clavaspirin diverges considerably from clavanins A-E, its precursor is almost identical to theirs.

ACKNOWLEDGMENTS

Dr. Lee performed this work with support from a Frederick B. Bang Scholarship from the American Association of Immunologists and a postdoctoral fellowship award from KOSEF, the Korea Science and Engineering Foundation. Dr. Lee's current address is: Department of Biochemistry, Hoseo University, 120-1 Anseo-dong Chonan, Chungnam, 330-180, Republic of Korea. We thank Dr. Audree Fowler for performing the amino acid sequencing, Ms. Birgitta Sjostrand for expert assistance and guidance with immunoelectron microscopy, and Professor Edwin Cooper for awakening our interest in tunicates.

REFERENCES

1. Lehrer, R.I., Bevins, C.L. and Ganz, T. Defensins and Other Antimicrobial Peptides. 1999. in Mucosal Immunology, Second Edition. Edited by P.L. Ogra, J. Mestecky, M.E. Lamm, W. Strober, J.R.. McGhee and J. Bienenstock. Academic Press, N.Y. pp 89-100.
2. Lee, I.H., CQ Zhao, Y. Cho, S.S.L. Harwig, E.L. Cooper and R.I. Lehrer. 1997. Clavanins, a-helical antimicrobial peptides from tunicate hemocytes. FEBS Lett. 400:158-162.
3. Lee, I.H., Cho, Y. and R.I. Lehrer. 1997. Styelins, broad-spectrum antimicrobial peptides from the solitary tunicate, *Styela clava.* Comp. Biochem. Physiol. B.118B: 515-521.
4. Lee, I.H., Cho, Y. and R.I. Lehrer. 1997. Effects of pH and salinity on the antimicrobial properties of Clavanins. Infect. Immun. 65:2898-2903.
5. Zhao, C., Liaw, L., Lee, I.H. and R.I. Lehrer. 1997. cDNA cloning of clavanins: antimicrobial peptides of tunicate hemocytes. FEBS Lett. 410:490-492.
6. Zhao, C., Liaw, L., Lee, I.H., and R.I. Lehrer. 1997. cDNA cloning of three cecropin-like antimicrobial peptides (Styelins) from the tunicate, *Styela clava.* FEBS Lett. 412:144-148.
7. Steinberg, D. and R.I. Lehrer. 1997. Designer Assays for Antimicrobial Peptides: Disputing the "One Size Fits All" Theory, in Methods in Molecular Biology, W. M. Shafer, editor, Humana Press, Totowa, N.J. pp. 169-187.
8. Turner, J., Cho, Y., Dinh, N., Waring, A.J. and Lehrer, R.I. 1998. Activities of LL-37, a cathelin-associated antimicrobial peptide of human neutrophils. Antimicrob. Agts Chemother 42:2206-2214.

THE DISCOVERY AND CHARACTERIZATION OF AN INTERLEUKIN 6 CYTOKINE FAMILY ANTAGONIST PROTEIN FROM A MARINE SPONGE, *Callyspongia sp*.

Jane V. Peppard, Patricia Loo, Matthew A. Sills, Lawrence Wennogle, *Amy Wright, *Shirley Pomponi and Maria Cueto

Novartis Institute for Biomedical Research and *Harbor Branch Oceanographic Institute

INTRODUCTION

The process of drug discovery in the pharmaceutical industry often begins with the setting up of a bioassay whereby many thousands of synthetic compounds, or less characterized substances such as natural product extracts, may be screened in high throughput mode. During the early 1990s, we started a search using this approach for inhibitors of the cytokine Interleukin 6 (IL-6). One of the sources for screening was

Phylogenetic Perspectives on the Vertebrate Immune System.
Edited by Beck *et al.*, Kluwer Academic/Plenum Publishers, 2001.

extracts of marine organisms provided to the former Ciba-Geigy organization by Harbor Branch Oceanographic Institute (HBOI). This chapter describes the finding of activity inhibiting IL-6 function from an extract of the marine sponge *Callyspongia* sp., and work which was carried out during the effort to characterize this activity more fully.

IL-6 is a cytokine with pleiotropic activities, eliciting several biological responses and acting on a wide range of tissues. According to the cell type or its state of maturation, IL-6 can drive differentiation or act as a growth factor. For example, IL-6 stimulates the proliferation of myeloma cells, T cells, keratinocytes, Kaposi's sarcomas and certain carcinomas. It can also stimulate B cells to terminally differentiate into immunoglobulin secreting cells, T cells to become cytotoxic, hepatocytes to increase production of acute-phase proteins, and it promotes osteoclastogenesis (1). Extended exposure to IL-6 through long-term supranormal production has been suggested to be contributory to the pathogenesis of several chronic diseases, such as myeloma, psoriasis, autoimmune disease, AIDS and osteoporosis (1-4). Reducing the activity of IL-6 should thus be of therapeutic value in such diseases.

IL-6 initiates cellular responses by binding with low affinity (~ 2 nM) to its specific receptor (IL-6R), a single chain transmembrane glycoprotein. The binding of IL-6 to IL-6R enables the formation of a high affinity binding site for IL-6 (*ca.* 70pM) through association of the IL-6/IL-6R complex with another transmembrane molecule, gp130 (5). This association is suggested to induce the formation of a hexameric receptor complex (Fig. 1; ref. 6) and disulfide-linked dimerization of gp130, which initiates signal transduction (5,7). Complex formation is followed rapidly by tyrosine phosphorylation of gp130 in the membrane proximal region of the cytoplasmic domain, by Jak-Tyk family kinases and others, and recruitment of Stat3. Stat3 also becomes tyrosine phosphorylated, dissociates from the complex, forms homo- or heterodimers and translocates to the nucleus. There it binds to cognate DNA response elements and activates gene transcription (reviewed in 8,9).

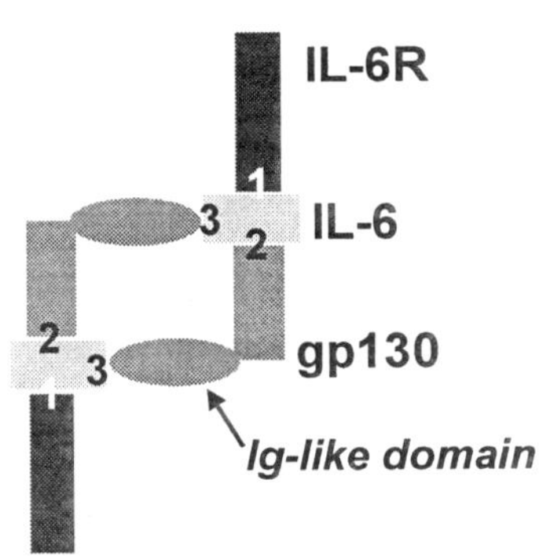

Figure 1. Model of gp130 interaction with IL-6 and IL-6R. IL-6 is proposed to require 3 sites of interaction with the complex for full signalling to occur .

Gp130 is a ubiquitous cell surface molecule and is utilized in the signal transduction process of several other cytokines which are structurally related to IL-6, including Oncostatin M (OSM), Ciliary Neurotrophic Factor (CNTF), Leukemia Inhibitory Factor

(LIF), Cardiotrophin-1 (CT-1) and IL-11. The consequent signalling process appears to be similar for all these molecules (8,9), and specificity appears to be conferred by the presence or absence on different cell types of ancilliary receptor molecules which specifically recognize individual ligands. These specific receptors are homologous to gp130 in both protein sequence and structure (Fig 2). Gp130 is also related to the IL-12R, leptin receptor (LEPR) and granulocyte colony-stimulating factor receptor (GCSR), the ligands of which do not trigger the same pathway as IL-6-type cytokines.

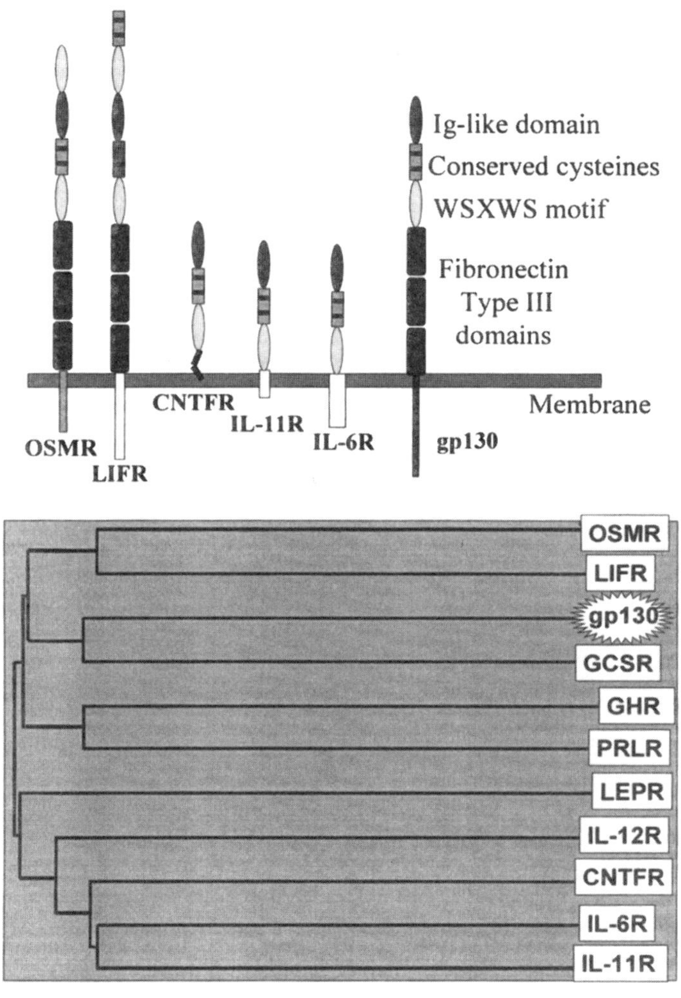

Figure 2. Structural and sequence homology of gp130 and cytokine receptors. **Top**: Predicted domain structures of receptors and co-receptors for the IL-6 cytokine family. **Bottom**: Phylogenetic tree derived from protein sequences related to gp130 using ClustalW and NJ methods.

THE SEARCH FOR IL-6 INHIBITORS

As is often the case when selecting novel targets for the discovery of drugs, little of the information on its mechanism of action now available was known when we started on the process of identifying inhibitors of IL-6. We therefore chose to look for inhibitors of IL-6 binding to cells, and selected a fast-growing human cell line, the B cell myeloma line U266 (10), which expresses high numbers (11,000/cell) of IL-6 receptors. This cell line was adapted for use in a high-throughput screen of synthetic compounds from the Ciba-Geigy library and extracts of marine organisms provided by HBOI. The inhibition of binding of ^{125}I-IL-6 to whole cells was used as the readout in the primary screening process. Compounds or extracts which inhibited binding were then evaluated in secondary screens to distinguish between antagonists and agonists of IL-6. Two functional assays were set up with human cell lines, using the IL-6 stimulated production of IgM by the myeloma SKW6.4, and the upregulation of haptoglobin production by the HepG2 hepatoma.

DISCOVERY AND PARTIAL CHARACTERIZATION OF AN IL-6 INHIBITOR

Following the process outlined above, IL-6 antagonistic activity was discovered in a water/alcohol extract from a marine sponge, *Callyspongia sp.* (11). This organism, of the family Callyspongiidae (Porifera, Desmospongiae, Haplosclerida; ref. 12) is most closely related to *C. simplex* and was collected off the coast of Madeira. Further testing of two other family members, *C. vaginalis* and *C. armigera*, collected at a different geographical location near the Bahamas, revealed that they contained similar activities, suggesting that this was a property of the sponge itself rather than of a commensurate organism. HBOI provided us with frozen samples of the organisms which we used to prepare freeze/thaw extracts into phosphate buffered saline. These extracts, stored at –20°C, were stable for many years.

INHIBITION OF BINDING AND IL-6 FUNCTION

Initially, the *Callyspongia* extracts were compared using U266 binding of ^{125}I-IL-6. For all three extracts, IL-6 binding was inhibited in a dose-dependent manner; the IC_{50} values for extract dilution were 1:5,880 for *Callyspongia sp.*, 1:1,563 *for C. armigera* and 1:3,226 for *C. vaginalis*, compared in the same experiment. The *Callyspongia* sp. extract had no effect on the binding of IL-1 to its receptor on EL4.6.1 cells, or of IL-4 to the IL-4R of Raji cells.

All further experiments were carried out using the *C.* sp. extract. Similar inhibition of IL-6 binding was seen with HepG2 cells (Fig. 3). These cells turned out to be exceptionally useful in functional testing of the extract. They produce haptoglobin constitutively and this can be stimulated several-fold after the addition of IL-6 (13). The constitutive production can be used as a measure of general toxicity of a substance for the cells, and the IL-6 stimulated production can be measured to assess IL-6 function. HepG2 cells also produce haptoglobin in response to other IL-6 family members, such as OSM or LIF. In addition, IL-1 stimulates haptoglobin secretion using a different signalling pathway.

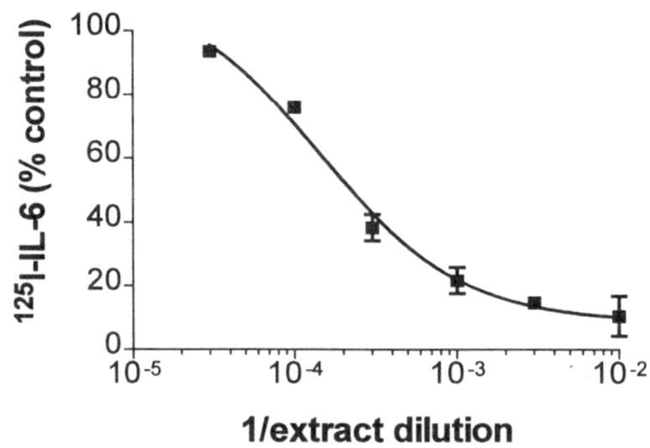

Figure 3. ^{125}I-IL-6 binding to HepG2 cells in a representative experiment. Results are expressed as % total binding (100% binding was 2,300cpm); IC_{50} value = 1:6,868.

For functional testing it was found that optimal results were obtained if the extract was not left in the cultures for extended periods. A method was developed (13) whereby HepG2 cells were stimulated for a relatively short period (1-4 hr) using medium containing IL-6 and extract, the medium was removed and and replaced with normal growth medium. The effects of the stimulation 'pulse' on haptoglobin production (measured by ELISA) were quantified 24 - 48 hr later. Fig. 4 shows the inhibition of IL-6 stimulation of haptoglobin production, though the basal level was unaffected. IL-1 stimulated haptoglobin to a smaller extent than IL-6, and this was not changed in the presence of extract. Thus, both specificity and a lack of toxicity was indicated.

WORK ON IDENTIFICATION OF COMPONENT RESPONSIBLE FOR ACTIVITY

The inhibition of ^{125}I-IL-6 binding to U266 cells was used to identify activity in fractions when various methods of separation of the extract components were tested. IL-6 inhibition was lost after heating the extract for 5 min at 100°C, and after treatment at 37°C with 1% trypsin for 10 min. It was found that activity was retained after dialysis using 12-14kDa cellulose membranes, concentration of the extract using ultrafiltration (by Centricon PM-10) or by 50% ammonium sulfate precipitation. These data indicated that the activity was likely due to a protein present in the sponge extract. Unfortunately, all fractionation methods resulted in a huge loss of activity, and most separation media tried (*e.g.* G-25, Sepharose, Ultrogel) removed activity entirely. Size-exclusion HPLC using a Bio-Sil 125 (Bio-Rad) column was somewhat successful (Fig. 5) and suggested that activity was due to a protein around 17kDa.

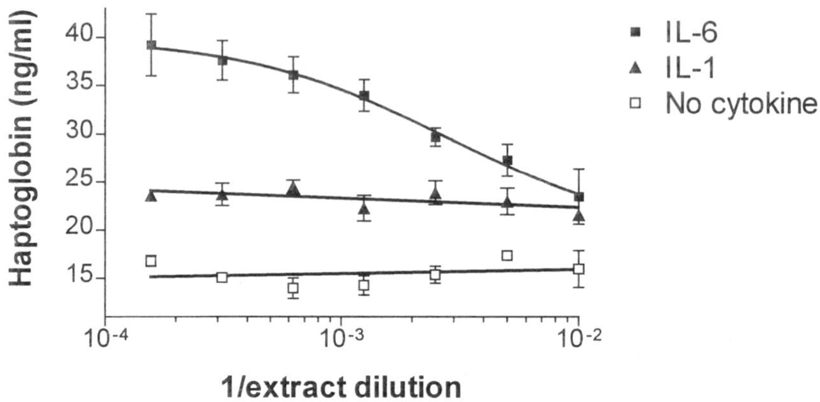

Figure 4. Functional testing in HepG2 cells. Inhibition of increased haptoglobin after IL-6 stimulation by sponge extract (EC_{50} = 1:430), but not of IL-1-induced upregulation or constitutive production.

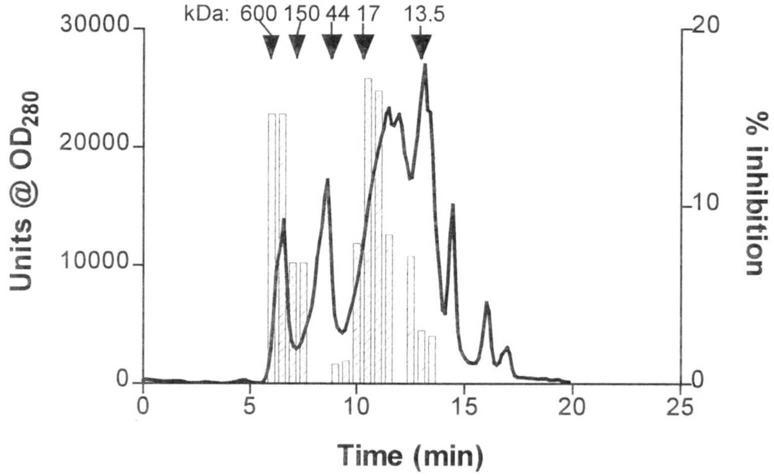

Figure 5. The extract was concentrated 10-fold using a Centricon-10 and 20ml was applied to a Bio-Sil 125 column; OD at 280nm was monitored (line) and 0.5min fractions were collected and assayed for activity (bar graph). In repeated runs, a consistent area of activity was located at 10.5-11.5 min, corresponding to a MW around 17kDa; a major portion of the activity was also seen in the exclusion peak (> 50kDa) at 6-7min. Integration of the ODs of the 6-7min and 10.5-11.5min peaks showed that altogether they contained ~ 22.5% of the total protein applied. The total recovery of activity compared to starting material was ~ 7%.

To characterize and purify the protein further, high performance reverse-phase chromatography was utilized (Fig. 6). Again, a small recovery of activity was observed and what remained eluted amongst many other components. Fractionation efforts were thus hampered by the loss of the active principle to such an extent that, for example, combining methods to achieve better purification was simply not feasible.

Because of this low recovery, we investigated whether fractionation had separated multiple components necessary for inhibition of binding. Active fractions from size-exclusion separation runs were each mixed with unfractionated extract at a dilution giving 25% inhibition of binding, and re-titrated in U266 binding. It was found that activity increased only additively, indicating that inhibitory activity could not be restored through synergy with other components.

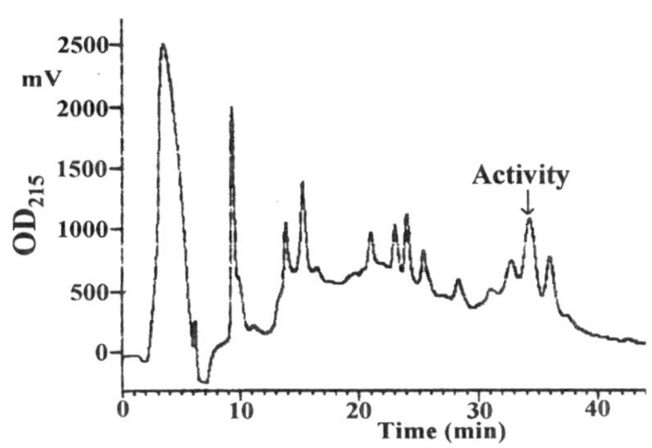

Figure 6. A typical run of 10ml whole extract using C-4 reverse-phase HPLC with a 10–90% acetonititrile gradient. A peak of activity (U266 binding inhibition) eluted in the region (arrow) around 50% acetonitrile. The activity recovered after pooling the active fractions was not more than 5% of that applied to the column. This material was also found to be correspondingly inhibitory for haptoglobin production by HepG2 cells.

The fractions which contained inhibitory activity from both chromatographic methods, *i.e.* reverse-phase and size-exclusion chromatography, were dried under vacuum and compared in reduced SDS-PAGE (Fig. 7). All showed poorly-resolved bands at ~25kDa. Furthermore, bands of similar MW were also seen in material fractionated by ammonium sulfate precipitation.

Taken together, these results suggested that the inhibitor was a protein of molecular weight around 20 kDa. However, the success of the purification efforts was such that we were not able to characterize this further in the time allowed.

MECHANISM OF ACTION OF THE *CALLYSPONGIA* INHIBITOR

In view of the increasing information which was becoming available at that time on the IL-6 cytokine family, we studied the effects of this extract on the stimulation of

HepG2 cells by IL-6 cytokine family members OSM and LIF. We found that the extract also inhibited these two cytokines, but that it was unable to reduce production to the extent that it did with IL-6 (Fig. 8).

We tested the individual components of the binding complex in immunoprecipitation experiments. In accord with the "stickiness" of the extract with separation media in general, we found that paramagnetic particles coated with a suitable antibody gave the

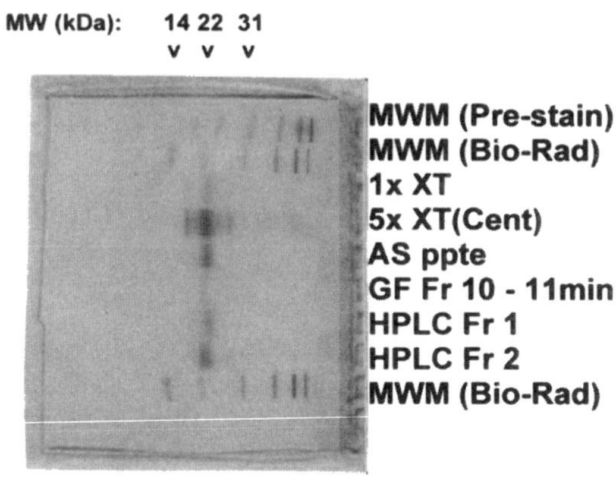

Figure 7. 16% Tris-Tricine SDS-PAGE of reduced samples of Callyspongia extract: MWM, molecular weight markers; 1X, untreated extract; 5X, Centricon-10 concentrated extract; AS ppte, 50% saturated NH_4SO_4-precipitated extract redissolved at 3x concentration; GF fraction; gel filtration active fraction, 10.5-11.5min (Fig 5); HPLC Fr 1, Fr 2, neighbouring active fractions from reverse-phase HPLC (Fig. 6).

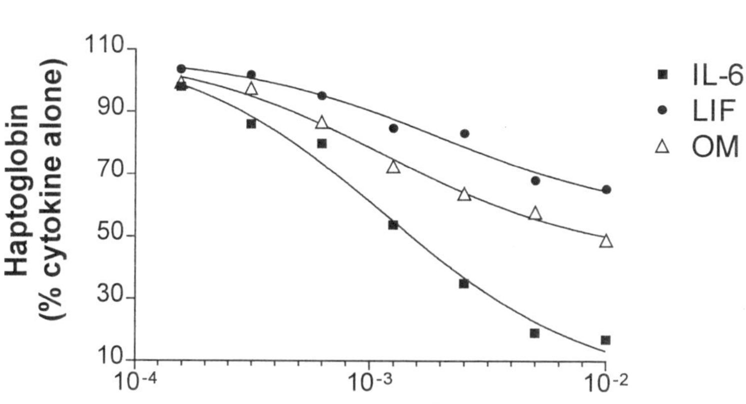

Figure 8. HepG2 stimulation by IL-6 compared with OSM and LIF. All were applied at 5ng/ml. The EC_{50} values obtained were 1:849 (IL-6), 1:513 (LIF) and 1: 865 (OSM).

lowest background binding of several methods tried. Using suitable antibodies and soluble IL-6R and gp130 moieties, we were able to demonstrate that the extract inhibited binding of IL-6 through interfering with the gp130 binding to the complex, but not with IL-6/IL-6R binding (Fig 9).

This observation may help to explain why the sponge protein is a better inhibitor of IL-6 than of LIF or OSM, since as described in the Introduction, IL-6 is the only one of these three which requires the dimerization of two molecules of gp130 to transduce its signal. Bearing in mind that all binding is transient, if the *Callyspongia* protein binds directly to gp130 at or near the sites which ligand/receptor complexes require, it can block two sites per signalling complex for IL-6 but only one site per complex for OSM or LIF, effectively increasing the avidity of its action.

Direct binding to gp130 also suggests a rationale for why the inhibitory acitvity is lost in extended culture (section 3.1), since gp130 is constitutively endocytosed (reviewed in 9). Presumably, the inhibitor is bound, then internalised with gp130 and broken down inside the cell, leaving IL-6 to activate gp130 once it is re-expressed on the cell surface several hours later.

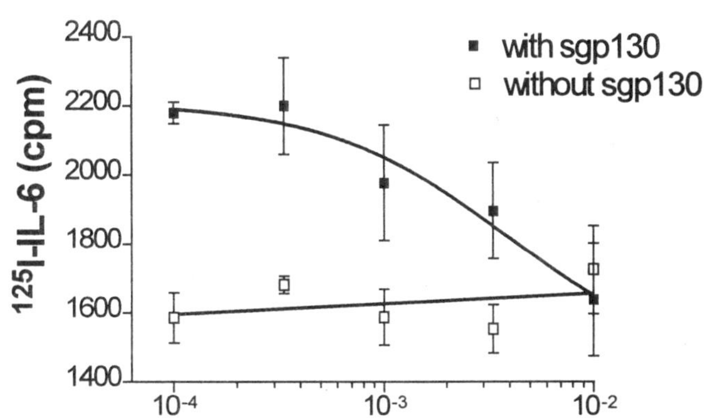

Figure 9. Immunoprecipitation of IL-6/IL-6R/gp130 complexes. ^{125}I-IL-6 was incubated with soluble IL-6R with (■) or without (□) soluble gp130 (sgp130) in the presence of extract dilutions. Immune complexes were pulled out of solution with mouse anti-IL-6R pre-bound to rabbit anti-mouse coated magnetic particles. Parallel determinations omitting the anti-IL-6R were carried out to determine non-specific binding and this had been substracted for each point. Soluble gp130 enhances binding of IL-6 in the complex and this enhancement is inhibited by extract, whereas it has no effect on IL-6 binding to IL-6R alone.

DISCUSSION

Gp130 is expressed by a wide variety of cell types and participates in the signalling for many cytokines. Gp130 knockout mouse embryos are not viable beyond 13 days (14) and postnatal inactivation of gp130 in mice results in defects of cardiac, hepatic, immunological, pulmonary and neurological systems (15). Gp130 has been suggested to be the prototype for the family of receptors which resemble it, in turn having evolved from a more primordial form of surface Ig domain (16); indeed, the Ig domain of gp130 remains essential for its function (Fig. 1) (6,17).

The discovery of a potent and apparently specific inhibitor protein in a marine sponge may suggest that a molecule with some structural resemblance to gp130, or to molecules which bind specifically to it, such as components of the IL-6/IL-6R complex, may exist also in sponges. Inhibitors of gp130/ligand complex formation which have been designed by humans include antibodies targeted to the IL-6, IL-6R or gp130 molecule (reviewed in 18), as well as altered cytokine molecules or peptides derived from them. However, the most successful antagonists have been produced by mutating IL-6 itself. For example, mutating IL-6 at sites required for binding to gp130 (Fig. 1) produces IL-6 antagonists with only limited agonist activity (19). In addition, super-antagonists of IL-6 have been designed with mutations at several sites to yield molecules with tighter binding to the signalling complex than IL-6 itself, yet which do not trigger function through gp130 (20). It is tempting to imagine that the activity in the *Callyspongia* extract, which appears to be due to a protein of MW around 20 kDa, a size close to that of IL-6 (~26kDa), may be some sort of IL-6 analogue. Until further characterization of the sponge protein, this can of course only be a matter for speculation.

ACKNOWLEDGMENTS

This work was particularly facilitated by the technical effort and lively scientific curiosity of Pat Loo, who sadly was killed in the crash of TWA 800. The authors would like to thank Deborah Munster, Siyuan Zhou, Michael Chin and Donna Yarwood for their technical help. Figures 3, 4, 8 and 9 are reprinted from the Journal of Biological Chemistry (ref. 11) with permission.

REFERENCES

1. Hirano, T. and Kishimoto, T. 1992. Molecular biology and immunology of interleukin-6 (46th Forum in Immunology). *Res. Immunol. 143:723*
2. Hirano, T. 1992. Interleukin-6 and its relation to inflammation and disease. *Clin. Immunol. & Immunopathol. 62:860.*
3. Breen, E.C., Rezai, A. R., Nakajima, K., Beall, G.N., Mitsuyasu, R.T., Hirano, T., Kishimoto, T. and Martinez-Maza, O. 1990. Infection with HIV is associated with elevated IL-6 levels and production. *J. Immunol. 144:480.*
4. Jilka, R.L., Hangoc, G., Girasole, G., Passeri, G., Williams, D.C., Abrams, J., Boyce, B., Broxmeyer, H. and Manolagas, S. 1992. Increased osteoclast development after estrogen loss: mediation by Interleukin-6. *Science 257:88.*
5. Taga, T., Hibi, M., Hirata, Y., Yamasaki, K., Yasukawa, K., Matsuda, T., Hirano, T. and Kishimoto, T. 1989. Interleukin-6 triggers the association of its receptor with a possible signal transducer, gp130. *Cell 58:573.*
6. Simpson, R.J., Hammacher, A., Smith, D.K., Matthews, J.M. and Ward, L.D. 1997. Interleukin-6: structure-function relationships. *Protein Sci. 6:929*
7. Murakami, M., Hibi, M., Nakagawa, N., Nagakawa, T., Yasukawa, K., Yamanishi, K, Taga, T. and Kishimoto, T. 1993. IL-6-induced homodimerization of gp130 and associated activation of a tyrosine kinase. *Science 260:1808.*
8. Hirano, T., Nakajima, K. and Hibi, M. 1997. Signaling mechanisms through gp130: a model of the cytokine system. *Cytokine & Growth Factor Rev. 8:241.*

9. Heinrich, P.C., Behrmann, I., Muller-Newen, G., Scahper, F. and Graeve, L. 1998. Interleukin-6-type cytokine signalling through the gp130/Jak/STAT pathway. *Biochem. J. 334:297.*
10. Taga, T., Kawanishi, K., Hardy, R.R., Hirano, T. and Kishimoto, T. 1987. Receptors for B cell stimulatory factor 2. Quantitation, specificity, distribution, and regulation of their expression. *J. Exp. Med. 166:967.*
11. Peppard, J.V., Loo, P., Sills, M.A.., Munster, D., Pomponi, S.A. and Wright, A. E. 1996. Characterization of an interleukin 6 cytokine family antagonist protein from a marine sponge, *Callyspongia sp. J. Biol. Chem. 271:7281.*
12. Pomponi, S.A. 1976. A cytological study of the *Haliclonidae* and the *Calyspongiidae* (Porifera, Desmospongiae, Haplosclerida). In: "Aspects of Sponge Biology", Eds. F.W. Harrison and R.R. Cowden, Academic Press.
13. Peppard, J.V., Smith, H.D. and Zhou, S. 1994. Kinetics of a cellular response to Interleukin 6. *Cell. Immunol. 155:253.*
14. Yoshida, K., Taga, T., Saito, M., Suematsu, S., Kumanogoh, A., Tanaka, T., Fujiwara, H., Hirata, M., Yamagami. T,, Nakahata, T., Hirabayashi, T., Yoneda, Y., Tanaka, K., Wang, W.Z, Mori, C., Shiota, K., Yoshida, N. and Kishimoto, T. 1996. Targeted disruption of gp130, a common signal transducer for the interleukin 6 family of cytokines, leads to myocardial and hematological disorders. *Proc. Natl. Acad. Sci. U.S.A. 93:407.*
15. Betz, U.A.K., Bloch, W., van den Broeck, M., Yoshida, K., Taga, T. Kishimoto, T., Addicks, K., Rajewsky, K. and Muller, W. 1998. Postnatally induced inactivation of gp130 in mice results in neurological, cardiac, hematopoietic, immunolgocial, hepatic and pulmonary defects. *J. Exp. Med. 188: 1955.*
16. Yamamori, T. and Sarai, A. 1992. Coevolution of cytokine receptor families in the immune and nervous system. Neurosci. Res. 15: 151.
17. Kallen, K-J., Grotzinger, J., Lelievre E., Vollmer, P., Aasland, D., Renne, C., Mullberg, J., Meyer zum Buschenfelde, K-H., Gascan, H. and Rose-John, S. 1999. Receptor recognition sites of cytokines are organized as exchangeable modules. *J. Biol. Chem. 274:11859.*
18. Huang, Y.W. and Vitetta, E.S. 1995 Immunotherapy of multiple myeloma. *Stem Cells 13:123.*
19. Brackenhoff, J.P.J., de Hon. F.D., Fontaine, V., ten Boekel, E., Schooltink, H., Rose-John, S., Heinrich, P.C., Content, J. and Aarden, L.A. 1994. Development of a human interleukin-6 receptor antagonist. *J. Biol. Chem. 269: 86.*
20. Sporeno, E., Savino, R., Ciapponi, L., Paonessa, G., Cabibo, A., Lahm, A., Pulkki, K., Sun, R-X., Toniatti, C., Klein, B. and Cilikberto, G. 1996. Human Interleukin-6 receptor super-antagonists with high potency and wide spectrum on multiple myeloma cells. *Blood 87:4510.*

PHYLOGENY OF VERTEBRATE CYTOKINES

C.J. Secombes, S. Bird, S. Hong, K.J. Laing, and J. Zou

Department of Zoology, University of Aberdeen, UK

INTRODUCTION

Considerable progress has been made in recent years in the sequencing of cytokine genes in non-mammalian vertebrates. This review will focus primarily on the recent advances made in isolating interleukin-1β (IL-1β), transforming growth factor-β (TGF-β) genes and chemokine genes, since they have been found in several vertebrate groups (typically fish, amphibians and birds) allowing some discussion of cytokine gene evolution. For recent advances on cytokine genes that have been isolated to date in only birds (outwith mammals), namely interferons and IL-2, the reader is referred to Schat & Kaiser (1997), Kaiser *et al.* (1998a,b), Sick *et al.* (1998), Michalski *et al.* (1999) and Kaiser & Mariani (1999).

INTERLEUKIN-1β

Over the last year or two, the complete cDNA sequence coding for IL-1β has been isolated from chicken (Weining *et al.*, 1998), *Xenopus* (Zou *et al.*, 2000), rainbow trout (Zou *et al.*, 1999) and carp (Fujiki *et al.*, 2000). Partial sequences have also

Phylogenetic Perspectives on the Vertebrate Immune System.
Edited by Beck *et al.*, Kluwer Academic/Plenum Publishers, 2001.

been obtained for plaice and turbot in our laboratory (Secombes *et al.*, 1999). The precursor molecules vary between 260 (trout) to 283 (*Xenopus*) amino acids (Table 1). Typically there is a short 5' untranslated region (UTR) of up to 100 bp and a 3' UTR of approximately 300-550 bp. As in mammals, all these sequences lack a signal peptide and with the exception of the chicken sequence, possess multiple copies of an mRNA instability motif (ATTTA) common to inflammatory molecules. However, in contrast to mammalian sequences they do not possess an obvious ICE cut site, despite a cloned ICE (caspase-1) cDNA in chickens (Johnson *et al.*, 1998), and it remains to be determined whether the precursor molecule is cleaved to release a biologically active "mature" peptide. Nevertheless, recombinant molecules made in chicken (Weining *et al.*, 1998) and trout (Fig. 1) that represent the predicted mature peptide are biologically active.

Table 1. Features of IL-1β genes in non-mammalian vertebrates.

	5' UTR	Open reading frame	3' UTR (excl. polyA)	No. potential glycosylation sites	No. ATTTA motifs	nucleotide identity to human IL-1β
chicken	32 bp	267 aa	271 bp	0	1	48%
Xenopus	64 bp	283 aa	519 bp	1	6	46%
trout	97 bp	260 aa	448 bp or 544 bp	3	7	48%
carp	53 bp	276 aa	329 bp	4	4	43%

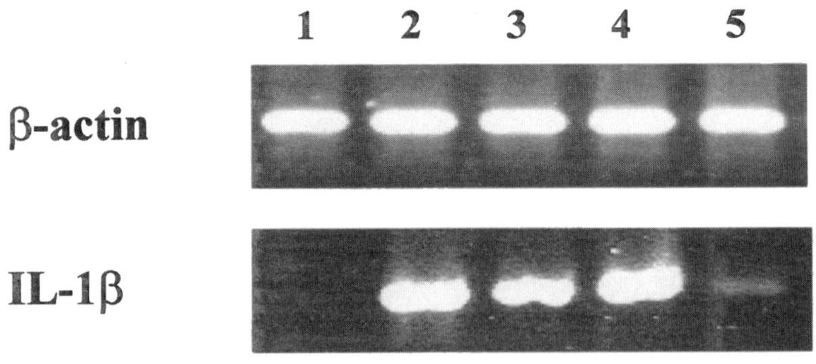

Figure 1. Expression of β-actin and IL-1β in trout head kidney leucocytes, determined by RT-PCR, in 1) control cells or cells stimulated for 4h *in vitro* with 2) 10 μg LPS/ml, 3) 50 ng recombinant trout IL-1β (rtIL-1β)/ml, 4) 25 ng rtIL-1β/ml, or 5) 12.5 ng rtIL-1β/ml.

Expression studies have shown that the IL-1β mRNA is inducible, with stimulants such as LPS, PHA and PMA being particularly effective *in vitro*. Injection of LPS or Gram negative bacteria also induce IL-1β expression *in vivo* (Zou *et al.*, 1999, 2000). A number of incompletely spliced transcripts of IL-1β have been detected in trout leucocytes poststimulation (Zou *et al.*, 1999), and may be involved with post-transcriptional regulation of protein expression. Multiple polyadenylation sites have also been identified in the trout transcripts, resulting in different sized 3' UTR (Secombes *et al.*, 1999). Whilst the gene organisation of trout IL-1β is known (Zou *et al.*, 1999), this has still to be determined for the other species above.

TRANSFORMING GROWTH FACTOR-β

It has been known for some time that in birds and mammals, three isoforms of TGF-β exist (β1, β2, β3), and that they possess a common 7 exon-6 intron gene organisation. More recently two isoforms have been found in amphibians (β2, β5) (Rebbert *et al.*, 1990; Kondaiah *et al.*, 1990), with the so-called TGF-β5 isoform having clear homology with TGF-β1 (Hardie *et al.*, 1998). The gene organisation of the *Xenopus* TGF-β5 has also been determined and is typical of the known homeotherm genes (Vempati & Kondaiah, 1998a), although the 5' flanking sequence appears to be unique (Vempati & Kondaiah, 1998b). In fish, isoforms equivalent to TGF-β1/5 (Hardie *et al.*, 1998; Harms *et al.*, 2000) and TGF-β2 (Sumathy *et al.*, 1997) have also been isolated, although in different species, and more recently it has become apparent that TGF-β3 does exist in both chondrostean (sturgeon) and teleostean (trout, eel, plaice) bony fish (Laing *et al.*, 1999, 2000), with some 81-84% nucleotide identity with known TGF-βs. Indeed, that all three isoforms are present within a species has been confirmed in plaice (Laing *et al.*, 2000).

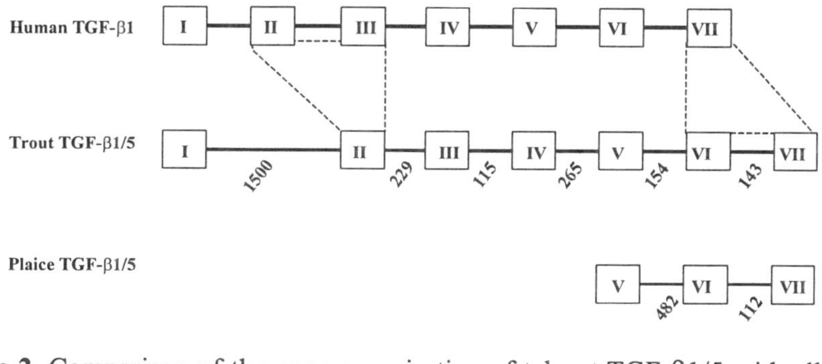

Figure 2. Comparison of the gene organisation of teleost TGF-β1/5 with all other known TGF-β genes (as illustrated with the human TGF-β1 gene).

One of the interesting recent findings is that the genomic organisation of trout TGF-β1/5 differs from that found in all other species (Daniels & Secombes, 1999) (Fig. 2). The trout gene appears to have an exon equivalent to exons 2 and 3 of birds, mammals and *Xenopus* and has an additional intron within exon 7. That this is not something peculiar to rainbow trout has been confirmed in part, since partial sequencing of the plaice TGF-β1/5 gene has revealed the additional intron at the 3' end of the gene (Fig. 2), with exon 6 corresponding exactly to the trout gene.

CHEMOKINES

A lot of progress has also been made with sequencing chemokines in nonmammalian vertebrates over recent years. The first chicken chemokine gene sequenced was 9E3/CEF4 (Martins-Green & Hanafusa, 1997), that codes for a 9 kDa CXC chemokine closely related to IL-8 (Fig. 3). Like other IL-8 related chemokines, it possess an "ELR" motif which identifies those chemokines able to attract neutrophils, use the CXC-R2 receptor and are angiogenic (Wuyts *et al.*, 1998). Whilst the recombinant molecule is both angiogenic and chemotactic, giving rise to the name chicken chemotactic and angiogenic factor (cCAF) for the protein, it attracts monocytes and lymphocytes rather than neutrophils. The angiogenic properties reside in the C-terminus of the molecule (Martins-Green & Kelly, 1998). Its expression is strongly induced by thrombin, and following wounding expression is found primarily in areas of neovascularisation. It is also highly expressed in transformed cells. A number of transcription factors have been shown to be important for expression of 9E3/CEF4, using reporter gene constructs containing the 5' flanking region (Li *et al.*, 1999).

A second chicken chemokine has been identified recently, from searches of EST-databases. It is most related to the "C" chemokine lymphotactin (Fig. 3), with which it has 28-33% amino acid identity (Rossi *et al.*, 1999). Northern blot analysis using RNA from a range of normal tissues shows strong hybridisation in the spleen, with a weaker signal in the colon and liver. The recombinant protein (produced in COS cells) induces B cell locomotion but in contrast to mammalian lymphotactin it has only a weak effect upon T cells. Thus, the possibility exists that this is a new member of the C chemokine subfamily.

An IL-8 homolog has also been sequenced in lamprey (Najakshin *et al.*, 1999), with highest homology to the chicken 9E3/CEF4 (40% identity) molecule. The 3' UTR possess multiple copies of the ATTTA motif and a motif present in "immediate-early" genes (TTTTGTA), suggesting it may be transiently expressed. Interestingly it lacks the "ELR" motif and whether the molecule is chemoattractive remains to be determined.

Two CC chemokines have also been sequenced in carp and trout recently. The carp sequence, which is in the EMBL database (Fujiki *et al.*, 1998a), codes for a molecule with highest identity to eotaxins and monocyte chemotactic proteins (MCP). It also has a copy of the TTTTGTA motif in the 3' UTR but lacks ATTTA. The trout sequence, CK-1, has been more thoroughly studied (Dixon *et al.*, 1998). It contains 6 cysteine residues in common with members of the C6-β subfamily of the CC chemokines, including MIP3A (also called LARC), leukotactin and MRP-1 and -2, with which it has highest homology (33-36% similarity). The gene contains four exons, as do the other C6-β chemokines although the organisation of the CK-1 gene is more similar to CXC chemokines. The gene has also been isolated from several other salmonid species (chinook salmon, coho salmon, brown trout) and in all cases has the same organisation.

That fish possess both CXC and CC chemokines is also borne out by the recent sequencing of CXC and CC receptors (Fujiki *et al.*, 1998b, c; Daniels *et al.*, 1999). It will be fascinating to determine whether theories of co-evolution of chemokines and their receptors (Hughes & Yeager, 1999) are supported by studies in lower vertebrates.

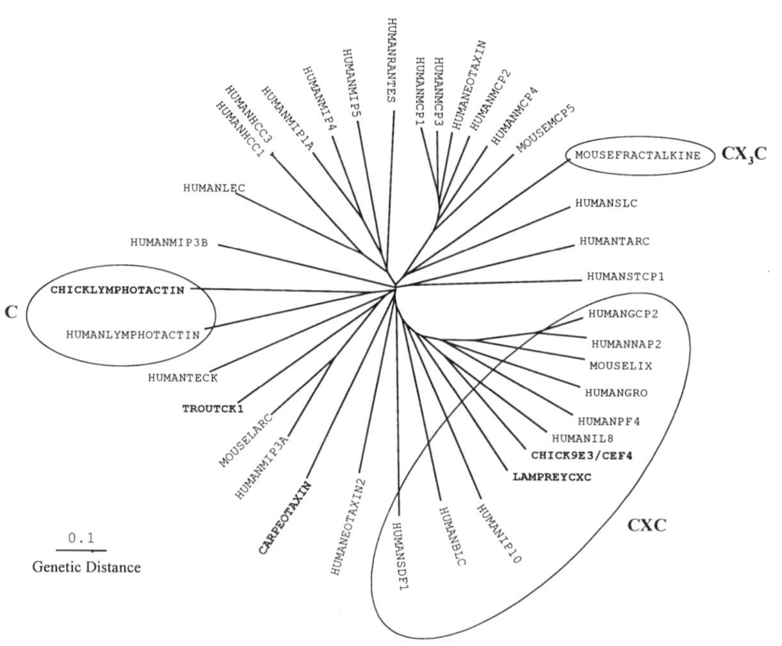

Figure 3. Phylogenetic tree of chemokines.

REFERENCES

Daniels GD, Secombes CJ. Genomic organisation of rainbow trout *Oncorhynchus mykiss* TGF-ß. Dev. Comp. Immunol. 1999;23:139-147.

Daniels GD, Zou J, Charlemagne J, Partula S, Cunningham C, Secombes CJ. Cloning of two chemokine receptor homologs (CXC-R4 and CC-R7) in rainbow trout *Oncorhynchus mykiss*. J. Leukocyte Biol. 1999;65:684-690.

Dixon B, Shum B, Magor KE, Hedrick RP, Muir DG, Parham P. CK-1, a putative chemokine of rainbow trout (*Oncorhynchus mykiss*). Imm. Rev. 1998;166:341-348.

Fujiki K, Nakao M, Shin D, Yano T. cDNA cloning of a carp CC chemokine homologous to mammalian eotaxins. EMBL accession No. AB010469, 1998a.

Fujiki K, Nakao M, Shin D, Yano T. cDNA cloning of a carp homologue-1 of mammalian interleukin-8 receptors. EMBL accession No. AB010468, 1998b.

Fujiki K, Shin D, Nakao M, Yano T. cDNA cloning of a carp homologue of mammalian interleukin-1 beta. EMBL accession No. AB010713, 1998c.

Fujiki K, Nakao M, Shin D, Yano T. Molecular cloning and expression analysis of carp (*Cyprinus carpio*) interleukin-1β high affinity immunoglobulin E, Fc receptor γ subunit and serum amyloid A. Fish and Shellfish Immunol. 200;10:229-242.

Hardie LJ, Laing KJ, Daniels GD, Grabowski PS, Cunningham C, Secombes CJ. Isolation of the first piscine transforming growth factor ß gene: Analysis reveals tissue specific expression and a potential regulatory sequence in rainbow trout (*Oncorhynchus mykiss*). Cytokine 1998;10:555-563.

Harms CA, Kennedy-Stoskopf S, Horne WA, Fuller FJ. Cloning and sequencing hybrid striped bass (*Morone saxatilis* x *M. chrysops*) transforming growth factor-ß (TGF-ß), and development of a reverse transcription quantitative competitive polymerase chain reaction (RT-qcPCR) assay to measure TGF-ß mRNA of teleost fish. Fish & Shellfish Immunol. 2000;10:61-85.

Hughes AL, Yeager M. Coevolution of the mammalian chemokines and their receptors. Immunogenetics 1999;49:115-124.

Johnson AL, Bridgham AL, Munks JT, Witty JP. Characterization of the chicken interleukin-1ß converting enzyme (caspase-1) cDNA and expression of caspase-1 mRNA in the hen. Gene 1998;219:55-62.

Kaiser P, Mariani P. Promoter sequence, exon-intron structure, and synteny of genetic location show that a chicken cytokine with T-cell proliferative activity is IL2 and not IL15. Immunogenetics 1999;49:26-35.

Kaiser P, Sonnemans D, Smith LM. Avian IFN-g genes: Sequence analysis suggests probable cross-species reactivity among galliforms. J. Interferon and Cytokine Res. 1998a;18:711-719.

Kaiser P, Wain HM, Rothwell L. Structure of the chicken interferon-g gene, and comparison to mammalian homologues. Gene 1998b;207:25-32.

Kondaiah P, Sands MJ, Smith JM, Fields A, Roberts AB, Sporn MB, Melton DA. Identification of a novel transforming growth factor-ß (TGF-ß5) mRNA in *Xenopus laevis*. J. Biol. Chem. 1990;265:1089-1093.

Laing KJ, Pilström L, Cunningham C, Secombes CJ. TGF-ß3 exists in bony fish. Vet. Immunol. Immunopathol. 1999;72:45-53.

Laing KJ, Cunningham C, Secombes CJ. Genes for three different isoforms of transforming growth factor-ß are present in plaice DNA. Fish & Shellfish Immunol. 2000;10:261-271.

Li QJ, Vaingankar SM, Green HM, Martins-Green M. Activation of the 9E3/cCAF chemokine by phorbol esters occurs via multiple signal transduction pathways that converge to MEK1/ERK2 and activate the Elk1 transcription factor. J. Biol. Chem. 1999;274:15454-15465.

Martins-Green M, Hanafusa H. The 9E3/CEF4 gene and its product the chicken chemotactic and angiogenic factor (cCAF): Potential roles in wound healing and tumor development. Cytokine & Growth Factor Rev. 1997;8:221-232.

Martins-Green M, Kelly T. The chicken chemotactic and angiogenic factor (9E3 gene product): Its angiogenic properties reside in the C-terminus of the molecule. Cytokine 1998;10:819-830.

Michalski WP, Shiell BJ, O'Neill TE, Beddome G, Lowenthal JW. Recombinant chicken interferon-g expressed in *Escherichia coli*: Analysis of Cterminal truncation and effect on biologic activity. J. Interferon and Cytokine Res. 1999;19:383-392.

Najakshin AM, Mechetina LV, Alabyev BY, Alexander V. Identification of an IL-8 homolog in lamprey (*Lampetra fluviatilis*): Early evolutionary divergence of chemokines. Eur. J. Immunol. 1999;29:375-382.

Rebbert ML, Bhatia-Dey N, Dawid IB. The sequence of TGF-ß2 from *Xenopus laevis*. Nucleic Acids Res. 1990;18:2185.

Rossi D, Sanchez-Garcia J, McCormack WT, Bazan JF, Zlotnik A. Identification of a chicken "C" chemokine related to lymphotactin. J. Leuk. Biol. 1999;65:87-93.

Schat KA, Kaiser P. "Avian cytokines." In *Cytokines in Veterinary Medicine*, VECJ Schijns, MC Horzinek, eds. Wallingford: CAB International, 1997.

Secombes CJ, Bird S, Cunningham C, Zou J. Interleukin-1 in fish. Fish & Shellfish Immunol. 1999;9:335-343.

Sick C, Schultz U, Munster U, Meier J, Kaspers B, Staeheli P. Promoter structures and differential responses to viral and nonviral inducers of chicken type I interferon genes. J. Biol Chem. 1998;273:9749-9754.

Sumathy K, Desai KV, Kondaiah P. Isolation of transforming growth factor-ß2 cDNA from a fish, *Cyprinus carpio* by RT-PCR. Gene 1997;191:103-107.

Vempati UD, Kondaiah P. Molecular organisation of the gene encoding *Xenopus laevis* transforming growth factor-ß5. Biochem. Molec. Biol. Int. 1998a;45:997-1003.

Vempati UD, Kondaiah P. Characterization of the 5' flanking region of the *Xenopus laevis* transforming growth factor-ß5 (TGF-ß5) gene. Gene 1998b;208:323329.

Weining KC, Sick C, Kaspers B, Staeheli P. A chicken homolog of mammalian interleukin-1ß: cDNA cloning and purification of active recombinant protein. Eur. J. Biochem 1998;258:994-1000.

Wuyts A, Proost P, Van Damme J. "Interleukin-8 and other CXC chemokines." In *The Cytokine Handbook*, A. Thomson, ed. London: Academic Press, 1998.

Zou J, Cunningham C, Secombes CJ. The rainbow trout *Oncorhynchus mykiss* interleukin-1ß gene has a different organisation to mammals and undergoes incomplete splicing. Eur. J. Biochem. 1999;259:901908.

Zou J, Bird S, Minter R, Horton J, Cunningham C, Secombes CJ. Molecular cloning of interleukin 1ß from *Xenopus laevis* and analysis of expression *in vivo* and *in vitro*. Immunogenetics 2000;51:332-338.

ANTIMICROBIAL PROTEINS IN CRUSTACEANS

Valerie J. Smith & June R.S. Chisholm

Comparative Immunology Group
Gatty Marine Laboratory
University of St Andrews
Fife, KY16 8LB Scotland, UK

The microbe is so very small, You cannot make him out at all
But many sanguine people hope, To see him through the microscope
Hilaire Belloc

INTRODUCTION

Crustacean shellfish along with their molluscan relatives have been an important source of food for humans for much of history. Shells of marine invertebrates have been found littered amongst the remains of primitive cooking utensils uncovered during twentieth century excavations of the neolithic settlements at Skara Brae on the Isle of Orkney, north of Scotland. Shrimps and prawns are listed amongst the crucial ingredients of a rich salty sauce, known as *liquamen* which was popular in Imperial

Rome. They are also known to have been a major source of protein for the early settlers in the Chesapeake Bay area of North America (hence contributing, in a small way, to the colonization of the New World by Europeans) Because of their abundance in the (then) clean waters of coasts and rivers, they have been a cheap and available form of nourishment for many of the poor in Britain from mediaeval times through to the beginning of the Industrial Revolution. Throughout other parts of the world, too, shellfish, have been, and in some countries still are, a prime form of sustenance for impoverished coastal villages and communities, as tragically revealed by the huge degree of human deprivation caused recently by hurricane damage in Honduras and which destroyed much of the country's small shrimp industry, upon which many thousands of people depended.

For the modern developed world, the consumption of shellfish, and especially crustaceans, has been elevated from the one-time coarse fare of the 'poor man', with some perceived prejudice as to its wholesomeness, to an epicurean delicacy, enjoyed with fine wines, champagne and other gastronomic *bonne-bouches,* by the consumer-culture of the modern middle classes. Demand for quality shellfish by the West continues to grow and is already beginning to outstrip the supply that can be obtained from the wild. For example, the world shrimp industry currently produces in excess of 660,200 metric tonnes, a value representing some 22% of total global seafood production (Rosenberry, 1997). In the UK alone, catches of crustacean shellfish are worth around £90,500,000 *pa*, accounting for around 16% of the total value of the fishery industry for the country (Ministry of Agriculture, Fisheries and Food report, 1994). Therefore it is not suprising that to meet this demand, artificial rearing of shellfish has become an expanding and profitable business venture.

Concomitant with the rise in aquaculture has been an increase in the problem of disease in the stock, either from specific pathogens or opportunistic microbial infections. The need to address the question of disease management has been one of the *raisons d'etre* for investigations into the antimicrobial defences of crustaceans, and it is notable that most scientific studies have focused on commercially important species or taxa. This is somewhat ironic given the central role some crustaceans (particularly the waterflea, *Daphnia*) played, historically, in the inception of the the discipline of immunology. Today, grant proposals and research papers tend to rest their justifications rather less on the phylogenetic significance of understanding crustacean immunity or on the intrinsic relevance of such knowledge *per se,* than on the economics of shellfish production.

Unfortunately, one of the consequences of this is that only a relatively small number of crustacean species have received attention, with the assumption that what we learn about these organisms applies generally to most, if not all, members of a large and highly diverse group of protostome invertebrates, without paying much heed to the life history, ecology or anatomy of the animal concerned. Another consequence of the preoccupation with the commercial value of crustaceans has been that findings are frequently reported in journals not regularly consulted by mammalian immunologists, evolutionary biologists or biomedical researchers. Thus crustacean immunology has tended to remain a small and specialised sub-field, with the phylogenetic or biomedical implications of discoveries made in this area frequently obscure to the rest of the scientific community. Thus, less effort seems to have been directed towards the biochemical and molecular biology of antibacterial proteins in crustaceans compared with their relatives, the insects and chelicerates.

It is therefore encouraging that the Federation of American Societies for Experimental Biology has recognised the broader ramifications of understanding the immune systems of lower organisms through its backing of the summer conference on phylogenetic perspectives of the vertebrate immune system, in which aspects of the antimicrobial defences of crustaceans have been permitted to take their rightful place.

This overview aims to consider the state of play as regards antimicrobial proteins in crustaceans with the hope that this will raise the profile of the subject and bring recent findings to the attention of a wider readership.

EARLY STUDIES

Early attempts to investigate the microbicidal properties of crustacean tissues date back to the 1960s. Since then an extensive literature has been produced describing antibacterial activity in crustacean haemolymph or tissue extracts. This has been reviewed previously by Cooper (1985), Ratcliffe *et al.* (1985) and Smith & Chisholm (1992). Much of this research was almost exclusively dedicated to decapods and the reports seldom give more than minimal physico-chemical characterisations of the factor(s) responsible (see Table 1). Virtually none go as far as purifying individual microbicidal proteins. The general picture that emerges from these papers is that decapod tissues or body fluids have potent cidal properties against a range of bacteria or fungi and that this activity is moderately heat stable, cation independent and distinct from agglutination (see review by Smith & Chisholm, 1992). In a few cases activity was thought to be enhanced by primary challenge of the host by bacteria, although this effect was found to be short-lived and non-specific (Smith & Chisholm, 1992). One observation of note was that lobster (*Homarus americanus*) haemolymph/serum seems to lack any activity against the lethal pathogen, *Gaffkya homari* (a Gram postive bacterium, now renamed as *Aerococcus viridans* var *homari*) (Cornick & Stewart, 1968a).

The site of synthesis and/or storage of the active factors is not made clear in most of these papers. Many authors tended to use the supernatant remaining after clotting as the test material. One paper by Mori & Stewart (1978a) reports bactericidal activity in hepatopancreas of lobster; another by White *et al.* (1985) describes activity in the gills of the shore crab, *Carcinus maenas*. However, the extent to which these phenomena are due to haemolymph within the tissues is unknown. The first unambiguous report that activity lies in the haemocytes, rather than the plasma, and more particularly in the granular haemocytes, was by Chisholm & Smith (1992), working with *C. maenas*. In this animal, antibacterial activity is effective against a wide range of Gram positive and Gram negative marine and non-marine bacteria, is freeze and heat stable (to 60°C), not dependent upon divalent cations and active at high titre (down to haemocyte lysate concentrations of ca 2 $\mu g\ ml^{-1}$) (Chisholm & Smith, 1992). Equivalent activity occurs in a number of temperate water crustaceans including, lobster, prawns, squat lobsters, and isopods (Chisholm & Smith, 1995). Interestingly, there are marked seasonal differences in antibacterial vigour by *C. maenas* haemocytes; activity is lower at midwinter and mid-summer, probably because at these times the animals are at the extremes of their physiologically tolerated range (Chisholm & Smith, 1994). Other studies have shown that antibacterial vigour in the shrimp, *Crangon crangon*, is unaffected by chronic exposure to pollutants, even though other blood parameters (eg. clotting, haemocyte number and phenoloxidase activity) are impaired (Smith *et al.*, 1995).

The mode of action of the antibacterial factors in *C. maenas* haemocytes was not elucidated in these early studies. It does not seem to be associated with phenoloxidase activation (Chisholm & Smith, 1992) but at least part of the activity against Gram negative bacteria may be mediated by LPS binding since pre-incubation of haemocyte lysate supernatants for 30 min with LPS at a final concentration of 1 mg ml results in loss of antibacterial activity against the marine bacterium *Psychrobacter immobilis* (Table 2). Similarly, 30 min pre-incubation with 10^9 Gram negative bacteria (Table 2) results in a partial loss of activity by the haemocytes (Chisholm, 1993).

From these baseline studies, we have a good picture of the potency, source, spectrum of activity and robustness of the antimicrobial factors in crustacean blood. The big questions which remain concern the biochemical nature of the active proteins involved, their mode of action and control of expression and bioprocessing.

ANTIBACTERIAL PROTEINS KNOWN FROM OTHER ANIMALS

Textbooks of mainstream immunology describe a battery of proteins that can be broadly regarded as antimicrobial, at least for mammals. They include such obvious molecules as immunoglobulins, complement factors and acute phase proteins which, directly or indirectly, neutralise or inactivate micro-organisms and prevent their replication in the host body. A more restricted group of antimicrobial molecules are those that actively kill, through one mechanism or another, invasive bacteria, fungi, rickettsias or protozoans. These constitute a diverse array of agents, many located in the blood or lymph, others on mucosal surfaces or within individual organs or tissues. Some are proteins that generate microbicidal substances, such as oxyradicals, hydroxy ions or nitric oxide, without necessarily being microbicidal in their inactive state. Hence the interpretation of what might be considered an antimicrobial protein is pretty much open.

This overview will restrict the definition to protein or glycoprotein molecules that directly inhibit the growth of microbes, especially bacteria and fungi. Within this narrow definition, three groups of proteins may be identified. These are: lytic enzymes, metal (or other nutrient) with-holding proteins, and a miscellany of cidal or growth inhibiting proteins, several of which have low molecular masses. Excluded are proteins or glycoproteins that act as agglutinins or precipitins. These are dealt with in detail elsewhere in this volume. Here, examples will be given of key microbicidal proteins that are likely to be found in crustaceans. Since invertebrates lack defined immunoglobulins and are incapable of true adaptive immunity, the repertoire of proteins associated with adaptive immunity, as mediated through lymphocytes, will not be considered further.

Table 1. Antimicrobial factors in crustacean haemolymph

Species	Comments	References
Panulirus argus *P. interruptus*	Thought to be inducible against Gram -ve bacteria	Evans *et al.*, 1968, 1969a,b; Weinheimer *et al.*, 1969
Homarus americanus	Tested against *Pseudomonas perolens* & *Aerococcus viridans* (var) *homari* Thought to be inducible	Acton *et al.*, 1969 Stewart & Zwicker, 1972; Mori & Stewart, 1978a,b
Callinectes sapidus	Tested against bacteriophages and poliovirus. 80 kDa polymer	McCumber *et al.*, 1979; McCumber & Clem, 1983
Carcinus maenas	6.5kDa peptide active against Gram +ve & Gram -ve bacteria	Schnapp *et al.*, 1996
	11.5 kDa protein active against Gram +ve bacteria Both present in granular cells	Relf *et al.,* 1999
Penaeus monodon	Thought to be inducible against Gram -ve bacteria	Adams, 1991
Penaeus vannamei	Penaeidins: 1 (5.48 kDa), 2 (5.52 kDa) and 3 (6.62 kDa)	Destoumieux *et al.,* 1997
Astacus astacus	Lysozyme acivity against *M.luteus* cell walls	Fenouil & Roch, 1991
A. leptodactylus,	"	"
Austropotamobius pallipes		
Pacifastacus leniusculus	Tested against *Aphanomyces astaci* Activity may be due to melanin or melanin precursors	Nyhlén & Unestam, 1980
	Lysozyme activity against *M. luteus* cell walls	Fenouil & Roch, 1991
Orconectes limosus	Lysozyme activity against *M. luteus* cell walls	Fenouil & Roch, 1991
Procambarus clarkii	"	"

Table 2. Antibacterial activity against *Psychrobacter immobilis* after pre-treatment of HLS[1] with LPS[2]

Sample	SI[6] (4h)
HLS-untreated	0.002 ± 0.001
HLS-LPS treated[3]	284 ± 43
HLS-bacteria treated[4]	15 ± 3
Control[5]	130 ± 6

[1] HLS: haemocyte lysate supernatant prepared in modified *Carcinus* saline pH 7.4 (Chisholm & Smith, 1992). [2] LPS: Lipopolysaccharide from *E. coli* 0111:B4 (Sigma). [3] HLS-pretreated with 1 mg ml^{-1} final concentration LPS for 30 min before ultracentrifugation to remove complexed material. [4] HLS: pretreated with the Gram - ve marine bacterium *Psychrobacter immobilis* at a concentration of $5x10^9$ before filtration through a 2µm Millipore filter. [5] Control: comprising equal parts modified *Carcinus* saline and marine broth (Difco). [6] SI: survival index calculated as:

$$SI = \frac{\text{Number of colonies at time t}}{\text{Number of colonies at time 0}} \times 100$$

LYSOZYME

Foremost amongst the non-specific proteins that are responsible for lysis of micro-organisms is the enzyme lysozyme. This was discovered by Fleming in the early nineteen twenties. Its abundance in human tears led to much satirical lampooning by the press, as evidenced by a cartoon by J.H. Dowd that appeared in the magazine Punch in 1922. It would be interesting to speculate whether lysozyme would have warranted as much media attention had it been discovered in an invertebrate!

Lysozyme is a very cationic 14 kDa protein with a pI of *ca* 9. Specifically, it enzymatically cleaves β-1,4 linkages of peptidoglycan, a polymer present in the cell walls of bacteria. In addition to being a component of neutrophil, monocyte and macrophage granules, it is extremely abundant in hen egg white and in saliva, tears and nasal secretions of mammals. Moreover, it has been found in a wide variety of vertebrate and invertebrate animals (Jolles, 1996; Prager, 1996; Yano, 1996; Ito *et al.*, 1999), so it has come to be regarded as a phylogenetically ancient molecule likely to be present in crustaceans. Lysozyme is now so well characterised and studied that it no longer generates excitement as a novel type of defence molecule, even though recent work has highlighted the diversity of lysozymes that have evolved in different organisms (Prager, 1996; Ito *et al.*, 1999) and thus begs questions about its primordial form and biological role.

METAL BINDING AND METAL WITH-HOLDING PROTEINS

That metal-binding proteins (cysteine-rich metallothioneins) occur in animal tissues is well established (Maroni, 1990) and, in vertebrates, metallothioneins are known to be induced by a variety of physiological stressors, including systemic infection (Kagi

& Schaffer, 1988). Whether or not metallothioneins have any antibacterial properties is not known, since most research has centred on their detoxifying or homeostatic functions.

As regards metal with-holding proteins, the best known are transferrin and lactoferrin (Weinberg, 1984, 1993). These are glycoproteins that chelate ferric iron which is required by many micro-organisms for normal growth. The genes for both transferrin and lactoferrin are highly homologous (Teng *et al.*,1987) and in vertebrates they bind reversibly to the metal and transfer it to the liver. Removal of iron from the immediate vicinity of a site of infection renders it unavailable to invading bacteria and therefore effectively halts microbial growth and pathogenicity (Weinberg, 1978, 1993). In mammals transferrin is a predominant serum globulin. Lactoferrin, however, is more frequently, though not exclusively, associated with mammary secretions and also occurs in neutrophil granules, serum, tears, semen and other secretions. Another form of transferrin, ovotransferrin, has been identified in eggs (Weinberg, 1993).

Iron binding is pH dependent, thus allowing ferric iron to be picked up and discharged according to 'local' conditions. The presence of iron with-holding proteins in the serum is up-regulated by IL-1. X-ray crystallography has shown that in mammals, transferrin and lactoferrin are folded into two lobes, which permit the strong but reversible binding of iron (Baker & Lindley, 1992). Transferrin has slightly different binding properties to lactoferrin, related to the chemical structure of its binding sites and possibly some functional differences between the folded lobes in the two forms (Baker & Lindley, 1992). Transferrin or transferrin-like molecules are known to be present in teleosts (Utter *et al.*, 1970; Yano, 1996), elasmobranchs and hagfish (see review by Yano, 1996) and were considered by Aisen & Listowsky (1980) to have arisen only in the Chordata. However, there is one report from Topham *et al.* (1988) which shows that the chelicerate *Limulus polyphemus* has an iron-binding protein, which has been purified and characterized. Since iron with-holding is a simple, non-adaptive defence strategy, suitable for any organism to use to protect itself against infection, it is is pertinent to ask whether such compounds are more widespread in invertebrates.

ACUTE PHASE PROTEINS

Acute phase proteins may be loosely defined as serum proteins whose levels increase during the early stages of infection or inflammation. The best known is C-reactive protein, a molecule which, in the presence of calcium, reacts with phosphorylcholine, a component of the cell walls of bacteria and fungi. It is present in the serum of most vertebrates, including fish, where it may also occur in mucus (Yano, 1996). It has also been reported for some invertebrates, particularly the marine chelicerate, *Limulus polyphemus* (Robey & Liu, 1981).

Another acute phase protein is LPS-binding protein (LBP). This forms complexes with liplopolysaccharide from Gram negative bacteria, transferring it to the CD14 receptor on the surface of leucocytes and other myeloid cells (see review by Ulevitch & Tobias, 1999). This results in the activation of the cells, which, amongst other activities, promotes phagocytosis and inflammation, possibly involving Toll or Toll-like receptors (Ulevitch & Tobias, 1999). Toll receptors and the associated signalling pathways of nuclear factor κB are thought to represent a very ancient form of universal host defence molecules (Kopp & Medzitov, 1999). Thus LBP seems to act to 'sense' the presence of Gram negative bacteria by their soluble product, LPS. Interestingly, LBP belongs to a group of lipid-binding proteins that are known to function in phospholipid transfer (Ulevitch & Tobias, 1999). At present, Toll-like receptors are assumed to occur in crustaceans but their role in antimicrobial defence of these animals has not been addressed.

Also of interest is the 55-60 kDa bacterial permeability increasing protein (BPI) which occurs in both plants and animals (Elsbach & Weiss, 1998). In vertebrates, this has ~45% sequence identity to LPB. Although the two molecules are regarded as homologous LPS-binding molecules and share certain similarities in structure, they interact with LPS in very different ways (see review by Ulevitch & Tobias, 1999).

ANTIMICROBIAL PEPTIDES

Of more recent interest is the ocurrence and biological role of antimicrobial peptides or other low molecular weight microbicidal proteins in animals. These agents have been defined by Boman (1995) as proteins of less than 10 kDa or 100 amino acids that kill micro-organisms in a stoichiometric, rather than a catalytic, manner. Their small size makes them easily synthesized without complex tissues or cells and they readily diffuse to sites of injury or infection (Boman, 1995). Moreover, they tend to have low toxicity towards eucaryotic cells and rarely cause problems of sepsis. It follows they must play a crucial role in innate protection against microbial infection. They were first described in insects, but a very large number (now over 100) have been found in a range of animal groups, including, for example, chelicerates, molluscs, ascidians, teleosts, amphibians and mammals (see reviews by Ganz & Lehrer, 1994; Boman, 1995; Lehrer & Ganz, 1999). Their mechanism of action is still largely unknown, although several seem to act by forming pores and making susceptible bacteria leaky (Elsbach & Weiss, 1998). A few of these proteins have attracted attention as potentially valuable chemotherapeutic agents for clinical use (Hancock & Lehrer, 1998) and currently there is much effort directed towards discovering new forms from different sources. It is thus within this category of antimicrobial proteins that recent research effort with crustaceans has been directed.

CRUSTACEAN ANTIMICROBIAL PROTEINS

Crustacean lytic enzymes

Lysis is arguably the easiest type of microbicidal activity to detect *in vitro*. Simply, it requires measurement of the decrease in absorbance of a bacterial or other microbial suspension caused by the test material (Salton, 1957; Smith *et al.*, 1995b). In the case of lysozyme, muramidase activity can be seen by evaluation of this change in absorbance using a suspension of *Micrococcus luteus* cell walls, since this microorganism is particularly rich in muramic acid residues and is highly sensitive to lysozyme (Salton, 1957). The only important criteria are the pH and ionic conditions, which for osmoconforming marine invertebrates might be expected to be close to those of the body fluids. It is therefore surprising that there is a paucity of information on the presence of lysozyme or lysozyme-like activity in crustaceans.

In contrast to the insects, molluscs and ascidians, for which there is a large literature on the presence, character and biological roles of lysozyme (Jolles, 1996), there appear to only very few reports for decapods (see Table 1) One, by Brown (1995) describes lysozyme activity in the gut of crayfish, which is believed to assist in digestion. Another by Fenouil & Roch (1991), reports muramidase activity in the haemolymph of four species of Astacidae (*Astacus astacus, A. leptodactylus, Austropotamobius pallipes* and *Pacifastacus leniusculus*) and two species of Cambaridae (*Orconectes limosus* and *Procambarus clarkii*). There is marked variability in the strength of activity between species, between individuals of the same species and blood fraction (serum, haemocytes or plasma) tested (Fenouil & Roch, 1991). Activity occurs only under weakly acidic conditions, (*ca* pH 6) and is only partially lost upon boiling (Fenouil & Roch, 1991). The speed of the response and its physico-chemical characterstics certainly point towards lysozyme as the active agent. However, as neither inhibition experiments with known inactivators of lysozyme, *e.g.* heparin, histamine or ultraviolet light (Prager, 1996), nor biochemical characterisations were performed, it is impossible to conclude unequivocally that lysozyme was responsible for the responses.

Certainly, with marine decapods, the available data are contradictory. Noga *et al.* (1996) state that there is no evidence of lysozyme-like activity in the haemolymph of the blue crab, *Callinectes sapidus*. By contrast, Chisholm (1993) has noted that a small amount of muramidase-like activity occurs in *C. maenas* haemocytes although the activity is very slow and is seen only with lysate supernatants prepared from separated hyaline cells (Fig 1). The hyaline haemocytes in *C. maenas* are phagocytic, so the presence of muramidase within the cytoplasm of these cells is consistent with their function. Again, until biochemical characterisation is performed, it is not possible to state categorically that lysozyme accounts for the effect of hyaline cell lysate supernatants on *M. luteus* cell walls.

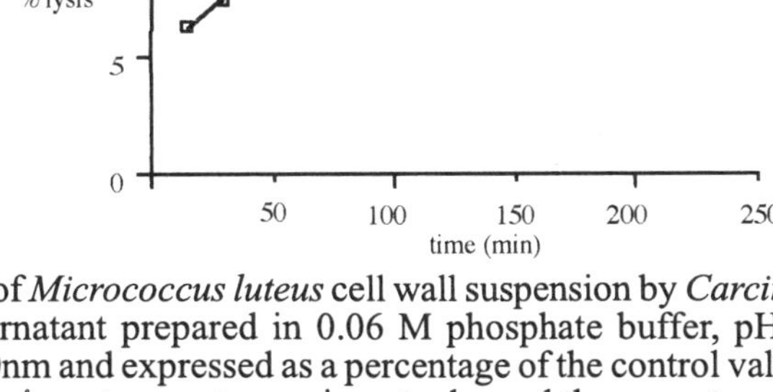

Figure 1. Lysis of *Micrococcus luteus* cell wall suspension by *Carcinus maenas* hylaine cell lysate supernatant prepared in 0.06 M phosphate buffer, pH 6.4. Absorbances were read at 570nm and expressed as a percentage of the control value. Data are related to a typical experiment; repeat experiments showed the same trend.

More recently a 14 kDa antibacterial protein has been found in mixed haemocyte populations of *C. maenas* by ion exchange chromatography (Schnapp *et al.*, 1996). The protein was not purified to homogeneity, but fractions containing it inhibit the growth of live *M. luteus* (Schnapp *et al.*, 1996). We have further purified this protein from separated granular haemocytes and confirmed that although it inhibits the growth of *M. luteus* cells it does not lyse *M. luteus* cell walls. Thus in spite of the similarity of molecular size, it seems unlikely that lysozyme is present in the granular cells. Nevertheless, the possibility remains that it is present in the hyaline cells.

Clearly the jury is still out on whether lysozymes exist in crustacean haemolymph. It would seem, from these few studies, that if lysozyme is present it is most obvious in freshwater species, perhaps associated with the higher probability of these animals encountering Gram positive bacteria than their marine relatives which inhabit a *milieu* rich in Gram negative organisms. Once its status as a constitutive lysin in crustaceans is confirmed, it will be interesting to see how far it resembles lysozymes in other invertebrates.

METAL-BINDING AND METAL WITH-HOLDING PROTEINS IN CRUSTACEANS

If the situation with respect to the presence of lysozyme in crustaceans is uncertain, it is even more enigmatic with regard to metal (or inorganic nutrient) with-holding proteins. There is a substantial literature relating specifically to metallothioneins in crustaceans (see for example, Engel, 1987; Engel & Brouwer, 1987, 1991; Landry & Topham, 1990; Rainbow, 1997; Pedersen *et al.*, 1994). These studies focus on copper, zinc, cadmium or other heavy metal chelation and their presence has been correlated with moulting (Engel & Brouwer, 1987; Engel, 1987; Landry & Topham, 1990) or tolerance of heavy metal pollution (Rainbow, 1997). No studies appear to have been undertaken to determine whether these small proteins have any direct antibacterial activity in the tissues of crustaceans

It might be expected that the advent of techniques to probe invertebrate blood and tissues for transferrin-like proteins or genes would have resolved the issue about their presence in crustaceans. Whilst iron-binding proteins have been found in invertebrates, (see for example Topham *et al.*, 1988; Lee *et al.*, 1978) there are scant reports of the existence of these proteins in decapods or other crustaceans. One study by Huebers *et al.* (1982) describes an invertebrate transferrin from *Cancer magister* and another by Landry & Topham (1990) reports on an iron-binding protein with a molecular mass of 155 kDa in the haemolymph of the blue crab (*Callinectes sapidus*). The presence of these molecules in crustaceans and other invertebrates supports the argument that transferrin and transferrin-like molecules are not newcomers to the evolutionary scene, but are of ancient development and not necessarily related to oxygen transport molecules. As well as having a bacteriostatic effect, they may serve to protect against the toxic effects free ionic iron (Landry & Topham, 1990).

In addition a recent paper by Huang *et al.* (1996) describes the purification and cDNA cloning of the iron-storage protein, ferritin, from the crayfish *Pacifastacus leniusculus,* which strengthens the theory that crustaceans have iron withholding, iron transport and iron storage mechanisms comparable to the vertebrate system. Indeed, Huang *et al.* (1996) found that the deduced amino acid sequence of the crustacean ferritin molecule bears a closer relationship to vertebrate ferritin heavy chains than to insect ferritin.

It is curious that this area of enquiry has been neglected for so long. It is a subject worthy of study and one which could explain the mystery of how crustaceans survive the vulnerable stages of moulting without succumbing to opportunistic microbial infections.

ACUTE-PHASE TYPE PROTEINS IN CRUSTACEANS

The literature is vague about the presence or not of acute-phase type proteins in crustaceans. That a variety of proteins appear in the plasma after antigenic challenge of the host is well established and much work has been performed by Söderhäll and others on those associated with phenoloxidase activation (reviewed elsewhere in this volume). Some of these factors bind to carbohydrates from the cell walls of bacteria and fungi (Söderhäll & Cerenius, 1998), others may include the release of microbicidal proteins from circulating haemocytes (see below). Because exocytosis is an integral part of cellular reactivity to micro-organisms in crustaceans *in vivo*, it is a matter of opinion as to the extent to which the appearance of haemocyte products in the plasma may be considered analogous to vertebrate-type acute phase proteins. At present there are few, if any, clear reports that molecules homologous to C reactive protein are present in true crustaceans (as opposed to their chelicerate relatives). Thus they will not be discussed further in this brief overview of microbicidal factors.

PEPTIDES AND OTHER LOW MOLECULAR WEIGHT PROTEINS IN CRUSTACEANS

Some effort has been made to identify, characterize and clone low molecular weight antimicrobial proteins in recent years. The first antibacterial protein to be purified from crustacean blood was by Schnapp *et al.* in 1996 working with *C. maenas*. The peptide was isolated from haemocyte lysate supernatants by Sep Pak C_{18} extraction, gel filtration and reverse phase HPLC. It has a molecular mass of ca 6.5 kDa and is active against both Gram positive and Gram negative bacteria (Schnapp *et al.*, 1996). Partial amino acid sequence analysis showed that it has a high proline content: 11 out of the first 30 amino acids are proline. As such it shares high degree of sequence identity (ca 60 %) with the mature form of a proline-rich 7 kDa antibacterial peptide (Bac 7, a bactenecin belonging to the cathelicidin group) which was identified from bovine neutrophils by Frank *et al.* (1990). Moreover, its titre of activity *in vitro* closely mimics that of Bac 7 (Schnapp *et al.*, 1996). As neither the full sequence nor the gene encoding this protein were obtained by Schnapp and co-workers, it is impossible to know whether or not the crab peptide has a precursor containing the cathelicidin domain signature (Zanetti *et al.*, 1995) which would place it firmly in this family of antibacterial proteins and assure its status as the first invertebrate cathelicidin to be recorded.

In the process of purifying the 6.5 peptide from *C. maenas*, Schnapp *et al.* (1996) noted the presence of at least four other bactericidal proteins in the haemocytes of this animal. These have molecular masses of ca 70 kDa, 45 kDa, 14 kDa and 11.5 kDa. Relf *et al.* (1999) have subsequently purified the 11.5 kDa protein to homogeneity by ion exchange chromatography on CM Sepharose, reverse phase HPLC on a C_{18} column and gel filtration. This protein (provisionally named Carcinin), unlike the 6.5 kDa peptide, is active only against marine Gram positive bacteria and appears to require high salinity for activity (Relf *et al.*, 1999). Interestingly it inhibits growth of the lobster pathogen, *Aerococcus viridans*, perhaps accounting for the lower susceptibility of crabs to this bacterium compared with lobsters (Cornick & Stewart, 1968b). Its effect on fungi was not tested. Carcinin is extremely stable and retains activity even after boiling (Relf *et al.*, 1999). Amino acid analysis shows that it is cysteine-rich and has 30% identity with some 4-disulphide core or whey acidic proteins (WAP) (Fig. 4). Proteins with a WAP motif frequently have serine proteinase inhibitory properties and may inhibit cell growth (Larsen *et al.*, 1998).

To date there is only one other detailed description of antibacterial peptides in crustacean haemolymph. This reports the isolation, characterisation and cloning of three proteins of 5.48, 5.52 and 6.62 kDa, from shrimp, *Penaeus vannamei* (Destoumieux *et al.* (1997). Accordingly, they have been termed Penaeidins 1,2 and 3, respectively (Destoumieux *et al.* (1997). Penaeidin 2 has strong cidal activity with a minimal inhibitory concentration (MIC) of 0.6-2.5 μM, whereas Penaeidin 3 is less potent with a MIC of > 5 μM and is bacteriostatic (Destoumieux *et al.*, 1997). Amino acid sequence analysis and deduction shows them to have a proline-rich N-terminus and a cysteine-rich C-terminus and are active against fungi and Gram positive bacteria (Destoumieux *et al.*, 1997). A comparison of the known characteristics of these antibacterial peptides can be found in Table 3.

Table 3. Low molecular mass microbicidal proteins in crustaceans

	Penaeidins	Crab 6.5 kDa	Crab 11.5 kDa (Carcinin)
Species	*Penaeus vannamei*	*Carcinus maenas*	*Carcinus maenas*
Molecular mass	5.5-6.5 kDa	6.5 kDa	11.5 kDa
Source	haemocytes	haemocytes	granular haemocytes
Amino acid composition	proline-rich (N) cysteine-rich	cysteine-rich (C)	proline-rich
Range of activity	Gram+ bacteria and fungi	Gram+ and G- bacteria	Gram+ marine bacteria
Induction	constitutive	constitutive	constitutive
Forms	3	1	1
Cloned	yes	no	partially
Reference	Destoumieux *et al.*, 1997	Schnapp *et al.*, 1996	Relf *et al.*, 1999

Additionally in a preliminary study, Majumdar *et al.* (1997) have found that a low molecular weight (4 kDa) lectin, Scyllin, from the edible crab *Scylla serrata*, exhibits antimicrobial activity against *Bacillus cereus* and *E. coli.* Previous reports of lectins in crustaceans have centred on their interactions with mammalian cells and micro-organisms (see review by Smith & Chisholm, 1992). Their agglutinating properties with respect to micro-organisms must inevitably assist in the sequestration of invasive organisms, but it appears that Scyllin acts via a pathway that inhibits bacterial respiration, although its biochemical character and the mechanism of action have not been fully elucidated (Majumdar *et al.*, 1997)

EXPRESSION OF CRUSTACEAN ANTIBACTERIAL PROTEINS

Molecular studies indicate that penaeidins are processed from mRNA to active products synthesised within the haemocytes, although the type of haemocyte responsible was not determined (Destoumieux *et al.*, 1997). It is likely that the antibacterial proteins in *C. maenas* are also synthesised by the circulating haemocytes and release from the haemocytes probably takes place through a regulated exocytosis, similar to that described for crayfish by Johansson & Söderhäll (1989). Certainly, in *C. maenas*, antibacterial proteins do not appear to reside normally within the plasma (Chisholm & Smith, 1992; Chisholm, 1993), although Schnapp (1996) has shown that antibacterial activity appears in the plasma following injection of *C. maenas* with Gram negative bacteria (Fig. 2a). A similar response occurs after injection with LPS and is evident within 30 min at 15°C (Fig 2b). A weaker effect is produced by saline injection, but unlike the plasma from bacteria or LPS-injected crabs the effect is short lived (Fig 2a & b). It is likely that the antibacterial activity seen in the plasma of bacteria/LPS-injected crabs is a result of release of antibacterial proteins by the circulating haemocytes through exocytosis in response to antigenic stimulation.

The antibacterial proteins in crustaceans differ significantly from those in insects in appearing to be fully constitutive. In insects, antibacterial peptides are inducible and released from the fat body into the plasma after wounding or bacterial injection (Lehrer & Ganz, 1999). Equivalent induction of fat body synthesis does not appear to occur in crustaceans. Indeed, the early reports of enhanced secondary bactericidal activity in lobsters and crayfish may be explained by the release of antimicrobial proteins into the plasma by the bacterial pre-treatment.

The extent to which antibacterial protein synthesis is affected by internal (physiological) or external (environmental) factors is at present unknown and awaits development of appropriate molecular or antibody probes. Observations of *C. maenas* by us, indicate that there are seasonal alterations in the amount of the 6.5 and 11.5 kDa proteins present in the granular haemocytes; levels are lower during the winter and early spring, although how far this maps to the moult cycle, water temperature or nutritional status of the host is not known. Similarly, the extent to which handling or other stresses impinge on microbicidal vigour is uncertain, although it is reasonable to predict that expression of antimicrobial proteins will be down-regulated by handling and other stresses, as stressed decapods tend to show decreased haemocyte counts

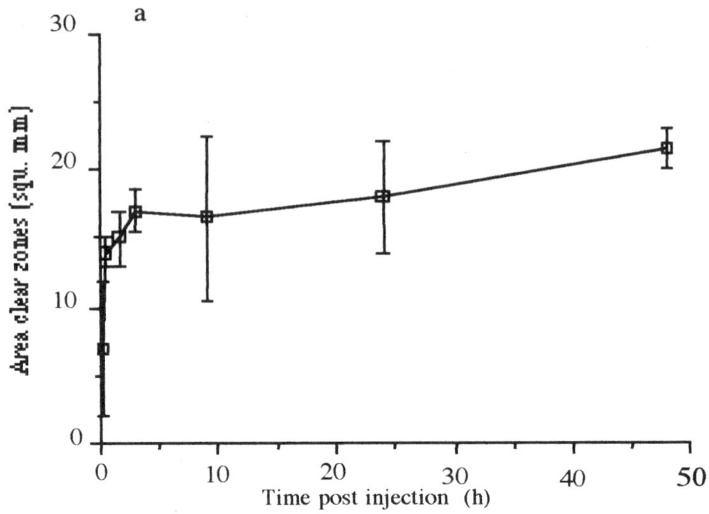

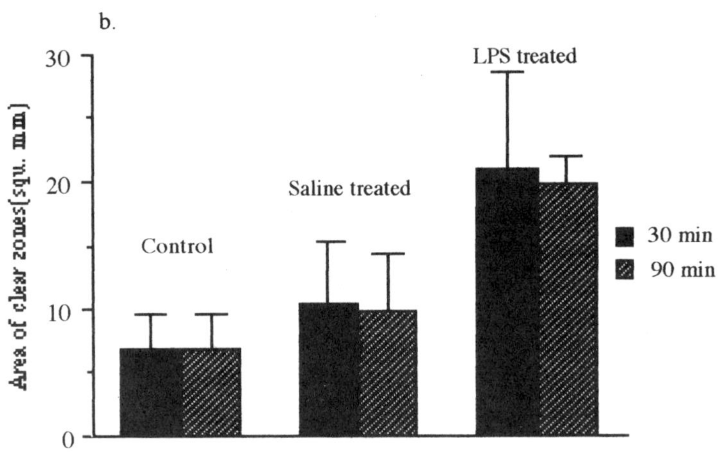

Figures 2a & 2b. Antibacterial activity in the plasma of *C. maenas* haemolymph challenge with a): an injection of 1 x 10^8 bacteria. and b): an injection of LPS (200 µl of a 0.2 mg ml^{-1} solution in *Carcinus* saline (CS). Controls were untreated crabs. Saline-treated crabs were injected with 200 µl modified CS. Antibacterial activity was determined against a) *P. immobilis* and b) *P. citreus* by radial diffusion assays (Lehrer *et al.*, 1991) on thin agarose plates. The method was modified for marine bacteria as described by Schnapp *et al.* (1996). Antibacterial activity was measured as the area of clear zones surrounding wells in the bacterial lawns.

(Hamann, 1975) and to suffer higher incidences of general bacteraemia or pathogenic infections (Stewart *et al.*, 1966, 1972; Welsh & Sizemore, 1985). These are important issues for crustacean aquaculture, where a constant high level of antimicrobial immune capability is required, throughout the entire developmental period, to minimise the risk of disease outbreak in the stock.

CONCLUSIONS AND FUTURE PERSPECTIVES

The most obvious conclusion to this brief overview of antimicrobial proteins in crustaceans is that there is a striking dearth of descriptions of indiviual microbicidal proteins and the small number of crustacean species that have been studied. Hand in hand with this is the lack of biochemical and molecular information about their synthesis, processing, regulation and mode of action. The work lags substantially behind that on antimicrobial proteins in insects and chelicerates. Since crustacean health is an important and timely area of research in relation to the aquaculture industry, this cannot be because crustaceans do not attract scientific interest. Rather it could be because research effort is responsive to the user needs, and understandably the shellfish producers require a cost effective 'quick fix' which is usually translated as the development of 'vaccines' or other preventative substances, which can be marketed for profitable use by shrimp or prawn farmers. The investment of time and money into finding native antibiotic substances and then to find ways of naturally promoting or maximizing their potential to reduce opportunistic infection in stock must seem too great to be risked. Likewise, the opportunity to use genetic modification for disease resistant transgenic animals, based on fundamental studies of constitutive antimicrobial proteins, must seem an unrealistic goal, especially given the current public concern about genetically modified crops and food derivatives. However, these reservations have to be overcome as it is probable that as a consequence of the very nature of immmunity in crustaceans, 'vaccination' type approaches will not work and genetic modification will become the way forward in the third millenium AD.

For such a bold strategy to be successful, we will need to understand a good deal more not only about the diversity of antimicrobial proteins within the Crustacea but also basic details of how the expression of these substances varies through development, during the moult cycle and in relation to changed environmental conditions. This can best be studied by taking advantage of molecular techniques, in which case, appropriate target genes (or gene products) need to be identified and suitable probes constucted. Antimicrobial peptides represent ideal 'markers' of immunity. They are produced by single genes, their small size simplifies their preparation, testing and use, and, undoubtedly, they play an key role in first line (surveillance) defence. Hopefully, through such endeavour, there might be serendipitous spin-offs which could benefit the the wider spheres of biomedicine, phylogeny, protein chemistry and cell signalling.

ACKNOWLEDGEMENTS

The work of Juliet Relf, Denni Schnapp and Robert Hughes, past and present members of the Comparative Immunology Group, St Andrews, performed some of the unpublished work described in this review. We would also like to thank Susan Ross, Comparative Immunology Group, for her valued technical input. Likewise, we acknowledge the help and advice on protein chemistry given by Dr Graham Kemp, Biomolecular Sciences, University of St Andrews. The original research work of the Comparative Immunology Group, St Andrews was supported by grants and studentships awards to VJS from the Natural Environment Research Council (NERC), UK. JRSC was supported by NERC under the DEMA initiative.

REFERENCES

Acton, R.T., Weinheimer, P.F. & Evans, E.E. 1969. A bactericidal system in the lobster *Homarus americanus. Journal of Invertebrate Pathology*, 13, 463-464.

Adams, A. 1991. Response of penaeid shrimp to exposure to *Vibrio* species. *Fish & Shellfish Immunology,* 1, 59-70.

Aisen, P. & Listowsky, I. 1980. Iron transport and storage proteins. *Annual Reviews in Biochemistry*, 49, 357-393.

Baker, E.N. & Lindley, P.F. 1992. New perspectives on the structure and function of transferrins. *Journal of Inorganic Biochemistry*, 47, 147

Boman, H. G. 1995. Peptide antibiotics and their role in innate immunity. *Annual Reviews in Immunology,* 13, 61-92.

Brown, P.B. 1995. Physiological adaptations in the gastro-intestinal-tract of crayfish. *American Zoologist,* 35, 20-27.

Chisholm, J.R.S. 1993. Antibacterial activity in the blood cells of Carcinus maenas (L), and other crustaceans. PhD thesis, University of St Andrews, Fife, Scotland pp 126-129.

Chisholm, J.R.S. & Smith, V.J. 1992. Antibacterial activity in the haemocytes of the shore crab, *Carcinus maenas* (L.). *Journal of the Marine Biological Association of the U.K.*, 72, 529-542.

Chisholm, J.R.S. & Smith, V.J. 1994. Variation of antibacterial activity in the haemocytes of the shore crab, *Carcinus maenas* with temperature. *Journal of the Marine Biological Association of the U.K.,* 74, 979-982.

Chisholm, J.R.S. & Smith, V.J. 1995. A comparison of antibacterial activity in the hemocytes of different marine crustaceans. *Comparative Biochemistry and Physiology,* 110A, 39-45.

Cooper, E.L. 1985. Overview of humoral factors in invertebrates. *Developmental and Comparative Immunology*, 9, 577-583.

Cornick, J.W. & Stewart, J.E. 1968a. Interaction of the pathogen *Gaffkya homari* with the natural defence mechanisms of *Homarus americanus. Journal of the Fisheries Research Board of Canada* 25, 695-709.

Cornick, J.W. & Stewart, J.E. 1968b. Red crab, *Geryon qinquedens,* and snow crab *Chinoectes opilio*; resistance to infection by the lobster pathogen, *Aerococcus viridans. Journal of the Fisheries Research Board of Canada* 32, 702-706

Destoumieux, D., Bulet, P., Loew, D., Van Dorsselaer, A., Rodriquez, J., & Bachere, E. 1997. Penaeidins, a new family of antimicrobial peptides isolated from the shrimp, *Penaeus vannamei* (Decapoda). *Journal of Biological Chemistry,* 272, 28398-28406.

Elsbach, P. & Weiss, J. 1998. Role of bactericidal/permability-increasing protein in host defence. *Current Opinion in Immunology*, 10, 45-49.

Engel, D.W. 1987. Metal regulation and molting in the blue crab, *Callinectes sapidus*: copper, zinc and metallothionein. *Biological Bulletin*, 172, 69-82.

Engel, D.W. & Brouwer, M. 1987. Metal regulation and molting in the blue crab, *Callinectes sapidus;* metallothionein function in metal metabolism. *Biological Bulletin*, 173, 237-249.

Engel, D.W. & Brouwer, M. 1991. Short-term metallothionein and copper changes in blue crabs at ecdysis. *Biological Bulletin*, 180, 447-452.

Evans, E. E., Painter, B., Evans, M. L., Weinheimer, P. F. & Acton, R. T. 1968. An induced bactericidin in the spiny lobster, *Panulirus argus. Proceedings of the Society for Experimental Biology and Medicine*, 128, 394-398.

Evans, E. E., Cushing, J. E., Sawyer, S., Weinheimer, P. F., Acton, R. T. & McNeely, J. 1969a. Induced bactericidal response in the California spiny lobster *Panulirus interruptus. Proceedings of the Society for Experimental Biology and Medicine*, 132, 111-114.

Evans, E.E., Weinheimer, P.F. Painter, B., Acton, R.T. & Evans, M. 1969b. Secondary and tertiary responses of the induced bactericidin from the West Indian spiny lobster, *Panulirus argus. Journal of Bacteriology,* 98, 934-946.

Fenouil, E. & Roch, P. 1991. Evidence and characterisation of lysozyme in six species of freshwater crayfishes from Astacidae and Cambaridae families. *Comparative Biochemistry and Physiology,* 99B, 43-49.

Frank, R.W., Gennaro, R. Schneider,K., Przbylski, M. & Romeo, D. 1990. Amino acid sequences of 2 proline rich bactenecins - antimicrobial peptides of bovine neutrophils. *Journal of Biological Chemistry,* 265, 18871-18874.

Ganz, T. & Lehrer, R.I. 1994. Defensins. *Current Opinion in Immunology,* 6, 584-589

Hamann, A. 1975. Stress induced changes in cell titre of crayfish hemolymph. *Z. Naturforsch*, 30, 850.

Hancock, R.E. & Lehrer, R.I. 1998. Cationic peptides; a new source of antibiotics. *Trends in Biotechnology*, 16, 82-88.

Huang, T.S., Law, J.H. & Söderhäll, K. 1996. Purification and cDNA cloning of ferritin from the hepatopancreas of the freshwater crayfish *Pacifastacus leniusculus*. *European Journal of Biochemistry*, 236(2), 450-456.

Ito, Y., Yoshikaw, A., Hotani, T., Fukuda, S., Sugimura, K. & Imoto, T. 1999. Amino acid sequences of lysozymes newly purified from invertebrates imply wide distribution of a novel class in the lysozyme family. *European Journal of Biochemistry*, 259, 456-461.

Johansson, M.W. & Söderhäll, K. 1989. A peptide containing the cell adhesion sequence RGD can mediate degranulation and cell adhesion of crayfish granular haemocytes *in vitro*. *Insect Biochemistry*, 19, 573-579

Jolles, P. (ed.). 1996. *Lysozyme: Model Enzymes in Biochemistry and Biology*. Birkhäuser-Verlag, Basel, Switzerland.

Kagi, J.H.R. & Schaffer, W. 1988. Biochemistry of metallothioneins. *Biochemistry*, 27, 8509-8515.

Kopp, E.B. & Medzhitov, R. 1999. The Toll receptor family and control of innate immunity. *Current Opinion in Immunology*, 11, 13- 18.

Landry, C.C. & Topham, R.W. 1990. Purification and characterisation of an iron-binding protein from blue crab (*Callinectes sapidus*). *Comparative Biochemistry and Physiology*, 97B, 831-836.

Larsen, M., Ressler, S.J., Lu, B., Gerdes, M.J., McBride, L., Dang, T.D. & Rowley, D.R. 1998. Molecular cloning and expression of ps20 growth inhibitor- A novel WAP-type four-disulphide core domain protein expressed in smooth muscle. *Journal of Biological Chemistry*, 273, 4574-4584.

Lehrer, R.I. & Ganz, T. 1999. Antimicrobial peptides in mammalian and insect host defence. *Current Opinion in Immunology*, 11, 23-27.

Lehrer, R.I., Rosenman, M., Harwig, S.S.L., Jackson, R. & Eisenhauer, P. 1991. Ultrasensitive assays for endogenous antimicrobial polypeptides. *Journal of Immunological Methods*, 137: 167-173.

McCumber, L.J. & Clem, L.W. 1983. Recognition of non-self in crustaceans. *American Zoologist*, 23, 173-183.

McCumber, L.J., Hoffmann, E.M. & Clem, L.W. 1979. Recognition of viruses and xenogeneic proteins by the blue crab, *Callinectes sapidus*: a humoral receptor for T_2 bacteriophage. *Journal of Invertebrate Pathology*, 33, 1-9.

Majumdar, M., Chattopadhyay, T. & Guha, A.K. & Chatterjee, B.P. 1997. Inhibition of bacterial respiration by a low molecular weight lectin, Scyllin, from *Scylla serrata* crab haemolymph. *Indian Journal of Biochemistry & Biophysics*, 34, 87-89.

Maroni, G. 1990. Animal metallothioneins. In : *Heavy Metal Tolerance in Plants: Evolutionary Aspects*. ed Shaw, A.J. RC Press, Boca, Raton, Florida, pp 215-232.

Mori, K. & Stewart, J.E. 1978a. Natural and induced bactericidal activities of the hepatopancreas of the American lobster, *Homarus americanus*. *Journal of Invertebrate Pathology*, 32, 171-176.

Mori, K. & Stewart, J.E. 1978b. The hemolymph bactericidin of American lobster (*Homarus americanus*): adsorption and activation. *Journal of The Fisheries Research Board of Canada*, 35, 1504-1507.

Noga, E.J., Arroll, T.A. & Fan, Z.Q. 1996. Specificity and some physiochemical characteristics of the antibacterial activity from the blue crab, *Callinectes sapidus*. *Fish & Shellfish Immunology*, 6, 403-412.

Nyhlén, L & Unestam, T. 1980. Wound reactions and *Aphanomyces astaci* growth in crayfish cuticle. *Journal of Invertebrate Pathology*, 36, 187-197

Pedersen, K.L., Pedersen, S.N., Hojrup, P., Andersen, J.S., Röpstorff, P., Knudsen, J. & Deplege, M. 1994. Purification and characterisation of a cadmium-induced metallothionein from the shore crab, *Carcinus maenas* (L). *Biochemical Journal*, 297-609-614.

Prager, E.M. 1996. Adaptive evolution of lysozyme: changes in amino acid sequence, regulation of expression and gene duplication. In: *Lysozyme: Model Enzymes in Biochemistry and Biology* (Jolles, P., ed.) pp 323-345. Birkhäuser-Verlag, Basel, Switzerland.

Rainbow, P.S. 1997. Trace metal accumulation in marine invertebrates: marine biology or marine chemistry. *Journal of the Marine Biological Association of the UK.*,77, 195-210.

Ratcliffe, N.A., Rowley, A.F., Fitzgerald, S.W. & Rhodes, C.P. 1985. Invertebrate immunity: basic concepts and recent advances. *International Review of Cytology*, 97, 183-350.

Relf, J.M., Chisholm, J.R.S., Kemp, G.D. & Smith, V.J. 1999. Purification and characterisation of a cysteine-rich 11.5 kDa antimicrobial protein from the granular haemocytes of the shore crab, *Carcinus maenas*. *European Journal of Biochemistry. In press*.

Robey, F.A. & Liu, T-Y. 1981. Limulin, a C- reactive protein from *Limulus polyphemus*. *Journal of Biological Chemistry*, 256, 969-975.

Rosenberry, B. (ed.) 1997. *World Shrimp Farming News Intl.*, San Diego, California.

Salton, M. R.J. 1957. The properties of lysozyme and its action on micro-organisms. *Bacteriological Reviews*, 21, 82-99.

Schnapp, D. 1996. Antiviral defnce and antibacterial proteins in the shore crab, *Carcinus maenas*. PhD thesis, University of St Andrews, Fife, Scotland.

Schnapp, D., Kemp, G.D. & Smith, V.J. 1996. Purification and characterization of a proline-rich antibacterial peptide, with sequence similarity to bactenecin 7, from the haemocytes of the shore crab, *Carcinus maenas*. *European Journal of Biochemistry*, 240, 532-539.

Smith, V.J. & Chisholm, J.R.S. 1992. Non-cellular immunity in crustaceans. *Fish & Shellfish Immunology*, 2, 1-31.

Smith, V.J., Swindlehurst, R.J., Johnston, P.A. & Vethaak, A. D. 1995a. Disturbance of host defence capability in the common shrimp, *Crangon crangon* by exposure to harbour dredge spoils. *Aquatic Toxicology,* 32, 43-58.

Smith, V.J., Findlay, C., Schnapp, D. & Hutton, D.M.C. 1995b. Detection of antibacterial activity in the tissues or body fluids of marine invertebrates. In: *Techniques in Fish Immunology 4. Immunology and Pathology of Aquatic Invertebrates* (J. S. Stolen, T.C. Fletcher, S.A. Smith, J.T. Zelikoff, S.L. Kaattari, R.S. Anderson, K. Söderhäll, & B.A. Weeks-Perkins, eds). SOS Publications, New Jersey, USA, 173-184.

Söderhäll, K. & Cerenius, L. 1998. Role of the prophenoloxidase -activiating system in invertebrate immunity. *Current Opinion in Immunology*, 10(1), 23-28.

Stewart, J.E. & Zwicker, B.M. 1972. Natural and induced bactericidal activities in the hemolymph of the lobster, *Homarus americanus:* products of haemocyte-plasma interaction. *Canadian Journal of Microbiology* 18, 1499-1509.

Stewart, J.E., Cornick, J.W., Spears, D.I. & McLeese, D.W. 1966. Incidence of *Gaffkya homari* in natural lobster (*Homarus americanus*) populations of the Atlantic region of Canada. *Journal of the Fisheries Research Board of Canada,* 23, 1325-1330.

Stewart, J.E., Zwicker, B.M., Arie, B. & Horner, G.W. 1972. Food and starvation as factors affecting the time to death of the lobster *Homarus americanus* infected with *Gaffkya homari*. *Journal of the Fisheries Research Board of Canada,* 29, 461-464.

Teng, C.T., Pentecost, B.T., Marshall, A., Solomon, A., Bowman, B.H., Lalley, P.A. & Naylor, S.L. 1987. Assignment of the lactotransferrin gene to human chromosome 3 and to mouse chromosome 9. *Somatic Cell Molecular Genetics*, 13, 689.

Topham, R., Cooper, B., Tesh, S., Godette, G, Bonaventura, C. & Bonaventura, J. 1988. Isolation, purification and characterisation of an iron-binding protein from the horseshoe crab (*Limulus polyphemus*). *Biochemical Journal*, 252, 151-157.

Ulevitch, R. & Tobias, P.S. 1999. Recognition of Gram negative bacteria and endotoxin by the innate immune system. *Current Opinion in Immunology,* 11, 19-22.

Utter, F.M., Ames, W.E., Hodgkins, H.O. 1970. Transferrin polymorphisms in Coho salmon. *Journal of the Fisheries Research Board of Canada*, 27, 2371-2373.

Weinberg, E.D. 1978. Iron and infection. *Microbioloogical Reviews*, 42, 45-66.

Weinberg, E.D. 1984. Iron-with-holding: a defence against intracellular infection and neoplasia. *Physiological Reviews*, 64, 65-102.

Weinberg, E.D. 1993. The iron with-holding defence system. *American Society for Microbiological News*, 59, 559-562.

Weinheimer, P.F., Acton, R.T., Sawyer ,S. & Evans, E.E. 1969. Specificity of the induced bactericidin of the West Indian spiny lobster, *Panulirus argus*. *Journal of Bacteriology*, 98, 947-948.

Welsh, P.C. & Sizemore, R.K. 1985. Incidence of bacteremia in stressed and unstressed populations of the Blue crab, *Callinectes sapidus*. *Applied and Environmental Microbiology*, 50, 420-425.

White, K.N., Ratcliffe, N.A. & Rossa, M. 1985. The antibacterial activity of haemocyte clumps in the gills of the shore crab, *Carcinus maenas*. *Journal of the Marine Biological Association of the UK*, 65, 857-870.

Yano, T. 1996. Non-specific immune system: humoral defense. In: *The Fish Immune System*. G. Iwama, & T. Nakanishi (eds). Academic Press, 106-159.

Zanetti, M., Gennaro, R. & Romeo, D. 1995. Cathelicidins; a novel protein family with a common pro-region and a variable C terminal antimicrobial domain. *FEBS Letters*, 374, 1-5.

FORM, FUNCTION AND PHYLOGENETIC RELATIONSHIPS OF MOSQUITO IMMUNE PEPTIDES

Carl A. Lowenberger

Animal Health and Biomedical Sciences
University of Wisconsin-Madison
1655 Linden Drive, Madison WI 53706

INTRODUCTION

Insects represent one of the earth's most successful groups of organisms, colonizing essentially every niche possible, with the exception of the oceans. One factor contributing to this success in exploiting such diverse ecological niches is their ability to defend themselves against harmful pathogens and parasites (1,2). Insects can mount an extremely effective response against invasion by prokaryotic and eukaryotic pathogens involving both cellular and humoral factors (3,4). These responses may include phagocytosis of bacteria, formation of nodules containing large aggregates of bacteria, the melanotic encapsulation of metazoan parasites (5,6) or the use of potent antimicrobial peptides (7,8).

Phylogenetic Perspectives on the Vertebrate Immune System.
Edited by Beck *et al.*, Kluwer Academic/Plenum Publishers, 2001.

These inducible responses, including those that result in the production of immune peptides, have been incorporated into what is termed innate immunity in both animals and plants (9-11). Immune peptides are quickly expressed *de novo* and delivered to the appropriate site to defend the organism in a manner that is neither learned or acquired, unlike the hallmark characteristics of classical immunology (9). The innate immune system of insects is a complex interactive system used extensively by all genera to protect themselves from infection from bacteria, fungi, and parasites, and although the insects' innate system lacks the sophistication of the mammalian system, there are similarities between different taxa that would suggest common origins of some immune molecules.

The role and goal of the innate immune system is a rapid and effective elimination of invading pathogens. We can find examples of innate immunity from primitive and advanced organisms; genes encoding immune molecules have been isolated from plants, mammals, birds, amphibians, fish and many classes of invertebrates, and we would anticipate that such genes are present in every living organism as a mechanism of defense against pathogens. Because very similar molecules have been isolated from such diverse taxonomic groups, it has been proposed that the efficacy of these ancient molecules has perpetuated their presence through evolutionary diversification. An assessment of the innate immune response of invertebrates may lead to understanding of the origins and evolution of immune systems in higher organisms including the identification of precursor molecules and progenitor responses of vertebrates.

It is now becoming apparent, as more and more related immune peptides are being isolated from very diverse taxa, that antimicrobial peptides are key, and highly conserved elements of the innate immune system in both animals and plants (12-14). Since the first cecropin molecule was identified and characterized in the laboratory of Hans Boman (15), over 170 immune peptides have been isolated from insects alone. This potent arsenal of antimicrobial peptides results in hemolymph concentrations from 1-100 mM within 24h of pathogen stimulation (7). These peptides have been classified into specific families by their physical structure and several recent reviews have described in detail these families of compounds, their range of activity and the taxa from which they have been identified (7,8,10,16-20).

It is interesting to note that closely related insects often have the same peptides in their antimicrobial repertoire, but do not rely on the same peptides as their major response to microbial invasion. Although the origins and evolution of these peptides in invertebrates may have arisen from common ancestors before the divergence of the insects, specific peptides are preferentially expressed by different species.

This review will address the antimicrobial peptides used by insects in general, and specifically by mosquitoes, and will review the growing literature on the role of these immune peptides in determining the competence of mosquitoes to transmit parasites. Mosquitoes are unquestionably the most important arthropod vectors of diseases to humans and are responsible for the transmission of malaria (500 million cases, 3 million deaths, (21)), lymphatic filariasis (100 million cases, (21)), and numerous arboviruses including yellow fever and Dengue fever (21).

Over the last few years, several immune peptides have been isolated and characterized from mosquitoes. These include defensins (1,22-25), cecropins (26-28), and a specific peptide active solely against Gram negative bacteria (Lowenberger and Bulet unpublished data). Other molecules found in mosquitoes that may have a role in antibacterial activity include insect transferrins (29). It generally is accepted that the site of production of antimicrobial peptides is the insect fat body, the functional equivalent of the mammalian liver. However, in some species immune peptides have been found in hemocytes (30,31) salivary glands (32), genital tracts, and the cuticle (33). As such many of the insect immune peptides may be produced as larger precursors that are transported to the hemolymph and subsequently to specific locations and activated by a cleavage of the precursor molecule.

INSECT DEFENSINS

Insect defensins were described first from the flesh fly *Sarcophaga peregrina* (19,20) and independently from *Phormia terranovae* (34). Since these initial reports, over 30 defensins have been reported from Diptera, Coleoptera, Hemiptera, Hymenoptera, Trichoptera, Odonata, and related molecules reported from scorpions and molluscs. Functional analogues also have been reported from mammals (35,36) and plants (14,37). Insect defensins contain 34-43 residues (with the exception of the defensins isolated from royal jelly (38) and the honey bee (39). A characteristic of the insect defensins is the presence of 6 cysteine residues that combine to form three intramolecular bridges to provide the characteristic 3-dimensional structure. The molecule consists of an N-terminal loop (residues 4-14, an α-helix structure, (residues 15-23), and a 2-stranded twisted antiparallel β-sheet (residues 27-31 and 35-39). The structure is stabilized by 2 of the disulphide bridges between the α-helix and the β-sheet (Cys2-Cys5, Cys3-Cys6) forming the "cysteine-stabilized α-β motif"described by Cornet et al. (40). This motif is present in other families of peptides, such as scorpion toxins (41) and plant g-thionins (40), and similarities have been described for the antifungal peptide, drosomycin (7). In insect defensins the disulphide bridges that maintain stability of the molecule are Cys1-Cys4, Cys2-Cys5, and Cys3-Cys6.

Defensins have been isolated from several orders of insects: from the higher insects such as the Diptera (*Drosophila* and mosquitoes), and Coleoptera, and also from ancient insects such as dragon flies (*Aeshna cyanea*), which evolved about 100 million years before the emergence of mosquitoes. Functional analogues also have been isolated from amoebae (42), nematodes (43), scorpions (44,45), molluscs (46,47) mammals (35,36) and plants (14) (Figures 1 and 2). The presence of defensin like molecules in such a diverse group of taxa suggests an ancient origin whose importance in protecting self against microbes has maintained the prominence of this peptide family through evolutionary divergence.

Insect defensins are active against many Gram positive, and a few Gram negative bacteria. Activity against Gram positive bacteria is carried out almost immediately with a lytic effect on the bacterial membranes. These membranes are permeabilized, resulting in a loss of cytoplasmic potassium, a depolarization of the inner membrane, reduced amounts of cytoplasmic ATP, and a reduction in respiration (48).

Family	Species	Sequence
A		
Insect	*P. terranovae*	--ATCDLLS----GTGINHSACAAHCL-LRGNRGGYCNG--KGVCVC-RN
Insect	*Ae. aegypti* A	--ATCDLLS----GFGVGDSACAAHCI-ARGNRGGYCNS--KKVCVC-RN
Insect	*Ae. aegypti* B	--ATCDLLS----GFGVGDSACAAHCI-ARGNRGGYCNS--QKVCVC-RN
Insect	*Ae. aegypti* C	--ATCDLLS----GFGVGDSACAAHCI-ARRNRGGYCNA--KKVCVC-RN
Insect	*An. gambiae*	--ATCDLAS----GFGVGSSLCAAHCI-ARRYRGGYCNS--KAVCVC-RN
Insect	*A. mellifera*	--VTCDLLS---FKGQVNDSACAANCL-SLGKAGGHCE---KGVCIC-RKTSFKDLWDKRF
Insect	*T. molitor*	--VTCDILSVEAKGVKLNDAACAAHCL-FRGRSGGYCNG--KRVCVC-R
Insect	*P. apterus*	--ATCDILSFQSQWVTPNHAGCALHCV-IKGYKGGQCKI---TVCHC-RR
Insect	*A. cyanea*	-GFGCPLD--Q--------MQCHRHCQTITGRSGGYCSGPLKLTCTCYR
Scorpion	*L. quinquestriatus*	-GFGCPLN--Q--------GACHRHCRSIR-RRGGYCAGFFKQTCTCYRN
Scorpion	*A. australis*	-GFGCPFN--Q--------GACHRHCRSIR-RRGGYCAGLFKQTCTCYR
Mollusc	*M. edulis*	-GFGCPND-----------YPCHRHCKSIPGRXGGYCGGRHRLRCTC
Mollusc	*M. galloprovinsialis*	-GFGCPNN-----------YQCHRHCKSIPGRCGGYCGGXHRLRCTCYRC
B		
Amoeba	*E. histolytica* 1	GEILCNLCTGLINTLENLLTTKGA-DKVKDYISSLCNKASGFIATLCTKVLDFGIDKLIQLIEDKVDANAICAKIHAC
Amoeba	*E. histolytica* 2	GAILCNLCKDTVKLVENLLTVDGA-QAVRQYIDNLCGKASGFLGTLCEKILSFGVDELVKLIENHVDPVVVCEKIHAC
Amoeba	*E. histolytica* 3	IPVLCPVCTSLVGKLIDLVLGGAV-DKVTDYLETLCAKADGLVETLCTKIVSYGIDKLIEKILEGGSAKLICGLIHAC
Nk-Lysin	*H. sapiens*	GYFCESCRKIIQKLEDMVGPQPNEDTVTQAASQVCDKL-KILRGLCKKIMRSFLRRISWDILTGKKPQAICVDIKICKE
Granulysin	*H. sapiens*	GRDYRTCLTIVQKLKKMVD-KPTQRSVSNAATRVCRTGRSRWRDVCRNFMRRYQSRVIQGLVAGETAQQICEDLRLCIPSTGPL
C		
Plant	*A. hippocastanum*	LCNERPSQTWSGNCGNTAHCDKQCQDWEKASHGACHKRENHWKCFCYFNC
Plant	*C. ternatea*	NLC-ERASLTWTGNCGNTGHCDTQCRNWESAKHGACHKR-GNWKCFCYFNC
Plant	*D. merckii*	ELC-EKASKTWSGNCGNTGHCDNQCKSWEGAAHGACHVRNGKHMCFCYFNC

Figure 1. Comparison of the amino acid sequences of defensin like molecules from invertebrates (**A**), amoebae and mammals (**B**), and plants (**C**). Sequences were aligned for similarities, and dashes used to maintain optimal alignment.

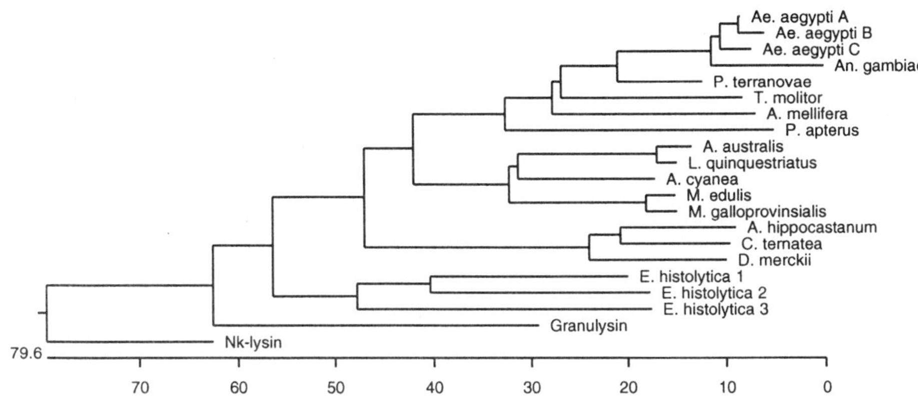

Figure 2. Phylogenetic analysis of the defensin amino sequences in Fig. 1. Constructions were done using the mature defensin sequences. The homology sequences were calculated using DNAStar software using the Jotun Hein method with PAM 250 residue table.

Three insect defensin isoforms have been isolated from *Ae. aegypti* (1,22,23), one from *An. gambiae* (25), and we have corresponding cDNAs from *Culex pipiens* (Lowenberger, Kamal, Solimen, Farid, and Christensen unpublished data). In *Ae. aegypti* the peptides were isolated from whole bodies of insects 24h after being inoculated with bacteria (23). Subsequent peptide analysis of the hemolymph of bacteria-inoculated mosquitoes confirmed that defensins are the predominant antibacterial peptide produced by this species.

Of the three isoforms of defensin isolated from *Ae. aegypti*, we believe that A and B are allelic variants of the same gene whereas isoform C likely is a separate gene. In the samples used in the initial study, the peptides corresponding to isoforms A and B were found in equal amounts in the insects whereas the peptide corresponding to isoform C was found at levels only 20% of A and B, and there are differences in the length and composition of the signal peptide region of the pre-pro defensins (1). Defensins are transcribed in the midguts of naïve mosquitoes, but only transcripts for isoform C were detected by RT-PCR, and these were up-regulated in the presence of a bloodmeal. The presence of lower amounts of isoform C in the mosquito, but its major presence in the midgut (1) and the much greater levels of isoforms A and B in the hemolymph (23) probably reflects tissue specific expression patterns. In immune activated mosquitoes, transcription for defensins continues for 21 days after inoculation with bacteria, and about 7-10 days after sterile injury (1).

If we consider the developmental stages of the mosquito, we only find transcription for defensin in the white or callow pupal stages, and no transcription in any of the larval stages (1). We thought initially that this may have been due to entry of bacteria through the soft white cuticle of the callow pupae during metamorphosis, or by a contamination of the hemocoel following the histolysis of the immature gut in the larva-pupa-adult transformation. However, because we do not see a similar transcriptional profile for cecropins (see below) in *Ae. aegypti*, which should occur if bacteria entered the hemocoel from the gut or through the cuticle, the transcription for defensin may be a developmentally regulated event independent of the presence of bacteria.

In *An. gambiae*, the major vector of malaria in sub-Saharan Africa, there is a similar pattern of transcription for defensins (25), but transcripts are short lived compared *to Ae. aegypti*, and there is not the similarly high level of defensin in the hemolymph of *An. gambiae* (1-5 μM) as is the case with *Ae. aegypti* (45 μM) (Table 1). Whether this reflects a more potent peptide in *An. gambiae* that is able to combat invading pathogens at a lower concentration, or the use of other, currently unknown, immune peptides is unknown. However, there are several differences in the transcription of insect defensins in these two mosquito species. In *Ae. aegypti* there is a low level of transcription for defensins in the midgut of naïve mosquitoes, that increases with the ingestion of a bloodmeal. In *An. gambiae*, an increase in transcription occurs when *Plasmodium* parasites are ingested in the bloodmeal, although not so with a non-parasite-infected bloodmeal (49). In addition, by RT-PCR, Dimopoulos et al. (32) demonstrated that defensin transcripts were found in the salivary glands 10 days after ingesting a *Plasmodium*-infected bloodmeal. In contrast, in *Ae. aegypti*, no transcripts were found by Northern analysis in whole body RNA extracts 10 days after ingesting a *P. gallinaceum*-infected bloodmeal. Whether these differences are due to different techniques used (northern analysis versus RT-PCR) or due to differences between mosquitoes of different genera remains to be elucidated.

CECROPINS

Cecropins were the first of the inducible antimicrobial peptides to be isolated and characterized (15). Approximately 20 cecropins have been reported from several Lepidoptera and Diptera, and from the intestine of pigs (50). These peptides are 31-39 residue proteins, devoid of cysteines, and generally amidated on the C-terminus. The helical structure of this molecule was described (51) as two helices joined by a Alanine-Glycine-Proline hinge. The N-terminal region forms an amphipathic α helix with equal hydrophobic and hydrophilic regions. The C-terminal region is more hydrophobic (8). Cecropins are active against Gr-negative and -positive bacteria: the mode of action seems to be through binding to the acidic components of the cell membrane, and inducing changes in cell permeability (52,53). Because of the amphipathic nature of the helices it has been proposed that the molecules bind to the cell membrane, inducing channel formation (54,55). In this "barrel-stave" mechanism trans-membrane amphiphilic α-helices form bundles that have outwardly facing hydrophobic surfaces that interact with membrane lipid regions, and inwardly facing hydrophilic regions that induce the pore (53,56-59).

Cecropins were originally isolated from Lepidoptera and, with the exception of cecropin D from *Bombyx mori* (60,61), all cecropins previously reported from insects have a tryptophan residue as the first or second amino acid in the mature peptide sequence. However, with the sequences we now have from mosquitoes (one obtained from immune-activated *Ae. aegypti* (26), one from a screen of a genomic library, and found expressed in immune activated mosquitoes (Lowenberger and Kamal unpublished data), 3 cDNA sequences found in bacteria exposed cells of an *Aedes albopictus* cell line (28,62), and one from *An. gambiae* (27)) no mosquito cDNA sequence has a tryptophan residue in this position. Another common feature of the cecropins is the blockage of the C-terminus by an amine group, a feature absent in the cecropins isolated from *Ae. aegypti,* but present in those isolated from *An. gambiae* (27) and *Ae. albopictus* (28).These features are considered significant because in *H. cecropia* the activity of a synthetic analogue deficient in the tryptophan residue was significantly reduced (63-65). Similarly in *S. peregrina*, an amidated cecropin molecule had a 4-fold greater antimicrobial activity than did the molecule with a free carboxylic group (66). Similar results implicating the importance of these two features have been reported by other authors (61,64,67,68). That the *Ae. aegypti* cecropin is not as active against bacteria as is the corresponding molecule from Drosophila (26), may reflect the lack of these two features in the mosquito cecropin.

From a phylogenetic perspective, cecropin like molecules are not limited to insects. Similar molecules have been isolated from tunicates, marine invertebrates belonging to the genus Chordata (69), from the intestine of a pig (50) and from bovine adrenal glands (70) once again exemplifying the diversity of taxa in which these immune peptides are found (Fig. 3). A comparison of the structure and phylogenetic relationships between cecropin like molecules identified from different organisms is presented in Figure 4.

TRANSFERRINS

Transferrins (TF) are molecules normally associated with transport of iron in vertebrates, and may play a role in limiting iron supplies to, and thus starving, parasites (71-74). Iron bound to lactoferrin found in mucosa and transferrin in blood serum and interstitial regions may lower the availability of free iron to levels below those needed by pathogens to survive. In insects transferrins have been isolated from *M. sexta* (75), *S. peregrina* (76), *D. melanogaster* (77) and the mosquito *Ae. aegypti* (29).The insect transferrins differ from those isolated from mammals in that the insects have the characteristic iron binding residues in the N-terminal region, but lack the iron binding domain in the C-terminal region (29). It has been speculated that this modification has arisen as a mechanism to reduce iron piracy by microorganisms (29). In *Ae. aegypti* there is an upregulation in TF transcripts after bacterial inoculation (29) or the injection of filarial worms (31). There is a similar increase in TF transcripts in *D. melanogaster* upon bacteria inoculation. When filarial worms were introduced into the hemocoel of *Ae. aegypti*, the transcripts were found in the hemocytes at the time when the mosquito was melanotically encapsulating the worms (31). However the process of injecting the worms also may have initiated the up regulation.

Family	Species	Sequence
Insect	*Ae. aegypti*	GGLKKLGKKLEGAGKRVFNAAEKALPVVA--G-AKALRK
Insect	*Ae. albopictus*	GGLKKLGKKLEGVGKRVFKASEKALPVAV--G-IKALG
Insect	*An. gambiae*	GRLKKLGKKIEGAGKRVFKAAEKALPVVA--G-VKAL
Insect	*D. melanogaster* A	GWLKKIGKKIERVGQHTRDATIQ-GLGIA--QQAANVAATAR
Insect	*D. melanogaster* B	GWLRKLGKKIERIGQHTRDASIQ-VLGIA--QQAANVAATAR
Insect	*D. melanogaster* C	GWLKKLGKRIERIGQHTRDATIQ-GLGIA--QQAANVAATAR
Insect	*S. peregrina* A	GWLKKIGKKIERVGQHTRDATIQ-GLGIA--QQAANVAATAR
Insect	*S. peregrina* B	GWLKKIGKKIERVGQHTRDATIQ-VIGVA--QQAANVAATAR
Insect	*C. capitata* A	GWLKKIGKKIERVGQHTRDATIQ-TIAVA--QQAANVAATARG
Insect	*H. cecropia* A	KW--KLFKKIEKVGQNIRDGIIKAGPAVAVVGQATQIAK
Insect	*H. cecropia* B	KW--KVFKKIEKMGRNIRNGIVKAGPAIAVLGEAKAL
Insect	*H. cecropia* C	-W--NPFKELEKVGQRVRDAVISAGPAVATVAQATAL
Insect	*A. pernyi* B	KW--KIFKKIEKVGRNIRNGIIKAGPAVAVLGEAKAL
Insect	*M. sexta* B2	-W--NPFKELERAGQRVRDAVISAAPAVATVGQAAAIAR
Insect	*M. sexta* B4	-W--NPFKELERAGQRVRDAIISAAPAVATVGQAAAIAR
Insect	*B. mori* A	RW--KLFKKIEKVGRNVRDGLIKAGPAIAVIGQAKSL
Insect	*B. mori* B	RW--KIFKKIEKMGRNIRDGIVKAGPAIEVLGSAKAI
Insect	*B. mori* D	G---NFFKDLEKMGQRVRDAVISAAPAVDTLAKAKALGQG
Tunicate	*S. clava* E	GWLRKAAKSVGKFYYKHKYYIKAAWKIGRHALGDMTDEEFQDFMKEVEQAREEELQSRQ
Tunicate	*S. clava* D	GWLRKAAKSVGKFYYKHKYYIKAAWQIGKHALGDMTDEEFQDFMKEVEQAREEELQSRQ
Tunicate	*S. clava* C	GWFGKAFRSVSNFYKKHKTYIHAGLSAATLLGDMTDEEFQEFMQDIEQAREEELLSRQ
Pig		SWLSKTAKKLENSAKKRISEGIAIAIQGGPR
Bovine		AMDLELQKIAEKFSGTRG

Figure 3. Comparison of the amino acid sequences of cecropin molecules from insects, tunicates, and mammals. Sequences were aligned for similarities, and dashes used to maintain optimal alignment.

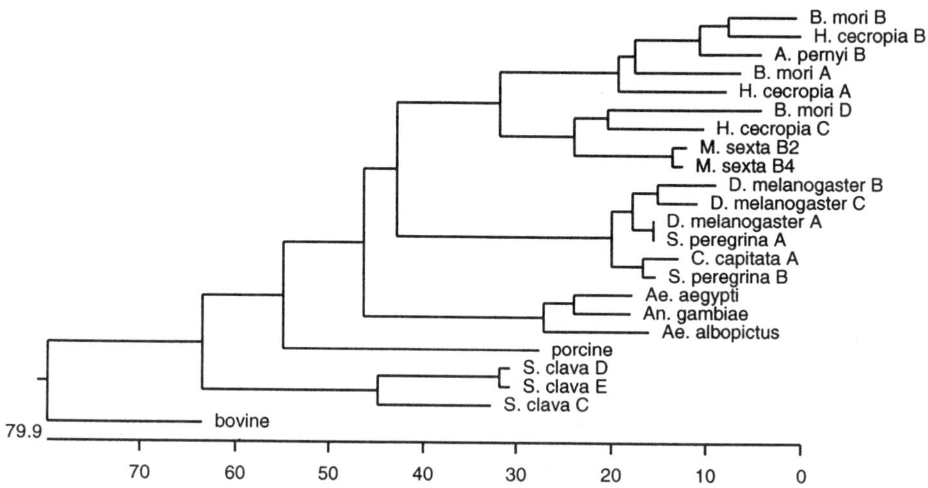

Figure 4. Phylogenetic analysis of the amino acid sequences of the mature cecropin sequences in Fig. 3. The homology sequences were calculated using DNAStar software using the Jotun Hein method with PAM 250 residue table.

We have been studying the immune response, including transferrin production, of another mosquito, *Armigeres subalbatus*, to infections with filarial worms. This is an ideal model to use because *Ar. Subalbatus* melanizes >95% of the microfilariae of *Brugia malayi* ingested in a blood meal. In this species the introduction of bacteria significantly increases transcription of TF, but TF transcripts also increase significantly during the melanotic encapsulation of *B. malayi* (Falk, Lowenberger and Christensen unpublished data). It is enticing to suggest that TF plays dual roles in the mosquito: that of iron transport in general, and as an immune response to pathogens or parasites to reduce their ability to establish and proliferate within the hemocoel during the period required for the production of other immune peptides.

One of the interesting aspects in looking at the different immune peptides in different insect species is the fact that even closely related Diptera do not produce the same molecules at the same concentration. Even making general references to the response of mosquitoes to a particular stimulus may be inappropriate because all species do not respond similarly. For instance, *Ae. aegypti* produces defensin in great amounts- about 45 μM in the hemolymph, and cecropin at about 1-5 μM, but *An. gambiae* produces defensin and cecropin at about 1mM, and *Drosophila* produces defensin at 1 μM and cecropin about 10 μM (Table 1). Whether these differences are related to the environments in which these different organisms live, and on the bacteria to which they have been exposed through evolutionary time, remains to be examined. When we examine the phylogenetic relationships of the cecropins and defensins solely within the insects, it is common to make comparisons between taxonomic groups. However, environmental factors and specific niches inhabited by particular insects also may have played a role in the evolution and preponderance of use of specific peptides based upon the selective pressures to which these organisms have been exposed. In the analysis of cecropins, the terrestrial Diptera and Lepidoptera arise from one branch whereas the Diptera with aquatic immature stages, the mosquitoes, arise from a different branch (1). Similarly with the defensins, the terrestrial Diptera, as well as other terrestrial insect orders arise from a different branch than do the mosquitoes (1). To test the hypothesis of environmental effects on the preferential expression of particular immune peptides will require a more detailed analysis encompassing more species from different environments, and the microbial fauna in these diverse locations, than is currently feasible.

Table 1. Known concentrations of immune peptides in different Diptera[1]

	Aedes aegypti	*Anopheles gambiae*	*Drosophila melanogaster*
Defensin	45	1-5	1
Cecropin	1-5	1-5	20
Transferrin	1-5	*	?
Drosomycin	*	*	100
Drosocin	*	*	40
Attacin	*	*	?
Metchnikowin	*	*	10

[1]An asterisk indicates that the peptide has not been identified in the species and a question mark indicates no known concentration for the peptide.

PEPTIDE-PARASITE INTERACTIONS

Insect immune peptides generally have been considered ineffective against eukaryotic cells (16). However, there is a growing literature on the effects of these immune peptides on parasites normally transmitted by mosquitoes (Table 2). Gwadz et al. (78) injected cecropins and magainins into mosquitoes, and reported a significant reduction in development of *Plasmodium* sp. Since this first report, several laboratories have reported similar results. Jaynes et al (79) used a modified synthetic cecropin, Shiva, to reduce the development of *Plasmodium* sp. Further modifications of the cecropin molecule subsequently were used to prevent *Plasmodium* development (80-82). Chalk et al. (83) co-injected cecropins with the nematode *Brugia pahangi* into *Ae. aegypti*,

Table 2. Reports of the effect of immune peptides or immune activation on parasite development.

Author	Peptide	Parasite
Gwadz *et al.*, 1989	cecropin / magainin	*Plasmodium*
Jaynes *et al.*, 1988	cecropin	*Plasmodium*
Jaynes *et al.*, 1988	cecropin	*Trypanosoma cruzi*
Shahabuddin *et al.*, 1998	defensin	*Plasmodium*
Lowenberger *et al.*, 1996	immune activation	*Brugia malayi*
Lowenberger *et al.*, 1999a	immune activation	*Plasmodium*
Chalk *et al.*, 1995	cecropin	*Brugia pahangi*
Albuquerque *et al.*, 1996	defensin	*Brugia pahangi*
Rodriguez *et al.*, 1995	cecropin / Shiva	*Plasmodium*
Possani *et al.*, 1998	cecropin / Shiva	*Plasmodium*
Boisbouvier *et al.*, 1998	cecropin / Shiva	*Plasmodium*

and Albuquerque et al. (84) co-injected mosquito defensins and *B. pahangi* into *Ae. aegypti.* Both studies reported a reduction in parasite development. Lowenberger et al. (85) reported that the inoculation of bacteria into the hemocoel (a process normally done to activate the immune system of mosquitoes) and then subsequent feeding of these mosquitoes on a gerbil infected with *Brugia malayi*, resulted in significant reductions in both parasite prevalence (percent of mosquitoes that became infected) and mean intensity of infection (mean number of parasites/infected mosquito). Depending on the parasite burden to which the mosquitoes were exposed, prevalence was reduced to 50-57% in immune activated mosquitoes while non-inoculated controls had a prevalence of 92-97%. Similarly the mean intensity of infection was reduced from 8-16 worms / mosquito in the controls to 2.3-2.5 worms/immune activated mosquito. When the mosquitoes were exposed to extremely high numbers of parasites the hosts apparently were overwhelmed in their efforts to eliminate the parasites, and numbers were no different from naïve controls. Interestingly, in this study the microfilariae that were ingested by the mosquito, and which subsequently penetrated the midgut and entered the hemocoel, were found dead in the immune-activated mosquitoes. This killing apparently occurred after entering the hemolymph, and before the nematodes could reach the thoracic musculature where they normally develop. However, there was no evidence of melanotic materials being deposited on the worms, thus killing them, as is in the case of some mosquito-nematode relationships.

In a subsequent study (2) we studied the effect of immune activation of *Ae. aegypti* on the development of *Plasmodium gallinaceum*. As was the case with *B. malayi*, the immune activation process had a significant effect on the number of parasites that developed within the mosquito. However, the timing of events was critical: Only those mosquitoes immune activated before, or immediately after bloodfeeding demonstrated this reduction in parasite development. Mosquitoes immune activated 1-5 days after bloodfeeding demonstrated no reduction in oocysts on the midgut as compared to naïve controls. These results implicated the ookinete as the susceptible parasite stage that was being killed. Approximately 24h after bloodfeeding, the ookinetes leave the blood bolus, migrate through the midgut wall, and establish on the hemolymph side of the midgut, beneath the basal lamina, where they form oocysts. As this study found very few oocysts, the authors assumed that the parasite killing factors had acted on the ookinetes. These data appear to conflict with those of Shahabuddin et al. (86) who injected purified defensins isolated from *P. terranovae* and *A. cyanea* into *Ae. aegypti*. In this study there was no effect of defensins *in vitro* or *in vivo* on the ookinetes or early oocysts, whereas later oocysts and sporozoites were killed by the defensins. However, whereas bacteria inoculation produced a significant reduction in prevalence, as well as in mean intensity of infection in one study (2) injection of purified defensins did not reduce the prevalence of infection in another study (86)The results of these two studies may differ because of the methodologies used: we (2) injected bacteria into the hemocoel, a process that turns on the expression of many immune peptides, some of which, alone or acting in concert, may have had an effect on the developing ookinetes, whereas in the study by Shahabuddin et al.(86) one individual peptide, defensin, was evaluated.

In order to determine what specific peptides are involved in the parasite killing effects reported, we need a mechanism by which we can express a particular peptide within the mosquito host, without expressing the complete arsenal of immune factors. The development of the Sindbis virus as a transient transducing vector for the expression of specific peptides has been developed for use by the AIDL at Colorado State University (87-89). Although the original viral construct had to be injected into the mosquitoes, the development of a double subgenomic construct that is orally infectious to *Ae. aegypti* (90) will prove invaluable in assessing the effects *in vivo* of particular peptides on developing parasites within mosquitoes.

There is no doubt that there are many more immune peptide molecules produced by mosquitoes and other insects, and the search continues to isolate, identify, and characterize them. With the recent announcement to sequence the entire genomes of *Ae. aegypti* and *An. gambiae* (91,92) future discoveries of genes coding for immune peptides will be able to probe these microchip libraries to find large flanking regions of these genes and such advances will no doubt increase the speed with which we can define the regulatory elements and promotor regions of specific genes. Similarly, with the ever expanding databases of sequences from diverse organisms, database searches will rapidly be able to discern the same genes within related species, members of the same gene family in diverse taxa, or even propose ancient precursor or ancestral genes that have evolved differently through different taxa.

However, gene sequence data alone will not allow us to determine what biological factors within mosquitoes act alone or in concert to induce or enhance the expression of particular peptides encoded by these genes or what signalling molecules are used for chemical communication between cells and tissues. The regulation of several immune peptides through Toll/ IMD pathways has been described in Drosophila (17,93,94), and these are, in all likelihood, similar in mosquitoes. However, the mechanisms, molecules, and means by which mosquitoes in particular, and insects in general, recognize non-self and then initiate the chain of events that results in the high concentrations of immune factors in the hemolymph is the holy grail of insect immunity today.

ACKNOWLEDGEMENTS

I thank B.M. Christensen for support and for reviewing an earlier version of this paper, and Sofie Kamal for invaluable technical assistance. Funding for this research was provided by NIH grants AI44966 to CAL and AI19769 and AI28781 to BMC.

REFERENCES

1. Lowenberger, C. A., Smartt, C. T., Bulet, P., Ferdig, M. T., Severson, D. W., Hoffmann, J. A., and Christensen, B. M. (1999) Insect immunity: molecular cloning, expression, and characterization of cDNAs and genomic DNA encoding three isoforms of insect defensin in *Aedes aegypti*. *Insect. Mol. Biol.* 8, 107-118
2. Lowenberger, C. A., Kamal, S., Chiles, J., Paskewitz, S., Bulet, P., Hoffmann, J. A., and Christensen, B. M. (1999) Mosquito-*Plasmodium* interactions in response to immune activation of the vector. *Exp. Parasitol.* 91, 59-69
3. Dunn, P. E. (1986) Biochemical aspects of insect immunology. *Annu. Rev. Entomol.* 31, 321-329
4. Lackie, A. M., and Vasta, G. R. (1988) The role of galactosyl-binding lectin in the cellular immune response of the cockroach *Periplaneta americana* (Dictyoptera). *Immunology* 64, 353-357
5. Christensen, B. M., and Forton, K. F. (1986) Hemocyte-mediated melanization of microfilariae in *Aedes aegypti*. *J. Parasitol.* 72, 220-225
6. Li, J. Y., and Christensen, B. M. (1990) Immune competence of *Aedes trivittatus* hemocytes as assessed by lectin binding. *J. Parasitol.* 76, 276-278
7. Bulet, P., Hetru, C., Dimarcq, J. L., and Hoffmann, D. (1999) Antimicrobial peptides in insects; structure and function. *Dev. Comp. Immunol.* 23, 329-344
8. Hetru C., H., D., and Bulet, P. (1998) Antimicrobial Peptides from Insects in *Molecular Mechanisms of Immune Responses in Insects* (Brey, P. T. and Hultmark, D. eds), pp. 40-66, Chapman and Hall, London
9. Boman, H. G. (1998) Gene-encoded peptide antibiotics and the concept of innate immunity: an update review. *Scand. J. Immunol.* 48, 15-25
10. Boman, H. G. (1991) Antibacterial peptides: key components needed in immunity. *Cell* 65, 205-207
11. Zasloff, M. (1992) Antibiotic peptides as mediators of innate immunity. *Curr. Opin. Immunol.* 4, 3-7
12. Boman, H. G. (1995) Peptide antibiotics and their role in innate immunity. *Annu. Rev. Immunul.* 13, 61-62
13. Ganz, T., and Weiss, J. (1997) Antimicrobial peptides of phagocytes and epithelia. *Semin. Hematol.* 34, 343-354
14. Broekaert, W. F., Terras, F. R., Cammue, B. P., and Osborn, R. W. (1995) Plant defensins: novel antimicrobial peptides as components of the host defense system. *Plant Physiol.* 108, 1353-1358
15. Steiner, H., Hultmark, D., Engstrom, A., Bennich, H., and Boman, H. G. (1981) Sequence and specificity of two antibacterial proteins involved in insect immunity. *Nature* 292, 246-248
16. Cociancich, S., Bulet, P., Hetru, C., and Hoffmann, J. A. (1994) The inducible antibacterial peptides of insects. *Parasitology Today* 10, 132-139
17. Hoffmann, J. A., Reachhart, J. M., and Hetru, C. (1996) Innate immunity in higher insects. *Curr. Opin. Immunol.* 8, 8-13
18. Hetru, C., Bulet, P., Cociancich, S., Dimarcq, J. L., Hoffmann, D., and Hoffmann, J. A. (1994) Antibacterial Peptides/Polypeptides in the insect host defense: a comparison with vertebrate antibacterial peptides/polypeptides in *Phylogenetic perspectives in immunity: the insect host defense* (Hofmann, J. A., Janeway, C. A., and Natori, S., eds), pp. 43-66, R.G. Landes, Austin
19. Matsuyama, K., and Natori, S. (1988) Molecular cloning of cDNA for sapecin and unique expression of the sapecin gene during the development of *Sarcophaga peregrina*. *J. Biol. Chem.* 263, 17117-17121
20. Matsuyama, K., and Natori, S. (1988) Purification of three antibacterial proteins from the culture medium of NIH-Sape-4, an embryonic cell line of *Sarcophaga peregrina*. *J. Biol. Chem.* 263, 17112-17116

21. WHO. (1999) *Disease Statistics. The World Health Report*, WHO, Geneva
22. Chalk, R., Albuquerque, C. M., Ham, P. J., and Townson, H. (1995) Full sequence and characterization of two insect defensins: immune peptides from the mosquito *Aedes aegypti*. *Proc. R. Soc. Lond. B* 261, 217-221
23. Lowenberger, C., Bulet, P., Charlet, M., Hetru, C., Hodgeman, B., Christensen, B. M., and Hoffmann, J. A. (1995) Insect immunity: isolation of three novel inducible antibacterial defensins from the vector mosquito, *Aedes aegypti*. *Insect. Biochem. Mol. Biol.* 25, 867-873
24. Cho, W. L., Fu, T. F., Chiou, J. Y., and Chen, C. C. (1997) Molecular characterization of a defensin gene from the mosquito, *Aedes aegypti*. *Insect. Biochem. Mol. Biol.* 27, 351-358
25. Richman, A. M., Bulet, P., Hetru, C., Barillas-Mury, C., Hoffmann, J. A., and Kafalos, F. C. (1996) Inducible immune factors of the vector mosquito *Anopheles gambiae*: biochemical purification of a defensin antibacterial peptide and molecular cloning of preprodefensin cDNA. *Insect. Mol. Biol.* 5, 203-210
26. Lowenberger, C., Charlet, M., Vizioli, J., Kamal, S., Richman, A., Christensen, B. M., and Bulet, P. (1999) Antimicrobial activity spectrum, cDNA cloning, and mRNA expression of a newly isolated member of the cecropin family from the mosquito vector *Aedes aegypti*. *J. Biol. Chem.* 274, 20092-20097
27. Vizioli, J., Bulet, P., Charlet, M., Lowenberger, C., Blass, C., Muller, H.-M., Dimopoulos, G., Hoffmann, J. A., Kafatos, F. C., and Richman, A. (2000) Cloning and analysis of a cecropin gene from the malaria vector mosquito, *Anopheles gambiae*. *Insect Mol. Biol.* In Press
28. Sun, D., Eccleston, E. D., and Fallon, A. M. (1998) Peptide sequence of an antibiotic cecropin from the vector mosquito, *Aedes albopictus*. *Biochem. Biophys. Res. Commun.* 249, 410-415
29. Yoshiga, T., Hernandez, V. P., Fallon, A. M., and Law, J. H. (1997) Mosquito transferrin, an acute-phase protein that is up-regulated upon infection. *Proc. Natl. Acad. Sci. U S A* 94, 12337-12342
30. Dimarcq, J. L., Zachary, D., Hoffmann, J. A., Hoffmann, D., and Reichhart, J. M. (1990) Insect immunity: expression of the two major inducible antibacterial peptides, defensin and diptericin, in *Phormia terranovae*. *Embo J.* 9, 2507-2515
31. Beerntsen, B. T., Severson, D. W., and Christensen, B. M. (1994) *Aedes aegypti*: characterization of a hemolymph polypeptide expressed during melanotic encapsulation of filarial worms. *Exp. Parasitol.* 79, 312-321
32. Dimopoulos, G., Seeley, D., Wolf, A., and Kafatos, F. C. (1998) Malaria infection of the mosquito *Anopheles gambiae* activates immune- responsive genes during critical transition stages of the parasite life cycle. *Embo J.* 17, 6115-6123
33. Brey, P. T., Lee, W. J., Yamakawa, M., Koizumi, Y., Perrot, S., Francois, M., and Ashida, M. (1993) Role of the integument in insect immunity: epicuticular abrasion and induction of cecropin synthesis in cuticular epithelial cells. *Proc. Natl. Acad. Sci. U S A* 90, 6275-6279
34. Lambert, J., Keppi, E., Dimarcq, J. L., Wicker, C., Reichhart, J. M., Dunbar, B., Lepage, P., Van Dorsselaer, A., Hoffmann, J., Fothergill, J., and et al. (1989) Insect immunity: isolation from immune blood of the dipteran *Phormia terranovae* of two insect antibacterial peptides with sequence homology to rabbit lung macrophage bactericidal peptides. *Proc. Natl. Acad. Sci. U S A* 86, 262-266
35. Ganz, T., and Lehrer, R. I. (1995) Defensins. *Pharmacol. Ther.* 66, 191-205
36. Evans, E. W., Beach, F. G., Moore, K. M., Jackwood, M. W., Glisson, J. R., and Harmon, B. G. (1995) Antimicrobial activity of chicken and turkey heterophil peptides CHP1, CHP2, THP1, and THP3. *Vet. Microbiol.* 47, 295-303
37. Penninckx, I. A., Eggermont, K., Terras, F. R., Thomma, B. P., De Samblanx, G. W., Buchala, A., Metraux, J. P., Manners, J. M., and Broekaert, W. F. (1996) Pathogen-induced systemic activation of a plant defensin gene in Arabidopsis follows a salicylic acid-independent pathway. *Plant Cell* 8, 2309-2323
38. Fujiwara, S., Imai, J., Fujiwara, M., Yaeshima, T., Kawashima, T., and Kobayashi, K. (1990) A potent antibacterial protein in royal jelly. Purification and determination of the primary structure of royalisin. *J. Biol. Chem.* 265, 11333-11337
39. Casteels, P., Ampe, C., Jacobs, F., and Tempst, P. (1993) Functional and chemical characterization of Hymenoptaecin, an antibacterial polypeptide that is infection-inducible in the honeybee (*Apis mellifera*). *J. Biol. Chem.* 268, 7044-7054
40. Cornet, B., Bonmatin, J. M., Hetru, C., Hoffmann, J. A., Ptak, M., and Vovelle, F. (1995) Refined three-dimensional solution structure of insect defensin A. *Structure* 3, 435-448

41. Bontems, F., Roumestand, C., Gilquin, B., Menez, A., and Toma, F. (1991) Refined structure of charybdotoxin: common motifs in scorpion toxins and insect defensins. *Science* 254, 1521-1523
42. Leippe, M. (1999) Antimicrobial and cytolytic polypeptides of amoeboid protozoa—effector molecules of primitive phagocytes. *Dev. Comp. Immunol.* 23, 267-279
43. Kato, Y., and Komatsu, S. (1996) ASABF, a novel cysteine-rich antibacterial peptide isolated from the nematode *Ascaris suum*. Purification, primary structure, and molecular cloning of cDNA. *J. Biol. Chem.* 271, 30493-30498
44. Cociancich, S., Goyffon, M., Bontems, F., Bulet, P., Bouet, F., Menez, A., and Hoffmann, J. (1993) Purification and characterization of a scorpion defensin, a 4kDa antibacterial peptide presenting structural similarities with insect defensins and scorpion toxins. *Biochem. Biophys. Res. Commun.* 194, 17-22
45. Ehret-Sabatier, L., Loew, D., Goyffon, M., Fehlbaum, P., Hoffmann, J. A., van Dorsselaer, A., and Bulet, P. (1996) Characterization of novel cysteine-rich antimicrobial peptides from scorpion blood. *J. Biol. Chem.* 271, 29537-29544
46. Charlet, M., Chernysh, S., Philippe, H., Hetru, C., Hoffmann, J. A., and Bulet, P. (1996) Innate immunity. Isolation of several cysteine-rich antimicrobial peptides from the blood of a mollusc, *Mytilus edulis*. *J. Biol. Chem.* 271, 21808-21813
47. Hubert, F., Noel, T., and Roch, P. (1996) A member of the arthropod defensin family from edible Mediterranean mussels (*Mytilus galloprovincialis*). *Eur. J. Biochem.* 240, 302-306
48. Cociancich, S., Ghazi, A., Hetru, C., Hoffmann, J. A., and Letellier, L. (1993) Insect defensin, an inducible antibacterial peptide, forms voltage- dependent channels in *Micrococcus luteus*. *J. Biol. Chem.* 268, 19239-19245
49. Richman, A. M., Dimopoulos, G., Seeley, D., and Kafatos, F. C. (1997) *Plasmodium* activates the innate immune response of *Anopheles gambiae* mosquitoes. *Embo J.* 16, 6114-6119
50. Lee, J. Y., Boman, A., Sun, C. X., Andersson, M., Jornvall, H., Mutt, V., and Boman, H. G. (1989) Antibacterial peptides from pig intestine: isolation of a mammalian cecropin. *Proc. Natl. Acad. Sci. U S A* 86, 9159-9162
51. Holak, T. A., Engstrom, A., Kraulis, P. J., Lindeberg, G., Bennich, H., Jones, T. A., Gronenborn, A. M., and Clore, G. M. (1988) The solution conformation of the antibacterial peptide cecropin A: a nuclear magnetic resonance and dynamical simulated annealing study. *Biochemistry* 27, 7620-7629
52. Shai, Y. (1995) Molecular recognition between membrane-spanning helices. *TIBS* 20, 460-464
53. Shai, Y. (1998) Mode of Action of Antibacterial Peptides in *Mode of action of antibacterial peptides in Molecular mechanisms of immune responses in insects* (Brey, P. T., and Hultmark, D., eds), pp. 111-134, Chapman and Hall, London
54. Christensen, B., Fink, J., Merrifield, R. B., and Mauzerall, D. (1988) Channel-forming properties of cecropins and related model compounds incorporated into planar lipid membranes. *Proc. Natl. Acad. Sci. U S A* 85, 5072-5076
55. Cruciani, R. A., Barker, J. L., Durell, S. R., Raghunathan, G., Guy, H. R., Zasloff, M., and Stanley, E. F. (1992) Magainin 2, a natural antibiotic from frog skin, forms ion channels in lipid bilayer membranes. *Eur. J. Pharmacol.* 226, 287-296
56. Ehrenstein, G., and Lecar, H. (1977) Electrically gated ionic channels in lipid bilayers. *Q. Rev. Biophys.* 10, 1-34
57. Inouye, M. (1974) A three-dimensional molecular assembly model of a lipoprotein from the *Escherichia coli* outer membrane. *Proc. Natl. Acad. Sci. U S A* 71, 2396-2400
58. Oiki, S., Danho, W., Madison, V., and Montal, M. (1988) M2 delta, a candidate for the structure lining the ionic channel of the nicotinic cholinergic receptor. *Proc. Natl. Acad. Sci. U S A* 85, 8703-8707
59. Lear, J. D., Wasserman, Z. R., and DeGrado, W. F. (1988) Synthetic amphiphilic peptide models for protein ion channels. *Science* 240, 1177-1181
60. Hara, S., and Yamakawa, M. (1995) Moricin, a novel type of antibacterial peptide isolated from the silkworm, *Bombyx mori*. *J. Biol. Chem.* 270, 29923-29927
61. Hara, S., Taniai, K., Kato, Y., and Yamakama, M. (1994) Isolation and alpha-amidation of the non-amidated form of cecropin-D from larvae of *Bombyx mori*. *Comp. Biochem. Physiol. B* 108, 303-308
62. Sun, D., Eccleston, E. D., and Fallon, A. M. (1999) Cloning and expression of three cecropin cDNAs from a mosquito cell line. *FEBS Lett* 454, 147-151

63. Andreu, D., Merrifield, R. B., Steiner, H., and Boman, H. G. (1983) Solid-phase synthesis of cecropin A and related peptides. *Proc. Natl. Acad. Sci. U S A* 80, 6475-6479
64. Andreu, D., Ubach, J., Boman, A., Wahlin, B., Wade, D., Merrifield, R. B., and Boman, H. G. (1992) Shortened cecropin A-melittin hybrids. Significant size reduction retains potent antibiotic activity. *FEBS Lett* 296, 190-194
65. Andreu, D., Merrifield, R. B., Steiner, H., and Boman, H. G. (1985) N-terminal analogues of cecropin A: synthesis, antibacterial activity, and conformational properties. *Biochemistry* 24, 1683-1688
66. Li, Z. Q., Merrifield, R. B., Boman, I. A., and Boman, H. G. (1988) Effects on electrophoretic mobility and antibacterial spectrum of removal of two residues from synthetic sarcotoxin IA and addition of the same residues to cecropin B. *FEBS Lett* 231, 299-302
67. Merrifield, R. B., Vizioli, L. D., and Boman, H. G. (1982) Synthesis of the antibacterial peptide cecropin A (1-33). *Biochemistry* 21, 5020-5031
68. Callaway, J. E., Lai, J., Haselbeck, B., Baltaian, M., Bonnesen, S. P., Weickmann, J., Wilcox, G., and Lei, S. P. (1993) Modification of the C terminus of cecropin is essential for broad- spectrum antimicrobial activity. *Antimicrob. Agents Chemother.* 37, 1614-1619
69. Zhao, C., Liaw, L., Lee, I. H., and Lehrer, R. I. (1997) cDNA cloning of three cecropin-like antimicrobial peptides (Styelins) from the tunicate, *Styela clava*. *FEBS Lett* 412, 144-148
70. Strub, J. M., Garcia-Sablone, P., Lonning, K., Taupenot, L., Hubert, P., Van Dorsselaer, A., Aunis, D., and Metz-Boutigue, M. H. (1995) Processing of chromogranin B in bovine adrenal medulla. Identification of secretolytin, the endogenous C-terminal fragment of residues 614-626 with antibacterial activity. *Eur. J. Biochem.* 229, 356-368
71. Britigan, B. E., Lewis, T. S., McCormick, M. L., and Wilson, M. E. (1998) Evidence for the existence of a surface receptor for ferriclactoferrin and ferrictransferrin associated with the plasma membrane of the protozoan parasite *Leishmania donovani*. *Adv. Exp. Med. Biol.* 443, 135-140
72. Svistunenko, D. A., Patel, R. P., Voloshchenko, S. V., and Wilson, M. T. (1997) The globin-based free radical of ferryl hemoglobin is detected in normal human blood. *J. Biol. Chem.* 272, 7114-7121
73. Wilson, M. E., Vorhies, R. W., Andersen, K. A., and Britigan, B. E. (1994) Acquisition of iron from transferrin and lactoferrin by the protozoan *Leishmania chagasi*. *Infect. Immun.* 62, 3262-3269
74. Wilson, M. E., and Britigan, B. E. (1998) Iron acquisition by parasitic protozoa. *Parasitology Today* 14, 348-353
75. Bartfeld, N. S., and Law, J. H. (1990) Isolation and molecular cloning of transferrin from the tobacco hornworm, *Manduca sexta*. Sequence similarity to the vertebrate transferrins. *J. Biol. Chem.* 265, 21684-21691
76. Kurama, T., Kurata, S., and Natori, S. (1995) Molecular characterization of an insect transferrin and its selective incorporation into eggs during oogenesis. *Eur. J. Biochem.* 228, 229-235
77. Yoshiga, T., Georgieva, T., Dunkov, B. C., Harizanova, N., Ralchev, K., and Law, J. H. (1999) *Drosophila melanogaster* transferrin. Cloning, deduced protein sequence, expression during the life cycle, gene localization and up-regulation on bacterial infection. *Eur. J. Biochem.* 260, 414-420
78. Gwadz, R. W., Kaslow, D., Lee, J. Y., Maloy, W. L., Zasloff, M., and Miller, L. H. (1989) Effects of magainins and cecropins on the sporogonic development of malaria parasites in mosquitoes. *Infect. Immunol.* 57, 2628-2633
79. Jaynes, J. M., Julian, G. R., Jeffers, G. W., White, K. L., and Enright, F. M. (1989) In vitro cytocidal effect of lytic peptides on several transformed mammalian cell lines. *Pept. Res.* 2, 157-160
80. Rodriguez, M. C., Zamudio, F., Torres, J. A., Gonzalez-Ceron, L., Possani, L. D., and Rodriguez, M. H. (1995) Effect of a cecropin-like synthetic peptide (Shiva-3) on the sporogonic development of *Plasmodium berghei*. *Exp. Parasitol.* 80, 596-604
81. Possani, L. D., Zurita, M., Delepierre, M., Hernandez, F. H., and Rodriguez, M. H. (1998) From noxiustoxin to Shiva-3, a peptide toxic to the sporogonic development of *Plasmodium berghei*. *Toxicon* 36, 1683-1692
82. Boisbouvier, J., Prochnicka-Chalufour, A., Nieto, A. R., Torres, J. A., Nanard, N., Rodriguez, M. H., Possani, L. D., and Delepierre, M. (1998) Structural information on a cecropin-like synthetic peptide, Shiva-3 toxic to the sporogonic development of *Plasmodium berghei*. *Eur. J. Biochem.* 257, 263-273
83. Chalk, R., Townson, H., and Ham, P. J. (1995) *Brugia pahangi*: the effects of cecropins on microfilariae in vitro and in *Aedes aegypti*. *Exp. Parasitol.* 80, 401-406
84. Albuquerque, C. M., and Ham, P. J. (1996) In vivo effect of a natural *Aedes aegypti* defensin on *Brugia pahangi* development. *Med. Vet. Entomol.* 10, 397-399

85. Lowenberger, C. A., Ferdig, M. T., Bulet, P., Khalili, S., Hoffmann, J. A., and Christensen, B. M. (1996) *Aedes aegypti*: induced antibacterial proteins reduce the establishment and development of *Brugia malayi*. *Exp. Parasitol.* 83, 191-201
86. Shahabuddin, M., Fields, I., Bulet, P., Hoffmann, J. A., and Miller, L. H. (1998) *Plasmodium gallinaceum*: differential killing of some mosquito stages of the parasite by insect defensin. *Exp. Parasitol.* 89, 103-112
87. Beaty, B. J., and Carlson, J. O. (1997) Molecular manipulation of mosquitoes. *Curr. Opin. Infect. Dis.* 10, 372-376
88. Rayms-Keller, A., Powers, A. M., Higgs, S., Olson, K. E., Kamrud, K. I., Carlson, J. O., and Beaty, B. J. (1995) Replication and expression of a recombinant Sindbis virus in mosquitoes. *Insect. Mol. Biol.* 4, 245-251
89. Olson, K. E., Higgs, S., Gaines, P. J., Powers, A. M., Davis, B. S., Kamrud, K. I., Carlson, J. O., Blair, C. D., and Beaty, B. J. (1996) Genetically engineered resistance to dengue-2 virus transmission in mosquitoes. *Science* 272, 884-886
90. Seabaugh, R. C., Olson, K. E., Higgs, S., Carlson, J. O., and Beaty, B. J. (1998) Development of a chimeric sindbis virus with enhanced per Os infection of *Aedes aegypti*. *Virology* 243, 99-112
91. NIAID. (1999) http://www.niaid.nih.gov/ncn/microbes.htm.
92. NIAID. (1999) http://www.niaid.nih.gov/facts/minutes/0599min1.htm.
93. Hoffmann, J. A., and Reichhart, J. M. (1997) *Drosophila* Immunity. *Trends Cell Biol.* 7, 309-316
94. Engstrom, Y. (1998) Insect Immune Gene Regulation in *Molecular mechanisms of immune responses in insects.* (Brey, P. T., and Hultmark, D., eds), pp. 211-244, Chapman and Hall, London

REACTIVE OXYGEN SPECIES AND ANTIMICROBIAL DEFENSES OF INVERTEBRATES: A BIVALVE MODEL

Robert S. Anderson

Chesapeake Biological Laboratory
University of Maryland, Box 38
Solomons, MD 20688-0038

INTRODUCTION

Reactive oxygen species (ROS) are cytotoxic agents produced by many phagocytic cells in response to membrane perturbations such as receptor-ligand interactions and phagocytosis. These agents are used to defend against infectious diseases by virtue of their antimicrobial properties. The initial ROS generated, the superoxide anion (O_2^-), is spontaneously or enzymatically converted to hydrogen peroxide (H_2O_2), which can give rise to even more toxic products, such as hydroxyl radical (·OH), hypochlorous acid (HOCl), and singlet oxygen ($^1O_2$). The ROS are toxic in their own right or may exert increased antimicrobial activity in concert with lysosomal hydrolases and/or reactive nitrogen species (RNS). Phagocyte activation triggers the assembly of a membrane-associated enzyme NADPH oxidase, which produces O from molecular oxygen and NADPH. Over-production of ROS by blood cells can overwhelm antioxidant defenses leading to various manifestations of oxidative stress. Human beings with genetic abnormalities in NADPH oxidase present with recurrent bacterial and

Phylogenetic Perspectives on the Vertebrate Immune System.
Edited by Beck *et al.*, Kluwer Academic/Plenum Publishers, 2001.

fungal infections, granulomatous infiltration of many organs, and early deaths characteristic of chronic granulomatous disease (Bridges, Berendes and Good, 1959). This clearly establishes the role of ROS in the antimicrobial mechanisms of vertebrates, and a large literature has been developed on the subject, as well as on the modulation of ROS production by environmental chemical stressors and disease. Much less is known about the physiological roles of hemocyte-generated ROS in invertebrates; however, it appears that this system has been highly conserved during metazoan evolution. Among the invertebrates, the defense mechanisms of commercially important bivalve mollusks, such as oysters, that have been decimated by parasitic diseases, have received considerable study. Research efforts on ROS production by molluscan hemocytes have been reviewed (Adema, van der Knaap and Sminia, 1991; Anderson, 1994, 1996); this chapter is an attempt to update these reviews, with emphasis on the modulation of ROS generation by exposure to xenobiotics, physical stressors and parasitic infection.

ROS GENERATION BY BIVALVE HEMOCYTES

When bivalve hemocytes are stimulated by phagocytosis of foreign particles, or by cell surface receptor interactions with agents such as phorbol myristate acetate (PMA), the hexose monophosphate shunt is activated, increasing the availability of NADPH. Typically, this is accompanied by an increase in O_2 uptake by the cells. The substrates O_2 and NADPH are converted to O_2^- by NADPH oxidase. If the level of reduced pyridine nucleotides is lowered by exposure of the cells to pentachlorophenol or chlorothalonil, O_2^- generation by oyster (*Crassostrea virginica*) hemocytes is much reduced (Baier-Anderson and Anderson 1997, and unpublished). The presence of bivalve NADPH oxidase has been shown by the use of specific inhibitors of its activity (Adema, van Deutekom-Mulder, van der Knapp and Smina, 1993; Noël, Bachère and Mialhe, 1993; Bramble and Anderson, 1999). The main difference from vertebrate NADPH oxidase seen with stimulated oyster hemocytes is the apparent lack of a typical respiratory burst, in that a large increment in O_2 uptake has not been recorded (Anderson, R.S., unpublished; Friedl, F. E.; Paynter, K.T., personal commun.). However, a burst of O_2^- production can be measured by chemiluminogenic probes, nitroblue tetrazolium reduction, ferricytochrome c reduction, fluorescent probes, and electron spin resonance spin trapping. The conversion of O_2^- to H_2O_2 in these cells has been quantified by chemiluminogenic probes, phenol red oxidation, homovanillic acid, and diaminobenzidine oxidation. A major pathway for H_2O_2 metabolism in the hemocytes appears to be the MPO-halide-H_2O_2 antimicrobial system. This can be shown by the presence of a strong luminol-augmented chemiluminescence (LuCL) response to stimuli by oyster and mussel hemocytes; LuCL is known to measure MPO-dependent ROS generation (Albrecht and Jungi, 1993). In addition, LuCL production in molluscan species is very sensitive to MPO inhibitors such as sodium azide (Noël, Bachère and Mialhe, 1993; Austin and Paynter, 1995) and salicyclhydroxamine acid (Anderson, unpublished). In mammalian leukocytes NADPH oxidase is inactive, unless specifically activated during the respiratory burst (Babior, 1984); however, a constant low-level luminol-CL production is typical of bivalve hemocytes, suggesting another minor difference from comparable cells of higher animals. Nevertheless, it seems reasonable to conclude that the ROS system of bivalve hemocytes closely resembles that seen in monocyte/macrophages and neutrophils.

MODULATION OF ROS PRODUCTION BY CHEMICAL AGENTS

If ROS of molluscan hemocytes, either directly or in concert with other cidal agents, contribute to protection against infectious disease, their modulation by environmental chemicals could influence the health and survival of exposed organisms. In fact, the available literature suggests that numerous aquatic contaminants, particularly when administered at relatively high (but sublethal) concentrations, will produce this kind of immunotoxicity. From this point on, the words "immune" and "defense" will be used interchangeably; in this paper, "immunity" refers to the phagocytic cells and nonimmunoglobulin serum molecules of bivalves analogous to comparable entities in the innate or nonspecific (nonadaptive) immune system of vertebrates.

In a pioneering study, Larson, Roberson, Hetrick (1989) reported that *C. virginica* hemocytes exposed to selected metals, pesticides, or other organic chemicals generally showed reduced LuCL. Similar results were obtained if the hemocytes were withdrawn from oysters previously exposed to these toxicants. Hormesis-like effects were produced by exposure to Cd, Al, Zn, dieldrin and naphthalene; that is, low doses of these chemicals sometimes made the cells more responsive to LuCL stimuli.

Upon exposure of *C. virginica* cells to noncytotoxic Cd levels (2-100 ppm) a dose-dependent suppression of zymosan-stimulated LuCL activity was seen, but basal LuCL was not affected (Anderson, Oliver and Jacobs, 1992). Increased xenobiotic sensitivity of stimulated LuCL vs basal LuCL appears to be a general characteristic of the immunotoxic response of bivalve hemocytes. It is possible that priming or activation of the cells by interaction with microbes produces the elevated ROS responses associated with protection from infection; therefore, reduction of that capacity might seriously impact host defense. In an attempt to reproduce in vivo the LuCL-suppressive effects seen after in vitro exposure, we exposed oysters to the highest sublethal Cd level known not to interfere with shell growth or condition index. However, no significant LuCL modulation was observed, possibly due to differences in in vitro/in vivo Cd bioavailability by the hemocytes, or by the protective effects of metal-binding proteins in vivo. The minimal intrahemocytic Cd level associated with LuCL inhibition was about 15μg/mg protein, based on the in vitro studies, but Cd levels in the hemocytes from in vivo exposed oysters were only about 2μg/mg protein. Mussels exposed to up to 400μg Cd/l for 1 wk showed significant Cd uptake in digestive gland tissues and cellular changes such as modulation of neutral red uptake and release of degradative enzymes; however, ROS release by hemocytes was not significantly altered (Coles, Farley and Pipe, 1995). Other studies in this laboratory report that bivalve hemocytes show reduced ROS production when exposed to numerous compounds such as copper (Anderson, Mora and Thomson, 1994b), brass (Anderson, Mora and Thomson, 1994a), pentachloropenol (Roszell and Anderson, 1992; Baier-Anderson and Anderson, 1996), and tributyltin (TBT) (Anderson, Brubacher, Ragone Calvo, Burreson and Unger, 1997). In those cases where examined, the suppression of LuCL could be at least partially reversed by washing the toxicant-exposed cells and returning them to toxicant-free media. Studies of ROS responses of bivalves collected from contaminated and reference field sites have yielded mixed results. Mussels (*M. galloprovincialis*) living in the Venice Lagoon are exposed to a complicated mixture of contaminants including energyrelated PAHs, lindane, DDT, PCBs, and a suite of heavy metals. Contaminant profiles vary from site to site, as well as with the season. ROS production by hemocytes from the mussels also showed considerable variability, but there was a general correlation between high levels of O production and low levels of organic contamination (Pipe, Coles, Thomas, Fossato and Pulsford, 1995). A major oil spill in Wales caused mass mortalities of *M. edulis*; surviving mussels showed significant modulations of hemocyte immune parameters, including decreased O_2- generation (Dyrynda, Law, Dyrynda, Kelly, Pipe, Graham and Ratcliffe, 1997). When the PAH levels in the mussels were reduced by depuration, there was a concomitant resolution of the impaired O_2-generating capacity. Mussels were also collected from other sites, where they were chronically exposed to mixtures of metals (Cu, Cd, Zn, Pb, and organotin) and other contaminants. Three contaminated sites were each paired with a nearby relatively clean reference site. In all three locations, extracellular O generation (by cytochrome c reduction) was elevated at the contaminated site, and intracellular O levels (by NBT reduction) were unchanged at two of the contaminated sites and elevated at the other one (Dyrynda, Pipe, Burt and Ratcliffe, 1998).

Clearly, the general utility of hemocyte ROS responses as biomarkers of effect for bivalves collected in the field has yet to be established. However, ROS suppression in hemocytes exposed to chemicals in vitro, may prove useful as a method to screen for potential immunomodulators, as has been well established in the case of mammalian macrophages (Tam and Hinsdill, 1990). In the few studies of this kind so far attempted, agent like metals, PAHs, and TBT that inhibit LuCL responses of *C. virginica* hemocytes after in vitro exposure, also enhance the severity of a major oyster disease (caused by *P. marinus*) when administered in vivo. These studies will be reviewed in the next section, as well as the ROS responses of the host elicited by this and other pathogens.

ENHANCEMENT OF INFECTIOUS DISEASE BY CHEMICAL AND PHYSICAL STRESSORS

An early example of this phenomenon was an unexpected byproduct of an attempt to produce neoplasia in *C. virginica* by exposure to the carcinogen nnitrosodiethylamine (Winstead and Couch, 1988). Although the study did not result in neoplastic lesions, an enhancement of already established *P. marinus* infections was noted, as well as diminished hemocyte infiltration in the area of the infections. The reduced cellular response to the foci of *P. marinus* infection was interpreted as a manifestation of immunotoxicity. Chu and Hale (1994) studied the effect of exposure of *C. virginica* to water-soluble fractions of PAHs and metals contaminated river sediments. They found that pollutant exposure enhanced preexisting *P. marinus* infections and increased susceptibility to experimental infections in a dose-dependent manner. Naphthalene, a major component of watersoluble fractions of PAH-rich sediments inhibited oyster hemocyte ROS responses (Larson, Roberson and Hetrick, 1989), as did certain exposures to sediments amended with PAHs common to polluted areas of the Chesapeake Bay (Chu, Volety, Lingenfelser and Hale, 1996). The antifouling agent TBT, when administered in vitro to *C. virginica* hemocytes, will also inhibit ROS responses (Fisher, Wishkovsky and Chu, 1990; Anderson, Brubacher, Ragone Calvo, Burreson and Unger, 1997). These authors also reported that experimental exposure of oysters to TBT could also promote the severity of *P. marinus* infections. Fisher, Oliver, Sutton, Manning, and Walker (1995) pre-exposed *C. virginica* to TBT for 8 wk prior to *P. marinus* challenge. Greater mortality, higher prevalence, and greater intensity of infection were seen in the TBT-treated oysters. Anderson, Unger, and Burreson (1996) exposed oysters to TBT for 30d prior to experimental *P. marinus* infection and continued the TBT exposure for the course of the study, with repeated sampling for 21 wk; their observations that TBT exacerbated the progression and lethality of this parasitic disease were essentially similar to those of Fisher, Oliver, Sutton, Manning, and Walker (1995). Subsequently, TBT was shown to enhance natural *P. marinus* infections in field-collected oysters and that hypoxia acted synergistically with TBT in this regard (Anderson, Brubacher, Ragone-Calvo, Unger and Burreson, 1998). These data suggest that the negative effects of multiple stressors on disease resistance may be more than additive.

An interesting attempt to measure modulation of molluscan defenses against an opportunistic bacterium was carried out by Pipe and Coles (1995). Hemocytes withdrawn from *M. edulis* exposed to *Vibrio tubiashi* for 1wk showed increased production of intracellular O_2^-, by NBT reduction. However, if the mussels were exposed to Cu for 7wk prior to the *V. tubiashi* challenge, the hemocytes showed reduced O_2^- production. This implied that Cu-exposure made the mussels unable to mount a typical antibacterial defense response, a conclusion supported by the observation that mortalities were consistently higher in *V. tubiashi*-exposed mussels that had been pre-exposed to Cu.

ROS RESPONSES ELICITED BY PATHOGENS

Immunomodulation caused by interactions of bivalve hemocytes with various pathogens has most commonly been measured by changes in LuCL. *Bonamia ostreae*, an intrahemocytic parasite of *Ostrea edulis,* was readily phagocytized by cells of either its typical host or another oyster, *C. gigas*. Hemocytes of both bivalves also took up zymosan particles, which triggered a strong LuCL response. However, ingestion of *B. ostreae* elicited no LuCL by the hemocytes of either species (Hervio, Bachère, Mialhe and Grizel, 1989). The lack of ROS production associated with phagocytosis of this parasite may partially explain how it avoids intracellular destruction, but does not provide an understanding of host specificity. In a similar fashion, a rickettsia-like organism, found in gill endothelial cells of scallop (*Pecten maximus*) populations suffering mass mortalities, was phagocytized by hemocytes of the host without producing LuCL (LeGall, Bachère, Mialhe and Grizel, 1989; LeGall, Bachère and Mailhe, 1991). The scallop hemocytes were quite capable of generating LuCL upon phagocytosis of zymosan or formalin-killed rickettsiae.

C. virginica hemocytes can contain large numbers of *P. marinus* organisms during heavy infections. In vitro studies have shown that *C. virginica* hemocytes failed to produce LuCL during phagocytosis of *P. marinus*, although showing good LuCL responses to zymosan or other stimuli (LaPeyre, Chu and Vogelbein, 1992; Volety and Chu, 1995). There is some electron microscopical evidence for limited intracellular destruction of *P. marinus* (LaPeyre, 1993; Bushek, Allen, Alcox, Gustafson and Ford, 1994), but mechanisms underlying such activity are currently unknown. In an attempt to identify changes in hemocyte-mediated immune mechanisms associated with *P. marinus* infection, we quantified several parameters, including zymosan-stimulated LuCL, in oysters with different levels of infection (Anderson, Paynter and Burreson, 1992). The most consistent findings were an increased hemocyte recruitment into the circulation and an increased LuCL responsiveness of hemocytes as the intensity of infection increased. The zymosan-stimulated LuCL responses of hemocytes withdrawn from oysters with moderate or heavy infections were higher than those of cells withdrawn from uninfected oysters (Anderson, Burreson and Paynter, 1995). Since the number of phagocytized *P. marinus* cells in the hemocytes was a function of the stage of the disease, it was suggested that the increased LuCL was a result of the priming or activation of the hemocytes by the presence of foreign particulates, *i.e.* the parasites. Although it was known that phagocytosis of *P. marinus* by cells from uninfected oysters did not produce ROS, it was possible that continued uptake of the parasites by already activated/primed hemocytes might result in a significant release of ROS. If so, this burst of ROS might either serve some protective function or might be a source of oxidative stress, partially underlying the pathogenesis of the disease. However, it was learned that phagocytosis of additional *P. marinus* was an ineffective ROS stimulus, regardless of the stage of *P. marinus* infection of the oyster from experimental hemocytes were drawn (Anderson, in press). Recently, we investigated the possibility that *P. marinus* actively produces factors that act as free radical scavengers or as inhibitors of ROS generation by *C. virginica* hemocytes (Anderson, 1999). When living *P. marinus* was added to *C. virginica* cells, a small (but statistically significant) reduction in LuCL was seen. However, the addition of osmotically killed, structurally intact, *P. marinus* triggered a large LuCL response from the hemocytes. When both zymosan and *P. marinus* were simultaneously incubated with oyster hemocytes, both particles were usually found in the same phagolysosomes. When a range of numbers of dead *P. marinus* plus a standardized dose of zymosan were incubated with aliquots from a common hemocyte pool, the LuCL activity increased with increasing numbers of parasites. This indicated that non-viable *P. marinus* elicited a LuCL response comparable to other foreign particles, such as zymosan. The zymosan produced a typical LuCL response and if dead *P. marinus* were included with the zymosan, the effects on LuCL were additive. However, if different amounts of living *P. marinus* plus the standard zymosan dose were phagocytized the resultant LuCL showed a *P. marinus* dose-dependent suppression. This suggested that living *P. marinus* not only fails to generate LuCL upon phagocytosis, but also can inhibit LuCL produced intracellularly in response to stimulatory agents such as zymosan. The identity of the agent(s) produced by *P. marinus* responsible for this apparent reduction in ROS generation remains unknown, but it is interesting to note that one of the functions of excretory/secretory proteins produced by the parasite *Schistosoma mansoni* is to suppress O_2- production by the hemocytes of its molluscan host, *Biomphalaria glabrata* (Connors and Yoshino, 1990; Connors, Lodes and Yoshino, 1991).

CONCLUSIONS

It appears that an ROS-generating system, akin to that typical of blood phagocytes of higher animals, can be demonstrated in the hemocytes of many invertebrates, including bivalve mollusks. Although some exceptions have been documented, it seems that this system involves an NADPH oxidase-like enzyme that produces O_2- and H_2O_2, and has the typical components of the MPO-halide-H_2O_2 antimicrobial system of mammalian neutrophils and monocyte/macrophages. Classical stimuli will activate this ROS pathway; however, the magnitude of the ROS response of bivalve hemocytes seems to be considerably lower than that of vertebrate (fish) macrophages stimulated under similar conditions (Bramble and Anderson, 1998). It is possible that hemocytic ROS act in concert with lysosomal enzymes or RNS to effect resistance to agents of infectious disease. A number of environmental contaminants have been shown to modulate ROS production by hemocytes; suppression of ROS generation in mammals has been shown to exacerbate the progression and lethality of microbial infections. Based on the very limited evidence available, it seems that exposure of *C. virginica* to chemicals shown to reduce ROS generation, when administered to the hemocytes in vitro, can augment the severity and lethality of *P. marinus* infections in vivo. However, attempts to quantify consistently altered immune functions, including significant changes in ROS production, in hemocytes withdrawn from xenobiotic-treated oysters with demonstrably impaired resistance to *P. marinus* have been generally unsuccessful (Oliver, Sutton and Fisher, 1995; Anderson, Unger and Burreson, 1996). There are reports of ROS modulation in bivalves collected from contaminated aquatic sites, but no direct correlations of these to measurements of disease susceptibility have been made. It appears that the intracellular presence of *P. marinus* primes or activates oyster cells, because hemocytes from infected animals show an elevated ROS response to zymosan. However, this phenomenon does not extend to *P. marinus* itself, which under no circumstances elicits detectable ROS when phagocytized. In fact, it is likely that *P. marinus* and other bivalve pathogens produce excretory/secretory products that suppress ROS production by hemocytes and/or serve as effective free radical scavengers.

REFERENCES

Adema C.M., van der Knaap W.P.W., Sminia T. Molluscan hemocyte-mediated cytotoxicity: the role of reactive oxygen intermediates. Rev. Aquat. Sci. 1991;4:201-223.

Adema C.M., van Deutekom-Mulder E.C., van der Knapp W.P.W., Smina T. NADPH-oxidase activity: the probable source of reactive oxygen intermediate generation in hemocytes of the gastropod *Lymnea stagnalis*. J. Leukoc. Biol. 1993;54:379-383.

Albrecht D., Jungi T.W. Luminol-enhanced chemiluminescence induced in peripheral blood-derived human phagocytes: obligatory requirement of myeloperoxidase exocytosis by monocytes. J. Leukoc. Biol. 1993;54:300-306.

Anderson R.S. "Modulation of blood cell mediated oxyradical production in aquatic species: implications and applications." In *Aquatic Toxicology: Molecular, Biochemical, and Cellular Perspectives,* D.C. Malins and G.K. Ostrander, eds. Boca Raton: Lewis Publishers, 1994.

Anderson R.S. "Production of reactive oxygen intermediates by invertebrate hemocytes: immunological significance." In *New Directions in Invertebrate Immunology,* K. Söderhäll, G. Vasta and S. Iwanaga, eds. Fair Haven, NJ: SOS Publications, 1996.

Anderson R.S. *Perkinsus marinus* secretory products modulate superoxide anion production by oyster (*Crassostrea virginica*) hemocytes. Fish Shellfish Immunol. 1999;9:51-60.

Anderson R.S. Lack of hemocyte chemiluminescence stimulation by *Perkinsus marinus* in oysters *Crassostrea virginica* with dermo disease. J. Aquat. Animal Health. In press.

Anderson R.S., Brubacher L.L., Ragone Calvo L.M., Burreson E.M., Unger M.A. Effect of in vitro exposure to tributyltin on generation of oxygen metabolites by oyster hemocytes. Environ. Res. 1997;74:84-90.

Anderson R.S., Brubacher L.L., Ragone-Calvo L., Unger M.A., Burreson E.M. Effects of tributyltin and hypoxia on the progression of *Perkinsus marinus* infections and host defence mechanisms in *Crassostrea virginica* (Gmelin). J. Fish Dis. 1998;21:371-379.

Anderson R.S., Burreson E.M., Paynter K.T. Defense responses of hemocytes withdrawn from *Crassostrea virginica* infected with *Perkinsus marinus*. J. Invertebr. Pathol. 1995;66:82-89.

Anderson R.S., Mora L.M., Thomson S.A. Exposure of oyster macrophages to particulate brass suppresses luminol-augmented chemiluminescence. Toxicologist 1994a;12:391.

Anderson R.S., Mora L.M., Thomson S.A. Modulation of oyster (*Crassostrea virginica*) hemocyte immune function by copper, as measured by luminol-enhanced chemiluminescence. Comp. Biochem. Physiol. 1994b;108C:215-220.

Anderson R.S., Oliver L.M., Jacobs D. Immunotoxicity of cadmium for the eastern oyster (*Crassostrea virginica* [Gmelin, 1791]): effects on hemocyte chemiluminescence. J. Shellfish Res. 1992;11:31-35.

Anderson R.S., Paynter K.T., Burreson E.M. Increased reactive oxygen intermediate production by hemocytes withdrawn from *Crassostrea virginica* infected with *Perkinsus marinus*. Biol. Bull. 1992;183:476-481.

Anderson R.S., Unger M.A., Burreson E.M. Enhancement of *Perkinsus marinus* disease progression in TBT-exposed oysters (*Crassostrea virginica*). Mar. Environ. Res. 1996;42:177-180.

Austin K.A., Paynter K.T. Characterization of the chemiluminescence measured in hemocytes of the Eastern oyster, *Crassostrea virginica*. J. Exp. Zool. 1995;273:461-471.

Babior B.M. Oxidants from phagocytes: agents of defense and destruction. Blood 1984;64:959-966.

Baier-Anderson C., Anderson R.S. The effect of pentachlorophenol on NADPH production in oyster hemocytes: immunomodulatory consequences. J. Shellfish Res. 1996;15:498-499.

Baier-Anderson C., Anderson R.S. The effect of pentachlorophenol on pyridine nucleotide production in oyster hemocytes: NADPH and immunomodulation. J. Shellfish Res. 1997;16:111-114.

Bramble L.H., Anderson R.S. A comparison of the chemiluminescent response of *Crassostrea virginica* and *Morone saxatilis* phagocytes to zymosan and viable *Listonella anguillarum*. Dev. Comp. Immunol. 1998;22:55-61.

Bramble L.H., Anderson R.S. Lack of involvement of reactive oxygen species in the bactericidal activity of *Crassostrea virginica* haemocytes in contrast to *Morone saxatilis* phagocytes. Fish Shellfish Immunol. 1999;9:109-123.

Bridges R.A., Berendes H., Good R.A. A fatal granulomatous disease of childhood. Amer. J. Dis. Child. 1959;97:387.

Bushek D., Allen S.K., Alcox K.A., Gustafson R., Ford S. Dose response of the eastern oyster, *Crassostrea virginica*, to cultured cells of *Perkinsus marinus*, the agent of Dermo disease. J. Shellfish Res. 1994;13:313.

Chu F.-L.E., Hale R.C. Relationship between pollution and susceptibility to infectious disease in the Eastern oyster, *Crassostrea virginica*. Mar. Environ. Res. 1994;38:243-256.

Chu F.-L.E., Volety A.K., Lingenfelser J.T., Hale R.C. Modulation of hemocyte activities in oysters (*Crassostrea virginica*) upon exposure to PAHs. SETAC Abstract Book, 17th Annual Meeting, SETAC Press, Pensacola, FL. p 150.

Coles J.A., Farley S.R., Pipe R.K. Alterations of the immune response of the common marine mussel *Mytilus edulis* resulting from exposure to Cadmium. Dis. Aquat. Org. 1995;22:59-65.

Connors V.A., Lodes M.J., Yoshino T.P. Identification of a *Schistosoma mansoni* sporocyst execretory-secretory antioxidant molecule and its effect on superoxide production by *Biomphalaria glabrata* hemocytes. J. Invertebr. Pathol. 1991;58:387-395.

Connors V.A., Yoshino T.P. In vitro effect of larval *Schistosoma mansoni* execretory-secretory products on phagocytosis-stimulated superoxide production in hemocytes from *Biomphalaria glabrata*. J. Parasitol. 1990;76:895-902.

Dyrynda E.A., Law R.J., Dyrynda P.E.J., Kelly C.A., Pipe R.K., Graham K.L., Ratcliffe N.A. Modulations in cell-mediated immunity of *Mytilus edulis* following the 'Sea Empress' oil spill. J. Mar. Biol. Assoc. UK 1997;77:281-284.

Dyrynda E.A., Pipe R.K., Burt G.R., Ratcliffe N.A. Modulations in the immune defenses of mussels (*Mytilus edulis*) from contaminated sites in the UK. Aquat. Toxicol. 1998;42:169-185.

Fisher W.S., Oliver L.M., Sutton E.B., Manning C.S., Walker W.W. Exposure of eastern oysters to tributyltin increases the severity of *Perkinsus marinus* disease. J. Shellfish Res. 1995;14:265-266.

Fisher W.S., Wishkovsky A., Chu F.-L.E. Effects of tributyltin on defense-related activities of oyster hemocytes. Arch. Environ. Contam. Toxicol. 1990;19:354-360.

Hervio D., Bachère E., Mialhe E., Grizel H. Chemiluminescent responses of *Ostrea edulis* and *Crassostrea gigas* to *Bonamia ostrea* (Ascetospora). Dev. Comp. Immunol. 1989;13:449.

LaPeyre J.F. (1993): Studies on the oyster pathogen *Perkinsus marinus* (Apicomplexa): interactions with host defenses of *C. virginica* and *C. gigas*, and in vitro propagation. Ph.D. Dissertation, The College of William and Mary, Gloucester Point, VA.

LaPeyre J.F., Chu F.-L.E., Vogelbein W.K. In vitro interaction of *Perkinsus marinus* with hemocytes from eastern and Pacific oysters, *Crassostrea virginica* and *Crassostrea gigas*. J. Shellfish Res. 1992;11:200.

Larson K.G., Roberson B.S., Hetrick F.M. Effect of environmental pollutants on the chemiluminescence of hemocytes from the American oyster *Crassostrea virginica*. Dis. Aquat. Org. 1989;6:131-136.

LeGall G., Bachère E., Mailhe E. Chemiluminescence analysis of the activity of *Pecten maximus* hemocytes stimulated with zymosan and host-specific Rickettsiales-like organisms. Dis. Aquat. Org. 1991;11:181-186.

LeGall G., Bachère E., Mialhe E., Grizel H. Zymosan and specific-rickettsia activation of oxygen free radicals production in *Pecten maximus* hemocytes. Dev. Comp. Immunol. 1989;13:448.

Noël D., Bachère E., Mialhe E. Phogocytosis associated chemiluminescence of hemocytes in *Mytilus edulis* (Bivalvia). Dev. Comp. Immunol. 1993;17:483-493.

Oliver L.M., Sutton E.B., Fisher W.S. Effects of tributyltin exposure on oyster (*Crassostrea virginica*) defense functions. J. Shellfish Res. 1995;14:274.

Pipe R.K., Coles J.A. Environmental contaminants influencing immune function in marine bivalve molluscs. Fish Shellfish Immunol. 1995;5:581-595.

Pipe R.K., Coles J.A., Thomas M.E., Fossato V.U., Pulsford A.L. Evidence for environmentally derived immunomodulation in mussels from the Venice Lagoon. Aquat. Toxicol. 1995;32:59-73.

Roszell L.E., Anderson R.S. Effects of pentachlorophenol on the chemiluminescent response of phagocytes from two estuarine species. Toxicologist 1992;12:392.

Tam P.E., Hinsdill R.D. Screening for immunomodulators: effects of xenobiotics on macrophage chemiluminescence in vitro. Fund. Appl. Toxicol. 1990;14:542-553.

Volety A.K., Chu F.-L.E. Suppression of chemiluminescence of eastern oyster (*Crassostrea virginica*) hemocytes by the protozoan parasite *Perkinsus marinus*. Dev. Comp. Immunol. 1995;19:135-142.

Winstead J.T., Couch J.A. Enhancement of protozoan pathogen *Perkinsus marinus* infections in American oysters *Crassostrea virginica* exposed to the chemical carcinogen n-nitrosodiethylamine (DENA). Dis. Aquat. Org. 1988;5:205-213.

A ROLE FOR PROTEASE INHIBITORS IN IMMUNITY OF LONG-LIVED ANIMALS

Peter B. Armstrong[1] and James P. Quigley[2]

[1]Department of Molecular and Cellular Biology, University of California, 1 Shields Ave, Davis, CA 95616 and Marine Biological Laboratory, 7 MBL Street, Woods Hole, MA 02543
[2]Department of Vascular Biology, Scripps Research Institute, 10550 Torrey Pines Rd, La Jolla, CA 92037 and Marine Biological Laboratory, 7 MBL Street, Woods Hole, MA 02543

INTRODUCTION

To reproduce successfully, all organisms from bacteria to man must survive the attacks of the myriad of parasites that would otherwise produce fatal episodes of disease and premature death. This is particularly true for long-lived species that require many years to reach reproductive maturity and, presumably, that are required to resist multiple episodes of parasitic infection prior to that happy day. Parasites are species that draw their food resources from the live bodies of a host species and that spend much or all of their life in association with that host species. Typically damage is done to the host by the parasite, resulting in a condition of pathogenesis and disease. Parasites may be unicellular or multicellular, prokaryote, eukaryote, or virus.

Phylogenetic Perspectives on the Vertebrate Immune System.
Edited by Beck *et al.*, Kluwer Academic/Plenum Publishers, 2001.

Parasites are faced with several tasks that must be accomplished to ensure their survival and reproduction. They must invade across the host integuments and then disseminate into the internal spaces of the host, procure the nutrients for growth and reproduction, escape from host immune defenses, and then disseminate their progeny to new, previously un-colonized hosts. Virulence factors are the unique molecular attributes of the pathogen used to facilitate the colonization and infection of the host and that contribute to the production of disease (1,2). These include the agents responsible for adhesion of the pathogen to and invasion of the integuments, invasins that promote uptake into cells of the host, toxins and cytolysins, and mediators that reduce the effectiveness of host immune defenses such as protective capsuls and soluble inactivators of host immune effectors (3). In the following, we will argue that secreted and surface-associated proteases play an important role in the survival and virulence of a variety of prokaryote and eukaryote parasites and that the proteins of the host that inhibit these proteases serve an important function in defense and immunity. Additionally, we will argue that the protease inhibitor, α_2-macroglobulin, is uniquely important in the clearance and inactivation of the proteases of invading pathogens and that the α_2-macroglobulin-based protease clearance pathway should be counted as one of a small number of evolutionarily conserved immune defense systems of higher animals.

Table 1. Proteases are virulence factors for pathogenic parasites

Role in pathogenesis	Pathogen	Protease	References
1. Invasion and dissemination in the host: Direct tissue destruction, fibrinolysis			
Corneal keratinitis	*Pseudomonas, Serratiais*	protease IV	(4,5)
Epidermal invasion	*S. stercoralis, S. mansoni*	Ss40	(6,7)
Peridontal disease	*P. gingivalis*	cysteine protease	(8)
Cytolysis (internal attack: proteases internalized as α_2M-protease complex)	*Serratia, Pseudomonas*	Serratia 56 K; Pseudomonas alkaline protease	(9,10)
Cytolysis (extracellular attack)	*E. histolytica*	cysteine proteases	(11,12)
Intracellular invasion	*P. chabaudi*	p68	(13)
Fibrinolysis	*S. pyogenes, Y. pestis, B. burgdorferi*	pla, OspA	(14-16)

2. Housekeeping functions			
Protein digestion	*S. mansoni; P. falciparum*	cysteine protease	(17-19)
Digestion of micronutrient sequestering proteins	*P. gingivalis*		(20)
Excystation	*Giardia, Paragonimus*	cathepsin B	(21,22)
Molting	*O. volvulus, T. cruzi*	cysteine protease	(23,24)
Processing of protein virulence factors	*P. gingivalis*		(8,25)
3. Immune evasion and misappropriation of host immune effectors			
a. Proteolytic destruction of defense mediators			
Depletion of protease inhibitors	*Pseudomonas, Serratia, Candida, C. salmositica*	elastase, Serratia 56 kDa protease	(26-28)
Inactivation of complement	*Streptococcus, Serratia, Psedomonas, Candida, E. histolytica*	Streptococcal C5a protease, Candida acid protease, Entamoeba cysteine protease	(29-31)
Cleavage of kininogen	*P. gingivalis*	Lys-gingivain	(32)
Cleavage of immunoglobulins	*Streptococci, Serratia, Candida, P. gingivalis*	IgA1 protease, Serratia 56 K, Candida protease,	(33,34)
b. Activation of host cytotoxic mediators			
Bradykinin cascade	*Serratia, Pseudomonas, Candida, Vibrio*	metalloproteases, aspartyl proteases, cysteine proteases	(35)
c. Activation of host protease zymogens			
Plasmin-catalyzed plasmin generation	*S. pyogenes, S. aureus*	Streptokinase-activated autoproteolytic conversion	(36)
Plasminogen activator-catalyzed plasmin generation	*E. coli, Y. pestis*	microbial plasminogen activators	(15,37)

(Continued)

Activation of host MMPs	*P. gingivalis*	40 kDa trypsin-like protease	(38,39)
4. Protease-dependent microbial toxin systems			
Proteolytic activation of microbial toxins	*B. anthracis, P. aeruginosa*	Host cell furin/PACE4	(40-42)
Microbial toxins that are proteases	*Botulinus, Tetanus*	Botulinus toxin, Tetanus toxin	(43,44)

PROTEASES AS VIRULENCE FACTORS

Secreted and cell surface proteases of the parasite play a variety of important roles in the discharge of the several key tasks faced by parasites and thus are candidates for service as virulence and survival factors (Table 1). Tissue barriers to invasion, such as the keratinocyte layers of the epidermis, the basal lamina, interstitial extracellular matrices, and the blood vessel walls have been shown to be susceptible to attack by parasite proteases (6,7). The blood clotting system plays an important role in the immobilization of invading parasites to prevent their systemic dispersal from the site of initial invasion in animals as diverse as man (45,46) and the horseshoe crab (47). The endogenous system of mammals for the proteolytic destruction of blood clots is well characterized (48). Pathogen-mediated fibrinolysis may involve proteases secreted by entrapped pathogens that have an inherent fibrinolytic action, in which case they destroy the clot directly, or they may be proteases or other proteins that activate the host's own plasminogen activator/plasminogen system. Many bacteria capture host plasmin at the cell surface. In this latter case, host-derived plasmin is the enzyme directly responsible for clot lysis. Alternatively, streptokinase and staphylokinase promote plasminogen autoproteolytic activation (36). So one potential function for proteases is invasion and dissemination.

Of major interest for the present study are the several ways that proteases secreted by pathogenic parasites function to inactivate or cause the inappropriate expression of host immune defense systems. As mentioned above, the blood clot serves as a device for immobilizing invading parasites at the site of initial penetration of the integument and fibrinolysis serves to free pathogens from this immobilization. In vertebrates, antibodies and the complement system are important mediators of immunity by contributing to outright lysis of foreign cells and by opsonizing foreign cells for subsequent phagocytosis by monocytes. As will be described below, pathogenic parasites present a variety of proteases that inactivate the complement and immunoglobuoin-mediated defense pathways (30,34).

In addition to a direct role as virulence factors, secreted proteases function in various housekeeping roles during parasitic infection. The proteolytic degradation of host proteins may serve important nutritive functions for the parasite, and in some organisms, such as *Porphyromonas gingivalis*, the anaerobic Gram-negative bacterium important in human peridontal disease, protein is the principal energy source. In some species, such as *Plasmodium*, nutritional proteolysis may occur in intracellular compartments

(18,49), but in *P. gingivalis*, this appears to be the product of extracellular proteases (8). Additionally, many animals attempt to limit the ability of parasites to thrive in the internal milieu by limiting the supply of essential micronutrients. For example, ferritin and the transferrin proteins bind iron and avidin binds biotin, limiting the availability of these nutrients for use by parasites (50). Proteolytic attack on these nutrient-binding proteins to free the bound nutrient could serve an important nutritional role for the parasite by liberating the sequestered micronutrients (20,51). In this situation, the protease could be considered as an agent of attack on a host defense function, and thus as a virulence factor rather than acting merely in the service of housekeeping functions. Similarly, extracellular proteases may play important roles in the expression of other types of virulence factors. In *P. gingivalis*, the expression of fimbriae is defective in protease-deficient mutant strains (8), possibly due to defective processing of the precursor of the FimA protein. The fimbriae mediate bacterial adherence to the oral epithelium and the tooth (51).

The characterization of proteases secreted by prokaryotic and eukaryotic parasites is represented by a voluminous literature (Table I). Some parasites produce a diverse array of secreted and surface-bound proteases. For example *P. gingivalis*, produces metalloproteases, aspartyl proteases, and serine proteases (8) and *Plasmodium falciparum*, one of the agents of malaria, produces serine, cysteine, and aspartyl proteases (52). While not all parasites have been shown to release the diverse array of proteases of *P. gingivalis* or *P. falciparum*, most that have been adequately studied appear to produce at least one or a few different proteases and, in sum, the ensemble of proteases secreted by all of the different bacterial (2,35,53) and eukaryote (52) parasites known to plague any individual species of higher animal is likely to include proteases of all different enzymatic classes.

Virulence has correlated positively with the expression of secreted proteases in a variety of systems where virulence can be modified genetically or in response to environmental cues (54). Similarly, expression of key proteases correlates with pathogenicity within genera of bacteria that contain pathogenic and non-pathogenic members. For example, mucosal pathogens in the genera *Neisseria, Haemophilus,* and *Streptococcus* express an extracellular protease that specifically cleaves human immunoglobulin A1 at Pro-Thr bonds in the heavy chain hinge region. Related species in these genera that are nonpathogenic lack the protease (33) and noninvasive isolates of *Neisseria meningitidis* express significantly lower levels of IgA1 protease than do invasive isolates (55). Immunoglobulin-A is the form of Ig that protects the mucosal surfaces against pathogenic attack and the mucosa is the specific attack site for these particular pathogens. The virulence of laboratory strains and clinical isolates of *Entamoeba histolytica* correlates well with the level of activity of a secreted cysteine protease (11,12). In some disease models, the regulated temporal progression of expression of different secreted proteases correlates with the development of progressively more profound tissue damage during disease progression (56). Saturation mutational screens for virulence factors have identified protease-encoding genes necessary for virulence in *Streptococcus pneumoniae* and *Salmonella* (57). Thus, in a variety of pathogens and disease models, and from a number of different experimental approaches, we find a strong correlation of virulence and protease expression.

A necessary role for proteases has been documented in a number of systems by showing a loss or reduction of virulence of pathogenic strains when the gene encoding the candidate protease virulence factor is inactivated by mutation or when the protease is inactivated by an appropriate inhibitor. For example, inactivation of the *pla* gene, which encodes a plasminogen activator of *Yersinia pestis*, the agent of plague, reduces the infectivity of bacteria injected subcutaneously by a factor of 10^6 (15). Strains of *Porphyromonas gingivalis* that are deficient in a 97 kDa C3-degrading protease bind complement more effectively and are more susceptible to phagocytosis by neutrophils than are the parent strain (30) and treatment of virulent strains of *P. gingivalis* with protease inhibitors similarly increased complement binding and susceptibility to phagocytosis (29). Strains of *Leishmania mexicana* generated by targeted gene disruption of the genes, *cpa* and *cpb*, which endode a pair of cysteine proteases, show attenuated virulence (58). Strains of *Pseudomonas aeruginosa* that are deficient in protease IV show reduced virulence and virulence of the protease-IVdeficient strain was augmented by cointroduction with purified protease IV (5). Inhibitors of serine proteases block entry of merozoites of *Plasmodium* into erythrocytes (13). Low molecular mass protease inhibitors directed against key proteases responsible for virulence have shown promise as therapeutic agents in a variety of conditions of parasitic disease (7,19,59).

In some cases, the toxins directly responsible for disease are proteases. For example, the neurotoxic actions of several of the clostridial neurotoxins (*e.g.*, tetanus toxin, botulinum toxins-B, -D, and -F) result from their proteolytic cleavage of synaptobrevin, a membrane protein of small synaptic vesicles (43). Cleavage prevents neurotransmitter release, resulting in paralysis of the host (60). These clostridial neurotoxins belong to a unique family of Zn proteases that specifically proteolyze synaptobrevin at the Gln-Phe, Gln-Lys, or Lys-Leu bonds (44). The Streptococcal pyrogenic exotoxin B (*SpeB*), a cysteine protease of *Streptococcus pyogenes,* is able to kill host leukocytes. *SpeB*-negative strains are nonpathogenic at doses that generate lethal dissemination of the wild-type bacterium (61).

As an alternative to producing their own proteases, a variety of pathogenic bacteria display surface receptors that allow them to capture host proteases, such as plasminogen, at their surface, which facilitate bacterial survival and invasion (16,36,62). The plasminogen receptor of *Borrelia burgdorferi,* the spirochete responsible for Lyme disease, has been identified as the surface protein OspA (16). In other cases, pathogenic bacteria stimulate the host tissues in the vicinity of the bacterial lesion to produce collagenases and other proteases that are the mediators of local tissue destruction (39). In sum, parasites have devised a diverse set of strategies to utilize proteases to contribute to their survival in the hostile environment of the internal milieu of an appropriate host organism. In an attempt to deny parasites the use of their proteases, their potential hosts have evolved an array of protein inhibitors of protease activity.

PROTEASE INHIBITORS ARE CONSTITUENTS OF THE INNATE IMMUNE SYSTEM

The immune system has evolved to thwart the nefarious designs of invading pathogens and is comprised of the array of tissues, cells, and effector molecules that operate to limit the severity of attack by pathogenic parasites that have gained access to the internal milieu. The primary defenses against pathogenic attack are the integuments, which include the skin and the mucosal linings of the digestive, respiratory, excretory, and reproductive systems. In many invertebrates, the skin and mucous membranes are supplemented by non-living surface layers, such as the cuticle of arthropods. These tissues limit the penetration of foreign organisms into the internal spaces of a potential host. But once inside, a second line of defense is mediated by the diverse array of cells and soluble effectors comprising the immune system. The immune system is the collection of cells and effector molecules that promote the inactivation or clearance of parasites and their toxic products. In coelomate animals, the bulk of the immune system is found in the blood, presumably because the blood has ready access to all parts of the body and is best prepared to concentrate defense effectors at a site of pathogenic invasion.

In addition to the proteases of invading pathogens, the endogenous proteases of the several zymogen cascades involved in blood clotting, clot resolution, and inflammation have the potential for undesirable destructive action should they survive to act beyond the limits of their intended target. With this in mind, it is not surprising that inhibitors of proteases have evolved as important elements in the system of host defenses against pathogens and as regulators of endogenous proteases. Inactivation of the secreted proteases of pathogens has the potential to restrict the steps of invasion, dispersal in the internal milieu, nutrition of those parasites that have successfully invaded, and parasite-mediated inactivation of other elements of host immunity. In mammals, ≈3-5% of the plasma proteins are protease inhibitors (63) and in our animal of choice, the American horseshoe crab, *Limulus polyphemus*, the plasma protease inhibitor, α_2-macroglobulin, is the third most abundant protein in the plasma (64).

THE α_2-MACROGLOBULINS ARE COMPONENTS OF THE INNATE IMMUNE SYSTEM

The protease inhibitors are of two basic functional classes, the active-site inhibitors, which bind to the active site of the target protease and inactivate its ability to hydrolyze all substrates, and the α_2-macroglobulins, which operate by the unique process of molecular entrapment. In this, the target protease is enclosed in a molecular cage of the α_2-macroglobulin polypeptide chain. Protease entrapment disables the enzyme's ability to proteolyze macromolecular substrates too large to fit into the α_2-macroglobulin cage, but leaves intact the ability of the entrapped enzyme to hydrolyze low molecular mass substrates small enough to enter the α_2-macroglobulin cage and interact with the active site of the protease. Protease entrapment is initiated by the proteolytic cleavage of a peptide bond in a specialized stretch of the α_2-macroglobulin polypeptide, which triggers the molecular compaction that entraps the protease molecule.

Most of the members of the α_2-macroglobulin family of protease inhibitors have a unique internal thiol ester bond that links the thiol of a cysteine residue with the γ carbonyl of a glutamyl residue three amino acids further along. The unique molecular trap mechanism of protease binding and the internal thiol ester form the basis for a suite of diagnostic features that allow the demonstration of the presence of protease inhibitors of the α_2-macroglobulin family in biological fluids (Fig. 1).

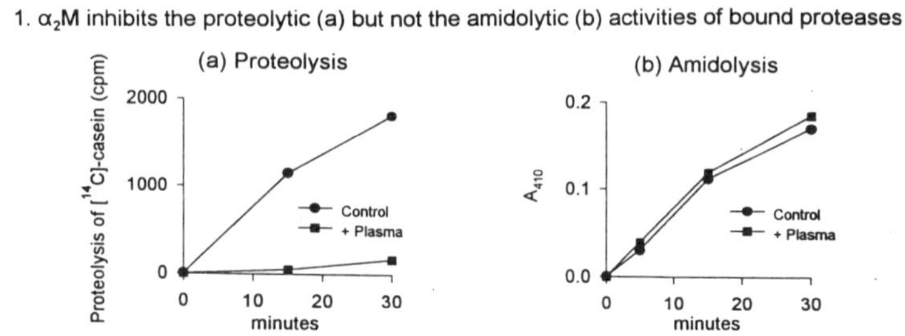

2. α_2M reacts with proteases of diverse catalytic activity: Hydrolysis of [^{14}C]-casein or [^{125}I]-fibrin

Protease	cpm - plasma	cpm + plasma
Trypsin	1957	194
Subtilisin	3187	270
Chymotrypsin	1384	132
Plasmin	15250	620
Elastase	22359	603
Thermolysin	2541	173
Papain	1747	227

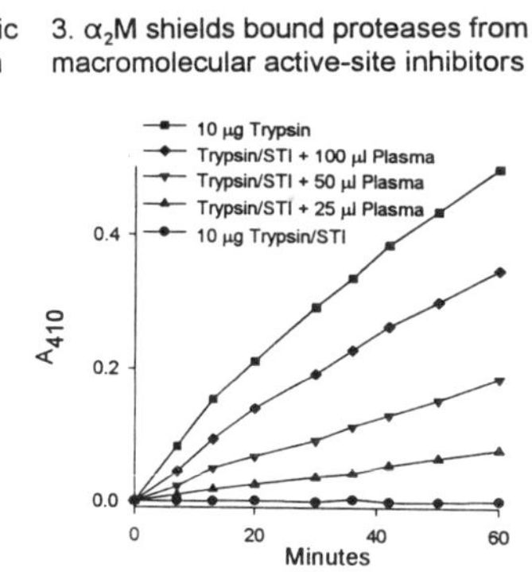

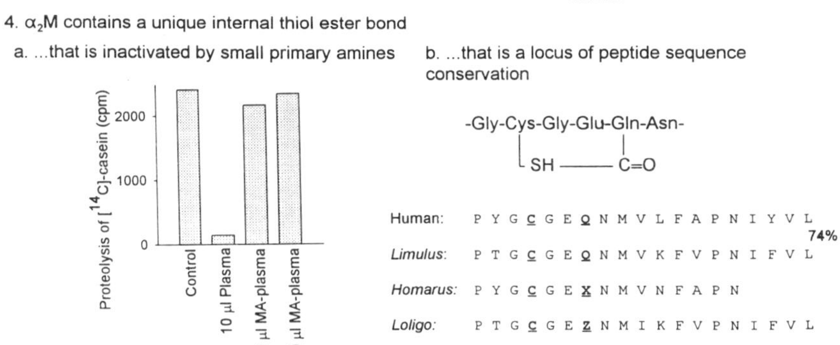

Figure 1. Identification of α_2-macroglobulin in the plasma of *Limulus*. A suite of unique properties can be employed to identify α_2-macroglobulin in a biological fluid, which, in this case is the plasma of the horseshoe crab, *Limulus*. Our first indication for α_2-macroglobulin in *Limulus* plasma was the demonstration that plasma could inhibit the ability of the protease trypsin to hydrolyze proteins, such as [^{14}C]-casein (top left), but without affecting the ability of trypsin to hydrolyze low molecular mass amide substrates, such as BA*p*NA (Nα-benzoyl-DL-arginine *p*-nitroanilide) (top right). An active-site trypsin inhibitor would be expected to eliminate both activities in parallel because it inactivates the enzymatic site of trypsin. By contrast, the active site of α_2-macroglobulin-bound trypsin is unaffected and is still capable of hydrolyzing low molecular mass substrates, such as BA*p*NA. Large molecular mass substrates, such as [^{14}C]-casein, which are unable to penetrate the α_2-macroglobulin cage, are protected from hydrolysis. We also observed that plasma (middle left) and our purified protein (not shown) could inhibit the proteolytic activity against a diverse array of proteases with very different catalytic mechanisms. Active-site protease inhibitors are typically fastidious in the spectrum of targeted proteases, whereas α_2-macroglobulin is notably

promiscuous. The unique protease-binding mechanism of α_2-macroglobulin both inhibits the ability of the bound protease to hydrolyze other proteins and protects the bound protease molecule from inactivation by macromolecular active-site protease inhibitors, such as soybean trypsin inhibitor (STI, molecular mass = 21.5 kDa). In this assay (middle right), trypsin was incubated with increasing amounts of plasma, then a 2-fold molar excess of STI was added to inactivate all unbound trypsin, and the quantity of bound trypsin was assayed for its hydrolysis of BA*p*NA. The addition of higher concentrations of plasma resulted in larger fractions of protected trypsin, as reflected by elevated rates of hydrolysis of BA*p*NA. The presence of a reactive internal thiol ester bond is diagnostic of α_2-macroglobulin (lower panel). Inactivation of the thiol ester with the low molecular mass primary amine methylamine (MA) inactivates the protease inhibitory activity of *Limulus* plasma (lower left). The small primary amines methylamine and ammonia are able to form amide bonds with the thiol esterified glutamine, thereby rupturing the thiol ester bond. Loss of protease inhibitory activity following exposure to methylamine is indicative of the thiol ester and, thus, of the presence of α_2-macroglobulin. The amino acid sequence of the thiol ester domain of the forms of α_2-macroglobulin from human and *Limulus* shows a 74% sequence identity (lower right).

The α_2-macroglobulins typically show a broad spectrum of reactivity with endopeptidases of diverse origin and diverse catalytic mechanism. This stands in marked contrast to the active site inhibitors, which are restricted in their activity to a limited subset of the world of proteases (63,65). The broad reactivity of α_2-macroglobulin is a product of the manner by which the α_2-macroglobulin trap is sprung on a reacting protease. As mentioned above, reaction of α_2-macroglobulin and target protease begins when the protease cleaves a peptide bond in a defined stretch of the α_2-macroglobulin molecule known as the bait region. In human and *Limulus* α_2-macroglobulins, the bait region comprises residues 666-706 and 719-771, respectively. Following proteolysis of the bait region, the α_2-macroglobulin chain experiences a rapid conformational change to enfold the hapless protease molecule. The sequence of amino acids in the bait region of the different members of the α_2-macroglobulin family show little conservation, but instead the bait regions typically include target peptide bonds that are susceptible to attack by most or all extant endopeptidases (66). The promiscuous character of the α_2-macroglobulins makes them ideal scavengers for the diverse array of proteases produced by the broad spectrum of parasites that might attack during the lifetime of any given animal.

The basic function of α_2-macroglobulin in protease clearance is to bind proteases of all diverse classes and convey them to a receptor-mediated endocytotic clearance pathway. Native, unreacted α_2-macroglobulin fails to bind its cell surface receptor, which recognizes and binds only the protease-reacted form. The reorganization of α_2-macroglobulin that follows proteolysis of the bait region not only activates the process of protease trapping, but also reorganizes a stretch at the carboxy terminus, the recognition domain, that can then bind protease-reacted α_2-macroglobulin to its receptor. The recognition domain of human α_2-macroglobulin is composed of the stretch from Val_{1299} to the carboxyl terminus at residue Ala_{1450} (67-69). In this regard, α_2-macroglobulin is a kind of opsonin that promotes the binding and endocytosis of a molecularly diverse array of endopeptidases to cell surface receptors that recognize as their cognate ligand a structure unique to the receptor-binding domain of the protease-reacted form of α_2-macroglobulin rather than the protease itself. Following binding, α_2-macroglobulin and its cargo of bound protease are degraded in secondary lysosomes.

A variety of mammalian cell types, including hepatocytes, fibroblasts, and monocyte/macrophages, bind protease-reacted α_2-macroglobulin. The best-characterized receptor for protease-reacted α_2-macroglobulin has been identified as low density lipoprotein related protein (LRP) and now is referred to as LRP/α_2M-R (70,71). LRP/α_2M-R is a member of the low-density lipoprotein receptor family of proteins. This 600 kDa protein with a single transmembrane domain shows binding activity for a variety of ligands in addition to protease-reacted α_2-macroglobulin (72,73).

COMPARATIVE BIOLOGY OF THE α_2-MACROGLOBULINS OF DIVERSE ANIMAL PHYLA

The suite of unique functional characteristics of α_2-macroglobulin allowed us to identify the first reported example of this protein in an invertebrate (74) (Fig. 1). We found a protease inhibitor in the plasma of the American horseshoe crab, *Limulus polyphemus*, that suppressed the proteolytic activity of a diverse array of proteases against the macromolecular substratum, [^{14}C]-casein, but that failed to reduce the amidolytic action of trypsin, using the low molecular mass synthetic substratum BA*p*NA (75,76). *Limulus* plasma protected bound trypsin from inactivation by the macromolecular active site inhibitor, soybean trypsin inhibitor (77,78). The protein was purified and shown to have the reactive internal thiol ester bond characteristic of most other members of the α_2-macroglobulin family. Its protease inhibitory activity was sensitive to the low molecular mass amine, methylamine (75,76). A new free thiol was produced upon reaction with protease (79). The protein was subject to peptide bond cleavage at the thiol ester by mild thermal denaturation (79). And, when the internal thiol ester region was sequenced, it showed significant amino acid sequence identity with human α_2-macroglobulin (80). The gene for *Limulus* α_2-macroglobulin has now been cloned and sequenced and functional domains can be identified by sequence similarity with human α_2-macroglobulin (Fig. 2). There is no doubt that the α_2-macroglobulins

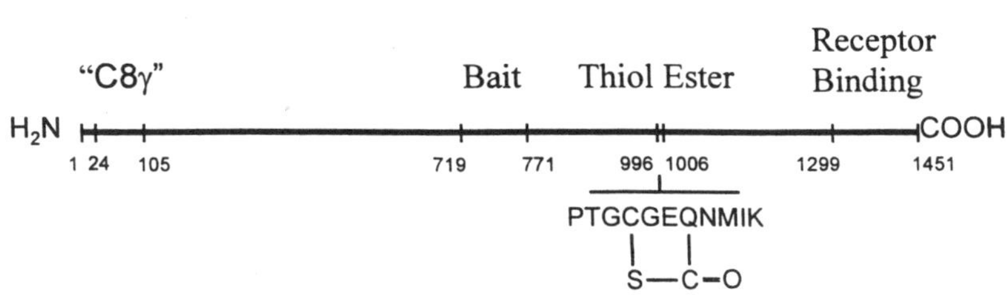

Figure 2. Domain structure of *Limulus* $\alpha_2$macroglobulin. Based on the amino acid sequence of *Limulus* α_2-macroglobulin (81), it is possible to identify important functional domains by comparison with the already established domain structure of human α_2-macroglobulin. The amino terminal end of *Limulus* α_2-macroglobulin has a unique stretch identified as the C8γ domain, based on modest sequence identity with the human defense protein C8γ of the mammalian complement pathway. The bait

region contains a sequence of amino acids that present target sites for nearly all proteases. Proteolysis of the bait region initiates the compaction of α_2-macroglobulin that is responsible for entrapment of the molecule of reacting protease. The thiol ester domain contains the unique thiol ester bond linking cysteine-999 and glutamine-1002. The receptor-binding domain at the carboxyl terminus of the polypeptide chain is exposed following molecular compaction and makes the protease-reacted form of α_2-macroglobulin a target for binding to the cell surface α_2-macroglobulin receptor.

of *Limulus* and mammals are molecular homologues, based on the sequence identity in key functional domains. Based on the criteria listed in Fig. 1 and on sequence information, members of the α_2-macroglobulin protein family have been found in nematodes (α_2-macroglobulin from *C. elegans* (EMBL: locus CEZK337, accession Z82090)), crustacea (78,82-85) mollusks (86-89), echinoderms (90), tunicates (91), and, of course, vertebrates. Thus it appears that α_2-macroglobulin and the related members of this protein family appeared in evolution prior to the great pre-Cambrian radiation that resulted in the diverse phyla of higher animals present today. Further, this protein has been important enough that it has been retained by members of disparate phyla, the members of which enjoy radically different life styles.

In parallel with sequence similarity, *Limulus* α_2-macroglobulin shares functional similarity at the biochemical and physiological levels with α_2-macroglobulin from humans (Table 2). Both forms of α_2-macroglobulin show the trap mechanism of protease binding. Both forms establish isopeptide bonds linking the γ-carbonyl residue of the thiol esterified glutamyl residue with ε amino groups of lysine residues, but with the difference that mammalian forms of α_2-macroglobulin cross link themselves to the bound protease molecule whereas *Limulus* α_2-macroglobulin forms intra-chain crosslinks, principally with lysine-254 (92). More importantly, α_2-macroglobulin appears to be the principle vehicle for clearance of proteases introduced into the plasma of *Limulus*. Fluoresceinated trypsin and thiol ester-activated forms of *Limulus* α_2-macroglobulin are cleared with the same kinetics, with a 10 min lag period for the injected protein to get pumped from the site of injection in the heart, followed by a 15 - 20 min clearance phase (Fig. 3). Enzymatically inactive trypsin and native, unreacted *Limulus* α_2-macroglobulin are not cleared from the blood. The blood cells appear to be one of the major cells that take up protease or protease-reacted α_2-macroglobulin because the fluorescein label appears with the blood cells during the clearance phase. Following clearance, fluorescein-labeled material reappeared in the plasma, which appears to be the proteolyzed fragments of the cleared protein (Fig. 3). During the clearance phase, the residual fluoresceinated trypsin partitions with an apparent molecular mass of approximately 400 kDa, consistent with its association with a high molecular mass protein, such as α_2-macroglobulin, while the material that reappears in the plasma following clearance has a molecular mass, as estimated by molecular filtration, of less than 10 kDa (93).

In addition to the structural and functional conservation of the α_2-macroglobulins of diverse animal phyla, we have the possibility that the other major molecular player in the α_2-macroglobulin-based protease clearance pathway, LRP/α_2M-R, may be conserved in evolution. The LRP protein family has at least two molecular homologues in the nematode, *Cenorhabditis elegans* (111,112). This animal also has the gene for

Table 2. Molecular and functional conservation of the α_2-macroglobulins

Character	Specific attribute	Ref.
1. Genetic identity		
Overall peptide sequence identity	28-29% identity of *Limulus* and mammalian α_2-macroglobulins	(81,94-97)
Bait region	Bounded by PXXC and EXXR consensus sequences in *Limulus* and mammalian α_2-macroglobulin	(66,81)
Thiol ester domain	>50% identity for α_2-macroglobulins of mammals,crustaceans,chelicerates, cephalopods, and gastropods	(64,80,81, 84,85,87, 89,94,98)
Receptor recognition domain	Conservation of two of the key Lys residues of human α_2-macroglobulin in *Limulus* α_2-macroglobulin; of 73 residues conserved amongst the mammals, 45 are conserved in *Limulus* α_2-macroglobulin	(69,81)
2. Biochemical homology		
Susceptibility of bait region to proteolysis by proteases of all catalytic mechanisms	Reactivity of purified α_2-macroglobulin of mammals, arthropods and mollusks to serine, cysteine, and metalloproteases	(64,76,82, 87,89,99)
Molecular compaction; non-covalent trapping of proteases	Molecular compaction of protease-reacted α_2-macroglobulin of mammals and *Limulus*	(100-103)
Covalent isopeptide bonding of thiol esterfied glutamyl residue	Covalent bonding to trapped protease in mammalian α_2-macroglobulin; covalent cross-linking to Lys-254 of partner chain of α_2-macroglobulin dimer in *Limulus*	(92,104-106)
Protease treatment exposes the recognition stretch for receptor-binding	Thiol ester-reacted *Limulus* α_2-macroglobulin binds the human α_2-macroglobulin receptor	(67,81, 107,108)
3. Functional similarity		
Protease clearance	Mammals and *Limulus* utilize α_2-macroglobulin for the clearance of proteases from the internal milieu	(93,109, 110)

α_2-macroglobulin (α_2-macroglobulin from *C. elegans* (EMBL: locus CEZK337, accession Z82090)), although there has been no functional characterization of α_2-macroglobulin in this animal and it has not been shown that either LRP family member can bind activated forms of α_2-macroglobulin. A family of proteins has been purified from detergent extracts of *Limulus* blood cells with characteristics reminiscent of mammalian LRP/α_2MR, based on their ability to bind human RAP. RAP (receptor-associated protein) is a 39 kDa cytoplasmic protein that functions as a chaperone for LRP/α_2M-R (113) and that can be used for the affinity purification of the receptor in the human system. Affinity matrices derivatized with the human version of RAP were used to isolate the candidate homologue of the receptor for protease-reacted α_2-macroglobulin from *Limulus* blood cell extracts (114). It must be emphasized that the *Limulus* protein has not been isolated in sufficient quantity to demonstrate that it has α_2-macroglobulin-binding activity, but its affinity for RAP makes it a serious candidate for that function. This is suggested further by the ability of thiol ester-reacted, but not native, forms of *Limulus* α_2-macroglobulin to bind purified human LRP/α_2M-R with a respectable affinity ($K_D \sim 20$ nM (81)).

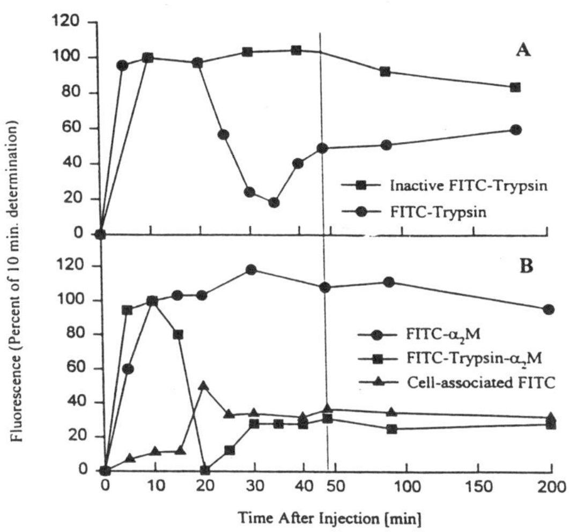

Figure 3. Clearance of proteases from the plasma of *Limulus* is mediated by α_2-macroglobulin. The role of α_2-macroglobulin in protease clearance is especially well illustrated in *Limulus* because α_2-macroglobulin is the only protease inhibitor in the blood (75). Fluoresceinated proteins were injected into the lumen of the heart and their concentration in samples of peripheral blood taken at various intervals after injection was measured with a fluorometer. Fluoresceinated protein reached the periphery by 5 - 10 min after injection. Enzymatically active trypsin, but not trypsin inactivated by treatment with the irreversible inhibitor phenylmethylsulfonylfluoride (PMSF), was substantially cleared from the blood by 30 min (Fig. 3A). The fluorescence that appeared at later times was of low molecular mass, and presumably consisted of peptide digest products of the injected protein. Protease-reacted, but not native, unreacted *Limulus* α_2-macroglobulin showed a similar pattern of clearance from the blood (Fig. 3B). During the period of clearance of fast-form α_2-macroglobulin from the plasma, fluorescence accumulated in the blood cells. This is interpreted to indicate that clearance of the introduced protease is mediated by its binding to α_2-macroglobulin and that the blood cells participate in the uptake and clearance of the α_2-macroglobulin-protease complex.

ROLE OF α_2-MACROGLOBULIN IN DISEASE

Evidence for an importance of α_2-macroglobulin in immunity is still fragmentary. The extraordinary structural and functional conservation of the α_2-macroglobulin system for protease clearance argues for its importance. On the one hand, it is clear that the α_2-macroglobulins of vertebrates (9,27,115,116) and invertebrates (117,118) react with a variety of the proteases of important parasites, but higher vertebrates have so diverse an array of protease inhibitors in the plasma (63,65) that it has been difficult to show in many cases that inhibition and clearance of any of these is unique to α_2-macroglobulin. The situation is clearer in *Limulus* for which α_2-macroglobulin appears to be the only protease inhibitor in the plasma (75) and in which the α_2-macroglobulin clearance pathway appears to be preeminent (93). In mammals, α_2-macroglobulin has been implicated as a key agent that mitigates bacterial elastase-induced septic shock (26,119) and members of the α_2-macroglobulin family have documented therapeutic potential in conditions where microbial proteases play a role in disease (120,121). The delivery of foreign proteins to macrophages by the α_2-macroglobulin-LRP/α_2M-R system (122) appears to enhance the ability of mammals to produce anti-protease antibodies to the major cysteine protease of *Trypanosoma cruzi*, the agent of Chagus' disease (123). α_2-Macroglobulin co-localizes with regions of myocardial lesions in experimental models of Chagus' disease (124).

The receptor, LRP/α_2M -R, appears to be an essential protein because the knockout mouse for this protein is an embryonic lethal (125), indicating that the receptor is required for development. However, the α_2-macroglobulin knockout mouse shows only minor phenotypic abnormalities (126). Interpretation of this latter observation is complicated by the aforementioned diversity of protease inhibitors in the blood of mammals. In addition to a diverse array of active site inhibitors, the mouse has several genes encoding different members of the α_2-macroglobulin family (127) so the potential for functional redundancy is significant. Additionally, it must be remembered that the selective pressure to preserve α_2-macroglobulin has been exerted on mice in the wild. Arguably immunity against proteases of pathogens seen by free-living mice may be the key unique function of $\alpha_2$macroglobulin. Knowing as little as we do about the diseases experienced by the free-living ancestors of the laboratory mouse, it would be a demanding task to test the α_2-macroglobulin knockout mouse for resistence to all of the relevant pathogens. Nevertheless, it would be potentially interesting to screen the α_2-macroglobulin knockout mouse for increased susceptibility against the many pathogens for which the mouse is an established disease model.

An interesting facet of the role of the α_2-macroglobulin-LRP axis in immunity are the situations where pathogens have coopted one or another player of this system to further their evil designs. We have argued that α_2-macroglobulin plays an important role in the immune protection from invading parasites but in these situations the pathogens have found ways to utilize α_2-macroglobulin or its receptor to contribute to the production of disease. The basic function of α_2-macroglobulin and its receptor is to deliver proteases to secondary lysosomes for proteolytic degradation. Some bacterial proteases are regenerated after lysosomal uptake, resulting in the proteolytic destruction of the cells involved in uptake of α_2-macroglobulin bearing these particular proteases (9,10). The important human pathogen, *Streptococcus pyogenes*, binds

α_2-macroglobulin via a surface receptor that recognizes specifically the native form of α_2-macroglobulin (27). The binding of α_2-macroglobulin correlates with bacterial virulence and functions to protect other surface receptors from proteolysis, including proteolysis mediated by a major cysteine protease that itself is an important virulence factor of the bacterium (27). LRP, the α_2-macroglobulin receptor, is the major agent for transport of *Pseudomonas* exotoxin A (128,129). Following transport, the exotoxin is proteolytically processed (130) to yield a fragment that translocates to the cytosol and inhibits protein synthesis (131).

CONCLUSION

Many of the elements of the innate immune system are superficially similar amongst the different phyla of higher animals. Blood clotting systems, phagocytosis, antimicrobial peptides, lysozyme, and the phenol oxidase system are widely distributed. It has been argued that most of the individual effectors of the innate immune system present in diverse phyla are not the products of evolutionary conservation but rather are the results of convergent evolution of different gene products into agents that serve similar immune functions. In a comparison of protein effectors of innate immunity in vertebrates and arthropods, Hughes (132) concluded that of the several systems studied, only lysozyme presented a picture consistent with evolutionary conservation from an ancient, and common precursor protein that itself was a mediator of immunity. Unfortunately, α_2-macroglobulin was not included in his analysis. As we have argued above, α_2-macroglobulin is properly included as an effector of immunity. It reacts with and binds the diversity of proteases released by invading pathogens that operate as virulence factors in many instances of parasitic attack. The α_2-macroglobulins are prominent components of the plasma of vertebrates, arthropods, and mollusks. As shown above, the α_2-macroglobulins of mammals and arthropods show significant sequence identity in key functional domains, including the thiol ester domain and, to a lesser extent, the receptor-binding domain. More tellingly, the α_2-macroglobulins of mammals and arthropods share strikingly similar modes of interaction with proteases and function identically as delivery agents to present their protease cargos to systems for receptor-mediated destruction. In sum, we have argued that the α_2-macroglobulins do indeed represent an evolutionarily conserved arm of the innate immune system whose ancestor arose prior to the great Precambrian radiation of the diverse phyla of higher animals and that has subsequently been preserved largely intact in the products of that divergence.

ACKNOWLEDGEMENTS

Supported by grant MCB-9726771 from the National Science Foundation.

REFERENCES

1. Mekalanos JJ. Environmental signals controlling expression of virulence determinants in bacteria. J Bacteriol 1992;174:1-7.
2. Lantz MS. Are bacterial proteases important virulence factors? J Periodontal Res 1997;32:126-132.
3. Salyers A.A., Whitt D.D. Bacterial Pathogenesis: A Molecular Approach. Washington, D.C. ASM Press, 1994:1-418.
4. Engel LS, Hill JM, Caballero AR, Green LC, O'Callaghan RJ. Protease IV, a unique extracellular protease and virulence factor from *Pseudomonas aeruginosa*. J Biol Chem 1998;273:16792-16797.
5. Engel LS, Hill JM, Moreau JM, Green LC, Hobden JA, O'Callaghan RJ. *Pseudomonas aeruginosa* protease IV produces corneal damage and contributes to bacterial virulence. Invest Ophthalmol Vis Sci 1998;39:662-665.
6. Brindley PJ, Gam AA, McKerrow JH, Neva FA. Ss40: the zinc endopeptidase secreted by infective larvae of *Strongyloides stercoralis*. Exp Parasitol 1995;80:1-7.
7. Cohen FE, Gregoret LM, Amiri P, Aldape K, Railey J, McKerrow JH. Arresting tissue invasion of a parasite by protease inhibitors chosen with the aid of computer modeling. Biochemistry 1991;30:11221-11229.
8. Kuramitsu HK. Proteases of *Porphyromonas gingivalis*: what don't they do? Oral Microbiol Immunol 1998;13:263-270.
9. Molla A, Oda T, Maeda H. Different binding kinetics of Serratia 56K protease with plasma a_2-macroglobulin and chicken egg white ovomacroglobulin. Journal of Biochemistry 1987;101:199-205.
10. Maeda H, Molla A, Oda T, Katsuki T. Internalization of serratial protease into cells as an enzyme-inhibitor complex with a_2-macroglobulin and regeneration of protease activity and cytotoxicity. J Biol Chem 1987;262:10946-10950.
11. Reed SL, Keene WE, McKerrow JH. Thiol proteinase expression and pathogenicity of *Entamoeba histolytica*. J Clin Microbiol 1989;27:2772-2777.
12. Reed S, Bouvier J, Pollack AS, Engel JC, Brown M, Hirata K, Que X, Eakin A, Hagblom P, Gillin F. Cloning of a virulence factor of *Entamoeba histolytica*. Pathogenic strains possess a unique cysteine proteinase gene. J Clin Invest 1993;91:1532-1540.
13. Breton CB, Blisnick T, Jouin H, Barale JC, Rabilloud T, Pereira da Silva LH. *Plasmodium chabaudi* p68 serine protease activity required for merozoite entry into mouse erythrocytes. Proc Natl Acad Sci U S A 1992;89:9647-9651.
14. Poon-King R, Bannan J, Viteri A, Cu G, Zabriskie JB. Identification of an extracellular plasmin binding protein from nephritogenic streptococci. J Exp Med 1993;178:759-763.
15. Sodeinde OA, Subrahmanyam YV, Stark K, Quan T, Bao Y, Goguen JD. A surface protease and the invasive character of plague. Science 1992;258:1004-1007.
16. Fuchs H, Wallich R, Simon MM, Kramer MD. The outer surface protein A of the spirochete *Borrelia burgdorferi* is a plasmin(ogen) receptor. Proc Natl Acad Sci U S A 1994;91:12594-12598.
17. Rosenthal PJ, McKerrow JH, Aikawa M, Nagasawa H, Leech JH. A malarial cysteine proteinase is necessary for hemoglobin degradation by *Plasmodium falciparum*. J Clin Invest 1988;82:1560-1566.
18. Francis SE, Sullivan DJ, Jr., Goldberg DE. Hemoglobin metabolism in the malaria parasite *Plasmodium falciparum*. Annu Rev Microbiol 1997;51:97-123.
19. Wasilewski MM, Lim KC, Phillips J, McKerrow JH. Cysteine protease inhibitors block schistosome hemoglobin degradation *in vitro* and decrease worm burden and egg production *in vivo*. Mol Biochem Parasitol 1996;81:179-189.
20. Carlsson J, Hofling JF, Sundqvist GK. Degradation of albumin, haemopexin, haptoglobin and transferrin, by black-pigmented *Bacteroides* species. J Med Microbiol 1984;18:39-46.
21. Ward W, Alvarado L, Rawlings ND, Engel JC, Franklin C, McKerrow JH. A primitive enzyme for a primitive cell: the protease required for excystation of *Giardia*. Cell 1997;89:437-444.
22. Chung YB, Kong Y, Joo IJ, Cho SY, Kang SY. Excystment of *Paragonimus westermani* metacercariae by endogenous cysteine protease. J Parasitol 1995;81:137-142.
23. Lustigman S, McKerrow JH, Shah K, Lui J, Huima T, Hough M, Brotman B. Cloning of a cysteine protease required for the molting of *Onchocerca volvulus* third stage larvae [published erratum appears in J Biol Chem 1997 Feb 14;272(7):4645]. J Biol Chem 1996;271:30181-30189.
24. Meirelles MN, Juliano L, Carmona E, Silva SG, Costa EM, Murta AC, Scharfstein J. Inhibitors of the major cysteinyl proteinase (GP57/51) impair host cell invasion and arrest the intracellular development of *Trypanosoma cruzi in vitro*. Mol Biochem Parasitol 1992;52:175-184.
25. Marsh JW, Taylor RK. Identification of the *Vibrio cholerae* type 4 prepilin peptidase required for cholera toxin secretion and pilus formation. Mol Microbiol 1998;29:1481-1492.
26. Khan MM, Shibuya Y, Nakagaki T, Kambara T, Yamamoto T. α_2-Macroglobulin as the major defence in acute pseudomonal septic shock in the guinea-pig model. International Journal of Experimental Pathology 1994;75:285-293.
27. Rasmussen M, Muller HP, Bjurck L. Protein GRAB of *Streptococcus pyogenes* Regulates Proteolysis at the Bacterial Surface by Binding α_2-Macroglobulin. J Biol Chem 1999;274:15336-15344.
28. Zuo X, Woo PT. Natural anti-proteases in rainbow trout, *Oncorhynchus mykiss* and brook charr, *Salvelinus fontinalis* and the in vitro neutralization of fish α_2-macroglobulin by the metalloprotease from the pathogenic hemoflagellate, *Cryptobia salmositica*. Parasitology 1997;114:375-381.
29. Cutler CW, Arnold RR, Schenkein HA. Inhibition of C3 and IgG proteolysis enhances phagocytosis of *Porphyromonas gingivalis*. J Immunol 1993;151:7016-7029.

30. Schenkein HA, Fletcher HM, Bodnar M, Macrina FL. Increased opsonization of a prtH-defective mutant of *Porphyromonas gingivalis* W83 is caused by reduced degradation of complement-derived opsonins. J Immunol 1995;154:5331-5337.
31. Reed SL, Keene WE, McKerrow JH, Gigli I. Cleavage of C3 by a neutral cysteine proteinase of *Entamoeba histolytica*. Journal of Immunology 1989;143:189-195.
32. Scott CF, Whitaker EJ, Hammond BF, Colman RW. Purification and characterization of a potent 70-kDa thiol lysyl-proteinase (Lys-gingivain) from *Porphyromonas gingivalis* that cleaves kininogens and fibrinogen. J Biol Chem 1993;268:7935-7942.
33. Plaut AG. The IgA1 proteases of pathogenic bacteria. Annu Rev Microbiol 1983;37:603-622.
34. Kilian M., Reinholdt J. Interference with IgA defense mechanisms by extracellular bacterial enzymes. In: Easmon C.S.F., Jeljaszewics J, eds. Medical Microbiology. London: Academic Press, 1986:173-208.
35. Maeda H, Yamamoto T. Pathogenic mechanisms induced by microbial proteases in microbial infections. Biol Chem Hoppe Seyler 1996;377:217-226.
36. Lottenberg R, Minning-Wenz D, Boyle MD. Capturing host plasmin(ogen): a common mechanism for invasive pathogens? Trends Microbiol 1994;2:20-24.
37. Leytus SP, Bowles LK, Konisky J, Mangel WF. Activation of plasminogen to plasmin by a protease associated with the outer membrane of *Escherichia coli*. Proc Natl Acad Sci U S A 1981;78:1485-1489.
38. Sorsa T, Ingman T, Suomalainen K, Haapasalo M, Konttinen YT, Lindy O, Saari H, Uitto VJ. Identification of proteases from periodontopathogenic bacteria as activators of latent human neutrophil and fibroblast-type interstitial collagenases. Infect Immun 1992;60:4491-4495.
39. DeCarlo AA, Grenett HE, Harber GJ, Windsor LJ, Bodden MK, Birkedal-Hansen B, Birkedal-Hansen H. Induction of matrix metalloproteinases and a collagen-degrading phenotype in fibroblasts and epithelial cells by secreted *Porphyromonas gingivalis* proteinase. J Periodontal Res 1998;33:408-420.
40. Gordon VM, Rehemtulla A, Leppla SH. A role for PACE4 in the proteolytic activation of anthrax toxin protective antigen. Infect Immun 1997;65:3370-3375.
41. Gordon VM, Klimpel KR, Arora N, Henderson MA, Leppla SH. Proteolytic activation of bacterial toxins by eukaryotic cells is performed by furin and by additional cellular proteases. Infect Immun 1995;63:82-87.
42. Inocencio NM, Moehring JM, Moehring TJ. Furin activates *Pseudomonas* exotoxin A by specific cleavage in vivo and in vitro. J Biol Chem 1994;269:31831-31835.
43. Schiavo G, Benfenati F, Poulain B, Rossetto O, Polverino de Laureto P, DasGupta BR, Montecucco C. Tetanus and botulinum-B neurotoxins block neurotransmitter release by proteolytic cleavage of synaptobrevin [see comments]. Nature 1992;359:832-835.
44. Montecucco C, Schiavo G. Tetanus and botulism neurotoxins: a new group of zinc proteases. Trends Biochem Sci 1993;18:324-327.
45. Dunn DL, Simmons RL. Fibrin in peritonitis. III. The mechanism of bacterial trapping by polymerizing fibrin. Surgery 1982;92:513-519.
46. Rotstein OD. Role of fibrin deposition in the pathogenesis of intraabdominal infection. Eur J Clin Microbiol Infect Dis 1992;11:1064-1068.
47. Bang F.B. Ontogeny and phylogeny of response to gram-negative endotoxins among the marine invertebrates. In: Cohen E. ed. Biomedical Applications of the Horseshoe Crab (Limulidae). New York, N.Y. Alan R. Liss, 1979:109-123.
48. Collen D, Lijnen HR. Molecular basis of fibrinolysis, as relevant for thrombolytic therapy. Thromb Haemost 1995;74:167-171.
49. Aikawa M. Fine structure of malaria parasites in the various stages of development. In: Wernsdorfer WH, McGregor I. eds. Malaria: Principles and Practices of Malariology. Edinburgh: Churchill Livingstone, 1988:97-129.
50. Crichton RR, Charloteaux-Wauters M. Iron transport and storage. Eur J Biochem 1987;164:485-506.
51. Cutler CW, Kalmar JR, Genco CA. Pathogenic strategies of the oral anaerobe, *Porphyromonas gingivalis*. Trends Microbiol 1995;3:45-51.
52. McKerrow JH, Sun E, Rosenthal PJ, Bouvier J. The proteases and pathogenicity of parasitic protozoa. Annu Rev Microbiol 1993;47:821-853.
53. Goguen JD, Hoe NP, Subrahmanyam YV. Proteases and bacterial virulence: a view from the trenches. Infect Agents Dis 1995;4:47-54.
54. Genco CA, Van Dyke T, Amar S. Animal models for *Porphyromonas gingivalis*-mediated periodontal disease. Trends Microbiol 1998;6:444-449.
55. Vitovski S, Reed RC, Sayers JR. Invasive isolates of *Neisseria meningitidis* possess enhanced immunoglobulin A1 protease activity compared to colonizing strains. FASEB Journal 1999;13:331-337.
56. Schaller M, Schafer W, Korting HC, Hube B. Differential expression of secreted aspartyl proteinases in a model of oral candidosis and in patient samples from the oral cavity. Mol Microbiol 1998;29:605-615.
57. Polissi A, Pontiggia A, Feger G, Altieri M, Mottl H, Ferrari L, Simon D. Large-scale identification of virulence genes from *Streptococcus pneumoniae*. Infect Immun 1998;66:5620-5629.
58. Alexander J, Coombs GH, Mottram JC. *Leishmania mexicana* cysteine proteinase-deficient mutants have attenuated virulence for mice and potentiate a Th1 response. J Immunol 1998;161:6794-6801.

59. Engel JC, Doyle PS, Hsieh I, McKerrow JH. Cysteine protease inhibitors cure an experimental *Trypanosoma cruzi* infection. J Exp Med 1998;188:725-734.
60. Pellizzari R, Rossetto O, Schiavo G, Montecucco C. Tetanus and botulinum neurotoxins: mechanism of action and therapeutic uses. Philos Trans R Soc Lond B Biol Sci 1999;354:259-268.
61. Lukomski S, Burns EH, Jr., Wyde PR, Podbielski A, Rurangirwa J, Moore-Poveda DK, Musser JM. Genetic inactivation of an extracellular cysteine protease (*SpeB*) expressed by *Streptococcus pyogenes* decreases resistance to phagocytosis and dissemination to organs. Infect Immun 1998;66:771-776.
62. Ullberg M, Kronvall G, Karlsson I, Wiman B. Receptors for human plasminogen on gram-negative bacteria. Infect Immun 1990;58:21-25.
63. Laskowski MJ, Kato I. Protein inhibitors of proteinases. Annu Rev Biochem 1980;49:593-626.
64. Enghild JJ, Thogersen IB, Salvesen G, Fey GH, Figler NL, Gonias SL, Pizzo SV. α_2-Macroglobulin from *Limulus polyphemus* exhibits proteinase inhibitory activity and participates in a hemolytic system. Biochemistry 1990;29:10070-10080.
65. Travis J, Salvesen GS. Human plasma proteinase inhibitors. Annu Rev Biochem 1983;52:655-709.
66. Sottrup-Jensen L, Sand O, Kristensen L, Fey GH. The α_2-macroglobulin bait region. Sequence diversity and localization of cleavage sites for proteinases in five mammalian α_2-macroglobulins. J Biol Chem 1989;264:15781-15789.
67. Sottrup-Jensen L, Gliemann J, Van Leuven F. Domain structure of human α_2-macroglobulin. Characterization of a receptor-binding domain obtained by digestion with papain. FEBS Lett 1986;205:20-24.
68. Holtet TL, Nielsen KL, Etzerodt M, Moestrup SK, Gliemann J, Sottrup-Jensen L, Thogersen HC. Receptor-binding domain of human α_2-macroglobulin. Expression, folding and biochemical characterization of a high- affinity recombinant derivative. FEBS Lett 1994;344:242-246.
69. Nielsen KL, Holtet TL, Etzerodt M, Moestrup SK, Gliemann J, Sottrup-Jensen L, Thogersen HC. Identification of residues in alpha-macroglobulins important for binding to the α_2-macroglobulin receptor/Low density lipoprotein receptor-related protein. J Biol Chem 1996;271:12909-12912.
70. Strickland DK, Ashcom JD, Williams S, Burgess WH, Migliorini M, Argraves WS. Sequence identity between the α_2-macroglobulin receptor and low density lipoprotein receptor-related protein suggests that this molecule is a multifunctional receptor. J Biol Chem 1990;265:17401-17404.
71. Kristensen T, Moestrup SK, Gliemann J, Bendtsen L, Sand O, Sottrup-Jensen L. Evidence that the newly cloned low-density-lipoprotein receptor related protein (LRP) is the α_2-macroglobulin receptor. Febs Letter 1990;276:151-155.
72. Krieger M, Herz J. Structures and functions of multiligand lipoprotein receptors: macrophage scavenger receptors and LDL receptor-related protein (LRP). Annu Rev Biochem 1994;63:601-637.
73. Strickland DK, Kounnas MZ, Argraves WS. LDL receptor-related protein: a multiligand receptor for lipoprotein and proteinase catabolism. FASEB J 1995;9:890-898.
74. Quigley JP, Armstrong PB, Gallant P, Rickles FR, Troll W. An endopeptidase inhibitor, similar to vertebrate α_2-macroglobulin, present in the plasma of *Limulus polyphemus*. Biol Bull (Woods Hole) 1982;163:402
75. Quigley JP, Armstrong PB. An endopeptidase inhibitor, similar to mammalian α_2-macroglobulin, detected in the hemolymph of an invertebrate, *Limulus polyphemus*. J Biol Chem 1983;258:7903-7906.
76. Quigley JP, Armstrong PB. A homologue of α_2-macroglobulin purified from the hemolymph of the horseshoe crab *Limulus polyphemus*. J Biol Chem 1985;260:12715-12719.
77. Ganrot PO. Determination of α_2-macroglobulin as trypsin-protein esterase. Clin Chim Acta 1966;14:493-501.
78. Armstrong PB, Rossner MT, Quigley JP. An α_2-macroglobulin-like activity in the blood of chelicerate and mandibulate arthropods. J Exp Zool 1985;236:1-9.
79. Armstrong PB, Quigley JP. *Limulus* α_2-macroglobulin. First evidence in an invertebrate for a protein containing an internal thiol ester bond. Biochem J 1987;248:703-707.
80. Sottrup-Jensen L, Borth W, Hall M, Quigley JP, Armstrong PB. Sequence similarity between α_2-macroglobulin from the horseshoe crab, *Limulus polyphemus*, and the proteins of the α_2-macroglobulin family from mammals. Comp Bioc B 1990;96:621-625.
81. Ikawi D, Kawabata S-I, Miura Y, Kato A, Armstrong PB, Quigley JP, Nielsen KL, Dolmer K. Molecular cloning of *Limulus* α_2-macroglobulin. Eur J Biochem 1996;242:822-831.
82. Hergenhahn H-G, Soderhall K. α_2-Macroglobulin-like activity in plasma of the crayfish, *Pacifastacus leniusculus*. Comp Biochem Physiol 1985;81B:833-835.
83. Hergenhahn HG, Hall M, Soderhall K. Purification and characterization of an α_2-macroglobulin-like proteinase inhibitor from plasma of the crayfish *Pacifastacus leniusculus*. Biochem J 1988;255:801-806.
84. Spycher SE, Arya S, Isenman DE, Painter RH. A functional, thioester-containing α_2-macroglobulin homologue isolated from the hemolymph of the American lobster (*Homarus americanus*). J Biol Chem 1987;262:14606-14611.
85. Stocker W, Breit S, Sottrup-Jensen L, Zwilling R. α_2-Macroglobulin from hemolymph of the freshwater crayfish *Astacus astacus*. Comp Biochem Physiol [B] 1991;98:501-509.
86. Armstrong PB, Quigley JP. Humoral immunity: α_2-macroglobulin activity in the plasma of mollusks. Veliger 1992;35:161-164.
87. Thogersen IB, Salvesen G, Brucato FH, Pizzo SV, Enghild JJ. Purification and characterization of an α_2-macroglobulin proteinase inhibitor from the mollusc *Octopus vulgaris*. Biochem J 1992;285:521-527.

88. Bender RC, Fryer SE, Bayne CJ. Proteinase inhibitory activity in the plasma of a mollusc - Evidence for the presence of α_2-macroglobulin in *Biomphalaria glabrata*. Comp Bioc B 1992;102:821-824.
89. Bender RC, Bayne CJ. Purification and characterization of a tetrameric α_2-macroglobulin protease inhibitor from the gastropod mollusc, *Biomphalaria glabrata*. Biochem J 1996;316:893-900.
90. Al-Sharif WZ, Sunyer JO, Lambris JD, Smith LC. Sea urchin coelomocytes spedifically express a homologue of the complement component C3. J Immunol 1998;160:2983-2997.
91. Baish MA, Lohr RL, Bartl S. Molecular evidence for complement and α_2-macroglobulin family members in the colonial ascidian, *Botryllus schlosseri*. Dev Comp Immunol 1997;21:147
92. Dolmer K, Husted LB, Armstrong PB, Sottrup-Jensen L. Localisation of the major reactive lysine residue involved in the self-crosslinking of proteinase-activated *Limulus* α_2-macroglobulin. FEBS Lett 1996;393:37-40.
93. Melchior R, Quigley JP, Armstrong PB. α_2-Macroglobulin-mediated clearance of proteases from the plasma of the American horseshoe crab, *Limulus polyphemus*. J Biol Chem 1995;270:13496-13502.
94. Sottrup-Jensen L, Stepanik TM, Kristensen T, Wierzbicki DM, Jones CM, Lonblad PB, Magnusson S, Petersen TE. Primary structure of human α_2-macroglobulin. V. The complete structure. J Biol Chem 1984;259:8318-8327.
95. Kan CC, Solomon E, Belt KT, Chain AC, Hiorns LR, Fey G. Nucleotide sequence of cDNA encoding human α_2-macroglobulin and assignment of the chromosomal locus. Proc Natl Acad Sci U S A 1985;82:2282-2286.
96. Gehring MR, Shiels BR, Northemann W, de Bruijn MH, Kan CC, Chain AC, Noonan DJ, Fey GH. Sequence of rat liver α_2-macroglobulin and acute phase control of its messenger RNA. J Biol Chem 1987;262:446-454.
97. Van Leuven F, Torrekens S, Overbergh L, Lorent K, De Strooper B, Van den Berghe H. The primary sequence and the subunit structure of mouse α_2-macroglobulin, deduced from protein sequencing of the isolated subunits and from molecular cloning of the cDNA. Eur J Biochem 1992;210:319-327.
98. Hall M, Soderhall K, Sottrup-Jensen L. Amino acid sequence around the thiolester of α_2-macroglobulin from plasma of the crayfish, *Pacifastacus leniusculas*. Febs Letter 1989;254:111-114.
99. Starkey PM, Barrett AJ. α_2-Macroglobulin, a physiological regulator of proteinase activity. In: Barrett AJ, ed. Proteinases in Mammalian Cells and Tissues. Amsterdam: Elsevier/North Holland Biomedical Press, 1977:663-696.
100. Barrett AJ, Starkey PM. The interaction of α_2-macroglobulin with proteinases. Characteristics and specificity of the reaction, and a hypothesis concerning its molecular mechanism. Biochem J 1973;133:709-724.
101. Barrett AJ, Brown MA, Sayers CA. The electrophoretically 'slow' and 'fast' forms of the α_2-macroglobulin molecule. Biochem J 1979;181:401-418.
102. Armstrong PB, Mangel WF, Wall JS, Hainfield JF, Van Holde KE, Ikai A, Quigley JP. Structure of α_2-macroglobulin from the arthropod *Limulus polyphemus*. J Biol Chem 1991;266:2526-2530.
103. Quigley JP, Ikai A, Arakawa H, Osada T, Armstrong PB. Reaction of proteinases with α_2-macroglobulin from the American horseshoe crab, *Limulus*. J Biol Chem 1991;266:19426-19431.
104. Sottrup-Jensen L, Petersen TE, Magnusson S. A thiol-ester in α_2-macroglobulin cleaved during proteinase complex formation. FEBS Lett 1980;121:275-279.
105. Sottrup-Jensen L, Hansen HF, Pedersen HS, Kristensen L. Localization of ε-Lysyl-gamma-glutamyl cross-links in 5 human α_2-macroglobulin-proteinase complexes - Nature of the high molecular weight cross-linked products. J Biol Chem 1990;265:17727-17737.
106. Jacobsen L, Sottrup-Jensen L. Localization of ε-lysyl-gamma-glutamyl cross-links in α_2-macroglobulin-plasmin complex. Biochemistry 1993;32:120-126.
107. Van Leuven F, Marynen P, Sottrup-Jensen L, Cassiman JJ, Van den Berghe H. The receptor-binding domain of human α_2-macroglobulin. Isolation after limited proteolysis with a bacterial proteinase. J Biol Chem 1986;261:11369-11373.
108. Enghild JJ, Thogersen IB, Roche PA, Pizzo SV. A conserved region in α_2-macroglobulins participates in binding to the mammalian a-macroglobulin receptor. Biochemistry 1989;28:1406-1412.
109. Davidsen O, Christensen EI, Gliemann J. The plasma clearance of human α_2-macroglobulin-trypsin complex in the rat is mainly accounted for by uptake into hepatocytes. Biochim Biophys Acta 1985;846:85-92.
110. Feldman SR, Rosenberg MR, Ney KA, Michalopoulos G, Pizzo SV. Binding of α_2-macroglobulin to hepatocytes: mechanism of *in vivo* clearance. Biochem Biophys Res Commun 1985;128:795-802.
111. Yochem J, Greenwald I. A gene for a low density lipoprotein receptor-related protein in the nematode *Caenorhabditis elegans*. Proc Natl Acad Sci U S A 1993;90:4572-4576.
112. Yochem J, Tuck S, Greenwald I, Han M. A gp330/megalin-related protein is required in the major epidermis of *Caenorhabditis elegans* for completion of molting. Development 1999;126:597-606.
113. Bu G. Receptor-associated protein: a specialized chaperone and antogonist for members of the LDL receptor gene family. Curr Opin Lipidol 1998;9:149-155.
114. Aimes RT, Quigley JP, Swarnakar S, Strickland D, Armstrong PB. Preliminary investigations on the scavenger receptors of the amebocyte of the American horseshoe crab, *Limulus polyphemus*. Biol Bull (Woods Hole) 1995;189:225-226.
115. Horvat RT, Clabaugh M, Duval-Jobe C, Parmely MJ. Inactivation of human gamma interferon by *Pseudomonas aeruginosa* proteases: elastase augments the effects of alkaline protease despite the presence of α_2-macroglobulin. Infection and Immunity 1989;57:1668-1674.

116. Giroux E, Vargaftig BB. Clostridiopeptidase B inhibition by plasma macroglobulins and microbial antiproteases. Biochimica Biophysica Acta 1978;525:429-437.
117. Fryer SE, Bender RC, Bayne CJ. Inhibition of cysteine proteinase from *Schsitosoma mansoni* larvae by α_2-macroglobulin from the plasma of *Biomphalaria glabrata.* J Parasitol 1996;82:343-347.
118. Srimal S, Armstrong PB. Interaction of immune defense systems of *Limulus polyphemus* with an extracorporeal protease secreted by the principal ectoparasite of *Limulus*, the triclad turbellarid worm, *Bdelloura candida.* Indian J Biol 1996;34:1081-1084.
119. Khan MM, Shibuya Y, Kambara T, Yamamoto T. Role of α_2-macroglobulin and bacterial elastase in guinea-pig pseudomonal septic shock. International Journal of Experimental Pathology 1995;76:21-28.
120. Neely AN, Law EJ, Holder IA. Increased susceptibility to lethal Candida infections in burned mice preinfected with *Pseudomonas aeruginosa* or pretreated with proteolytic enzymes. Infection and Immunity 1986;52:200-204.
121. Miyagawa S, Kamata R, Matsumoto K, Okamura R, Maeda H. Therapeutic intervention with chicken egg white ovomacroglobulin and a new quinolone on experimental *Pseudomonas* keratitis. Graefes Arch Clin Exp Ophthalmol 1994;232:488-493.
122. Chu CT, Pizzo SV. Receptor-mediated antigen delivery into macrophages. Complexing antigen to α_2-macroglobulin enhances presentation to T cells. J Immunol 1993;150:48-58.
123. Morrot A, Strickland DK, Higuchi Md, L, Reis M, Pedrosa R, Scharfstein J. Human T cell responses against the major cysteine proteinase (cruzipain) of Trypanosoma cruzi: role of the multifunctional α_2-macroglobulin receptor in antigen presentation by monocytes. Int Immunol 1997;9:825-834.
124. Coutinho C.M.L.M., van Leuven F., Araujo-Jorge T.C. Detection of α-macroglobulin in the heart of mice infected with *Trypanosoma cruzi.* Parasitol Res 1999;85:249-255.
125. Herz J, Clouthier DE, Hammer RE. LDL receptor-related protein internalizes and degrades uPA-PAI-1 complexes and is essential for embryo implantation [published erratum appears in Cell 1993 May 7;73(3):428]. Cell 1992;71:411-421.
126. Umans L, Serneels L, Overbergh L, Lorent K, Van Leuven F, Van den Berghe H. Targeted inactivation of the mouse α_2-macroglobulin gene. J Biol Chem 1995;270:19778-19785.
127. Overbergh L, Hilliker C, Lorent K, Van Leuven F, Van den Berghe H. Identification of four genes coding for isoforms of murinoglogulin, the monomeric form of mouse α_2-macroglobulin; characterization of the exons coding for the bait region. Genomics 1994;22:530-539.
128. Kounnas MZ, Morris RE, Thompson MR, FitzGerald DJ, Strickland DK, Saelinger CB. The α_2-macroglobulin receptor/low density lipoprotein receptor- related protein binds and internalizes *Pseudomonas* exotoxin A. J Biol Chem 1992;267:12420-12423.
129. FitzGerald DJ, Fryling CM, Zdanovsky A, Saelinger CB, Kounnas M, Winkles JA, Strickland D, Leppla S. Pseudomonas exotoxin-mediated selection yields cells with altered expression of low-density lipoprotein receptor-related protein [published erratum appears in J Cell Biol 1995 Aug;130(4): 1015]. J Cell Biol 1995;129:1533-1541.
130. Gu M, Gordon VM, FitzGerald DJ, Leppla SH. Furin regulates both the activation of Pseudomonas exotoxin A and the Quantity of the toxin receptor expressed on target cells. Infect Immun 1996;64:524-527.
131. Fitzgerald D, Pastan I. Pseudomonas exotoxin and recombinant immunotoxins derived from it. Ann N Y Acad Sci 1993;685:740-745.
132. Hughes AL. Protein phylogenies provide evidence of a radical discontinuity between arthropod and vertebrate immune systems. Immunogenetics 1998;47:283-296.

SUPPRESSION OF THE *Drosophila* CELLULAR IMMUNE RESPONSE BY *Ganaspis xanthopoda*

Hsiling Chiu[1], Richard Paul Sorrentino[1,2], and Shubha Govind[1,2]

[1]The City College of City University of New York
[2]Graduate School and University Center of City University of New York

INTRODUCTION

The cellular immune response of larvae of *Drosophila melanogaster* is mediated by two types of hemocyte. Plasmatocytes are small, round, and phagocytic hemocytes that engulf small particles, such as bacteria. When large objects, such as eggs of parasitoid wasps, are introduced into the larval hemocoel, plasmatocytes differentiate into larger, flat, and adhesive lamellocytes. Lamellocytes adhere to one another and wrap around the foreign body, forming capsules. These capsules are often melanized by the second type of hemocyte, called crystal cells, which house enzymes for melanization reactions (Rizki and Rizki, 1984). The capsules render the parasite inert and allow the host to survive. In the larva, plasmatocytes and crystal cells originate in

Phylogenetic Perspectives on the Vertebrate Immune System.
Edited by Beck *et al.*, Kluwer Academic/Plenum Publishers, 2001.

the lymph gland, which is the hematopoietic organ in *Drosophila.* (Shrestha and Gateff, 1982).

Capsules, similar to the ones around parasitic wasp eggs, are also found in many *D. melanogaster* mutants, often referred to as "melanotic tumor" mutants. An example of one such mutant is *hopscotch*$^{Tumorous\text{-}lethal}$ (*hop*$^{Tum\text{-}l}$; Hanratty and Ryerse, 1981). *hop*$^{Tum\text{-}l}$ is an X-linked, temperature-sensitive semidominant mutation in the Janus Kinase protein of *Drosophila* (Binari and Perrimon, 1994; Hanratty and Dearolf, 1993). This mutation causes hypertrophy of the lymph gland that leads to an overabundance of both plasmatocytes and lamellocytes (Luo *et al.*, 1995). The large number of lamellocytes causes encapsulation of self-tissue. These capsules become melanized and are clearly visible through the larval cuticle. As such, they are referred to as melanotic capsules (MC).

Endoparasitic wasps have developed strategies by which they can either evade or suppress the host cellular immune response (Carton and Nappi, 1997; Rizki and Rizki, 1990, 1994; Whitfield, 1990). *Leptopilina heterotoma* is a highly successful parasitoid wasp of *Drosophila* spp. When wild-type *D. melanogaster* hosts are infected by *L. heterotoma*, over 90% of the parasitoid embryos escape encapsulation (Nappi *et al.*, 1984). During oviposition, *L. heterotoma* females inject virus-like-particles (VLPs), along with their eggs into the host. Formed in the accessory long gland associated with the ovaries, VLPs specifically target lamellocytes for destruction (Rizki and Rizki, 1984, 1990, 1991). Interestingly, VLPs of *L. heterotoma* do not appear to affect lamellocyte differentiation that is induced upon infection (Rizki and Rizki, 1992).

G. xanthopoda, another parasitoid wasp of *Drosophila* spp. (Carton *et al.*, 1986; Melk and Govind, 1999), also exhibits high virulence, and an encapsulation response in wild-type *D. melanogaster* hosts is never observed. To determine if this lack of encapsulation is due to immune suppression or evasion, we have studied the effect of *G. xanthopoda* parasitization on mutant *hop*$^{Tum\text{-}l}$ hosts. To compare the effects of *G. xanthopoda* infection with that of *L. heterotoma* on *hop*$^{Tum\text{-}l}$ larvae, we quantified the reduction of melanotic capsule formation and lamellocyte abundance. Here, we present evidence that, like *L. heterotoma, G. xanthopoda* can also suppress melanotic encapsulation by *hop*$^{Tum\text{-}l}$, by reducing the incidence of melanotic capsules and percentage of lamellocytes in circulation. Our results also show that in this *in vivo* assay of immune suppression, the degree of suppression by *G. xanthopoda* is stronger than that by *L. heterotoma.*

MATERIALS AND METHODS

Insect stocks and encapsulation capacity assay

y v hop$^{Tum\text{-}l}$*/Basc* stock (a description of markers can be obtained by visiting the Flybase website at http://Flybase@flybase.bio.indiana.edu) was allowed to lay eggs on yeast-agar medium for 9-24 hr. Approximately 48 hr after the initiation of egglay, second-instar larvae were exposed to 4-20 female representatives of either *Leptopilina heterotoma* or *Ganaspis xanthopoda* for 6 hr. Control larvae were not exposed to wasps. Six days after the initiation of egglay, third-instar larvae were inspected for the presence of egg encapsulations (EC) and melanotic capsules (MC). Larvae exhibiting both ECs and MCs were included in the EC class. Hemi- and homozygous *hop*$^{Tum-l}$ mutants were identified by their yellow mouthhooks. All other siblings (*hop*$^{Tum-l}$*/Basc*, *Basc/Basc*, *Basc/Y*) had black mouthhooks. All experiments took place at 25°C.

Hemolymph samples and lymph gland dissections

Egglays were performed as described above, except second-instar larvae were exposed to wasps for 24 hr. Six days after initiation of egglays, third-instar larvae were collected, washed in PBS (0.14 M NaCl, 0.003 M KCl, 0.01 M Na_2PO_4, 0.002 M KH_2PO_4), then washed in 95% ethanol, and allowed to air-dry. Hemolymph samples were obtained by opening *Drosophila* larvae with forceps and allowing hemolymph to collect on ethanol-sterilized slides. The same larvae were transferred to drops of PBS on slides, and their lymph glands were removed and mounted. Both hemolymph samples and lymph glands were allowed to air-dry for 20-30 min, fixed in 4% formaldehyde in PBS for 3 minutes, stained with 10 mg/ml of Hoechst nuclear dye (Molecular Probes) in PBS for 5 minutes, washed three times with PBS, and mounted in 50% glycerol in PBS. Lymph gland samples were photographed on a Zeiss Axioplan compound fluorescent microscope. Larvae were identified as described above, and males were identified by presence of testes.

RESULTS

The melanotic capsules of *hopscotch*$^{Tum-l}$ (*hop*$^{Tum-l}$) larvae are composed primarily of lamellocytes (Nappi and Streams, 1969). If the *G. xanthopoda* survival mechanism is based on suppression of the host cellular immune response, then this mechanism should be able to reduce the number and size of melanotic capsules in *hop*$^{Tum-l}$ larvae. However, if *G. xanthopoda* relies on an evasive mechanism, then it should be able to avoid egg encapsulation while having no effect on the *hop*$^{Tum-l}$ melanotic capsule phenotype. In order to distinguish between these two possibilities, and to be able to compare the effects of *G. xanthopoda* to those of *L. heterotoma*, *hop*$^{Tum-l}$ larvae that had been parasitized by either wasp were observed for both the frequency of egg encapsulation (EC), as well as penetrance of the melanotic capsule (MC) phenotype (an example of an encapsulation of an *L. heterotoma* egg by a *Basc/Y* larva is presented in Figure 2A). At 25°C, 96.0% of unparasitized third-instar *hop*$^{Tum-l}$ (yellow-mouthhooked) larvae exhibit MCs, most of which tend to be large (Figures 1, 2B). The strength of this phenotype was severely reduced in *hop*$^{Tum-l}$ larvae parasitized by *L. heterotoma*, of which only 7.6% exhibited an EC, and 53.4% had MCs; 39.0% of larvae had no ECs or MCs (Figures 1, 2C). A greater reduction was observed in *hop*$^{Tum-l}$ larvae parasitized by *G. xanthopoda*, however, which allowed none of the larvae to form ECs, and only 27.3% to form MCs; the remaining 72.7% of larvae had neither ECs nor MCs (Figures 1, 2D). As expected, wasps were able to produce similar, albeit weaker, results for black-mouthhooked sibling larvae (*hop*$^{Tum-l}$*/Basc*, *Basc/Basc*, *Basc/Y*; see Figure 1).

The suppression of egg encapsulation and melanotic capsule formation can be effected by the destruction of host lamellocytes by the *L. heterotoma* VLPs (Rizki and Rizki, 1990). In unparasitized six-day-old *hop*$^{Tum-l}$*/Y* larvae raised at 25°C, lamellocytes represent over 70% of the circulating hemocyte population (unpublished observations).

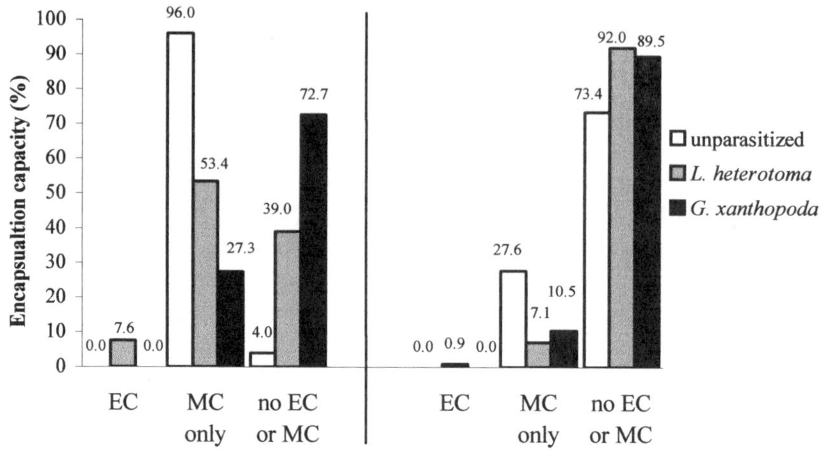

Figure 1. Percentage of parasitized third-instar larvae exhibiting an egg encapsulation (EC) or melanotic capsules (MC) after parasitization by *L. heterotoma* or *G. xanthopoda*. Left half of figure shows data for hemi- and homozygous hop^{Tum-l} larvae (yellow-mouthhooked). Right half of figure shows data for hop^{Tum-l}/*Basc*, *Basc/Basc*, and *Basc/Y* larvae (black-mouthhooked). Numerical percentage values are atop each bar. Sample sizes: yellow-mouthhooked: unparasitized, 25; *L. heterotoma*, 280; *G. xanthopoda*, 118; black-mouthhooked: unparasitized, 123; *L. heterotoma*, 336; *G. xanthopoda*, 112.

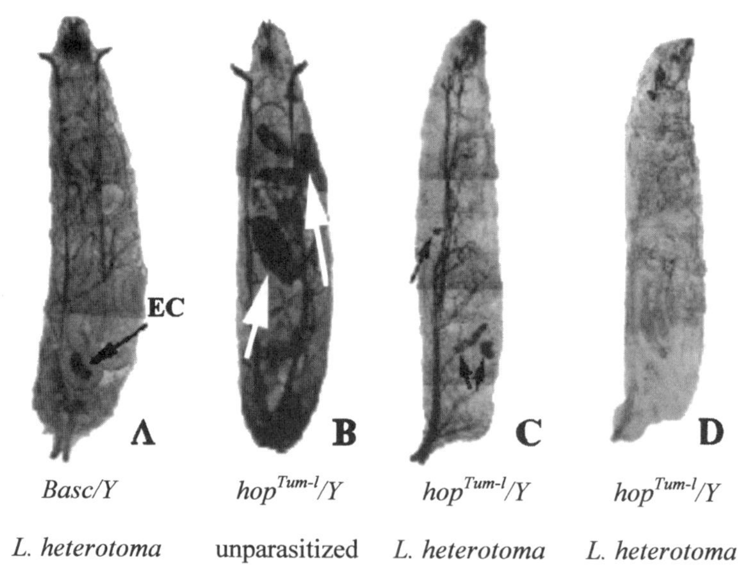

Figure 2. Wild-type and hop^{Tum-l} mutants with or without parasitization. Suppression of melanotic capsules can be seen in larvae parasitized by either *L. heterotoma* or *G. xanthopoda*. Unlabeled arrows indicate melanotic capsules, except in **(A)**, where only an egg encapsulation (EC) can be seen.

Three days after parasitization by *L. heterotoma*, fixed hemolymph samples from *hop*$^{Tum\text{-}l}$*/Y* larvae show a lamellocyte population of less than 5%, and are similar in appearance to samples from *Basc/Y* siblings. In order to determine whether the *G. xanthopoda* suppression mechanism could produce a similar effect on the abundance of circulating lamellocytes, we observed hemolymph samples from *hop*$^{Tum\text{-}l}$*/Y* and *Basc/Y* larvae three days after parasitization. Like their counterparts that were parasitized by *L. heterotoma*, larvae parasitized by *G. xanthopoda* showed a striking reduction in lamellocyte abundance (there was again a lamellocyte population of less than 5%). This result strongly suggests that like *L. heterotoma*, *G. xanthopoda* can also suppress the host immune response by acting on *Drosophila* lamellocytes.

In order to confirm the activation of the host cellular immune response, the lymph glands of *Basc/Y* (*i.e.*, wild-type) larvae were examined three days after parasitization by either *L. heterotoma* or *G. xanthopoda*. Figure 3A depicts the lymph gland of an unparasitized *Basc/Y* larva; note the prominent pair of anterior lobes (al) and the multiple pairs of posterior lobes (pl). Strikingly, after parasitization by *L. heterotoma* (Fig. 3B) or *G. xanthopoda* (Fig. 3C), the anterior lobes are absent, and only vestiges of the posterior lobes remain. Thus even though we could not see an appreciable difference in lamellocyte percentage between parasitized and unparasitized larvae, both wasps clearly were able to elicit a cellular immune response, as indicated by lymph gland morphologies.

DISCUSSION

In general, endoparasites must be able to effectively negotiate host immune responses. Two common tactics used by endoparasites are suppression and evasion (reviewed by Strand and Pech, 1995; Vinson, 1993). Evidence presented in this paper is consistent with *G. xanthopoda* making use of a mechanism that suppresses the host cellular immune response, and with this mechanism being stronger than that of *L. heterotoma*. We have demonstrated that *G. xanthopoda* is able to completely prevent the encapsulation of its eggs in wild-type hosts. Even more striking is the observation that it can also do so in *hop*$^{Tum\text{-}l}$*/Y* larvae, which have far greater cellular immune capacity (Nappi *et al.*, 1984). Additionally, in almost three-quarters of *hop*$^{Tum\text{-}l}$*/Y* larvae parasitized by *G. xanthopoda*, melanotic capsule formation around self-tissue did not occur. The remaining larvae had fewer and smaller capsules than did larvae parasitized by *L. heterotoma* or unparasitized controls. Examination of hemolymph samples from parasitized *hop*$^{Tum\text{-}l}$*/Y* larvae showed that both *G. xanthopoda* and *L. heterotoma* produced similar effects: a reduction of the lamellocyte population to less than one-fourteenth its normal size, rendering hemolymph samples almost indistinguishable from those of parasitized wild-type larvae. Finally, it is apparent that the wild-type lymph gland is affected by the presence of either *G. xanthopoda* or *L. heterotoma*, and that the reduction in the *hop*$^{Tum\text{-}l}$ lamellocyte population is not due to a failure of the hematopoietic system to respond to parasitization.

The exact nature of the *G. xanthopoda* mechanism is still unknown. However, it is possible that despite the evolutionary distance between these two wasp genera (Schilthuizen *et al.*, 1998), *G. xanthopoda*, like *L. heterotoma*, employs VLPs to mediate immune suppression. Possession of an additional evasion mechanism by *G. xanthopoda* cannot yet be ruled out. Further studies are required to characterize the *G. xanthopoda* immune suppressive mechanism, and its action on host tissues.

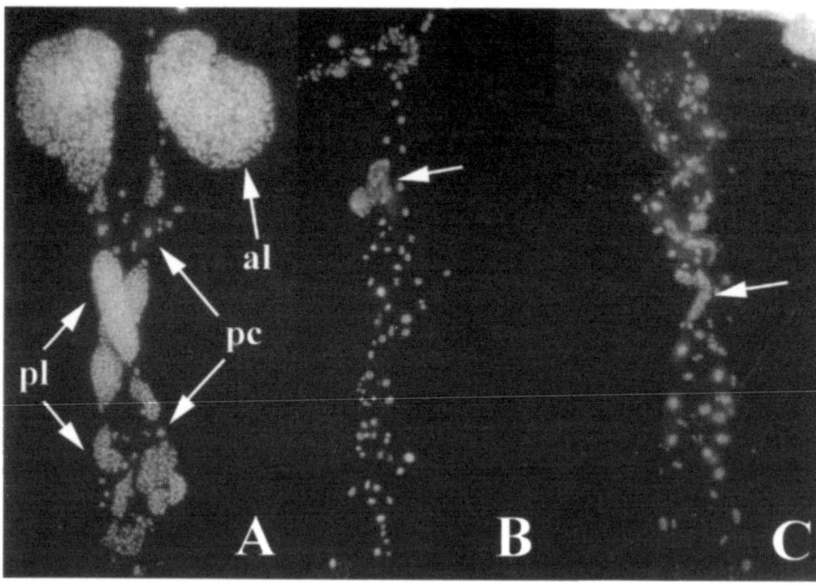

Figure 3. Lymph glands of unparasitized and parasitized *Drosophila* larvae stained with Hoechst nuclear dye. (**A**). Lymph gland from unparasitized larva; note prominent anterior and posterior lobes (al, pl) as well as presence of pericardial cells (pc). (**B**) and (**C**). Note absence of anterior lobes from larvae parasitized by either (**B**) *L. heterotoma* or (**C**) *G. xanthopoda*. Only a vestige of what may be either anterior or posterior lobes is present in (**B**) (arrow), and what appear to be vestigial posterior lobes are present in (**C**) (arrow).

ACKNOWLEDGEMENTS

Both authors (H.C. and R.P.S.) contributed equally to this paper. This work was supported by grants from the American Heart Association, Heritage Affiliate, Inc., and American Cancer Society RPG 98-228-01DDC. Support from the NIG-RCMI RR03060, NIH-MBRS, and the PSC-CUNY Awards Program is gratefully acknowledged.

REFERENCES

Binari R and Perrimon N (1994). *Genes Dev.* 8, 300–312.

Carton Y, Bouletreau M, vanLenteren JC, vanAlphen JCM (1986). In: *The Genetics and Biology of Drosophila* (Ed. Ashburner M, Carson HL, Thompson JN), New York: Academic Press, 347 - 394.

Carton Y and Nappi AJ (1997). *Parasitol. Today* 13(6), 218-227.

Hanratty WP and Dearolf CR (1993). *Mol. Gen. Genet.* 238, 33 – 37.

Hanratty WP and Ryerse JS (1981). *Dev. Biol.* 83, 238 – 249.

Luo H, Hanratty WP, and Dearolf CR (1995). *EMBO J.* 14(7), 1412 - 1420.

Melk JP and Govind S (1999). *J. Exp. Biol.*, 202, 1885 - 1896.

Nappi AJ, Kmiecik J and Silvers M (1984). *J. Inver. Pathol.* 44, 220 - 227.

Nappi AJ and Streams FA (1969). *J. Insect Physiol.* 15, 1551-1566.

Rizki RM and Rizki TM (1990). *Proc. Natl. Acad. Sci. USA* 87, 8388 – 8392.

Rizki RM and Rizki TM (1991). *J. Exp. Zoology* 257, 236 – 244.

Rizki TM and Rizki RM (1984). In: *Insect ultrastructure* (Ed. King RC), New York: Plenum Publishing Co., 2, 574 - 604.

Rizki TM and Rizki RM (1992). *Develop. Comp. Immunol.* 16, 103 – 110.

Rizki TM and Rizki RM (1994). *Ann. New York Acad. Sci.* 712, 178 – 194.

Schilthuizen M, Nordlander G, Stouthamer R, and vanAlphen JJM (1998). *Systematic Entomology* 23, 253 - 264

Shrestha R and Gateff E (1982). *Develop., Growth, and Differ.* 24(1), 65 – 82.

Strand MR and Pech LL (1995). *Annu. Rev. Entomol.* 40, 31 - 56.

Vinson SB (1993). In: *Insect Immunity* (Ed. Pathak JPN), Norwell: Kluwer Academic Publishers, 103 – 112.

Whitfield JB (1990). *Parasitol. Today* 6, 381-384.

ANNELID HUMORAL IMMUNITY: CELL LYSIS IN EARTHWORMS

Edwin L. Cooper[1], Ellen Kauschke[2] and Andrea Cossarizza[3]

[1]Laboratory of Comparative Immunology, Department of Neurobiology, School of Medicine, University of California Los Angeles, California, USA 90095–1763
[2]Zoological Institute and Museum University of Greifswald, J.-S.-Bach-Strasse 11/12 17489, Germany
[3]Department of Biomedical Sciences, Section of General Pathology, University of Modena, Via Campi 287 41100 Modena, Italy

INTRODUCTION

Earthworms and other organisms have deployed several strategies essential for maintaining and perpetuating species and one is the immune system that functions effectively against microbes. Effector activity against experimental antigens such as cancer cells is mediated by leukocytes and molecules that they synthesize and secrete (Cooper, *et al.*, 1995; Cossarizza *et al.*, 1995; 1996; Quaglino, *et al.*, 1996; Cooper *et al.*, 1992; 1999; Bilej *et al.*, 1995; Kauschke *et al.*, 1997). Current work emphasizes

Phylogenetic Perspectives on the Vertebrate Immune System.
Edited by Beck *et al.*, Kluwer Academic/Plenum Publishers, 2001.

the capacity of coelomic fluid to effect lysis by means of a protein which has been referred to as Eiseniapore (Lange *et al.*, 1997). Clearly the earthworm's importance in understanding this aspect of invertebrate humoral immunity is significant as indicated by the recent increase in molecular studies relevant to lysis, (Lassegues *et al.*, 1997; Milochau *et al.*, 1997; Sekizawa, *et al.*, 1997; Yamaji *et al.*, 1998), its apparent regulation by serine proteases (Roch *et al.*, 1997) and its possible relation to humoral agglutinins (H1, H2, H3), (Eue et al., 1997) and those antimicrobial peptides that are not hemolytic (Cho *et al.*, 1998). These research groups propose the names fetidin, lysin, and now Eiseniapore, and perhaps perforin (Komiyama *et al.,* 1997). From molecular analyses, these groups have cloned cDNAs which encode the cytolytic proteins and confirmed the activities of recombinant proteins. Certain common characteristics seem to be shared. As an example, their work reveals that the 40kDa fetidin has at least four isoforms with different isoelectric points and lysenin represents two isoforms of molecular masses of 41 and 42 kDa. This suggests that coelomic fluid of *Eisenia fetida* contains several sphingomyelin-binding cytolytic proteins with molecular masses around 40 kDa. Thus, when all results are viewed together, a certain bit of confusion emerges. There is the need to reach a consensus among these several laboratories. This is required for further characterization using identical methods. Expected results should clarify in order to codify the lytic components of *Eisenia* and to distinguish them functionally from the humoral agglutinins.

This works suggests neurophysiological bridges that may obliterate connections to immunological boundaries. For example, cross communication is essential in the formation of ion channels during nerve transmission (Gordon *et al.*, 1996, Mazumder *et al.*, 1996, Pennington *et al.*, 1996) and to characteristics related to allergic and autoimmune diseases (Nastuk, 1971, Brumback, 1981, Kaminski *et al.*, 1997). With respect to immunological responses of earthworms, formation of holes (ion channels?) may play a role in the mechanism by which non-self cells are destroyed. As an extension, the earthworm model could therefore apply to the destruction and ultimate elimination of pathogens or antigens that are naturally occurring or that are introduced experimentally (*e.g.* erythrocytes). Taken together these two systems that require channel formation in target cell membranes, clearly underscore a unifying motif at the cellular level with respect to survival.

PERFORIN IS INVOLVED IN THE DESTRUCTION OF TARGETS DURING CYTOTOXICITY IN MAMMALS

We now turn to the mammalian model to aid in analyzing putative mechanisms of cytotoxicity in earthworms. Perforin (cytolysin; pore-forming protein) is thought to be a major mediator in the cytolytic properties of cytotoxic T-lymphocytes and NK cells (Walsh *et al.*, 1994; Henkart, 1985; Podack *et al.*, 1991). When cytoplasmic granules of CTLs recognize aberrant or foreign "target" cells, any perforin stored in CTLs (along with a range of other products, such as proteolytic enzymes) are released directly into the target cell. Perforin monomers appear to form polymeric pore structures in target cell plasma membranes. These pores, in conjunction with other granule components, induce rapid, apoptosis-like cell death in target cells (Young, 1989;

Tschopp and Nabholz, 1990). Kägl *et al.* (1994) have suggested several mechanisms that may account for lymphocyte-mediated cytotoxicity (LMC), including granule exocytosis and apoptosis (Russell; 1983, Rouvier *et al.*, 1993), and the action of serine esterases (Jenne and Tschopp, 1988), tumor necrosis factors and other analogs. Morphological and biochemical criteria (Dourmashkin et al., 1980; Dennert and Podack, 1983; Young *et al.*, 1986) show that granule exocytosis involves release of granule contents when target cells are recognized and bound as conjugates to cytolytic lymphocytes (Henkart, 1985; Podack, 1985).

Perforin, which forms pores when polymerized (Podack, 1985), is a granular molecule that is responsible for cell lysis (Acha-Orbea *et al.*, 1990; Shiver *et al.*, 1992). The functional similarity between perforin and complement C9 is also supported by the homology of their amino-acid sequences (Lichtenheld *et al.*, 1988; Shinkai *et al.*, 1988). Perforin expressed in both cytotoxic T lymphocytes (CTL) and NK cells (Shinkai *et al.*, 1988; Podack *et al.*, 1988; Garcia-Sanz *et al.*, 1987; Kawasaki *et al.*, 1992), has been detected histologically in immunopathologic conditions where cytolytic lymphocytes are involved, such as during graft rejection and antiviral immune responses (Mueller *et al.*, 1988, 1989; Young *et al.*, 1990).

PORE-FORMING FACTORS ARE PRESENT IN EARTHWORMS AND OTHER INVERTEBRATES

Because of increased clarity with respect to perforin's function in mammals, it is reasonable to present material on lytic mechanisms causing destruction and found in earthworms that could be related to perforin (Bilej *et al.*, 1995). According to Canicatti (1990), invertebrates possess molecules that lyse vertebrate erythrocytes. Hemolytic activity depends on proteins that possess a wide range of reactivity and that are generally calcium-dependent and heat-labile, although calcium-independent and heat-stable hemolysins have also been detected. The molecules interact with sugars or lipids that may represent the membrane receptors by which lesions on target membranes are produced. Based upon certain analogies with vertebrate lytic molecules, it is conceivable that hemolysins evolved from a common ancestral gene that may have been a precursor of vertebrate pore-forming proteins. Canicatti's review (1990) is most informative since certain crucial topics are considered including: effects of cations, physico-chemical and kinetic properties, isolation of lytic components, cells responsible for synthesis, surface binding and mechanisms of action, role in the lysis of bacteria and erythrocytes, a comparison of hemolysins and complement, and finally a consideration of regulatory mechanisms including the involvement of proteases.

According to Beschin, *et al.* (1998), coelomic fluid of *Eisenia foetida* earthworms contains a 42-kDa protein named coelomic cytolytic factor 1 (CCF-1) that was described previously to be involved in cytolytic, opsonizing, and hemolytic properties of the coelomic fluid. Cloning and sequencing of CCF-1 (Figure 1) reveal significant homology with the putative catalytic region of β-1,3- and β-1,3-1,4-glucanases. CCF-1 also displays homology with coagulation factor G from *Limulus polyphemus* and with Gram-negative bacteria-binding protein of *Bombyx mori* silkworm, two proteins involved in invertebrate defense mechanisms. This group should show that CCF-1 efficiently binds both β-1,3-glucan and lipopolysaccharide and is involved in the activation of pro-PO. Moreover, CCF-1 participates in the activation of prophenoloxidase cascade via recognition of yeast and Gram-negative bacteria cell wall components (Figure 2). These results suggest that the 42-kDa CCF-1 protein of *E. foetida* coelomic fluid likely plays a role in the protection of earthworms against microbes.

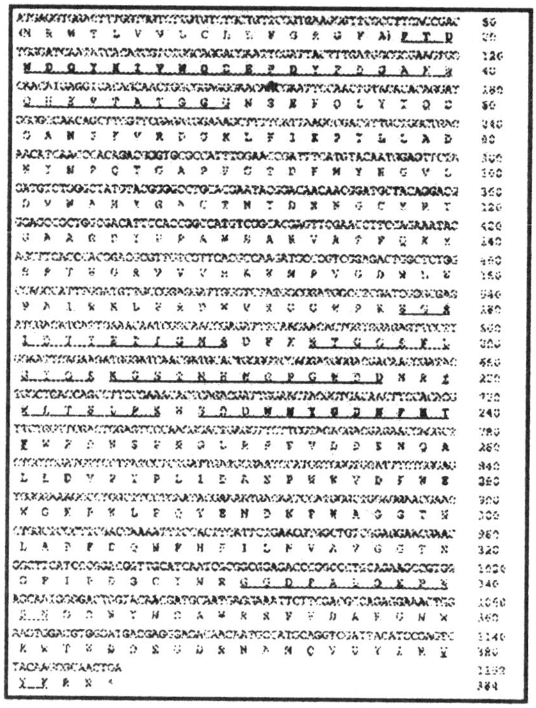

Figure 1. Nucleotide and amino acid sequence of CCF-1 (GenBankTM accession number AF030028). Parentheses indicate the signal peptide of 17 amino acids. An asterisk marks the TGA stop codon. Underlined amino acid residues were identified by Edman degradation. Amino acid residues in boldface were used for the synthesis of degenerated primers. (from Beschin, *et al.*, 1998).

To follow this work more specifically, (Pagliara *et al.*, 1993) examined the structure and enzyme content of sea urchin cytolytic granules. Amoebocytes of the sea urchin *Paracentrotus lividus* are responsible for cytotoxic activity. These cells possess a single, usually spheroidal, nucleus with a single nucleolus. The cytoplasm contains abundant rough endoplasmic reticulum, mitochondria and smooth-walled vesicles designated as lucid granules that are empty with circular and irregular in profiles. Dense granules are also scattered throughout the cytoplasm and phagosomes with ingested material have been frequently observed. The cytotoxic event requires contact between effector and target cells and calcium-dependent hemolysins, causing disintegration of target cells.

It is highly likely that natural cytotoxic cells which resemble mammalian NK cells are also present in invertebrates (Cooper *et al.*, 1992). The active effector cells of echinoderms, however, differ from mammalian counterparts with respect to morphology and biochemistry. Large granular lymphocytes (LGL), in fact, are distinguishable from coelomocytes (C) since LGL are non-adherent, non-phagocytic, peroxidase-negative cells (Ferrarini and Grossi, 1985). However, both LGL and C possess lysosomal granules which are probably associated with a lytic mechanism based on a stimulus-secretion model. (Porchet-Henneré *et al.*, 1992) have analyzed natural killer cells in the polychaete annelid, *Nereis diversicolor*. In this marine worm, they identified globular, non-adherent coelomocytes called "G3 granulocytes" that display spontaneous cytotoxicity. These cells kill invertebrate and vertebrate target cells by a contact-dependent cytolytic process and within the cytoplasm, organelles orient toward contact sites, similar to what has been observed in mammalian cells, (Kupfer *et al.*, 1985).

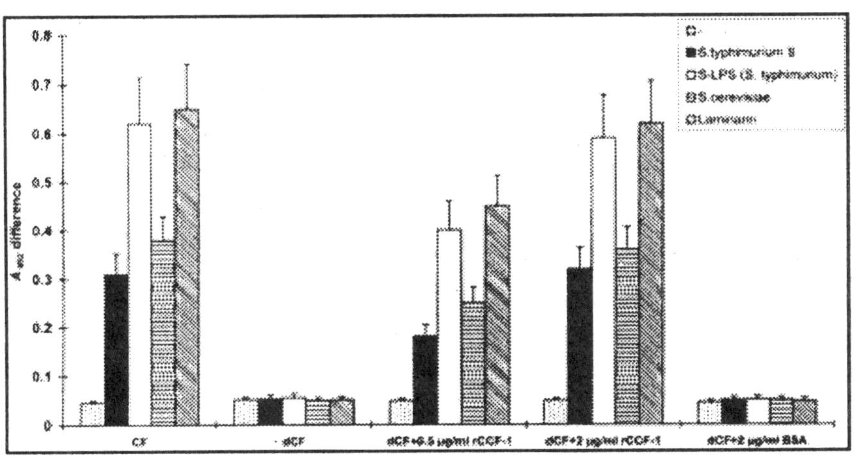

Figure 2. Involvement of CCF-1 in the activation of pro-PO. Levels of L-DOPA oxidation are expressed as the mean ± S.D. of A492 values difference of the sample without and with protease inhibitor. CF, entire coelomic fluid; dCF, CCF-1-depleted coelomic fluid; dCF + 0.5 µg/ml rCCF-1 or dCF + 2 µg/ml rCCF-1, CCF-1-depleted coelomic fluid supplemented, respectively, with 0.5 or 2.0 µg/ml rCCF-1; (dCF + 2 µg/ml BSA, CCF-1-depleted coelomic fluid supplemented with 2.0 µg/ml BSA. Bacteria *S. typhimurium* (S), yeast *S. cerevisiae*, laminarin, or S-LPS from S. *typhimurium* were given at 50 ng/ml. (Beschin, *et al.*, 1998)

There is pertinent information with respect to the role of proteases. Three serine proteases have been isolated from *Eisenia foetida andrei coe*lomocyte lysates, one of which was a trypsin-type (Roch *et al.*, 1991). Similarly, serine protease activity has been reported in *Lumbricus terrestris* (Leipner, *et al.*, 1993; Kauschke, *et al.*, 1997). According to interactions observed principally in insects, such intracellular proteolytic enzymes are regulated by extracellular serine protease inhibitors (see Polanowski and Wilusz, 1996 for review).

EARTHWORMS:PRELIMINARY EVIDENCE FOR PERFORIN

Preliminary work exists on the question of perforin existence in earthworms. Komiyama, *et al.* (1997) have used the earthworm coelomocyte cytotoxic response against K562 which requires cell to cell contact as has been observed for human NK cell interaction with targets. In their preliminary study, the perforin gene transcript and its protein in the coelomocytes has been analyzed. They employed immunohistochemistry and Western blot analysis using mouse perforin antibody. RT-PCR and southern blot analysis were performed by perforin specific primer and probe synthesizing based on the human perforin cDNA sequence. Perforin antibody reacted with cytoplasmic granules in only small but not in large coelomocytes. These coelomocytes also reacted with Thy-1 antibody. Molecular size of earthworm perforin analyzed by Western blot was about 70kDa. The perforin mRNAs from coelomocytes, mouse spleen cells and human PBL revealed the same size which was identified by RT-PCR and southern blot analysis. These data indicated that perforin may be associated with earthworm coelomocytes.

SEVERAL RELATED FACTORS MEDIATE LYSIS ARE SYNTHESIZED AND SECRETED BY EARTHWORM: RELATED TO PERFORIN?

Fetidin

According to Milochau, *et al.* (1997), The coelomic fluid of the earthworm *Eisenia fetida andrei* exhibits antibacterial, hemolytic and hemagglutinating activities. These activities are mainly mediated by two proteins, named fetidins, of apparent molecular mass 40 kDa and 45 kDa, respectively. For the first time, the two proteins have been purified to homogeneity from dialysed coelomic fluid by means of anion-exchange chromatography. Three peaks had hemolytic activity. The first fraction was found to correspond to the 40 kDa fetidin, the second to mixed 40 and 45 kDa fetidins, the last one to the 45 kDa fetidin. Both purified proteins still exhibited their hemolytic and antibacterial activities as dialysed coelomic fluid did. In this study, the amino-acid sequence of purified proteins is compared to the amino-acid sequence predicted by cDNA. This cDNA was isolated by screening an expression cDNA library from earthworm total tissues (unpublished data).

Using another approach, Lassegues *et al.* (1997), have characterized fetidins that are 40-kDa and 45-kDa hemolytic and antibacterial glycoproteins present in the coelomic fluid of the earthworm *Eisenia fetida andrei*. By screening a cDNA library with a polyclonal antifetidin serum, they cloned a cDNA that encoded the 40-kDa fetidin. The clone contains an insert of 1.44 kb encoding a protein of 34 kDa, which corresponds to the size of deglycosylated fetidins. The recombinant protein inhibits *Bacillus megaterium* growth. Restriction fragment polymorphisms were observed on Southern blots and correspond to a known protein polymorphism. The sequence of the cDNA contains a peroxidase signature and fetidins from earthworm coelomic fluid have peroxidase activity. The 40-kDa and 45-kDa fetidins therefore represent two related polymorphic defense factors in invertebrates.

Roch, *et al.* (1998) have purified and characterized the serine protease inhibitor activity contained in the coelomic fluid of the earthworms, *Eisenia*. Serine protease inhibitor activity was stable between pH3 and 9.5, not flocculable by pH 3.0 and resistant to 100 degrees C for 15 min. or to 4 degrees C for 24 h. Ten μl of coelomic fluid was sufficient to inhibit *in vitro* the protease activity of 0.12μg of trypsin. Injection of living bacteria into earthworms resulted in increased serine protease activity 1-2 days post-injection, and increased serine protease inhibitor activity on day 4, suggesting that serine protease inhibitor is responsible for serine protease neutralization. Purified to homogeneity by affinity chromatography on trypsin, the serine protease inhibitor of *Eisenia* is a monomer of 14 kDa. Its partial NH_2 amino acid sequence revealed a basic hydrophobic fragment which shared 68-75% homologies and 47-60% identities with several plant serine protease inhibitors. *Eisenia* cytotoxic activity due to the two fetidins of 40 and 45 kDa was stimulable in vitro by several serine proteases. Incubation with soybean trypsin inhibitor variant a (STIa) resulted in less cytotoxicity. The inhibitory effect occurred only when STIa was added before cell disruption. Interpretative cytotoxic scheme involving the release of intracellular cytotoxic proteins, intracellular trypsin-like activator and extracellular serine protease inhibitor suggests regulatory mechanisms for cellular/humoral immune system of earthworms.

Lysenin

According to Sekizawa *et al.*, (1997), lysenin, which causes contraction of rat vascular smooth muscle, is a protein that was isolated from *Eisenia foetida*. A cDNA encoding lysenin was isolated by using a partial cDNA probe that had been generated by the PCR with a primer designed by reference to an internal peptide sequence of lysenin. The recombinant lysenin was produced in *Escherichia coli.* This clone had an ORF encoding 297 amino acid residues. The amino acid sequence deduced from the cDNA revealed the absence of any significant homology to those of previously characterized vasoactive substances.This protein and native lysenin isolated from the earthworm had similar contractile activities when tested on rat aorta. Northern blot analysis of the RNA from various tissues of the earthworm indicated that lysenin is produced by coelomocytes. As a parenthetical remark, the authors indicated that the physiological function of lysenin is unknown, it may be involved in defense systems, since it exudes from the dorsal pore by mechanical stimulation of the body. This single statement, that this component also has hemolytic activity is apparently not supported by experimental data and seems to require further analysis.

Actually, this group has extended the above work in order to support the argument of a hemolytic capacity (Yamaji *et al.*, 1998). Lysenin, a novel 41-kDa protein purified from coelomic fluid of the earthworm *Eisenia foetida*, induced erythrocyte lysis (Figure 3). Preincubation of lysenin with vesicles containing sphingomyelin inhibited lysenin-induced hemolysis completely, whereas vesicles containing phospholipids other than sphingomyelin showed no inhibitory activity, suggesting that lysenin bound specifically to sphingomyelin on erythrocyte membranes (Figure 4). The specific binding of lysenin to sphingomyelin was confirmed by enzyme-linked immunosorbent assay, TLC immunostaining, and liposome lysis assay. In these assays, lysenin bound specifically to sphingomyelin but showed no cross-reaction with other phospholipids including sphingomyelin analogs such as sphingosine, ceramide, and sphingosylphosphorylcholine, indicating that it recognized a precise molecular structure of sphingomyelin. Incorporation of cholesterol into the sphingomyelin membrane significantly increased the total amount of lysenin bound to the membrane, whereas it did not change the kinetic parameters of the lysenin-membrane interaction. This suggests that lysenin specifically recognized sphingomyelin and cholesterol incorporation changed the topological distribution of sphingomyelin in the membranes, thereby increasing the accessibility of sphingomyelin to lysenin. This study demonstrated that lysenin bound specifically to sphingomyelin on cellular membranes and should be a useful tool to probe the molecular motion and function of sphingomyelin in biological membranes. This includes those with clinical application such as fibroblasts derived from patients with Nieman-Pick disease type A (Figure 5). The results are relevant to those recently published by Lange *et al.* (1997) and point toward the need for a concerted evaluation of properties of the various lytic components that are found in the earthworm's coelomic fluid.

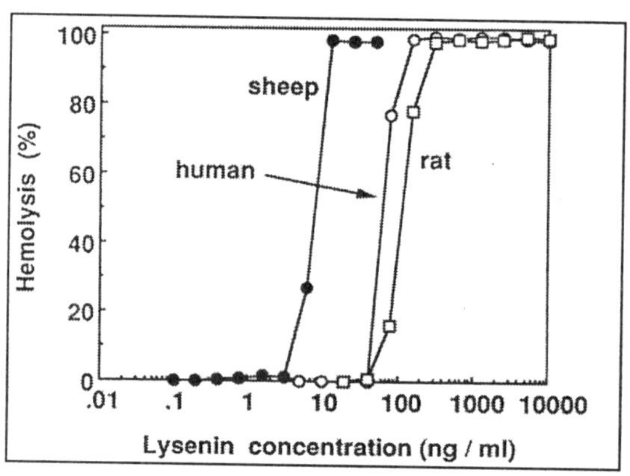

Figure 3. Lysenin-induced hemolysis. Erythrocytes (3 x 10^7 cells/ml) from various species (sheep; human; rat) were incubated with various concentrations of lysenin at 37 °C for 30 m (Yamaji *et al.*, 1998)

Hemolysin

In another related work, according to Eue, *et al.* (1998), hemolytic activity in coelomic fluid of *Eisenia fetida* (ECF) is due to three proteins H1, H2, H3 with molecular weights of 46, 43 and 40 kD. Isolated by preparative PAGE, H1 and H2 were stable in SDS and α-2-ME whereas H3 splits into two fragments with molecular weights of 18 and 21 kD. IEF indicates that each protein consists of different isoforms with pIs between 5.1 and 6.2. H3 was found to be a bifunctional protein that can lyse and agglutinate erythrocytes. At 56°C, hemolytic activity of all three proteins was inactivated, but the agglutination activity of H3 was stable. Intracoelomic injection of erythrocytes reduced the number of hemolysins from three to two. Monospecific antisera were raised against the isolated hemolysins H1, H2 and H3. Using these antibodies and carbohydrates as inhibitors of biological activity of the molecules demonstrates close structural relationship of agglutinins and hemolysins in the CF of *E. fetida.*

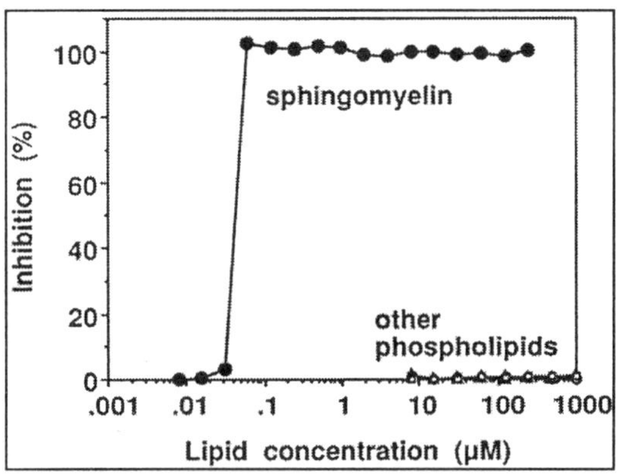

Figure 4. Inhibition of lysenin-induced hemolysis by sphingomyelin. Lysenin was preincubated with various concentrations of vesicles composed of 50 mol % each phospholipid and 50 mol % ch (Yamaji *et al.,* 1998).

LUMBRICIN I: AN ANTIMICROBIAL PEPTIDE IN LUMBRICUS

Cho, *et al.* (1998), have isolated and characterized a novel antimicrobial peptide from the earthworm, *Lumbricus rubellus*. The antimicrobial peptide was purified to homogeneity by a heparin-affinity column and C18 reverse-phase HPLC, and named lumbricin I. Lumbricin I is a proline-rich antimicrobial peptide of 62 amino acids (15% proline in molar ratio; molecular mass, 7231 Da), whose complete sequence was determined by a combination of peptide sequence and cDNA analysis. The peptide and cDNA sequence analysis revealed that lumbricin I was produced as a precursor form consisting of 76 amino acids, with 14 residues in a presegment and 62 residues in mature lumbricin I. Lumbricin I showed antimicrobial activity *in vitro* against a broad spectrum of microorganisms without hemolytic activity. In addition, a 29-amino acid peptide, named lumbricin I (6-34), which was derived from residues 6-34 of lumbricin I, showed marginally stronger antimicrobial activity than lumbricin I. Northern blot analysis on total RNA revealed that expression of lumbricin I gene was not induced by bacterial infection, but was constitutively expressed. Furthermore, the expression of lumbricin I gene was specific in adult *L. rubellus*: Lumbricin I mRNA was detected only in adult *L. rubellus*, but not in eggs and young *L. rubellus.*

IS THERE AN APPROPRIATE MODEL TO EXPLAIN DESTRUCTION OF TARGETS FOLLOWING RECOGNITION AND ADHESION?

In an attempt to understand the mechanism of synthesis and secretion of a lytic agent in the earthworm, Lange *et al.* (1997) have defined a novel new system. Lytic activityin the coelomic fluid of earthworm (*Eisenia foetida*) has been ascribed to

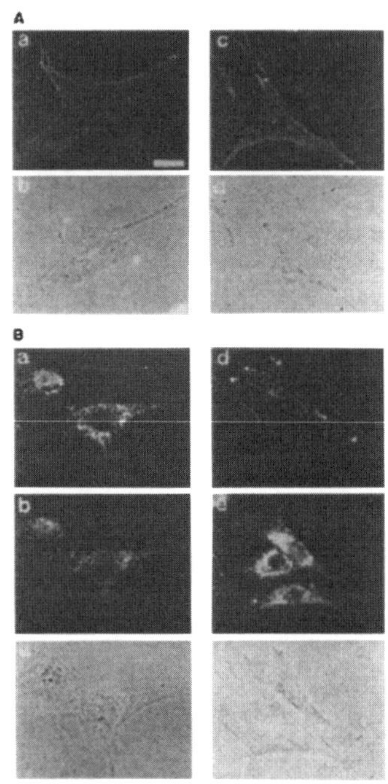

Figure 5. Immunofluorescence staining of human fibroblasts by lysenin. A, fibroblasts from a patient with NP-A (a and b) or a normal subject (c and d) were fixed and stained by lysenin. Bound lysenin was detected by sequential incubation with anti-lysenin antiserum and tetramethylrhodamine isothiocyanate-conjugated anti-rabbit IgG. Phase contrast micrograph (b and d) and lysenin staining (a and c) of the same specimens are shown. (Bar = 20 μm.) B, fibroblasts from a patient with NP-A (a-c) or a normal subject (d-f), which had been cultured in the medium containing 5 mg/ml fluorescein-labeled dextran for 24 h, were fixed and permeabilized. Staining by lysenin was performed as described in A. Phase contrast micrograph (c and f) and lysenin staining (a and d) and dextran fluorescence (b and e) of the same specimens are shown. (Yamaji *et al.*, 1998)

eiseniapore a hemolytic protein of 38 kDa (Figure 6). Since receptors for eiseniapore on target cell membranes are not known, we used lipid vesicles of various composition to determine whether specific lipids may serve as receptors. Lytic activity of eiseniapore was probed by the relief of fluorescence dequenching from the fluorophore 8-aminonaphthalene-1,3,6-trisulfonic acid originally incorporated into the vesicles lumen as a complex with *p*-xylene-bis-pyridinium bromide. Hemolysin binds to and disturbs the lipid bilayer only when distinct sphingolipids consisting of a hydrophilic head group as phosphorylcholine or galactosyl as well as the ceramide backbone, *e.g.* sphingomyelin, are present. Cholesterol enhances eiseniapore lytic activity toward sphingomyelin-containing vesicles probably due to interaction with sphingomyelin. Leakage of vesicles was most efficient when the lipid composition resembled that of the outer leaflet of human erythrocytes. Presumably, an oligomeric protein pore formed by six monomers is responsible for leakage of sphingomyelin-containing vesicles. The secondary structure of eiseniapore did not change upon binding to lipid membranes. The lytic activity of eiseniapore was completely abolished after its denaturation or after preincubation with polyclonal antibodies. Our results suggest two conclusions: 1) the presence of specific sphingolipids is sufficient to mediate lytic activity of eiseniapore; 2) membrane-(glyco)proteins, as receptors for eiseniapore, are not required as long as specific sphingolipids are present.

Further analyses by Lange *et al.* (1999) revealed interesting characteristics with respect to eiseniapore and erythrocyte membranes as viewed by electron microscopy (Figure 7). They observed several ring shaped structures (pores) with a central channel containing outer (10 nm) and an inner (3 nm) diameters. This channel complex consists of six monomers as deduced from the molecular weight of 228 kD. Functional evidence of pore formation by EP was revealed since protection of lysis by carbohydrates occurred at an effective diameter above 3nm. From these results there are two suggested conclusions: 1) the existence of a channel or a "barrel stave model" for vesicle and cell interaction of eiseniapore; 2) a plausible explanation for the mechanism by which components of the earthworm immune system destroy non-self cells.

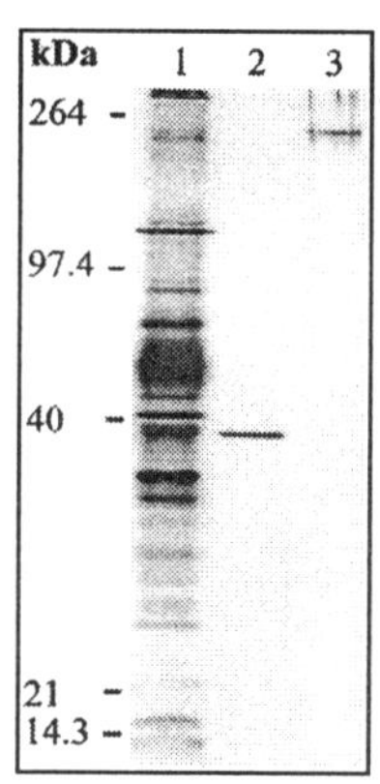

Figure 6. Formation of stable oligomers of Eiseniapore in the presence of egg-PC/Chol/SM (1:1:2) liposomes. Incubation of the Eiseniapore-vesicle suspension was done at pH 7.4, 25 °C. The L/P was 250:1. Subsequently, analytical SDS-PAGE with gels of 5-10% acrylamide was performed (silver stain). Lane 1, total coelom fluid content (25 μg); lane 2, Eiseniapore (2.1 μg); lane 3, Eiseniapore after preincubation in the presence of egg-PC/Chol/SM liposomes (1.4 μg). (Lange *et al.*, 1997).

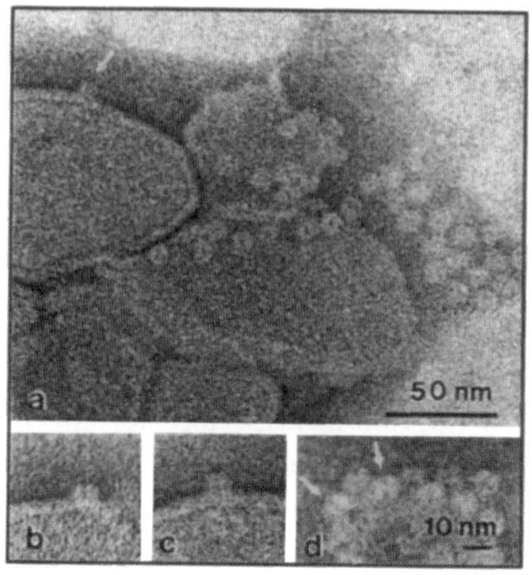

Figure 7. Electron micrographs of negatively stained egg-PC/Chol/SM liposomes in the presence of eiseniapore. Liposomes were incubated with eiseniapore for 15 min at 25 °C at an L/P = 125:1, fixed with gluteraldehyde, and negatively stained with phosphotungstate. Arrows point to side views of Eiseniapore complexes. A, general view: b-d, selected images of side views. Magnification 360,000x (A) and 480,000x (b-d). (Lange *et al.*, 1997)

PERSPECTIVES ON RECOGNITION, ADHESION, LYSIS AND AGGLUTINATION IN THE IMMUNE SYSTEM OF EARTHWORMS

Here we have focused on one aspect of the earthworm's humoral immune system. In all instances the emphasis has been on those molecules which lyse target cells of various types primarily in *Eisenia*. In the case of *Lumbricus* and in *Eisenia*, agglutinins can be detected depending upon the assay system. In the earthworm's immune system, recognition precedes cell adhesion to a target and this is followed by as yet a series of undefined signalling events; putative receptors on leukocytes are essential for recognition of self *vs* non-self. Encounters with foreign antigens are a normal event in the environment, replete with microbial pathogens that pose a threat to survival. To better understand events when an effector cell first recognizes a foreign target followed by its adhesion to it, isolated leukocytes, in sufficient quantities, has been beneficial. *In vitro* approaches when accompanied by biochemical, immunological and molecular technologies, have opened up new vistas concerning the immune response of earthworms and other invertebrates. The most recent discovery includes the preliminary identification of cell differentiation (CD) markers that play vital roles in recognitive and adhesive events. Three challenges for the future face invertebrate immunologists focusing on earthworms as a model. First, it is essential to determine the molecular structure of CD markers. Second, once their molecular and biochemical characteristics have been analyzed, their roles in cellular and humoral mechanisms must be clarified with respect to effector cell products and resulting immune responses. Third, distinctions must be elucidated between the several lytic factors which will most probably reveal multiple families with corresponding functional differentiations. Finally the role of outer membrane components such as sphingomyelin in mediating cell destruction must be clarified. There is the need to distinguish between those pathways that lead to either apoptosis or necrosis.

ACKNOWLEDGEMENT

The senior author extends a special note of gratitude to Rose Marie Norton (Waimanalo, Hawaii) who has recently expressed support for his research efforts on earthworm immunity. We also acknowledge one undergraduate volunteer in the UCLA Student Research Program, Andrew Kan, who helped prepare the manuscript. ELC is the recipient of a visiting professorship (German Academic Exchange Program) and the Alexander von Humboldt Prize.

REFERENCES

Acha-Orbea, H., Scarpellino, L., Hertig, S., Dupuis, M., and Tschopp J. (1990) Inhibition of lymphocyte mediated cytotoxicity by perforin antisense oligonucleotides. EMBO J., 9: 3815-3819.

Beschin A, Bilej M, Hanssens F, Raymakers J, Van Dyck E, Revets H, Brys L, Gomez J, De Baetselier P, and Timmermans M. (1998) Identification and cloning of a glucan- and lipopolysaccharide-binding protein from Eisenia foetida earthworm involved in the activation of prophenoloxidase cascade. J Bio Chem 273:24948-24954.

Bilej, M., Brys, L., Beschin, A., Lucas, R., Vercauteren, E., Hanusova, R., and De Baetselier, P. (1995) Identification of a cytolytic protein in the coelomic fluid of Eisenia foetida earthworms. Immunol. Lett. 45:123-128.

Brumback, R.A. (1981) The neuralmuscular junction. Part 1: physiology and the effects of drugs and toxins. Am. Fam. Physician 23, 188-192.

Canicatti, C. (1990) Hemolysis: Pore-forming proteins in invertebrates. Experientia 46: 239-244.

Cho J. H., Park, C. B., Yoon, Y. G., and Kim, S. C. (1998) Lumbricin I, a novel proline-rich antimicrobial peptide from the earthworm: purification, cDNA cloning and molecular characterization. Biochem Biophys Acta. 1408:67-76.

Cooper, E. L., Cossarizza, A. Suzuki, M. M., Salvioli, S., Capri, M., Quaglino, D., and Franceschi, C. (1995) Autogeneic but not allogeneic earthworm effector coelomocytes kill the mammalian tumor target K562. Cell. Immunol. 166:113-122.

Cooper, E.L., and Roch, P. (1992). The capacities of earthworms to heal wounds and to destroy allografts are modified by polychlorinated biphenyls (PCB). J. Invert. Pathol. 60: 59-63.

Cooper, E.L., Cossarizza, A., Kauschke, E., and Franceschi, C. (1999) Cell adhesion and the immune system: a case study using earthworms. Micro. Res. Tech. 44, 237-253.

Cooper, E.L., Rinkevich, B., Uhlenbruck, G., and Valembios, P. (1992) Invertebrate immunity: another viewpoint. Scand. J. Immunol. 35, 247-26.

Cossarizza, A., Cooper, E. L., Quaglinio, D., Salvioli, S., Kalachnikova, G., and Franceschi, C., (1995) Mitochondrial mass and membrane potential in coelomocytes from the earthworm Eisenia foetida: Studies with fluorescent probes in single intact cells. Biochem. Biophys. Res. Comm. 214: 503-510.

Cossarizza, A., Cooper, E. L., Suzuki, M. M., Salvioli, S., Capri, M., Gri, G., Quaglino, D., and Franceschi, C. (1996) Earthworm leukocytes that are not phagocytic and cross-react with several human epitopes can kill human tumor cell lines. Exptl. Cell Res. 224: 174-182.

Dennert, G. and Podack, E.R. (1983) Cytolysis by H-2-specific T killer cells: Assembly of tubular complexes on target membranes. J. Exp. Med. 157: 1483-1495.

Dourmashkin, R.R., Deteix, P., Simone C.B., and Henkart, P. (1980) Electron microscopic demonstration of lesions in target cell membranes associated with antibody-dependent cellular cytotoxicity. Clin. Exp. Immunol. 42: 554-60.

Eue, I., Kauschke, E., Mohrig, W., and Cooper, E. L. (1998) Isolation and characterization of earthworm hemolysins and agglutinins. Dev. Comp. Immunol., 22: 13-25.

Ferrarini, M. and. Grossi, C.E. (1985). Ultrastructure and cytochemistry of the human large granular lymphocytes. In: Immunobiology of natural killer cells. Lotzová, E. and Herbermann, R. B. (eds) CRC Press, Boca Raton, FL. 1: 33-43.

Garcia-Sanz, J.A., Plaetinck, G., Velotti, F., Masson, D., Tschopp, J., MacDonald, H.R., and Nabholz, M. (1987) Perforin is present only in normal activated Lyt2+ T-lymphocytes and not in L3T4+ cells, but serine protease granzyme A is made by both subsets. EMBO J. 6: 933-938.

Gordon, D., Martin-Eauclaire, M.F., Cestele, S., Kopeyan, C., Carlier, E., Khalifa, R.B., Pelhate, M., and Rochat, H. (1996) Scorpin toxins affecting sodium current inactivation bind to distinct homologous receptor sites on rat brain and insect sodium channels. J. Biol. Chem. 271, 8034-8045.

Henkart, P.A. (1985) Mechanisms of lymphocyte-mediated cytotoxicity. Ann. Rev. Immun. 3: 31-58.

Jenne, D.E and Tschopp, J. (1988) Granzymes a family of serine proteases released from granules of cytolytic T lymphocytes upon T cell receptor stimulation. Immunol. Rev. 103: 53-71.

Kägl, D., Ledermann, B., Burkl, K., Seller, P., Odermatt, B., Olsen, K. J., Podack, E. R., Zinkernagel, R. M., and Hengartner, H. (1994) Cytotoxicity mediated by T cells and natural killer cells is greatly impaired in perforin-deficient mice. Nature. 369: 31-37.

Kaminski, H.J., Suarez, J.I., and Ruff, R.L. (1997) Neuromuscular junction physiology in myasthemia gravis: isoforms of the acetylcholine receptor in extraocular muscle and the contribution of sodium channels to the safety factor. Neurology 48, 8-17.

Kauschke, E., Pagliara, P., Stabili, L., and Cooper, E. L. (1997) Characterization of proteolytic activity in coelomic fluid of Lumbricus terrestris. Comp. Biochim. Physiol. 116B: 235-242.

Kawasaki, A., Shinkai, Y., Yagita, H., and Okumura, E. (1992) Expression of perforin in murine natural killer cells and cytotoxic T lymphocytes in vivo. Eur. J. Immun. 22: 1215-1219.

Komiyama, K., Yoshimura, M., Iwase, T., Sato, J, Okumura, K., Cooper, E. L., and Moro, I. (1997) Identification of perforin gene and its protein in the earthworm coelomocytes. Dev. Comp. Immunol. 21:115.

Kupfer, A., Dennert, G., and Singer, S.J. (1985) The reorientation of the Golgi apparatus and the microtubule-organizing center in the cytotoxic effector cell is a prerequisite in the lysis of bound target cells. J. Mol. Cell. Immunol. 2: 37-49.

Lange, S., Kauschke, E., Mohrig, W., and Cooper, E. L. (1999) Biochemical characteristics of eiseniapore, a pore forming protein in the coelomic fluid of earth worms. Eu. J. Biol. Chem. 262: 1-11.

Lange, S., Nübler, Kauschke, E., Lutsch, G., Cooper, E. L., and Herrmann, A. (1997) Interactions of earthworm hemolysin with lipid membranes requires sphingolipids. J. Biol. Chem. 272: 20884-20892.

Lassegues, M. Milochau, A. Doignon, F., Du Pasquier, L., and Valembois, P. (1997) Sequence and expression of an Eisenia-fetida-derived cDNA clone that encodes the 40-kDa fetidin antibacterial protein Eur. J. Biochem. 246: 756-762.

Leipner, C., Tuckova, L., Rejnek, J., and Langner, J. (1993) Serine proteases in coelomic fluid of annelids Eisenia fetida and Lumbricus terrestris. Comp. Biochim. Physiol. 105B: 637-641).

Lichtenheld, M. G., Olsen, K. J., Ping, L., Lowrey, D.M., Hameed, A., Hengartner, H., and Podack, E.R. (1988) Structure and function of human perforin. Nature 335: 448-451.

Mazumder, P.K. and Dube, S. N. (1996) Marine toxins as molecular probes for biological interactions: a review. Indian J. Physiol. Allied Sci. 50, 34-47.

Milochau, A., Lassegues, M., and Valembois, P. (1997) Purification, characterization and activities of two hemolytic and antibacterial proteins from coelomic fluid of the annelid Eisenia fetida andrei. Biochem. Biophys, Acta. 1337: 123-132.

Mueller, C., Gershenfeld, H.K., Lobe, C.G., Okada, C.V., Bleackley, R.C., and. Weissman, I.L. (1988) A high proportion of T lymphocytes that infiltrate H-2-incompatible heart allografts in vivo express genes encoding cytotoxic cell-specific serine proteases, but do not express the MEL-14-defined lymph node homing receptor. J. Exp. Med. 167: 1124-1136.

Mueller, C., Kagi, D., Aebischer, T., Odermatt, B., Held, W. Podack, E.R., Zinkernagel, R.M., and Hengartner, H. (1989) Detection of perforin and granzyme A mRNA in infiltrating cells during infection of mice with lymphocytic choreomeningitis virus. Eur.J. Immunol. 19: 1253-1259.

Nastuk, W.L. (1971) Mechanisms of neuromuscular blockade. Ann. N.Y. Acad. Sci. 183, 171-182

Pagliara, P., Canicatti, C., and Cooper, E.L. (1993) Structure and enzyme content of sea urchin cytolytic granules. Comp. Biochim. Physiol. 106B: 813-818.

Pennington, M.W., Mahnir, V.M., Krafte, D.S., Zaydenberg, I., Byrens, M.E., Khaytin, I., Crowler, K., and Kem, W.R. (1996) Identification of three separate binding sites of SHK toxin, a potent inhibitor of voltage dependent potassium channels in human T-lymphocytes and rat brain. Biochem Biophys. Res. Commun. 219, 696-701.

Podack, E.R. (1985) The molecular mechanism of lymphocyte-mediated tumor cell lysis. Immunol. Today 6: 21-27.

Podack, E.R., Hengartner, H., and Lichtenheld, M.G. (1991) A central role of perforin in cytolysis. Ann. Rev. Immunol. 9: 129-157.

Podack, E.R., Lowrey, D.M., Lichtenheld, M., Olsen, K.J., Aebischer, T., Binder, D., Rupp, F., and Hengartner, H. (1988) Structure, function and expression of murine and human perforin 1 (P1). Immun. Rev. 103: 203-211.

Polanowski, A. and Wilusz, T. (1996) Serine proteinase inhibitors from insect hemolymph. Acta Biochim Polinica 43: 445-454).

Porchet-Henneré, E., Dugimont, T., and Fischer, A. (1992) Natural killer cells in a lower invertebrate, Nereis diversicolor. Eur. J. Cell Biol. 58: 99-107.

Quaglino, D., Cooper, E. L., Salvioli, S., Capri, M., Suzuki, M. M., Pasquali-Ronchetti, I., Franceschi, and C. Cossarizza, A. 1996 Earthworm coelomocytes in vitro: cellular features and "granuloma" formation during cytotoxic activity against the mammalian tumor cell target K562. Eu. J. Cell Biol. 70:278-288.

Roch, P., Ville, P., and Cooper, E. L. (1998) Characterization of a 14 kDa plant-related serine protease inhibitor and regulation of cytotoxic activity in earthworm coelomic fluid. Dev. Comp. Immunol. Dev. Comp. Immunol. 22: 1-12.

Roch, P., Ville, P., and Cooper, E. L. (1998) Characterization of a 14 kDa plant-related serine protease inhibitor and regulation of cytotoxic activity in earthworm coelomic fluid. Dev. Comp. Immunol. Dev. Comp. Immunol. 22: 1-12.

Roch, Ph., Stabili, L., and Pagliara, P. (1991) Purification of three serine proteases from the coelomic cells of earthworms (Eisenia foetida). Comp. Biochim. Physiol. 98B:597-602).

Rouvier, E., Luciani, M.F., and Golstein, P. (1993) Fas involvement in Ca2+-independent T cell-mediated cytotoxicity. J. Exp. Med. 177: 195-200.

Russell, J.H. (1983) Internal disintegration model of cytotoxic lymphocyte-induced target damage. Immunol. Rev. 72: 97-118.

Sekizawa, Y., Kubo, T., Kobayashi, H., Nakajima, T., and Natori, S. (1997) Molecular cloning of cDNA for lysenin, a novel protein in the earthworm Eisenia foetida that causes contraction of rat vascular smooth muscle. Gene 191: 97-102.

Shinkai, Y., Takio, K., and Okumura, K. (1988) Homology of perforin to the ninth component of complement (C9). Nature 334: 525-527.

Shiver, J.W., Su, L., and Henkart, P.A. (1992) Cytotoxicity with target DNA breakdown by rat basophilic leukemia cells expressing both cytolysis and granzyme A. Cell 71: 315-321.

Tschopp, J. and Nabholz, M. (1990) Perforin mediated target cell lysis by cytolytic T lymphocytes. Annu. Rev. Immunol. 8: 279-302.

Walsh, C.M. Matloubian, M., Liu, C., Ueda, R., Kurahara, C., Christensen, J., Huang, M.T.F., Young, J.D., Ahmed, R., and Clark, W.R. (1994) Immune function in mice lacking the perforin gene. Proc. Natl. Acad. Sci. USA 91: 10854-10858.

Yamaji A, Sekizawa Y, Emoto K, Sakuraba H, Inoue K, Kobayashi H, and Umeda M. (1998) Lysenin, a novel sphingomyelin-specific binding protein. J. Biol Chem 273: 5300-5306.

Young, J.D.E. (1989) Killing of target cells by lymphocytes: A mechanistic view. Physiol. Rev. 69: 250-314.

Young, J.D.E., Hengartner, H., Podack, E.R., and Cohn, Z.A. (1986) Purification and Characterization of a cytolytic pore-forming protein from granules of cloned lymphocytes with natural killer activity. Cell 44: 849-859.

Young, L.H., Joag, S.V., Zheng, L.M., Lee, C.D., Lee, Y.S., and Young, J.D.E. (1990) Perforin mediated myocardial damage in acute myocarditis. Lancet 336: 1019-1021.

HORSESHOE CRAB HEMOCYTE-DERIVED LECTIN RECOGNIZING SPECIFIC O-ANTIGENS OF LIPOPOLYSACCHARIDES

K. Inamori[1], T. Saito[1], D. Iwaki[1], T. Nagira[1], S. Iwanaga[1], F. Arisaka[2], S. Kawabata[1]

[1]Department of Biology, Kyushu University, Fukuoka, 812-8581 Japan [2]Department of Life Science, Faculty of Bioscience and Bioengineering, Tokyo Inst. of Technology, Yokohama 226-8501, Japan

INTRODUCTION

In the Japanese horseshoe crab, *Tachypleus tridentatus*, one of the major defense systems is carried by hemolymph that contains granular hemocytes comprising 99% of the total hemocytes. These granular hemocytes are filled with two populations of secretory granules, named large (L) and small (S) granules. These granules selectively store defense molecules, such as clotting serine protease zymogens, a clottable protein coagulogen, protease inhibitors, lectins, and antimicrobial peptides (1-3). The

Phylogenetic Perspectives on the Vertebrate Immune System.
Edited by Beck *et al.*, Kluwer Academic/Plenum Publishers, 2001.

hemocytes are highly sensitive to LPS and these defense molecules stored in both granules are secreted by exocytosis after stimulation with LPS. This response is important for the host defense related to engulfing and killing invading microbes, in addition to preventing the leakage of hemolymph.

We reported evidence of several types of hemocyte-derived lectins, named tachylectins, which may play in a functional role of the innate immunity of this animal. Tachylectin-1, identical to an L-granule-derived protein L6 (4), binds to both S- and R-types of LPS, and inhibits the growth of Gram-negative bacteria (5). Tachylectin-2, identical to another L-granule-derived protein L10 (4), exhibits binding specificity to *N*-acetyl-D-glucosamine (GlcNAc) or *N*-acetyl-D-galactosamine (GalNAc) and agglutinates a certain strain of Staphylococcus (6). Tachylectin-4 has affinity for L-fucose and binds to S-type LPS from several Gram-negative bacteria through O-specific polysaccharides (O-antigens) (7). We have recently identified a new type of the hemocyte-derived lectin, named tachylectin-3, which differs structurally and functionally from those of the tachylectins mentioned before. Tachylectins-3 has specificity for a blood group A antigen and recognizes specific O-antigens of bacterial lipopolysaccharides.

RESULTS AND DISCUSSION

The purified tachylectin-3 isolated from the hemocyte lysate was used for subsequent experiments (8). Tachylectin-3 agglutinated human A-type erythrocytes, but not human B- and O-types of erythrocytes (8). Any monosaccharides, including GlcNAc and GalNAc did not inhibit the hemagglutinating activity of tachylectin-3, even at 100 mM. However, a blood group A-pentasaccharide completely inhibited the hemagglutination at the minimum inhibitory concentration (MIC) of 0.16 mM. On the other hand, a blood group A-trisaccharide had no inhibitory effect at 2.5 mM, although it did inhibit the hemagglutinating activity of tachylectin-2 (6) at MIC of 0.63 mM.

LPS from various Gram-negative bacteria were also examined, as shown in Table 1. S-type LPS were potent inhibitors, whereas R-type LPS were not inhibitory. LPS derived from *Escherichia coli* O111:B4 and several strains strongly inhibited the hemagglutination at MIC of 0.006 μg/ml. However, LPS isolated from rough mutants of *Escherichia coli* O111:B4, such as *Escherichia coli* J5 (Rc), showed no inhibitory effect, suggesting that the O-antigen is important for carbohydrate recognition of tachylectin-3. Lipoteichoic acid (LTA), a cell wall component of Gram-positive bacteria, also inhibited the hemagglutinating activity of tachylectin-3, but the MIC were significantly higher than that of LPS from *Escherichia coli* O111:B4. These results indicate that tachylectin-3 recognizes specific O-antigens of LPS derived from several Gram-negative bacteria.

Comparison of the chemical structures of the A-pentasaccharide and the O-antigens suggests a potential ligand for tachylectin-3 (Figure 1, underlined). A disaccharide of Gal α1,3GalNAc or GalNAc α1,3GalNAc in the repeating units of O-antigens is shared in these O-antigens. In addition to this fact, the GalNAc α1,3Gal structure of the A-pentasaccharide is very similar to those of the repeating units of O-antigens. However,

Table 1 Inhibition of hemagglutinating activities of tachylectins (TL) by LPS and LTA

Source	TL-2	TL-3	TL-4[a]
		MIC (μg/ml)[b]	
LPS			
Escherichia coli O111:B4 (S)	100	0.006	0.1
Escherichia coli O55:B5 (S)	25	0.006	-
Escherichia coli O128:B12 (S)	NI	0.012	-
Escherichia coli O127:B8 (S)	NI	0.006	-
Escherichia coli O26:B6 (S)	100	6.3	-
Escherichia coli EH100 (Ra)	NI	NI	-
Escherichia coli J5 (Rc)	NI	NI	NI
*Escherichia coli*F583 (Rd)	NI	NI	NI
Salmonella minnesota (S)	12.5	6.3	15.6
Salmonella minnesota R7 (Rd)	NI	NI	NI
Salmonella minnesota R595 (Re)	NI	NI	NI
Salmonella typhimurium (S)	NI	3.1	7.8
Salmonella typhimurium TV119 (Ra)	25	NI	NI
Salmonella typhimurium SL684 (Rc)	NI	NI	NI
Vibrio cholerae O22 (S)	NI	0.006	-
Vibrio cholerae O139 (S)	NI	25	-
Vibrio cholerae Inaba 569B (S)	NI	12.5	-
Klebsiella pneumoniae (S)	NI	6.3	-
Pseudomonas aeruginosa 10 (S)	100	NI	-
Shigella flexneri 1A (S)	100	6.3	-
LTA			
Staphylococcus aureus	6.3	3.1	31.3
Streptococcus faecalis	NI	100	31.3
Streptococcus mutans	50	100	62.5
Streptococcus pyogenes	25	50	125
Streptococcus sanguis	12.5	25	62.5
Bacillus subtilis	25	NI	500

a) Data from Saito et al. (7).
b) Minimum concentration required for inhibition of two hemagglutinating titers of tachylectins against human A-type erythrocytes.
-, not tested.
NI, not inhibited at 100 μg/ml (TL-2 and TL-3) or 1 mg/ml (TL-4).

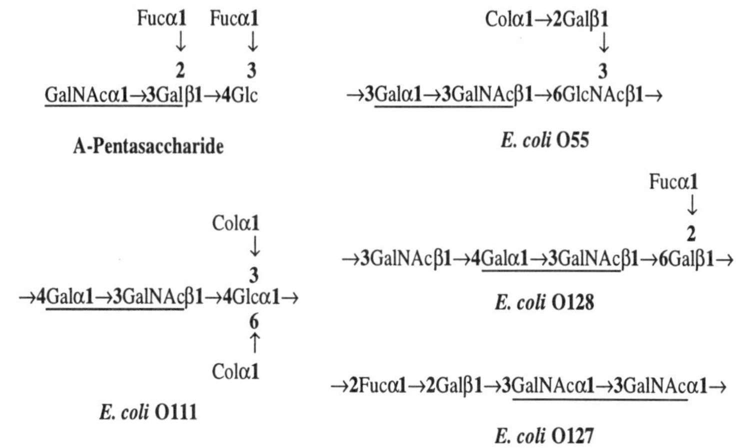

Figure 1. Chemical structures of the A-pentasaccharide and the repeating unit of O-antigens. Structures of the A-pentasaccharide and the repeating unit of O-antigens of LPS possessing potent inhibitory activity for the hemagglutinating activity of tachylectin-3.

the potential candidate cannot explain the inability of the A-trisaccharide (GalNAc α1,3 [Fuc α1,2] Gal) to inhibit the hemagglutinating activity of tachylectin-3, since the A-trisaccharide also contains the disaccharide of GalNAc α1,3Gal. Lima bean (Phaseolus lunatus) lectin binds to the A-trisaccharide and the A-tetrasaccharide of GalNAc α1,3 [Fuc α1,2] Gal β1,4Glc (9). Furthermore, the lectin more strongly binds to the A-hexasaccharide of GalNAc α1-3 [Fuc α1,2] Gal β1,3GlcNAc β1,3Gal β1,4Glc than the A-trisaccharide. Galectin-3 with affinity for β-galactosides, such as lactose (Gal β1,4Glc) and lactosamine (Gal β1,4GlcNAc), also has an affinity for the A-tetrasaccharide, and the affinity is about 30-fold higher than that of lactose (10, 11). Like these lectins, tachylectin-3 may recognize an annexed structure, in addition to the disaccharide of the O-antigens. Tachylectin-3 probably contains interaction sites for the extended oligosaccharides in addition to the primary binding sites, termed as subsite multivalency (12). Furthermore, tachylectin-3 forms a dimer in solution and this dimeric form may lead to the increase in affinity for appropriate multivalent ligands, such as O-antigens of LPS, termed as subunit multivalency (12).

Tachylectins isolated from hemocytes, such as tachylectin-1 and tachylectin-2, are located in L-granules of hemocytes (5, 6). Figure 2A indicates that tachylectin-3 is also located in L-granules of hemocytes. To determine whether or not tachylectin-3 could be released from hemocytes in response to external stimuli, hemocytes were treated with the calcium ionophore A23187 that induces exocytosis of horseshoe crab hemocytes (13, 14). Tachylectin-3 was indeed released into the extracellular fluid, as detected by immunoblotting and ELISA (Figure 2B). Similarly, the release of other hemocyte-derived lectins, including tachylectin-1 and tachylectin-2 were also confirmed, using specific polyclonal antibodies (data not shown).

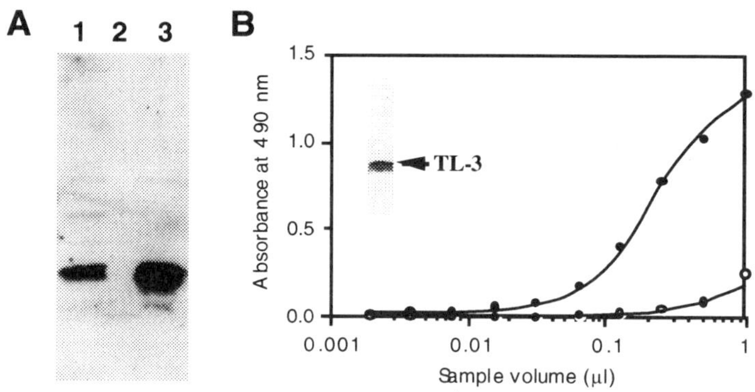

Figure 2. Subcellular localization and release of tachylectin-3 from hemocytes. A, subcellular localization of tachylectin-3. One μg each of large and small granule samples were treated with 2% SDS at 100°C for 3 min and subjected to SDS-PAGE under reducing conditions for immunoblotting. Lane 1, large granules; lane 2, small granules; lane 3, purified tachylectin-3. B, fresh hemocytes were incubated with 10 mM ionophore A23187 and the exocytosed fluid was subjected to ELISA using anti-tachylectin-3 antibody. Closed and open circles represent presence and absence of the ionophore A23187, respectively. The inset shows Western blot analysis on the exocytosed fluid, using anti-tachylectin-3 antiserum.

Mammalian lectins have been suggested to play an important role in host-pathogen interactions by specific recognition with cell surface substances of bacteria (15). For example, galectin-3, expressed in the cytosol of activated macrophages or epithelial cells, is secreted in response to inflammatory stimuli through non-classical pathways (16). Furthermore, galectin-3 binds to LPS and is likely to mediate interactions between host and pathogens (17). In the horseshoe crab, four types of lectins have been identified in the circulating hemocytes, and released from hemocytes in response to the external stimuli (Figure 2). Hence, they probably have a major functional role in host defense, including recognition, opsonization, and killing of invading microbes.

REFERENCES

1. Muta, T., and Iwanaga, S. (1996) Curr. Opin. Immunol., 8, 41-47
2. Kawabata, S., Muta, T., and Iwanaga, S.(1996) in New Directions in Invertebrate Immunology (Söderhäll, K., Iwanaga, S., and Vasta, G. R., eds) pp. 255-284, SOS Publications, Fair Haven, NJ
3. Iwanaga, S., Kawabata, S., and Muta, T. (1998) J. Biochem. (J. B. Review) 123, 1-15
4. Shigenaga, T., Takayenoki, Y., Kawasaki, S., Seki, N., Muta, T., Toh, Y., Ito, A., and Iwanaga, S. (1993) J. Biochem. 114, 307-316
5. Saito, T., Kawabata, S., Hirata, M., and Iwanaga, S. (1995) J. Biol. Chem., 270, 14493-14499
6. Okino, N., Kawabata, S., Saito, T., Hirata, M., Takagi, T., and Iwanaga, S. (1995) J. Biol. Chem., 270, 31008-31015
7. Saito, T., Hatada, M., Iwanaga, S., and Kawabata, S. (1997) J. Biol. Chem., 272, 30703-30708

8. Inamori, K., Saito, T., Iwaki, D., Nagira, T., Iwanaga, S., Arisaka, F., and Kawabata, S.(1999) J. Biol. Chem., 274, 3272-3278
9. Sikder, S. K., and Kabat, E. A. (1986) Carbohydr. Res. 151, 247-260
10. Leffler, H., and Barondes, S. H. (1986) J. Biol. Chem. 261, 10119-10126
11. Sparrow, C. P., Leffler, H., and Barondes, S. H. (1987) J. Biol. Chem. 262, 7383-7390
12. Rini, J. M. (1995) Annu. Rev. Biophys. Biomol. Struct. 24, 551-577
13. Armstrong, P. B., and Rickles, F. R. (1982) Exp. Cell Res. 140, 15-24
14. Miura, Y., Kawabata, S., Wakamiya, Y., Nakamura, T., and Iwanaga, S. (1995) J. Biol. Chem. 270, 558-565
15. Mandrell, R. E., Apicella, M. A., Lindstedt, R., Leffler, H. (1994) Methods Enzymol. 236, 231-254
16. Sato, S., and Hughes, R. C. (1994) J. Biol. Chem. 269, 4424-4430
17. Mey, A., Leffler. H., Hmama, Z., Normier, G., and Revillard, J.-P. (1996) J. Immunol. 156, 1572-1577

A FAMILY OF C-TYPE LECTINS IN *Manduca sexta*

Xiao-Qiang Yu and Michael R. Kanost

Department of Biochemistry, Kansas State University, Manhattan, KS 66506

INTRODUCTION

Insects have a rapid and effective immune system against microbial infections (Gillespie *et al.*, 1997) that shares common characteristics with the innate immune system of vertebrates. It also consists of humoral and cellular responses. In the cellular responses such as phagocytosis, nodule formation and encapsulation, hemocytes play important roles. In insect humoral responses, many hemolymph proteins are involved. These proteins include a group of antimicrobial proteins/peptides, lectins, and cell adhesion proteins.

In insect defense against microbial infections, non-self recognition molecules may play important roles in recognizing and clearing invading microbes. Pattern recognition molecules or receptors can recognize certain patterns on the surface of microorganisms (Janeway Jr., 1992). Such microbial patterns are lipopolysaccharide (LPS) and peptidoglycan from bacterial cell walls, and β-1,3-glucan and zymosan from fungal cell walls.

Phylogenetic Perspectives on the Vertebrate Immune System.
Edited by Beck *et al.*, Kluwer Academic/Plenum Publishers, 2001.

Lectins are sugar-binding proteins that participate in the innate immune system. Many animal C-type lectins (calcium-dependent lectins) have been reported be important in the innate immune system of vertebrates (Weis *et al.*, 1998). A subgroup of the C-type lectin superfamily, collectins, play a key role in the first line of defense. Members of the collectins all contain a carbohydrate recognition domain (CRD) connected to a collagen domain. The most extensively studied example is the serum mannose-binding protein (MBP). MBP can activate the complement system through the lectin pathway. Activation of the complement system by MBP is associated with C1r/C1s-like proteases.

C-type lectins have also been isolated from a few insect species. These insect lectins had different functions in the innate immune system, including hemocyte nodule formation (Koizumin *et al.*, 1999), prophenol oxidase activation (Yu *et al.*, 1999; Chen *et al.*, 1995), and opsonization (Jomori and Natori, 1992).

RESULTS AND DISCUSSION

In the tobacco hornworm, *Manduca sexta*, we have discovered a new family of C-type lectins, each of which consists of two tandem carbohydrate recognition domains (CRDs). This new family includes three lectins isolated so far from *M. sexta*, immulectin (Yu *et al.*, 1999), lipopolysaccharide (LPS)-binding lectin (MsLBL), and N-acetylgalactosamine-binding lectin (GalNac-BL). Members of this family were also isolated from two other insect species, including the LPS-binding protein (LBP) from the silkworm *Bombyx mori* (Koizumin *et al.*, 1999), and a putative lectin Hdd15 from the fall webworm *Hyphantria cunea* (Shin *et al.*, 1998). Figure 1 shows the alignment of *M. sexta* immulectin (Ms-IL) with *B. mori* LPS-binding protein (BmLBP) and *H. cunea* Hdd15.

M. sexta immulectin is synthesized in fat body upon challenge with bacteria or yeast, and the protein is secreted into hemolymph (Yu *et al.*, 1999). Immulectin has a predicted ligand binding specificity for mannose or glucose. Immulectin can agglutinate gram-negative, gram-positive bacteria, and yeast, in a calcium-dependent manner (Yu *et al.*, 1999). Agglutination of bacteria by immulectin can be inhibited by LPS, but not by laminarin (β-1,3-glucan). It seems that immulectin has a ligand specificity for a complex carbohydrate.

We also found that immulectin can stimulate prophenol oxidase activation. Immulectin alone can activate phenol oxidase activity to a certain level. However, immulectin complexed with LPS can activate phenol oxidase more rapidly and to a higher level (Figure 2). This indicates that immulectin may function as a non-self recognition protein in the immune system of *M. sexta*.

Recently, we have cloned two more new C-type lectins, LPS-binding lectin (MsLBL) and N-acetylgalactosamine binding lectin (GalNac-BL) from *M. sexta*. These two lectins have the same two-CRD organization as in immulectin. MsLBL showed 27% identity in amino acid sequence to immulectin, 39% to GalNac-BL, 55% to BmLBP, and 47% to Hdd15. It has a ligand binding specificity for xylose/glucose and for LPS and mannan. Unlike immulectin, MsLBL is present at a low constitutive level in hemolymph in naive larvae, and its synthesis is induced only after injection of gram-negative bacteria or LPS. Therefore, MsLBL has more restricted ligand binding specificity for LPS or gram-negative bacteria. MsLBL can agglutinate gram-negative bacteria at low physiological concentration. The agglutination was both concentration and calcium dependent. We also observed that MsLBL complexed with LPS can stimulate prophenol oxidase activation in hemolymph.

```
                   R  DY Y       K                C  EG  L  P
Ms-IL     PSPYQTKQYR SDYVYNKKTD AFYKLHIEGK RGWQVQKLCE VEGAKLMVPT
BmLBP     ......QQFR YDYTYMRDIN GWLKLQEIPA TWQEARLRCH LEGSLLASPL
Hdd15     ......VKFR CDYEYSDI.. GWFKYHEIPA TWDEARLRCH LEGAVLASPT
                     L                                PL
Ms-IL     T...QLDIIQ LHSMFKRFPD LGNYVWVAED GHNHESAEEQ PLIVLTPNPE
BmLBP     DDALKSGMLS LIKNKKTSCG IFTGIHATFS KGDYRSVEGV PLAKIPHDWA
Hdd15     TDKMKSIMLK LF....CKPE VFTGIAAIFS KGDYYTIDGI PLTEIHHEWA

             P         C      G       C        C              CG
Ms-IL     DSQPRDTWHS A.CDVVTRTG EVETYPCYRE LPFMCKVDAR DAPYDNHCGV
BmLBP     DYEPDNAGGD ENCILMNPDG NFADVNCTET FQYVCYKKKT ATLAMASCGS
Hdd15     QCEPDNKNND ENCTALSSDG KLSDVRCDAP RPYICYRE.Y SKVDVNLCGT

              Y         CYK              A  C  EG  L  IN E E
Ms-IL     YARDYEYIES VGSCYKIPRV VYPWNQAYAE CQAEGAHLVV INSEAEMLAV
BmLBP     VDSEYVLSKD TGNCYKFHKV PRTWSRAYMA CSAEGGYLTI INNEKEATFL
Hdd15     PDPDYHFETQ TNTCYKFHTK ARNFERAHMV CSSEGAHLAI INSEEEAKVI

                        G          G                  T I        AG
Ms-IL     KNII..NTKP SVLGATTSYF FFAGFRAEPA QDGKPKVFKT IFNQTLEEAG
BmLBP     RDLFAKNPAG QMIGSFWKDV AFIGFH.... DWNERGEWLT INGEKLQEAG
Hdd15     AQIFARYPKE KVVGSPHPDI AVIGYK.... YWDLNLEWTT IQGQPIQKAG

          Y      P  N    E CG          ND  C     FI CEK
Ms-IL     YSQWSPEPN NFDNKEDCGT LFKNDGNFND VICSHPYAFI CEKEVHL    (1-291)
BmLBP     YEKWSGEPS NATTGEYCGS IYRS.ALLND LWCEKPAPFI CEKEPRS   (20-305)
Hdd15     YAKFAPQPD NFKNHEYCGT VFRT.GLLND GDCDVKYPFI CEKKPDF   (19-297)
```

Figure 1. Alignment of insect C-type lectins consisting of two carbohydrate recognition domains (CRDs). The mature polypeptide sequences of three insect lectins are aligned. Residues conserved in all the three lectins are placed on the top of the alignment. Ms-IL: *M. sexta* immulectin; BmLBP: *B. mori* LPS-binding protein; Hdd15: putative lectin from *H. cunea*. The two residues that are important for the determination of ligand binding specificity are underlined.

The new family of C-type lectins discovered in the tobacco hornworm, *M. sexta*, may function as non-self recognition proteins for microorganisms with different specificities. Since there is no antibody-mediated immunity in insects, multiple lectins with diverse specificities may function as non-clonal effectors in the insect immune system. So far, multiple lectins have been isolated from four insect species: the tobacco hornworm *M. sexta*, the American cockroach *Periplaneta americana*, the silkworm *B. mori* and the West Indian leaf cockroach *Blaberus discoidalis* (Wilson *et al.*, 1999). These insect lectins may contain one CRD or two tandem CRDs. The organization of CRDs in single polypeptide and multimerization of CRDs between polypeptides may form different binding sites for microbial components.

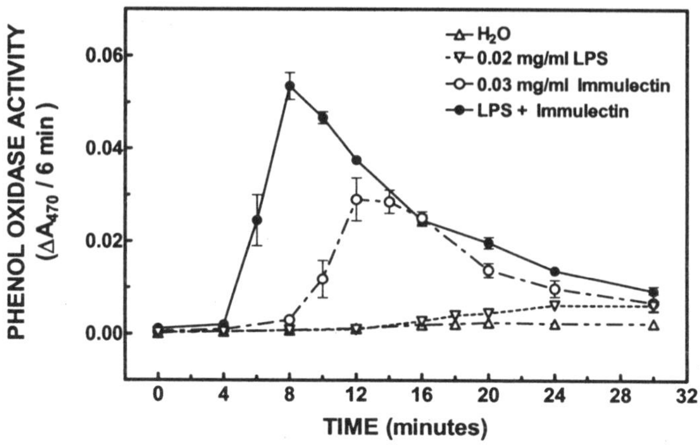

Figure 2. Activation of phenol oxidase by immulectin. Forty µl of cell free hemolymph collected from a naive fifth instar day 2 larva was incubated with 1.0 mg of LPS, 1.5 mg of recombinant immulectin (from baculovirus expression) or a mixture of LPS and immulectin at room temperature. At time intervals, an aliquot of hemolymph was removed and added to 2 mM dopamine solution in 50 mM sodium phosphate, pH 6.5. Absorbance at 470 nm was measured over 6 min. The points represent the mean of four individual measurements (except for the point at 6 minutes for the mixture of LPS plus immulectin, which was from two individual measurements). The bar represents the standard error of the mean. (From Yu *et al.*, 1999).

REFERENCES

Chen, C., Durrant, J., Newton, R. P. and Ratcliffe, N. A. (1995) A study of novel lectins and their involvement in the activation of the prophenoloxidase system in *Blaberus discoidalis*. *Biochem. J.* **310**, 23-31.

Gillespie, J. P., Kanost, M. R. and Trenczek, T. (1997) Biological mediators of insect immunity. *Annu. Rev. Entomol.* **42**, 611-643.

Janeway Jr., C. A. (1992) The immune system evolved to discriminate infectious nonself from noninfectious self. *Immunol. Today* **13**, 11-16.

Jomori, T. and Natori, S. (1992) Function of the lipopolysaccharide-binding protein of *Periplaneta americana* as an opsonin. *FEBS Letters* **296**, 283-286.

Koizumi, N., Imamura, M., Kadotani, T., Yaoi, K., Iwahana, H. and Sato, R (1999) Thelipopolysaccharide-binding protein participating in hemocyte nodule formation in the silkworm *Bombyx mori* is a novel member of the C-type lectin superfamily with two different tandem carbohydrate-recognition domains. *FEBS Letters* **443**, 139-143.

Shin, S. W., Park, S.-S., Park, D.-S., Kim, M. G., Kim, S. C., Brey, P. T. and Park, H.-Y. (1998) Isolation and characterization of immune-related genes from the fall webworm, *Hyphantria cunea*, using PCR-based differential display and subtractive cloning. *Insect Biochem. Molec. Biol.* **28**, 827-837.

Weis, W. I., Taylor, M. E. and Drickamer, K. (1998) The C-type lectin superfamily in the immune system. *Immunol. Rev.* **163**, 19-34.

Wilson, R., Chen, C. and Ratcliffe, N. A. (1999) Innate immunity in insects: the role of multiple, endogenous serum lectins in the recognition of foreign invaders in the cockroach, *Blaberus discoidalis*. *J. Immunol.* **162**, 1590-1596.

Yu, X.-Q., Gan, H. and Kanost, M. R. (1999) Immulectin, an inducible C-type lectin from an insect, *Manduca sexta*, stimulates activation of plasma prophenol oxidase. *Insect. Biochem. Molec. Biol.* In press.

ROLE OF TACHYLECTINS IN HOST DEFENSE OF THE JAPANESE HORSESHOE CRAB *TACHYPLEUS TRIDENTATUS*

S. Kawabata[1], H. -G. Beisel[2], R. Huber[2], W. Bode[2], S. Gokudan[1], T. Muta[1], R. Tsuda[1], K. Koori[1], T. Kawahara[1], N. Seki[1], Y. Mizunoe[3], S. N. Wai[3], S. Iwanaga[1]

[1]Department of Biology, Kyushu University Fukuoka, 812-8581 Japan, [2]Max-Planck-Institute fur Biochemie, D-82152 Martinsried, Germany, [3]Department of Bacteriology, Graduate School of Medical Sciences, Kyushu University Fukuoka, 812-8582 Japan

INTRODUCTION

Immunity to pathogens is mediated by two general systems, innate and acquired immunity. Acquired immunity is found only in vertebrates and it is the system of B and T lymphocytes, which produces a multitude of specific antigen receptors and

Phylogenetic Perspectives on the Vertebrate Immune System.
Edited by Beck *et al.*, Kluwer Academic/Plenum Publishers, 2001.

antibodies by somatic gene rearrangement. Innate immunity is phylogenetically older than acquired immunity and is present in all multicellular organisms. These proteins pre-exist in hosts, or are rapidly induced within hours of infection. With this immunity polysaccharides or other substances on pathogens are detected through pattern recognition (1,2). The innate immunity of invertebrates is triggered by polysaccharide derivatives, such as lipopolysaccharides (LPS), proteoglycans, and β-1,3-glucans. Examples are hemolymph coagulation in horseshoe crabs (3-5) and phenoloxidase-mediated melanization in crustaceans and insects (6).

In the Japanese horseshoe crab, *Tachypleus tridentatus,* one of the major defense systems is carried by the hemolymph that contains granular hemocytes comprising 99% of the total hemocytes. The granular hemocytes are filled with two populations of secretory granules, named large and small granules. The granules selectively store defense molecules, such as clotting factors, a clottable protein coagulogen, protease inhibitors, lectins, and antimicrobial peptides (3-5). The hemocytes are highly sensitive to LPS and the defense molecules stored in granules are secreted by exocytosis in response to the stimulation of LPS. This response is very important for host defense involving the engulfing and killing invading microbes, in addition to preventing the leakage of hemolymph. Non-self recognizing lectins, named tachylectins, have been identified in *Tachypleus tridentatus*, four from hemocytes (tachylectins-1 to 4) and two (tachylectins-5A and -5B) from plasma, as shown in Table 1. Tachylectin-1 interacts with Gram-negative bacteria through KDO (2-keto-3-deoxyoctonate) and lipid A of LPS (7). Tachylectin-3 (8) and tachylectin-4 (9) specifically bind to S-type LPS found in several Gram-negative bacteria with a certain sugar moiety on O-specific polysaccharides. Here, we review recent functional and structural studies of tachylectin-2 and tachylectins-5A and -5B, and discuss their importance in the innate immunity of this invertebrate.

TACHYLECTIN-2

Tachylectin-2 agglutinates human A-type erythrocytes and it binds specifically to *N*-acetylglucosamine (GlcNAc) with Kd of 0.05 mM (10). Detailed sugar binding analysis indicated that the acetamide group at the C-2 position and a free OH group at the C-4 position of GlcNAc are required for recognition by tachylectin-2. Tachylectin-2 also recognizes several of LPS and lipoteichoic acids of Gram-positive bacteria.

Tachylectin-2 consists of 236 amino acid residues and the most interesting feature of the sequence is the presence of five-tandem repeats of 47 amino acids. The sequence identity between the repeats ranges from 49 to 68%. Tachylectin-2 contains no cysteine and no *N*- and *O*-linked sugars. The amino acid sequence exhibits no significant sequence similarity with other known proteins. Tachylectin-2 is present as a monomer in solution, as judged from ultracentrifugal analysis, and therefore at least two sugar binding sites must be present in the molecule for the expression of agglutinating activity. Recently, the X-ray structure of tachylectin-2 was solved at 2.0 Å resolution and the X-ray structure of the complex with GlcNAc was also determined (11).

Tachylectin-2 is a 5-fold β-propeller structure and it exhibits the shape of a rather flat pentagonal torus (Fig. 1). The single chain of tachylectin-2 is organized in five β-sheets, which are arranged in consecutive order with 5-fold symmetry around a central tunnel. Each twisted β-sheet is built up by four antiparallel β-strands. Protein structures with a 4- to 8-fold β-propeller geometry have been reported, and tachylectin-2 is the first example of a 5-fold β-propeller structure. The X-ray structure of tachylectin-2 in the complex with GlcNAc reveals five equivalent binding sites, with virtually identical occupancy and geometry in the crystal, perfectly reflecting the 5-fold symmetry of tachylectin-2. The GlcNAc points its acetamide group to the bottom of the binding pocket and the C6-OH group points outside into solvent.

Table 1. Tachylectins (TLs) found in hemocytes and hemolymph plasma of the Japanese horseshoe crab.

TLs	Molecular weight	Specificity	Localization
TL-1	27,000 (SP) 24,383 (Se)	LPS (KDO), LTA	Hemocyte (L)
TL-2	27,000 (SP) 26,757 (Se) 27,000 (Uc)	GlcNAc, LTA	Hemocyte (L)
TL-3	15,000 (SP) 13,723 (Se) 28,850 (Uc)	LPS (O-Antigen), LTA	Hemocyte (L)
TL-4	30,000 (SP) 26,625 (Se) 470,000 (Gf)	LPS (O-Antigen), LTA	Hemocyte
TL-5A	41,000 (SP) 31,185 (Se) 210,000/300,000 (Gf)	Acetyl group	Plasma
TL-5B	41,000 (SP) 33,069 (Se) 160,000 (Gf)	Acetyl group	Plasma

LPS, lipopolysaccharides; KDO, 2-keto-3-deoxyoctonic acid; LTA, lipoteichoic acid; L, large granule of hemocytes; SP, SDS-PAGE, Se, sequence; Uc, ultracentrifugation; Gf, gel filtration.

A binding site geometry of tachylectin-2 in complex with GlcNAc is shown in Figure 2. Carbonyl oxygens and NH-groups of the protein backbone accomplish most hydrogen bonds between the protein and the carbohydrate. Of the sugar functional groups, the acetamide group at the C2 position and C3-OH and C4-OH are used for ligand recognition. The acetamide methyl group interacts with the indole group of Trp 134 through hydrophobic interaction.

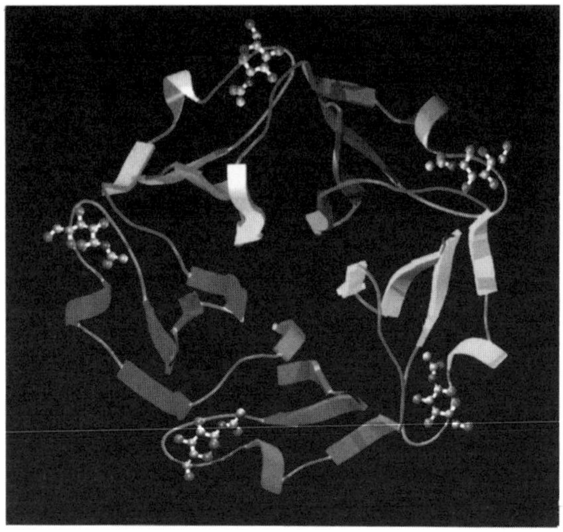

Figure 1. Ribbon plot of the 5-fold tachylectin-2 β-propeller structure showing bound GlcNAc as a ball-and-stick model.

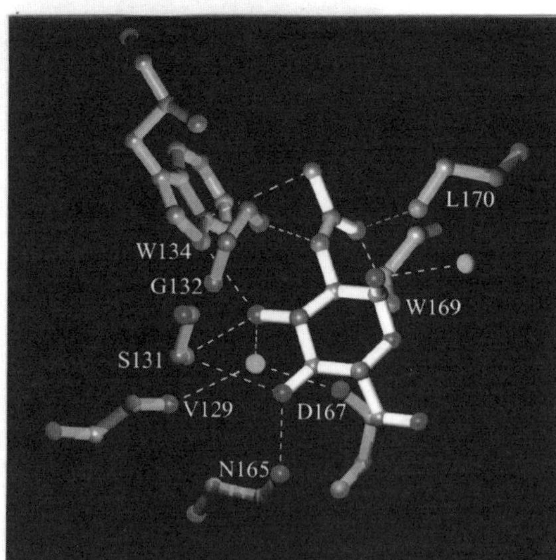

Figure 2. A binding site geometry (blade IV) of tachylectin-2 in complex with GlcNAc.

Tachylectin-2 shows virtually no change in main or side chain conformation on binding of GlcNAc. The deviation (root mean square) between the free tachylectin-2 and the complex with the ligands is only 0.18 Å, suggesting no cooperativity for ligand binding. The central tunnel is not a ligand-binding site but it is filled with water molecule. The water molecules are arranged in a pentagonal dodecahedron. This structural element of water molecules is also found in clathrate hydrates. However, the water cage in tachylectin-2 is empty and no guest species is observed in the cage. The water dodecahedron is framed by the polar main chain groups of the five innermost β-strands and by the benzyl side chains of the conserved Phe residues. This water cage is not affected by the crystal packing of tachylectin-2, and it is expected to exist also in solution. It may contribute to the stabilization of the protein fold.

The nature of the binding pocket explains the strict specificity of tachylectin-2 for GlcNAc. The exoskeleton of horseshoe crab contains a ubiquitous polysaccharide of β-1,4-linked GlcNAc, chitin. However, tachylectin-2 shows no significant binding affinity to chitin, mainly because there is no free 4-OH group in the β-1,4-linked GlcNAc units except for the non-reducing end. The mechanism of self/non-self distinction is reinforced by a distance of two individual binding sites of 25 Å and 40 Å, respectively, which seems to prevent multiple binding to distant terminal GlcNAc units of chitin. Therefore, the high valency of five binding sites of the single molecule suggests the recognition of GlcNAc units on pathogens with a fairly high ligand density, as shown in figure 3. Multiple binding of the five binding sites to the repetitive structures on pathogens will generate very tight interactions.

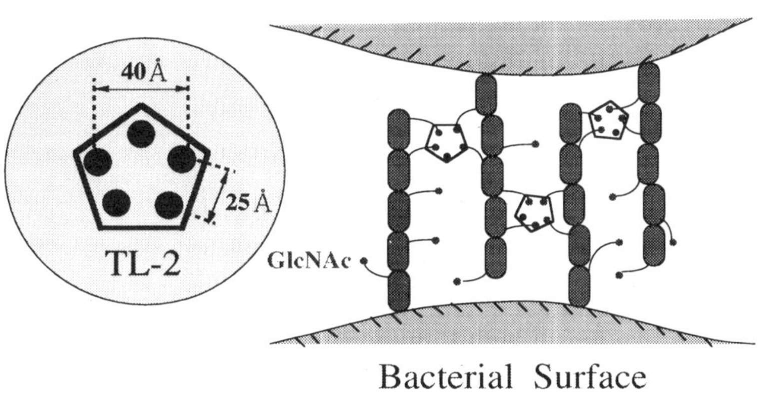

Figure 3. Molecular pattern on pathogen recognized by tachylectin-2.

TACHYLECTINS-5A AND-5B

The hemolymph plasma has been also reported to exhibit hemagglutinin activity. However, much less is known of plasma-derived lectins. We have found that the hemagglutinin activity in the hemolymph plasma is adsorbed mostly to an *N*-acetyl group-immobilized resin (12).

The acetyl group was coupled with a noncarbohydrate resin TOYOPEARL, a hydrophilic vinyl polymer. When hemolymph plasma was applied to this column, about two thirds of the total hemagglutinating activity bound to the resin. Two protein peaks with hemagglutinating activity were obtained by stepwise elution, using different concentrations of GlcNAc. The first lectin eluted by 25 mM GlcNAc gave a smear band of 72-74 kDa on SDS/PAGE, under nonreducing conditions. In contrast, the second lectin eluted by 250 mM GlcNAc gave a single band of 72 kDa. After reduction, both protein bands respectively shifted to a smear band of 40-42 kDa and a single band of 41 kDa, suggesting that each hemagglutinin consists of two identical subunits connected by disulfide linkage(s). We named these isolated proteins tachylectins-5A and -5B for first and second lectins, respectively.

Tachylectins-5A and -5B agglutinated all types of human erythrocytes, indicating that the primary recognition substance is not a blood group antigen. Tachylectin-5A had stronger hemagglutinating activity than did tachylectin-5B. The hemagglutinating activities of tachylectins-5A and -5B were completely inhibited by 5 mM EDTA, and this inhibition was overcome by adding an excess amount of $CaCl_2$.

Mono- and disaccharides containing the acetyl group, such as GlcNAc and GalNAc, inhibited the hemagglutination, but Glc and Gal did not. Furthermore, *N*-acetyl derivatives of amino acids and other acetyl group-containing substances including acetylsalicylic acid, acetylcholine, and acetyl coenzyme A were also effective competitors. One of the simplest chemicals containing the acetyl group, sodium acetate or acetamide, strongly inhibited the activity even at millimolar levels. Propionamide and benzamide also inhibited the hemagglutination, but 4 to 8-fold higher concentrations were required. On the other hand, sodium formate, as well as methyl-α-glucoside and methyl-β-glucoside were without effect. Therefore, acetyl group is a necessary and sufficient condition.

Tachylectins-5A and-5B also agglutinated Gram-negative and Gram-positive bacteria, and the activity was more efficient for Gram-negative than for Gram-positive bacteria. As in the case of hemagglutinating activity, tachylectin-5A exhibited stronger bacterial agglutinating activity than did tachylectin-5B, particularly against Gram-positive bacteria.

Expression of tachylectin-5A was detected in heart and intestine by immunoblotting. In contrast, tachylectin-5B was detectable only in hemocytes. cDNA clones for tachylectins-5A and -5B were isolated from the cDNA libraries of heart and hemocytes, respectively. The overall sequence identity between tachylectins-5A and -5B was 45%. A homology search revealed significant sequence similarity of the COOH-terminal regions of tachylectins-5A and -5B with those of vertebrate fibrinogens and their related proteins, consisting of 210-250 amino acid residues.

Pairwise comparisons of the COOH-terminal domains of tachylectins-5A and -5B with the 37 fibrinogen-like domains were made and normalized alignment scores (NAS) were calculated, according to the method of Doolittle (13). Based on the relationship between lengths of aligned sequences and NAS, the domains with NAS over about 250 points are homologous. NAS for fibrinogen β- (NAS = 340-374) and γ- (NAS = 279-368) chains derived from various species were sufficiently high to be homologous. In contrast, fibrinogen α-chain had only 15-17% identity with low NAS of 38-83. Amoung the proteins containing the fibrinogen-like domain, mammalian ficolins showed strikingly sequence identity of up to 51%. Ficolins are composed of an NH2-terminal cysteine-rich segment followed by a collagen-like region and a COOH-terminal fibrinogen-like globular domain (14). Interestingly, a collagenous domain found in ficolins is missing in the corresponding regions of tachylectins-5A and -5B.

Electron microscopy of negatively stained tachylectins-5A and -5B provided higher resolution images of their oligomeric structures. Tachylectin-5A formed a three- or four-bladed propeller structure, and tachylectin-5B formed a two-bladed propeller structure. If the ligand binding site(s) is located in each blade, tachylectin-5A could have more avidity against polymeric ligands than tachylectin-5B. This may explain why tachylectin-5A exhibits higher hemagglutinating and bacterial agglutinating activities than does tachylectin-5B.

It has been suggested that nonvertebrates may have fibrinogen, perhaps present inside blood cells. The horseshoe crab is equipped with a unique functional homologue of vertebrate fibrinogen, coagulogen, as the target protein of the clotting cascade. However, crystal analysis indicated that coagulogen is a structural homologue of nerve growth factor (15). Our observations show here that the horseshoe crab has fibrinogen-related molecules in hemolymph plasma and that they function as non-self recognizing lectins. An ancestor of fibrinogen-like proteins may have functioned as a non-self recognizing protein.

ACKNOWLEDGEMENTS

A part of this work was supported by a Grant-in-Aid for scientific research from Ministry of Education, Science, Sports and Culture of Japan (S.K.) and CREST of Japan Science and Technology Corporation (S.K.).

REFERENCES

1. Janeway, C. A., Jr. Approaching the asymptote? Evolution and revolution in immunology. *Cold Spring Harbor Symp. Quant. Biol.* **54**, 1-13 (1989).
2. Medzhitov, R. and Janeway, C. A. Jr. Innate immunity: The virtues of a nonclonal system of recognition. *Cell* **91**, 295-298 (1997).
3. Muta, T. and Iwanaga, S. The role of hemolymph coagulation in innate immunity. *Curr. Opin. Immunol.* **8**, 41-47 (1996).
4. Kawabata, S., Muta, T. and Iwanaga, S. Clotting cascade and defense molecules found in hemolymph of horseshoe crab. in *New directions in invertebrate immunology* (Söderhäll, K., Iwanaga, S. and Vasta, G. R., eds.) SOS Publications, Fair Haven, NJ., 1996, pp. 255-284.
5. Iwanaga, S., Kawabata, S. and Muta, T. New types of clotting factors and defense molecules found in horseshoe crab hemolymph: their structures and functions. *J. Biochem.* **123**, 1-15 (1998).
6. Söderhäll, K. and Cerenius, L. Role of the prophenoloxidase-activating system in invertebrate immunity. *Curr. Opin. Immunol.* **10**, 23-28 (1998).
7. Saito, T., Kawabata, S., Hirata, M. and Iwanaga, S. A novel type of limulus lectin-L6: purification, primary structure and antibacterial activity. *J. Biol. Chem.* **270**, 14493-14499 (1995).
8. Inamori, K., Saito, T., Iwaki, D., Nagira, T., Iwanaga, S., Arisaka, F. and Kawabata, S. A newly identified horseshoe crab lectin with specificity for blood group A antigen recognizes specific O-antigens of bacterial lipopolysaccharides. *J. Biol. Chem.* **274**, 3272-3278 (1999).
9. Saito, T., Hatada, M., Iwanaga, S. and Kawabata, S. A newly identified horseshoe crab lectin with binding specificity to O-antigen of bacterial lipopolysaccharides. *J. Biol. Chem.* **272**, 30703-30708 (1997).

10. Okino, N., Kawabata, S., Saito, T., Hirata, M., Takagi, T. and Iwanaga, S. Purification, characterization, and cDNA cloning of a 27-kDa lectin (L10) from horseshoe crab hemocytes. *J. Biol. Chem.* **270**, 31008-31015 (1995).
11. Beisel, H.-G., Kawabata, S., Iwanaga, S., Huber, R. and Bode, W. Tachylectin-2: crystal structure of a specific GlcNAc/GalNAc-binding lectin involved in the innate immunity host defense of the Japanese horseshoe crab *Tachypleus tridentatus*. *EMBO J.* **18**, 2313-2322 (1999).
12. Gokudan, S., Muta, T., Tsuda, R., Koori, K., Kawahara, T., Seki, N., Mizunoe, Y., Wai, S. N., Iwanaga, S. and Kawabata, S. Horseshoe crab acetyl group-recognizing lectins involved in innate immunity are structurally related to fibrinogen. *Proc. Natl. Acad. Sci. USA* (1999) in press.
13. Doolittle, R. F. in *Of Urfs and Orfs: a primer on how to analyze derived amino acid sequence*. University Science Books, Mill Valley, CA, 1987, pp. 10-15.
14. Lu, J. and Le, Y. Ficolins and the fibrinogen-like domain. *Immunobiol.* **199**, 190-199 (1998).
15. Bergner, A., Oganessyan, V., Muta, T., Iwanaga, S., Typke, D., Huber, R. and Bode, W. Crystal structure of coagulogen, the clotting protein from horseshoe crab: a structural homologue of nerve growth factor. *EMBO J.* **15**, 6789-6797 (1996).

HEMOLYTIC CAPACITY, PROPERTIES, AND ACTIVATION REQUIREMENTS OF COBRA VENOM FACTOR-TREATED CHICKEN SERUM

T. L. Koppenheffer

Trinity University, Biology, 715 Stadium Dr.
San Antonio, TX 78212

INTRODUCTION

In contrast to the depletion of complement (C') activity which occurs following the addition of cobra venom factor (CoVF) to mammalian serum, we have found that CoVF-treated chicken serum retains its lytic potential. We report herein some of the lytic and physiochemical properties of CoVF-treated chicken serum, and compare its activity to that of similarly treated turtle and bullfrog sera.

MATERIALS & METHODS

Pooled serum was obtained by bleeding White Leghorn chickens, bullfrogs (*Rana catesbeiana*), and turtles (*Pseudemys scripta*) and stored at –70°. Rat, mouse and

Phylogenetic Perspectives on the Vertebrate Immune System.
Edited by Beck *et al.*, Kluwer Academic/Plenum Publishers, 2001.

chicken blood were collected into PBS-EDTA and used within a day, while rabbit, sheep, horse, bovine, and goose blood were purchased from the Dept. of Veterinary Sciences, The University of Texas M.D. Anderson Cancer Center, Bastrop, TX. Sera were adsorbed twice with one-fifth volumes of washed target cells before use. Isotonic phosphate-buffered saline (PBS), veronal buffered saline (VBS), VBS containing 0.6 mM $MgCl_2$ and 0.6 mM $CaCl_2$ (VBS-M^{2+}), and VBS containing 10 mM EDTA (VBS-EDTA) were used as buffers. Prior to use erythrocytes (RBC) were washed three times in VBS and brought to 1% with an appropriate buffer. Hemolytic activity was measured by a standard microhemolytic test in which 0.1 ml of 1% target RBC were added to 0.5 ml of varying serum dilutions and incubated for 60-90 min. After this time 1.9 ml of cold PBS was added to each tube, the mixtures centrifuged, and the absorbance of the supernatants measured at 412 nm. Lytic activity was quantified and expressed as the percentage specific lysis.

Cobra venom factor was purchased from Calbiochem, and upon receipt was portioned into small volumes and stored at –70°. In a typical experiment 25 μl of serum was mixed with 0.5 U CoVF and diluted to a final volume of 50 ml with VBS-M^{2+}. Occasionally multiples of these quantities were employed. Thereafter, the mixtures were incubated at 37° for 30 min. and tested for lytic activity as previously described. To determine divalent cation requirements for hemolytic complex formation, 25 μl of serum was chelated by adding 2 μl of 0.1 M EDTA (pH 7.4) and incubating for 10 min. at room temp. An excess of Mg^{2+}, Ca^{2+}, or both divalent cations was then added and the mixtures brought to a final volume of 50 μl, incubated for 30 min. at room temp., and tested for hemolytic activity. Divalent cation requirements for hemolytic activity were determined by testing CoVF-treated serum diluted in VBS-EDTA.

RESULTS

The lytic activity of CoVF-treated chicken serum against RBC of various species is shown in Figure 1. The highest level of activity occurs against horse RBC, and the lowest against mouse RBC. This activity far exceeds that which occurs by the alternative C' pathway (Fig. 2) or Ab-dependent mechanisms (Fig. 3).

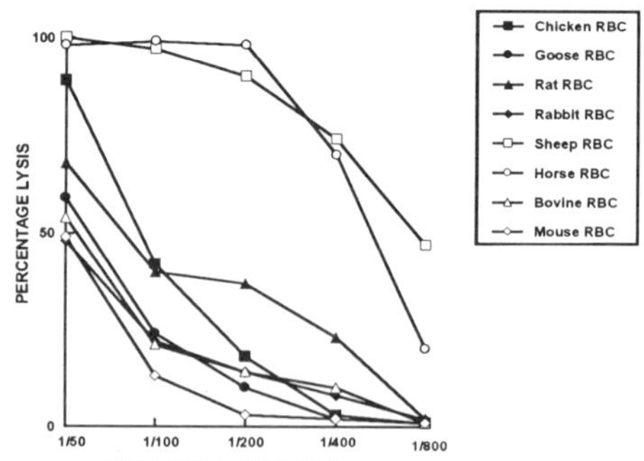

Figure 1. Hemolytic activity of CVF-treated chicken serum against various vertebrate erythrocytes.

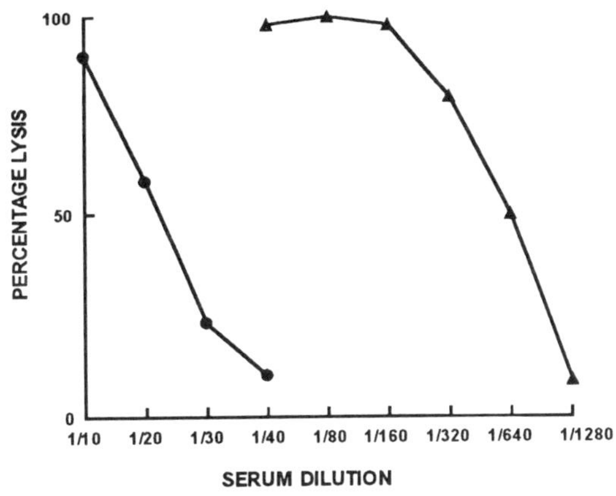

Figure 2. Lysis of unsensitized horse RBC by normal (•) and CoVF-treated (Δ) (0.5 U per 25μl serum) chicken serum.

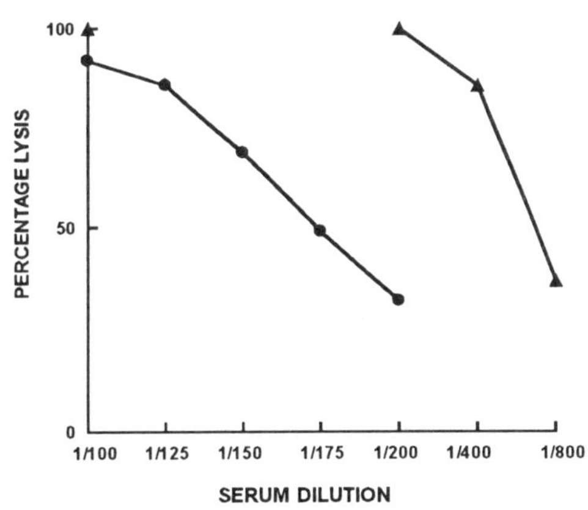

Figure 3. Lysis of sensitized sheepRBC by normal chicken serum (•) and unsensitized sheep RBC by CoVF-treated (0.5 U per 25μl serum) chicken serum (Δ).

The hemolytically active complex formed by the action of CoVF is very stable, showing only a slight loss in activity after heating at 56° or storage (data not shown). The initial interaction involving CoVF requires Ca^{2+} ions, but once formed, the hemolytic complex or factor lyses target cells in their absence (Table I). In contrast to CoVF-treated chicken serum, similarly treated bullfrog (Fig. 4) and turtle (Fig. 5) sera are depleted of lytic activity.

DISCUSSION

We originally carried out experiments with CoVF-treated chicken serum in an attempt to selectively inhibit alternative C' pathway activity as has been reported by (Ohta et al., 1984; 1986). To the contrary, our observations indicate that CoVF interacts with one or more components of chicken serum to produce a complex or factor that exhibits unusually high lytic activity. At present it is not clear whether CoVF interacts with

TABLE I. Divalent Cation Requirements for

Hemolysis by CoVF-treated serum+ in:	Percentage Lysis
VBS-M^{2+}	100
VBS-EDTA	100
Activation by CoVF* in buffer containing:	
M^{2+}	100
Ca^{2+}	100
Mg^{2+}	<0
–	<0

+Twenty-five μl of chicken serum was diluted with 25 μl VBS-M^{2+} containing 0.5 U CoVF and incubated at 37° for 30 min. After this time serum was diluted to 1/50 and tested for hemolytic activity against horse RBC using VBS-M^{2+} and VBS-EDTA as diluents.

*Twenty-five μl volumes of chicken serum were treated with 2 μl of 0.1M EDTA and incubated at room temp for 10 min. These mixtures were then brought to 50 μl with VBS containing 5.5 mM $MgCl_2$ and $CaCl_2$, or VBS containing 11 mM of either divalent cation, and 0.5 U of CoVF. After incubation at 37° for 30 min., sera were diluted to 1/50 and tested for hemolytic activity against unsensitized horse RBC.

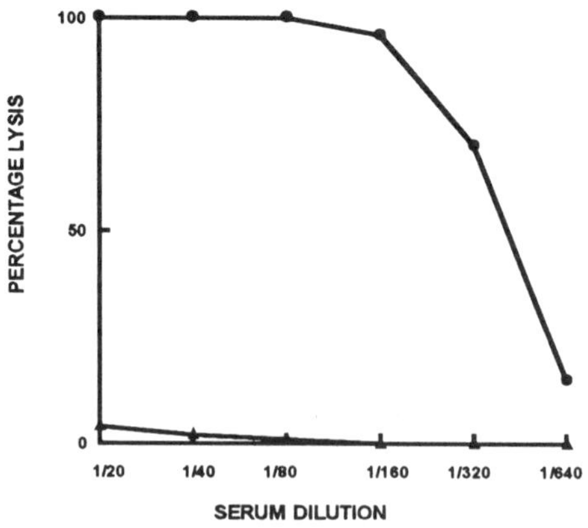

Figure 4. Lysis of rabbit RBC by normal (•) and CoVF-treated (Δ) (1.0 U per 25μl serum) bullfrog serum.

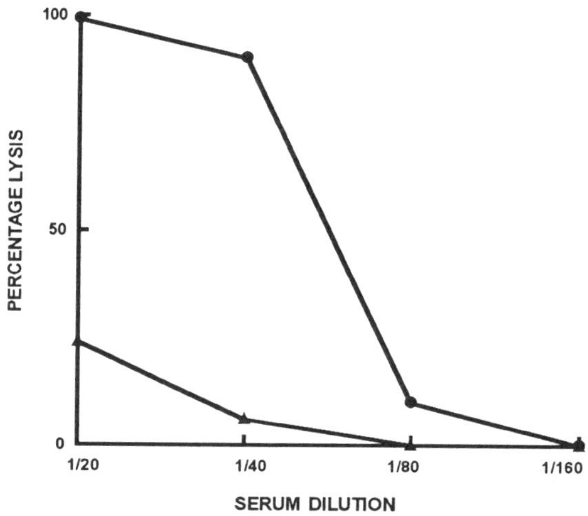

Figure 5. Lysis of rabbit RBC by normal (•) and CoVF-treated (Δ) (1.0 U per 25μl serum) turtle serum.

one or more C' components, or other serum constituents. The findings that formation of the hemolytic complex or factor requires Ca^{2+} ions and that hemolytic activity is resistant to heating at 56° suggest the possibility that CoVF interacts with a non-C' component. Moreover the elution pattern of hemolytic activity upon gel filtration (data not shown) indicates that it is not due to formation of a putative membrane attack complex. On the other hand, CoVF alone has no known lytic activity and it did not lyse any of the target cells used in this study. These observations, together with the loss of C' activity which occurred in CoVF-treated bullfrog and turtle sera, leave the interesting possibilities that either the chicken, and perhaps avian, C' system has structural and functional properties which are quite different from those of other vertebrate groups, or that chicken serum contains substances with lytic potential that are unrelated to the known components of the C' system.

REFERENCES

Ohta H, Yoshikawa Y, Kai C, Yamanouchi K, Okada H. Lysis of horse red blood cells mediated by antibody-dependent activation of the alternative pathway of complement. Immunology, 1984; 49:29-35.

Ohta H, Yoshikawa Y, Kai C, Yamanouchi K, Taniguchi H, Komine K-I., Ishijima Y, Okada H. Effect of complement depletion by cobra venom factor in fowlpox virus infection in chickens and chicken embryos. J. Virology, 1986; 57:670-673.

MOLECULAR STUDIES OF THE MOLLUSCAN RESPONSE TO *DIGENEAN* INFECTION

Eric S. Loker[1] and Christopher J. Bayne[2]

[1]Deparment of Biology University of New Mexico
Albuquerque, New Mexico 87131
[2]Department of Zoology Oregon State University
Corvallis,Oregon 97331

INTRODUCTION

The Mollusca is one of the largest, most diverse, and structurally complex of the animal phyla. Molluscs are of considerable importance in aquaculture, as intermediate hosts of several parasites of medical and veterinary significance, and because many molluscs are rare and endangered. Recent years have witnessed the increasing application of modern techniques to molluscan immunobiology, and there is a growing understanding of the molecules involved in molluscan defense responses. Among the notable approaches are 1) dissection of signal transduction pathways in molluscan hemocytes; 2) characterization of molecules involved in adherence of molluscan cells; 3) the identification of markers segregating with resistance of snails to digenetic trematodes; and 4) molecular characterization of a family of hemolymph proteins from planorbid snails that are induced following exposure to digeneans. The latter proteins are lectins and have similarity to fibrinogen and have been termed fibrinogen-related proteins, or

Phylogenetic Perspectives on the Vertebrate Immune System.
Edited by Beck *et al.*, Kluwer Academic/Plenum Publishers, 2001.

FREPs. FREPs are comprised of a C-terminal domain with homology to fibrinogen and an N-terminal domain with homology to V type loops of the immunoglobulin superfamily (IgSF). Other members of the IgSF occur in molluscs and other fibrinogen-related proteins lacking Ig domains have been found in external body mucus of snails and are believed to play roles in defense.

MOLLUSCS: PROMINENT REPRESENTATIVES OF A MAJOR ANIMAL LINEAGE

Although resolution of patterns of relatedness among the extant phyla of animals remains a very active field of investigation, a number of recent studies have provided agreement on a framework for a metazoan phylogeny (Eernisse *et al.*, 1992; Aguinaldo *et al.*, 1997; Giribet and Ribera, 1998; Valentine *et al.*, 1999), one that reveals four major groups of animals (Figure 1). From the standpoint of comparative immunobiology, what emerges from this phylogeny is that much of our current effort to understand the basic functioning of internal defense systems is focused on only two of the four groups: the Deuterostomia as represented by vertebrates and the Ecdysozoa, primarily represented by insects particularly *Drosophila* (*e.g.* Hoffmann *et al.*, 1999).

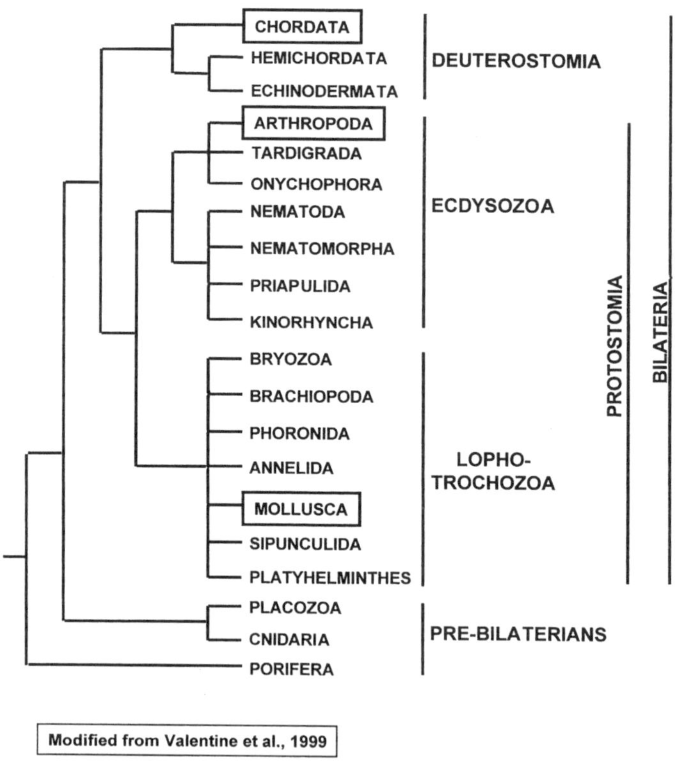

Figure 1. A metazoan phylogeny based on small subunit rRNA data. Note that the Mollusca is in a different major lineage from either the Arthropoda or Chordata.

The remaining two groups of animals, the Lophotrochozoa and the "Pre-Bilaterians" remain relatively understudied from an immunological point of view. Included among the Lophotrochozoa is one of the most prominent of the animal phyla, the Mollusca. Molluscs are remarkable for their present day diversity, a very conservative estimate being 50,000 extant species, and for the fact that the largest and most complex invertebrates, the cephalopods, are found in this phylum. Many members of the phylum have considerable significance in aquaculture and as outlined below, as intermediate hosts of persistent parasites of humans and domestic animals. Many, such as the unionid mussels of North America and the land snails of oceanic islands, are endangered.

WHY THE IMMUNOBIOLOGY OF MOLLUSCS IS IMPORTANT

Because molluscs do not have particularly close affinities to arthropods or vertebrates, their solutions to immunological problems might be expected to be quite distinctive. This is not to say that advances made with insects and vertebrates will not illuminate molluscan defense responses, but rather, that some real immunobiological novelties might be expected among molluscs and other Lophotrochozoans. Although the rate of discovery of these novelties might be slowed by a relative shortage of workers and because no Lophotrochozoan has achieved the same prominence as a "model organism" as *Drosophila* or the mouse, the results of such investigations are likely to be rewarding. This is likely to be particularly true for the cephalopods, a group that remains a nearly untapped resource for the comparative immunologist.

Another aspect of molluscan biology, the direct presentation to the environment of a body surface largely covered with a moist epithelium, no doubt has required development of unique defensive strategies for ubiquitous pathogens like viruses, bacteria and fungi. The molluscan solution to these problems may prove to be of considerable relevance to health care workers concerned with what is an increasingly recognized problem, the control of persistent bacterial populations (biofilms) on moist membranes (Costerson *et al.*, 1999). In contrast to the majority of invertebrate taxa, the gastropods have – like the Amphibia – made a successful transition to terrestrial life and aerial respiration. The muco-ciliary epithelia of the body surface have therefore met and successfully dealt with the novel defensive challenges associated with life out of water.

Additionally, some molluscs, particularly snails, serve as hosts for uniquely-adapted molluscan parasites, the digenetic trematodes, and thus present opportunities to investigate how this important group of molluscan and vertebrate pathogens has driven the evolution of distinctive defense strategies. Knowledge of such defenses will provide insights into how to control pervasive snail-transmitted parasites such as schistosomes and liver flukes.

DIGENETIC TREMATODES - A LINEAGE OF ENTIRELY DEDICATED MOLLUSCAN PARASITES

There are more than 100 families and probably >10,000 species of digenetic trematodes (Platt and Brooks, 1997), also known as digeneans, trematodes or, more informally, as "flukes." With the rare exception of a few species that parasitize marine annelids (almost certainly secondarily derived from molluscan parasites) (Køie, 1982), all digeneans undergo an obligatory period of larval development in intimate contact with molluscan tissues. Most digenean species use snails for this purpose although many also develop in bivalve molluscs and a few infect scaphopods or tusk shells (Gibson and Bray, 1994). Digeneans do not parasitize all molluscs: cephalopods and chitons are not infected.

Most digenean species are relatively specific with respect to the mollusc they colonize (Adema and Loker, 1997). During the course of infection, they undergo clonal asexual reproduction, occupy large volumes of their host's body, and in the process, the host is typically castrated (deJong-Brink, 1990). The adult stage in the digenean life cycle usually parasitizes a vertebrate host. There are numerous variations on this basic theme of asexual reproduction in molluscs followed by sexual reproduction in vertebrates. This provides a remarkable example of how a successful evolutionary innovation, namely the ability of digeneans to proliferate asexually within the molluscan host, has lead to a radiation in the number of species of flukes (Brooks and McLennan, 1993).

Digeneans have probably parasitized snails for at least 200-400 million years (Gibson, 1987). They are common in Nature (Kuris and Lafferty, 1994) and their prevalence can sometimes reach high levels (Curtis, 1996). In some circumstances, such as with estuarine and salt marsh snails, larval digeneans can reduce snail population densities (Lafferty, 1993). Some relatively long-lived marine snails retain digenean infections for decades (Curtis, 1995). Given their pronounced effects on host reproduction, it is likely that digeneans have had a major effect on the evolution of molluscan defenses.

Digeneans are relatively large, specialized metazoan parasites that must live for a prolonged period (usually at least a month) in intimate contact with molluscan tissues before they can release infective progeny. The selective pressures exerted on molluscs by digeneans would thus be expected to be quite different from those exerted by smaller, rapidly dividing, more ubiquitous, generic pathogens like bacteria or fungi. Accordingly, it might also be expected that the defensive repertoire evolved to meet the challenge of this ancient, specialized group of molluscan parasites would include unique attributes.

SOME RECENT STUDIES OF MOLLUSCAN IMMUNOBIOLOGY RELEVANT TO DIGENEAN INFECTIONS

Studies on Hemocytes

The ability or inability of a snail to destroy an invading digenean is faithfully replicated in vitro by the snail's amebocytic hemocytes, even in the absence of humoral plasma components. Hemocytes will bind to digenean larvae and encapsulate them, resulting in the destruction of the parasite. Conversely, in those host-digenean parasite systems in which infections are successful, the ability of the hemocytes to respond to the digenean larva appears to have been successfully abrogated by the parasite. At least with sporocysts of the schistosome *Schistosoma mansoni*, this is not apparently due to an aggressive counter-attack by the parasite against the cells of the host snail

Biomphalaria glabrata, neither is digenean survival assured by building a defensive barrier. Rather it appears from ultrastructural and physiological evidence that the host is hardly aware of the schistosome's presence. The ability of a relatively large (>200 mm long), multicellular worm to achieve this anonymity in the face of the mollusc's potent and broadly reactive immune system (Adema and Loker, 1997) parallels the evident absence of immunological attack on the adult schistosomes that live for years in mammalian blood vessels, and begs for a mechanism to be demonstrated.

Among attributes that are likely to be relevant in this relationship, the intramolluscan schistosome parasite mimicks host antigens: rabbit antisera raised to pure snail antigens recognize schistosome antigens, and *vice versa* (Bayne and Stephens, 1983; Yoshino and Bayne, 1983; Yoshino and Boswell, 1986). Molecular mimicry has been thought to be a passive means to help parasites avoid detection by a host. We considered it likely that molecular mimicry plays a more active role in ensuring survival of schistosomes.

Many immunocytes express both positive-signaling and negative-signaling receptors that, respectively, elicit and inhibit aggressive responses to non-self or to self markers. For example, among vertebrates, the failure of natural killer (NK) cells to attack healthy somatic cells is ensured by the expression of self markers (MHC molecules) by those somatic cells and the presence, on the NK cells, of negative signalling receptors for these markers (Karre, 1995). Similar mechanisms may occur among invertebrates (Humphreys and Reinherz, 1994). A picture is emerging that positive signalling is mediated by way of kinases, whereas negative signalling is mediated by way of early acting phosphatases (Long, 1999).

We were struck by the powerful, theoretical potential of a parasite's ability to switch off the aggressive responses of its host, thereby achieving freedom from immunological attack. If parasites are to exploit the negative signalling pathways of immunocytes, an obvious means of doing this is by mimicry of host ligands for these receptors that trigger early phosphatase activity. This possibility was investigated by Ing Wei Khor in the Bayne laboratory.

To test kinase and phosphatase activation in response to various stimuli, it was first necessary to obtain snail hemocytes (from *B. glabrata*) in a resting condition. The high level of vigilance of the cells, however, makes this difficult: as soon as they contact glass or plastic they adhere and spread – typical signs of response to non-self. Such adherent cells will ignore nearby parasites. Two surfaces decrease the avidity with which hemocytes adhere and spread, thereby allowing them to crawl around and respond normally to incompatible parasites (*i.e.* encapsulate and kill them): agarose (Bayne *et al.*, 1980) and poly-L-lysine (Boemhler *et al.*, 1996). In at least the latter case, this appears due to the failure of a plasma protein to first coat the surface (S.E. Fryer, personal communication). In contrast to these 'calming' surfaces, the neoglycoprotein mannose-BSA rapidly activates snail hemocytes to undergo an oxidative burst (U.E. Hahn, personal communication). We coated plastic cell culture dishes with poly-L-lysine, agarose or mannose-BSA, and placed freshly collected hemolymph in the wells. The cells were allowed 30 minutes to rebound from the trauma of bleeding. Then, to replicate wells, we added either physiological saline (CBSS, Chernin, 1963) or a strong phagocytic stimulus: β-1,3-glucan-enriched yeast particles (Macrogard, Makzymal, Tromso, Norway). The extracted cytosol of the cells, when electrophoresed and probed (Western blot) with a monoclonal antibody to phospho-tyrosine proteins, revealed several such antigens. These were minimal in cells retrieved from agarose, and predominant in cells retrieved from poly-L-lysine and mannose-BSA. Encounter with the phagocytic stimulus for 10 minutes appeared to induce no tyrosine kinase in cells on agarose, but to do so strongly in cells on poly-L-lysine (Figure 2). Cells on mannose-BSA had abundant phospho-tyrosine proteins, even without the phagocytic stimulus; indeed quantities were less after the addition of the stimulus. This membrane was stripped and re-probed with a polyclonal rabbit antiserum to snail hemocyte antigens. This revealed that differences in antigen loading was a possible reason for some but not likely all the differences seen. The phospho-proteins are not abundant or strong antigens in the hemocyte extracts.

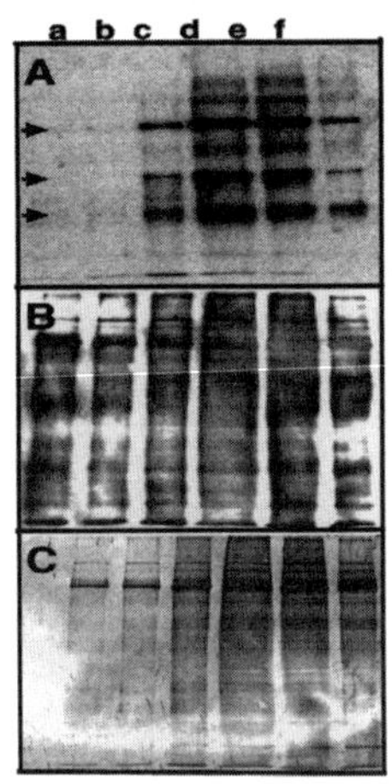

Figure 2. Cytosol proteins extracted from hemocytes of *Biomphalaria glabrata*. Panel A: probed with MoAb specific for phospho-tyrosine; arrows indicate 3 dominant phospho-tyrosine proteins of mass 120 kDa, 75 kDa and 50 kDa. Panel B: probed with polyclonal antiserum to *B. glabrata* hemocyte antigens. Panel C: probed with colloidal gold (total protein stain). Cells were held in wells of a 48-well tissue culture plate in which wells had been coated with 0.1% agarose (lanes a and b), poly-L-lysine (lanes c and d), or mannose-BSA (lanes e and f). The extracts were separated on a 7.5% reducing SDS-PAGE gel, and electroblotted onto PVDF membrane. The second lane of each treatment (lanes b, d and f) was loaded with extracts from cells stimulated with Macrogard in buffered saline, while the first (lanes a, c, and e) received buffered saline alone. Less cell protein was retrieved from the agarose-treated wells, reflecting the weak adhesion of hemocytes on this surface. Panel B shows variable but sufficient antigen for easy detection with the rabbit antiserum, and Panel C confirms variable antigen loading; yet anti-phospho-tyrosine detected essentially no antigen in the 'agarose' lanes. Agarose may fail to stimulate kinases or may stimulate phosphatases in these cells.

Many experiments over several months in which plastic-adherent hemocytes were rested for various times up to 12 hr in the hope of obtaining cells with minimal levels of phospho-proteins revealed that the presence of phospho-proteins did not require specific stimulation. Furthermore, heat-killed yeast (*Saccharomyces cerevissiae*), which is avidly phagocytosed (Fryer and Bayne, 1990), and mannose-BSA which stimulates the oxidative burst in these cells (U.E. Hahn, pers.comm.) did not induce phospho-protein levels above those seen in control cell populations given just CBSS. Cell responses were inconsistent to the β-1,3-glucan-particles. In vertebrate phagocytes, kinases are used for signal transduction following stimuli such as those used here. Since both yeast and Macrogard induce behavioral responses in the snail cells, it appears that some receptors and second messenger systems in snail hemocytes differ from their vertebrate counterparts when encountering such stimuli.

In contrast, the bacterial peptide *formyl*-methionine-leucine-phenylalanine (*f*-MLP), which stimulates the oxidative burst in mammalian neutrophils and is a chemoattractant, induced new phospho-proteins (27, 32 and 33 kDa in reducing SDS PAGE) in snail hemocytes in 3 of 4 independently run assays. Following exposure of hemocytes to 10^{-6} M *f*MLP, phospho-threonine and phospho-serine, but not phospho-tyrosine, were induced.

Affinity chromatography with rabbit antisera against snail antigens was used to purify cross-reactive antigens from schistosome extracts. Some of these antigens were themselves phospho-proteins. The effects of parasite antigens on cell responses were inconsistent. If host mimicry by parasite antigens serves to protect the invader by activating negative signaling receptors in hemocytes, the evidence is elusive.

STUDIES OF MOLLUSCAN CELL ADHESION MOLECULES

Characterization of molecules involved in cell adhesion and spreading are also relevant to understanding the molluscan defense against digeneans because such molecules may play a role in binding to parasite surfaces, and initiating encapsulation responses. Recently, two different molecules relevant to cell adhesion have been identified and sequenced from *B. glabrata*. A cDNA encoding a 36 kDa polypeptide with sequence similarity to selectins of the cell adhesion molecule superfamily was recently characterized by Duclermortier *et al.* (1999). This molecule has a C-terminal domain with similarity to C-type lectins. It is expressed in both hemocytes and in a cell line (*Bge* cells) derived from this snail species. Further studies are underway to characterize the interaction of this molecule with sporocysts of *S. mansoni*.

The arginine-glycine-aspartic acid motif occurs in a number of cell adhesion molecules (Hynes, 1992). Cells of the *Bge* line use an RGD-dependent mechanism for cell spreading, as do hemocytes of *B. glabrata* (Davids and Yoshino, 1998; Davids *et al.*, 1999). An integrinβ subunit from *Bge* cells has recently been cloned and sequenced by Davids *et al.* (1999). Further study is needed to determine if this particular β subunit, or others that might be present in *B. glabrata,* is involved in hemocyte adhesion and motility. These workers are now searching for the presence of an α integrin subunit and will use the tools developed from this study to explore the interactions between hemocytes and *S. mansoni* sporocysts.

SEARCHING FOR RESISTANCE GENES

One of the most intensively studied aspects of digenean-snail interactions has been to understand why some snails are resistant to digenean infection. Using the *B. glabrata*-

S. mansoni model system, it has been found that both the level of host resistance and the degree of parasite infectivity are controlled genetically and that that resistance of adult snails is controlled by a single gene locus (Richards, 1975; Richards, 1984; Richards *et al.*, 1992). Knight, Richards, Lewis and co-workers recently reported results of crosses between susceptible and resistant snail stocks that were undertaken to identify markers associated with resistance (Knight *et al.*, 1999). They screened progeny using a large panel of arbitrary primers in conjunction with RAPDs analysis and found two markers, of 1.0 and 1.2 kb, that segregated with resistance. The 1.2 kb band was a repetitive sequence with no homologs in the data base. The 1.0 kb band was also a repetitive sequence with an imperfect open reading frame with homology to retroviral-encoded *gag* polypeptides. Further studies are underway to define additional sequence associated with these markers.

FREPS, A DISTINCTIVE FAMILY OF PLASMA PROTEINS, AS AN EXAMPLE OF THE SNAIL RESPONSE TO DIGENEAN INFECTION

One of the major goals in the Loker laboratory has been to characterize the IDS responses of *B. glabrata* following exposure to either *S. mansoni* or the relatively distantly related digenean, *Echinostoma paraensei* (Loker and Adema, 1995). The two species differ in their mode of intramolluscan development: the former species has at least two generations of sporocysts, whereas the latter has a sporocyst generation followed by at least two successive redial generations. Rediae differ from sporocysts in having a mouth, oral sucker and a gut.

This discussion will focus on changes occurring in the snail's soluble blood (plasma) proteins of >10 kDa. Following exposure to *S. mansoni*, there is little if any indication of altered plasma composition as a consequence of infection. By contrast, examination of plasma from snails exposed to *E. paraensei* (Figure 3) revealed the presence on SDS-PAGE gels of at least three different groups of diffusely staining molecules, of 180-200, 80-120 and 65 kDa (Adema *et al.*, 1997a,b). These are aggregated in their native configuration into large complexes of ~1,300-1,600 kDa. These bands are consistently broad in appearance, and not easy to describe concisely. In other words, complex profiles of plasma polypeptides appear following exposure of *B. glabrata* to *E. paraensei*.

Deliberate wounding or injection of gram positive or gram negative bacteria did not stimulate production of these molecules by *B. glabrata*. When the related planorbid snail *Helisoma trivolvis* was exposed to echinostomes, it responded in a manner similar to that seen with *B. glabrata* (Adema *et al.*, in press). This suggested that production of complex and potentially diverse arrays of plasma proteins may be a common feature of the response of planorbids, and possibly other snails, to infection with at least some trematodes.

These particular plasma proteins became more interesting to us when we realized that if secretory-excretory products (SEP) derived from cultured *E. paraensei* larvae were added to plasma from echinostome-infected snails, conspicuous precipitates would form (Figure 4), and the precipitates were enriched in the broad, diffuse bands noted previously. A similar response did not occur in plasma from unexposed control snails.

Parasite antigens could also be observed in the precipitates. Precipitate formation could not be prevented by addition of protease inhibitors nor did addition of proteases provoke precipitate formation. Furthermore, by changing the ratios of plasma and parasite SEP, we could change the composition of precipitates and precipitates could be dissolved by addition of excess SEP. These features suggested the precipitates formed in a manner analogous to antigen-antibody precipitation reactions. Precipitate formation was calcium-dependent and inhibitable by addition of L-fucose (Adema *et al.*, 1997a,b).

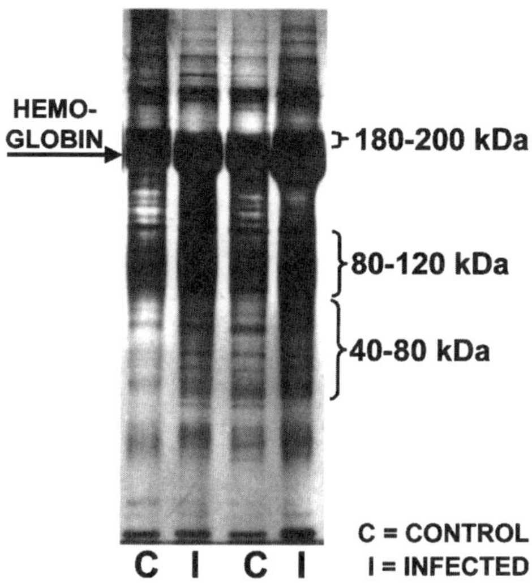

Figure 3. SDS-PAGE profiles of plasma proteins in control (C) snails and snails infected (I) with *Echinostoma paraensei.* Note the complex response to infection, consisting of increased abundance of broad, diffuse plasma bands.

The precipitation reaction was also fortuitous from the standpoint of allowing us to collect adequate quantities of the target bands for microsequencing efforts. We selected material from the very broad, snail-derived band of 65 kDa and following tryptic digestion obtained 4 peptides. Two peptides, one of which had homology to human fibrinogen (FBG), obtained directly from the 65 kDa band were used to design PCR primers that were then used to amplify a 354 bp fragment from M line *B. glabrata* genomic DNA. This amplicon was sequenced, and sequence internal to the primers was also found to have homology to FBG. The same primers were also used to amplify similar sized PCR products from other strains of *B. glabrata*, species of *Biomphalaria*, genera of planorbid snails and pulmonate families, some of which were sequenced and also found to have homology to FBG (Adema *et al.*, 1997a).

When used as a probe in genomic Southern blots, the PCR product revealed conservatively 6 bands, suggesting a family of FBG-encoding sequences is present in *B. glabrata*. A cDNA library was then prepared using mRNA from the head-foot of *B. glabrata* harboring 4-day infections with *E. paraensei*. Again using the 354 bp PCR fragment as a probe, we have identified over 35 different positive clones from this cDNA library and have found 7 distinct cDNAs (Figure 5). These encode FBG-related polypeptides or FREPs (Adema *et al.*, 1997a; Léonard *et al.*, in preparation). Two of these seven FREPs have been found to contain sequence matching one of the four peptides originally obtained by microsequencing. Thus the broad 65 kDa plasma band is comprised of at least two different FREPs.

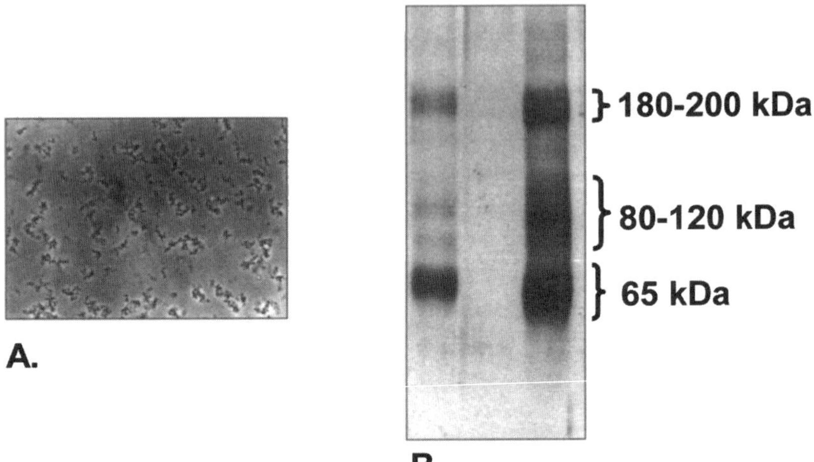

Figure 4. Precipitation of broad, induced plasma bands by addition of parasite antigens. A. Microscopical appearance of precipitates. B. Composition of precipitates as revealed on SDS-PAGE gel lanes. Note that these plasma bands are reactive with parasite antigens.

Using the same original primers to amplify FREPs from pooled *B. glabrata* genomic DNA, we have identified 4 additional FREPs, bringing the total number known for M line *B. glabrata* to 11 (Léonard *et al.*, in preparation). We presently do not know how many different FREP genes an individual snail possesses, but the minimum number of FREPs expressed by one snail is at least five (Adema *et al.*, 1997a).

FREPs have a tripartite structure: an N-terminal domain with similarity to sequences of IgSF intra-chain loops; an interceding region of variable length with no obvious similarities to other known molecules; and a C-terminal FBG β/γ domain (Adema *et al.*, 1997a; Léonard *et al.*, in preparation). The FREP Ig domains are unusual because they most resemble V-type Ig domains which are rarely encountered in invertebrates (DuPasquier and Flajnik, 1999). Most IgSF members from invertebrates contain C-type Ig domains and are involved in embryological development or cell-cell interactions. However, hemolin from moths binds to bacteria, promotes interactions among hemocytes and is an IgSF member (Sun *et al.*, 1990; Ladendorff and Kanost, 1991; Gillespie *et al.*, 1997). V-type Ig domains play an important role in specific antigen recognition in vertebrates and are generated by gene rearrangements (DuPasquier and Flajnik, 1999). There is no evidence at this time that gene rearrangements are involved in *B. glabrata* FREP Ig domains.

The *B. glabrata* FBG domains align well with FBG domains present in both vertebrates and other invertebrates (Doolittle, 1992). As *B. glabrata* plasma does not clot in any conventional sense, the FBG domain would seem to be responsible for the known lectin properties of FREPs. FBG domains have been implicated in the calcium-dependent carbohydrate binding shown by P35, an opsonic human serum lectin (Matsushita *et al.*, 1996).

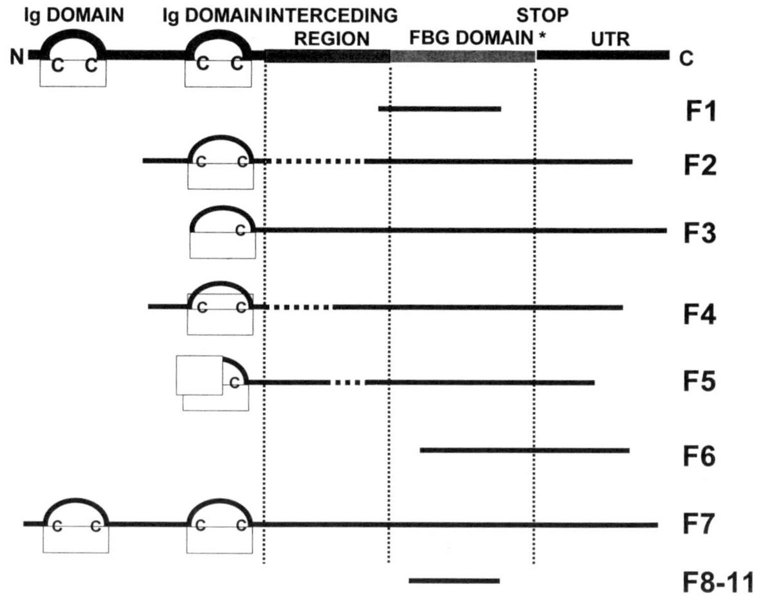

Figure 5. Schematic view of the known portions of 11 different M line fibrinogen-related proteins, or FREPs. At the top of the figure is indicated an overview of known FREP structure, with all parts thus far known labeled. Note that two FREPs F3 and F7, are known to have two Ig loops. FREPs 1-7 were obtained from a cDNA library. FREPs 8-11 were obtained by PCR amplification from genomic DNA.

The cDNAs examined clearly comprise a family of related, but different, sequences. A representative stretch of 111 amino acids residues shows 67-73% identity and 87-92% similarity between 4 of the FREPs. Some of the cDNAs are remarkable in having more than one IgSF intra-chain loop. We believe we have the complete cDNA for at least one of the known FREPs (FREP 2). We are actively seeking additional N-termini by examining genomic FREP sequences. RT-PCR using redundant primers has been used to document upregulation of FREP production and to demonstrate that at least some FREPs are produced in hemocytes, the circulating defense cells of *B. glabrata* (Adema *et al.*, 1997a).

Recent studies of the slug *Limax* reveal that its body mucus contains 3 different sialic acid-binding lectins, all of which are comprised primarily of single FBG domains (Kurachi *et al.*, 1998). Although unquestionably similar to the FBG domains of *B. glabrata* FREPs, the *Limax* sequences are also different, in lacking Ig domains. It is intriguing that both *B. glabrata* FREPs and the *Limax* mucus lectins occur in families, raising the possibility that diversity in FBG domains promotes diversity in carbohydrate recognition capability. Another recent example of an invertebrate lectin with a FBG domain has been described by Abe *et al.* (1999) from the ascidian *Halocynthia roretzi*. Collectively these studies suggest that FBG domains may have been used in carbohydrate recognition molecules since prior to the divergence of protostomes and deuterostomes.

Another molluscan IgSF member has been reported from *Lymnaea stagnalis* (Hoek *et al.*, 1996). It has C instead of V-type Ig domains. A preliminary analysis suggests that the selectin-like molecule from *B. glabrata* reported by Duclermortier *et al.* (1999) may also contain Ig-like sequences (Adema, personal communication). A recent phylogenetic analysis of V-domains has included the *B. glabrata* FREPs and concludes that the available molluscan sequences provide new opportunities for understanding the evolution of V-type domains across animal phyla (DuPasquier and Flajnik, 1999).

Another goal is to understand the genomic organization of FREP-encoding genes, because this will help us understand how FREP expression is regulated and how FREP genes may have evolved and diversified. We have recently isolated 3 genomic clones containing FREP sequences, all of which are different from one another. Two correspond to known FREP cDNAs. Some intron-exon boundaries have been revealed by this analysis (Léonard *et al.*, in preparation).

In conclusion, FREPs are the first molecules known to share Ig and FBG domains, and the first molecules with V-type Ig loops from invertebrates functioning in a likely context of internal defense. Among the important questions yet to be resolved are: 1) How many FREP genes does a single snail possess and how diverse is the overall FREP repertoire? 2) Do FREP expression patterns vary between snails susceptible or resistant to digenean infection? If so, what is their functional role? 3) How have FREP genes diversified and how widely do FREPs occur among molluscs and possibly other phyla? Much additional work remains to be done. We feel that FREPs are good candidates as non-self recognition molecules in *B. glabrata* and merit additional study.

SUMMARY AND CONCLUSIONS

A classical body of knowledge regarding molluscan immunobiology has developed over the years that indicates an important role for both soluble hemolymph proteins and hemocytes in internal defense. The incorporation of powerful molecular approaches into the toolkit of the molluscan immunobiologist holds considerable promise for allowing a more precise description of the molecules and processes involved. Recent discoveries indicate that the molluscan internal defense system has components that are simultaneously both tantalizingly similar to, and different from, molecules functioning in a defense context in other phyla. Study of molecular components of the internal defense systems of molluscs promises to be a rewarding and surprising field.

ACKNOWLEDGMENTS

The authors gratefully acknowledge the support provided by grants AI24340 and AI16137 from the National Institutes of Health.

REFERENCES

Abe Y, Tokuda M,Ishimoto R,azumi K, Yokosawa H A unique primary structure, cDNA cloning and function of a galactose-specific lectin from ascidian plasma. Eur. J. Biochem. 1999;261;33-9.

Adema CM, Hertel LA, Loker ES. Evidence from two planorbid snails of a complex and dedicated response to digenean (echinostome) infection. Parasitology, in press.

Adema CM, Hertel LA, Miller RD, Loker ES. A family of fibrinogen-related proteins that precipitates parasite-derived molecules is produced by an invertebrate after infection. Proc Natl Acad Sci USA 1997a;94:8691-96

Adema CM, Hertel LA, Loker ES. "Infection with *Echinostoma paraensei* (Digenea) induces parasite-reactive polypeptides in the hemolymph of the gastropod host *Biomphalaria glabrata*" In *Parasites and Pathogens: Effects on Host Hormones and Behavior.* N.E. Beckage, ed. Chapman and Hall, 1997b

Adema CM, Loker ES. Specificity and immunobiology of larval digenean-snail interactions. In. *Advances in trematode biology*, B. Fried TK Graczyk, eds. Boca Raton, CRC Press, 1997.

Aguinaldo AMA, Turbeville JM, Linford LS, Rivera MC, Garey JR, Raff R, Lake JA. Evidence for a clade of nematodes, arthropods and other moulting animals. Nature 1997;387:489-93

Bayne CJ, Buckley PM, DeWan PC. Macrophage-like hemocytes of resistant *Biomphalaria glabrata* are cytotoxic for sporocysts of *Schistosoma manoni in vitro.* J. Parasitol. 1980;66:413-9.

Bayne CJ, Stephens JA. Schistosoma mansoni and *Biomphalaria glabrata* share epitpoes: antibodies to sporocysts bind snail hemocytes. J. Invertebr. Pathol. 1983;42:221-3.

Boemhler A, Fryer SE, Bayne, CJ. Killing kinetics of *Schistosoman mansoni* sporocysts by *Biomphalaria glabrata* hemolymph in vitro: alterationof hemocyte behavior after poly-L-lysine treatment of plastic and the kinetics of killing by different strains. J. Parasitol. 1996;82:332-5.

Brooks DR, McLennan DA. *Parascript – Parasites and the Language of Evolution.* Washington: Smithsonian, 1993.

Chernin E. Observations on hearts explanted *in vitro* from the snail *Australorbis glabratus.* J. Parasitol. 1963;49:353-64.

Costerson JW, Stewart PS, Greenberg, EP. Bacterial biofilms: a common cause of persistent infections. Science 1999; 284:1318-22

Curtis LA. Growth, trematode parasitism, and longevity of a long-lived marine gastropod (*Ilyanassa obsoleta*). J. Mar. Biol. Assoc. U.K.1995;75:913-25

Curtis LA. The probability of a marine gastropod being infected by a trematode. J. Parasitol. 1996;82:830-33

Davids BJ, Xiao-Jun W, Yoshino TP. Cloning of a b integrin subunit cDNA from an embryonic cell line derived from the freshwater mollusc, *Biomphalaria glabrata.* Gene. 1999; 228:213-23.

Davids BJ, Yoshino TP. Integrin-like RGD-dependent binding mechanism involved in the spreading response of circulating molluscan phagocytes. Dev. Comp. Immunol. 1998;22:39-53.

DeJong-Brink M. How trematode parasites interfere with reproduction of their intermediate hosts, freshwater snails. J.Med. Appl. Malacol. 1990;2:101-33.

Doolittle RF. A detailed consideration of a principal domain of vertebrate fibrinogen and its relatives. Protein Sci. 1992;1:1563-77

Duclermortier P, Lardans V, Serra E, Trottein F, Dissous C. *Biomphalaria glabrata* embryonic cells express a protein with a domain homologous to the lectin domain of mammalian selectins. Parasitol Res. 1999;85:481-6.

DuPasquier L, Flajnik M. "Origin and evolution of the vertebrate immune system." In *Fundamental Immunology*, W.E. Paul, ed. Philadelphia, PA: Lippincott-Raven Publishers, 1999.

Eernisse DJ, Albert JS, Anderson FE. Annelida and Arthropoda are not sister taxa: A phylogenetic analysis of spiralian metazoan morphology. Syst Biol 1992;4:305-30

Fryer SE, Bayne CJ. *Schistosoma mansoni* modulation of phagocytosis in *Biomphalaria glabrata.* J. Parasitol. 1990;76: 45-52.

Gibson DI. Questions in digenean systematics and evolution. Parasitology 1987;95: 429-60.

Gibson DI, Bray, RA. The evolutionary expansion and host-parasite relationships of the digenea. Int. J. Parasitol. 1994;24:1213-26

Gillespie JP, Kanost MR, Trenczek T. Biological mediators of insect immunity. Annu. Rev. Entomol. 1997;42: 611-43.

Giribet G, Ribera C. The position of arthropods in the Animal Kingdom: in search for a reliable outgroup for internal arthropod phylogeny. Mol Phylogen Evol 1998;9:481-88

Hoek RM Smit AB, Frings H, Vink JM, DeJong-Brink M, Geraerts WPM. A new Ig-superfamily member, molluscan defense molecule (MDM) from *Lymnaea stagnalis*, is down-regulated during parasitosis. Eur. J. Immunol. 1996;26:939-44.

Hoffmann JA, Kafatos FC, Janeway CA, Jr, Ezekowitz RAB. Phylogenetic perspective in innate immunity Science 1999;284;1313-1318.

Humphreys T, Reinherz EL. Invertebrate immune recognition: natural immunity and the evolution of positive selection. Immunol. Today 1994;15:316-20.

Hynes RO. Integrins: versatility modulation and signaling in cell adhesion. Cell 1992;69:11-25.

Karre K. Express yourself or die: peptides, MHC molecules and NK cells. Science 1995;267:978-9.

Knight M, Miller AN, Patterson CN, Rowe CG, Michaels G, Carr D, Richards CS, Lewis FA. The identification of markers segregating with resistance to *Schistosoma mansoni* infection in the snail *Biomphalaria glabrata*. Proc Natl Acad Sci USA 1999;96:1510-5.

Køie M. The redia, cercaria and early stages of *Aporocotyle simplex* Odhner, 1900 (Sanguinicolidae) – a digenetic trematode which has a polychaete annelid as the only intermediate host. Ophelia 1982;21:115-45.

Kurachi S, Song Z, Takagaki M, Yang Q, Winter HC, Kurachi K, Goldstein IJ. Sialic-acid-binding lectin from the slug *Limax flavus*. Cloning, expression of the polypeptide, and tissue localization. Eur. J. Biochem. 1998;254:217-222.

Kuris AM, Lafferty KD. Community structure: larval trematodes in snail hosts. Annu. Rev. Ecol. Syst. 1994;25:189-217

Ladendorff NE, Kanost MR. Bacteria-induced protein-p4 (hemolin) from *Manduca sexta*: a member of the immunoglobulin superfamily which can inhibit hemocyte aggregation. Arch. Insect Biochem. Physiol. 1991;18: 285-300.

Lafferty KD. Effects of parasitic castration on growth, reproduction and population dynamics of the marine snail *Cerithidea californica*. Mar. Ecol. Progr. Ser. 1993;96:229-37.

Loker ES, Adema CM. Schistosomes, echinosotmes and snails: comparative immunobiology. Parasitol. Today 1995;11:120-4.

Long EO. Regulation of immune responses through inhibitory receptors. Annu. Rev. Immunol. 1999;17:875-904.

Matsushita M, Endo Y, Taira S, Sato Y, Fujita T, Ichikawa N, Nakata, M., Mizuochi T. A novel human serum lectin with collagen-like and fibrinogen-like domains that functions as an opsonin. J. Biol. Chem.1996;271:2448-54.

Platt TR, Brooks DR. Evolution of the schistosomes (Digenea, Schistomatoidea): the origin of dioecy and colonization of the venous system. J. Parasitol. 1997;83:1035-44.

Richards CS. Genetic studies on variation in infectivity of *Schistosoma mansoni*. J. Parasitol. 1975;61:233-36.

Richards CS. Influence of snail age on genetic variation in susceptibility of *Biomphalaria glabrata* for infection with *Schistosoma mansoni*. Malacologia 1984;25:493-502.

Richards CS, Knight M, Lewis FA. Genetics of *Biomphalaria glabrata* and its effect on the outcome of *Schistosoma mansoni* infection. Parasitol. Today 1992;8:171-4.

Sun S, Lindstrom I, Bowman HG, Faye I, Schmidt O. Hemolin: an insect-immune protein belonging to the immunoglobulin superfamily. Science 1990;250:1729-32.

Valentine JW, Jablonski D, Erwin DH. Fossils, molecules and embryos: new perspectives on the Cambrian explosion. Development 1999;126:851-59.

Yoshino TP, Bayne CJ. Mimicry of snail host antigens by miracidia and primary sporocysts of *Schistosoma mansoni*. Parasite Immunol. 1983;5:317-28.

Yoshino TP, Boswell CA. Antigen sharing between larval trematodes and their snail hosts: How real a phenoomenon in immune evasion? Symp. Zool. Soc. London 1986;56:221-38.

INSECT LECTINS AND INNATE IMMUNITY

S. Natori

The Institute of Physical and Chemical Research
2-1 Hirosawa, Wako-shi, Saitama 351-0198, Japan

INTRODUCTION

Proteins capable of recognizing carbohydrates are called lectins. Lectins are widely distributed in the animal as well as the plant kingdom and function in various ways. Lectins are assumed to play roles in mediating cellular and cell-substate interactions during development and immune reactions. Sometimes, they have been supposed to relay signals to the immune system, which allow an organism to distinguish self-determinants from non-self or modified-self determinants.

We have been studying the humoral lectins of *Sarcophaga peregrina* and *Drosophila melanogaster*. So far, we have isolated two lectins from *Sarcophaga* and one from *Drosophila* (1-3). Of these lectins, *Sarcophaga* lectin has been shown to play a role in the innate immunity of this insect (4). Throughout studies on these lectins, we noticed that the *Sarcophaga* lectin plays two independent roles in defense and development. Namely, the *Sarcophaga* lectin is not simply a defense molecule, but also an effector molecule in the differentiation of imaginal discs into adult structures (5). In this chapter, I would like to focus on the roles of *Sarcophaga* lectin in innate immunity.

Phylogenetic Perspectives on the Vertebrate Immune System.
Edited by Beck *et al.*, Kluwer Academic/Plenum Publishers, 2001.

LECTINS FROM *SARCOPHAGA* AND *DROSOPHILA*

Immunized larval hemolymph of *Sarcophaga* was found to agglutinate sheep red cells in the presence of calcium and this agglutinating activity was inhibited by galactose or lactose, whereas no appreciable hemagglutinating activity was detected for the hemolymph of naive larvae. These findings indicated that a galactose-specific C-type lectin was induced in the larval hemolymph when larvae were infected with bacteria or their body walls were simply injured. We purified this lectin from the hemolymph of immunized larvae to homogeneity by affinity chromatography on Sepharose (a galactose polymer) (1). This lectin is a homohexamer of a subunit protein with a molecular mass of 30 kDa, and 4 of the 6 subunits are glycosylated. We isolated the cDNA and found that this subunit consists of 260 amino acid residues and contains possible N-glycosylation sites (6). The carbohydrate recognition domain (CRD) is located in its amino-terminal half. We named this lectin *Sarcophaga* lectin. *Sarcophaga* lectin was found to be synthesized by fat bodies in response to bacterial infection or body injury, and to be secreted into the hemolymph. The function of this lectin in the innate immunity of *Sarcophaga* was extensively studied. When sheep red cells were injected into the abdominal cavities of *Sarcophaga* larvae, they were lysed gradually and finally disappeared from the body fluid. When antibody against *Sarcophaga* lectin or galactose was injected with sheep red cells however, these cells were not lysed, remaining in the body cavity. Injection of control IgG or glucose did not interfere with the elimination of the red cells (7). These results clearly indicated that *Sarcophaga* lectin is essential for the clearance of foreign cells introduced into the body cavity. The precise mechanisms by which *Sarcophaga* lectin participates in the elimination of sheep red cells are not clear. As *Sarcophaga* lectin itself has no hemolytic activity, it seems to activate a cascade system to finally lyse invading foreign cells or it may induce hemocytes to phagocytize these cells. We analyzed expression of the *Sarcophaga* lectin gene. This gene was activated when larvae were infected with bacteria or the larval body wall was simply injured with a hypodermic needle. Except for this acute-phase expression, the *Sarcophaga* lectin gene was found to be activated transiently at the early embryonic stage and the early pupal stage during the normal development of *Sarcophaga*. The function of this lectin in the development of *Sarcophaga* was described elsewhere (8).

Recently, we isolated another C-type lectin from NIH-Sape-4 cells derived from *Sarcophaga*. Subsequently, we found that this lectin was constitutively synthesized by granulocytes in naive third instar larvae and then secreted into the hemolymph (2). So, we named this lectin granulocytin. Granulocytin is a homotrimer of a subunit consisting of 151 amino acid residues. This subunit contains two possible N-glycosylation sites and in fact it is glycosylated. Granulocytin caused the agglutination of rabbit red cells, but no appreciable agglutination was detected with sheep, pig, chicken, horse, guinea pig, bovine or human red cells. Thus, the cell specificity of granulocytin is quite stringent and different from that of *Sarcophaga* lectin. Moreover, we could not identify its specific hapten sugar in monosaccharides. However, bovine submaxillary gland mucin inhibited its agglutinating activity, indicating that granulocytin recognizes polysaccharide chains rather than monosaccharides. Concerning the expression pattern of the granulocytin gene, no significant expression was detected from the embryonic stage to the second instar larval stage, but this gene was expressed in third instar larvae and the expression continued until the early pupal stage. Then this gene was turned off during the rest of the pupal stage and was activated again in the newly emerged adult flies. These expression patterns are quite different from those of the *Sarcophaga* lectin gene. Contrary to the *Sarcophaga* lectin gene, expression of the granulocytin gene on the acute phase is not clear, because this gene is always expressed in the hemocytes of naive insects (background level expression), but its expression tends to be enhanced in the hemocytes of immunized larvae. At the present time, nothing is known about the biological role of granulocytin in innate immunity.

We found a galactose-binding C-type lectin in a pupal lysate of *Drosophila melanogaster* and purified it to homogeneity (3). Since we reported the purification of *Sarcophaga* lectin, many people, including ourselves, have tried to isolate the *Drosophila* counterpart of *Sarcophaga* lectin for its genetic analysis. In the beginning, it seemed to be relatively easy to isolate the *Drosophila* counterpart of the *Sarcophaga* lectin, since cDNA and a specific antibody for the *Sarcophaga* lectin are available. However, it turned out that it is not so easy. We could not detect clear signals on Northern blotting of total RNA isolated from immunized *Drosophila* larvae with *Sarcophaga* lectin cDNA as a probe. No positive signal was detected either on immunoblotting of a lysate of immunized *Drosophila* larvae with antibody against *Sarcophaga* lectin. However, as *Sarcophaga* lectin was found to play a crucial role in the innate immunity of *Sarcophaga*, we assumed that the situation is the same for *Drosophila,* because *Sarcophaga* and *Drosophila* are phylogenetically related dipteran insects. In fact, *Drosophila* cecropin A was found to be exactly the same protein as sarcotoxin IA, which was originally isolated from the hemolymph of immunized *Sarcophaga* larvae.

As we know that *Sarcophaga* lectin is transiently synthesized at the early pupal stage of *Sarcophaga*, we examined if a pupal extract of *Drosophila* contains hemagglutinating activity toward sheep red cells, but the results were negative. After a long struggle, we finally found that such an extract only causes agglutination of trypsin-treated and glutalaldehyde-fixed bovine red cells. Calcium was required for their agglutination, which was inhibited by galactose. These results indicated that *Drosophila* pupae contain a galactose-binding C-type lectin. In order to purify this lectin, we adopted a novel strategy, which was very effective for its purification. We added trypsin-treated and glutalaldehyde-fixed bovine red cells to the pupal extract and then mixed them well. Then we recovered the cells by centrifugation and washed them with a galactose solution to elute the attached lectin. In this way, we were able to purify a galactose-specific C-type lectin named *Drosophila* lectin to homogeneity. The purified *Drosophila* lectin gave two bands corresponding to molecular masses of 17 kDa and 14 kDa, respectively, in a density ratio of 1:2, on SDS polyacrylamide gel electrophoresis. However, when the lectin was treated with N-glycanase, the 17 kDa band was converted to the 14 kDa band. The molecular mass of the intact lectin was estimated to be 41 kDa by molecular sieve column chromatography. Thus, we assume that the 17 kDa protein is a glycosylated form of the 14 kDa protein, and *Drosophila* lectin is a trimer of these proteins in a molar ratio of 1:2. We isolated cDNA for *Drosophila* lectin, and found that it is a protein consisting of 163 amino acid residues. The expression pattern of the *Drosophila* lectin gene was very similar to that of the granulocytin gene. Thus, *Drosophila* lectin appears to be the *Drosophila* counterpart of *Sarcophaga* granulocytin, judging from their amino acid sequences, molecular masses, subunit structures and gene expression patterns. However, carbohydrate specificities of these two lectins are different.

SARCOPHAGA LECTIN AS AN ACTIVATOR OF MAMMALIAN MACROPHAGES

Of the three lectins so far purified from dipteran insects, only *Sarcophaga* lectin was suggested to participate in innate immunity. If *Sarcophaga* lectin activates hemocytes to ingest non-self sheep red cells, there is a possibility that it also activates mammalian macrophages, because large parts of the innate immunity are assumed to be common between vertebrates and invertebrates. Therefore, we examined the effect of *Sarcophaga* lectin on the murine immune system. We found that mouse syngeneic tumor MM46 cells were killed by peritoneal macrophages only in the presence of *Sarcophaga* lectin *in vitro*. We extensively studied this killing mechanism and found that *Sarcophaga* lectin activates murine macrophages to produce TNF-α, that kills tumor cells selectively. A subsequent study revealed that both macrophages and MM46 cells have receptors for *Sarcophaga* lectin on their surface (9, 10). *Sarcophaga* lectin is a divalent molecule and one molecule contains two carbohydrate-binding sites. Possibly, *Sarcophaga* lectin forms a bridge between macrophages and tumor cells by binding to these receptors. Macrophages are activated when *Sarcophaga* lectin binds to their receptors and secrete TNF-α to kill the adherent tumor cells.

We purified both macrophage and tumor cell receptors and raised antibodies against them. These receptors seem to be different molecules, although both of them exhibit affinity to *Sarcophaga* lectin, because they did not cross-react immunologically with each other. Both of these receptors were found to consist of two subunits. Therefore, their molecular structures seem to be similar. Antibody against the macrophage receptor inhibited the *Sarcophaga* lectin-dependent macrophage-mediated tumor cell killing reaction. Similarly, antibody against the tumor cell receptor also inhibited this killing reaction. Therefore, both receptors are important for the tumor killing reaction, and *Sarcophaga* lectin has to form a bridge between macrophages and tumor cells to kill these tumor cells.

It is noteworthy that receptors for insect lectins are present on the surface of murine cells. We are interested in the macrophage receptor for *Sarcophaga* lectin. Macrophages are known to be activated by bacterial lipopolysaccharide (LPS). For the activation of macrophages by LPS, it is generally believed that an LPS-binding protein carrying LPS interacts with CD14, that couples with TLR4 (a Toll-like receptor), that contains a Toll/IL-1 receptor homology domain (TIR). We assume that CD14, and thus TLR4, is not directly involved in the macrophage activation by *Sarcophaga* lectin. Because although human intestinal macrophages, that lack CD14, are not activated by LPS, we found that they could be activated by *Sarcophaga* lectin, as shown in Fig. 1. This finding is extremely interesting, because it is known that intestinal macrophages of patients with Crohn's disease are activated to produce TNF-α, which is assumed to worsen the intestinal inflammation. However, the mechanism of the activation of intestinal macrophages is totally unknown. If the *Sarcophaga* lectin receptor participates in the activation of macrophages in Crohn's disease, analysis of this receptor may provide a clue as to drugs for treating Crohn's disease. We believe that the *Sarcophaga* lectin receptors of mammalian macrophages are intimately related to innate immunity.

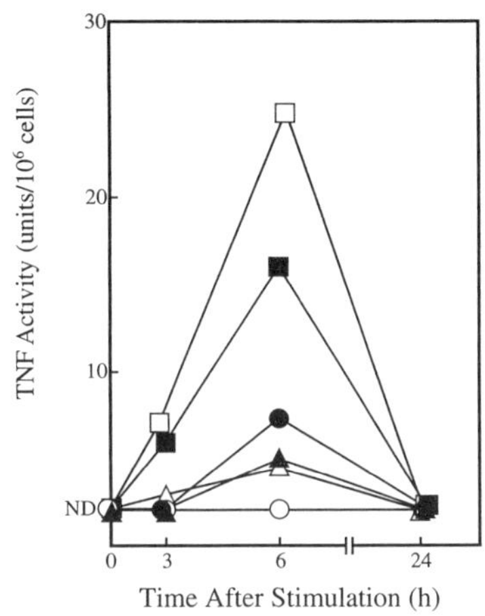

Figure 1. Stimulation of human intestinal macrophages by various lectins. Control (○), lipopolysaccharide (10 μg/ml) (●), pokeweed mitogen (10 μg/ml) (Δ), *Sarcophaga* lectin (10 μg/ml) (▲), *Sarcophaga* lectin (1.0 μg/ml) (□), *Sarcophaga* lectin (100 ng/ml) (■),

REFERENCES

1. H. Komano, D. Mizuno and S. Natori. A possible mechanism of induction of insect lectin. ***J. Biol. Chem. (1981) 256, 7087-7089***
2. S. Haq, T. Kubo, S. Kurata, A. Kobayashi and S. Natori. Purification, characterization, and cDNA cloning of a galactose-specific C-type lectin from *Drosophila melanogster*. ***J. Biol. Chem. (1996) 271, 20213-20218***
3. Y. Fujita, S. Kurata, K. Homma and S. Natori. A novel lectin from *Sarcophaga*. ***J. Biol. Chem. (1998) 273, 9667-9672***
4. S. Natori and T. Kubo. Role of lectins in development and morphogenesis in insects. In ***New Directions in Invertabrate Immunology*** (K. Soederhall, S. Iwanaga and G. R. Vasta eds) pp. 175-187, SOS Publications, Fair Haven (1996)
5. N. Kawaguchi, H. Komano and S. Natori. Involvement of *Sarcophaga* lectin in the development of imaginal discs of *Sarcophaga peregrina* in an autocrine manner. ***Dev. Biol. (1991) 144, 86-93***
6. H. Takahashi, H. Komano, N. Kawaguchi, N. Kitamura, S. Nakanishi and S. Natori. Cloning and sequencing of cDNA of *Sarcophaga peregrina* humoral lectin induced on injury of the body wall. ***J. Biol. Chem. (1985) 260, 12228-12233***
7. H. Komano and S. Natori. Participation of *Sarcophaga peregrina* humoral lectin in the lysis of sheep red blood cells injected into the abdominal cavity of larvae. ***Devel. Comp. Immunol. (1984) 9, 31-40***
8. S. Natori. Defense molecules in host-defense and development of insects. In ***Phylogenetic Perspectives in Immunity: the Insect Host Defense*** (J. A. Hoffman, C. A. Janeway and S. Natori eds) pp.67-75, R.G. Landes Company, Austin (1994)

9. Y. Ohkuma, H. Komano and S. Natori. Identification of target proteins participating in a lectin-dependent macrophage-mediated cytotoxic reaction. ***Cancer Res. (1984) 45, 288-292***
10. Y. Ohkuma, H. Komano and S. Natori. Comparison of binding proteins on the surface of murine tumor cells for two lectins active in the lectin-dependent macrophage-mediated cytotoxic reaction. ***Cancer Res. (1985) 45, 4397-4400***

COLLAGENOUS LECTINS IN TUNICATES AND THE PROTEOLYTIC ACTIVATION OF COMPLEMENT

David Raftos, Peter Green, Deepika Mahajan, Rebecca Newton[1], Sarina Pearce, Rodney Peters, Jennifer Robbins and Sham Nair

Dept of Biological Sciences, Macquarie University, North Ryde, NSW, 2109, Australia
[1]Department of Anatomy, University of California San Francisco, San Francisco, CA

INTRODUCTION

Complement activation in vertebrates has traditionally been associated with the classical (antibody mediated) or alternative (spontaneously hydrolytic) pathways. Recently, a new mechanism of complement activation has been identified in which a collagenous lectin (collectin), mannose binding protein (MBP) and its associated serine protease (MASP) activate the central component of the complement system, C3 (Epstein, *et al.*, 1996, Holmskov, *et al.*, 1994, Hoppe and Reid, 1994, Lu, 1997). Two components of this lectin-mediated complement pathway have already been identified in invertebrates. Echinoderms express a C3 homologue and a number of other complement

Phylogenetic Perspectives on the Vertebrate Immune System.
Edited by Beck *et al.*, Kluwer Academic/Plenum Publishers, 2001.

components, whilst a MASP-like serine protease and a C3 homologue have been identified in tunicates (Al-Sharif, *et al.*, 1998, Azumi, *et al.*, 1993, Nonaka, *et al.*, 1998, Smith, *et al.*, 1996, Smith, *et al.*, 1998).

The existence of C3 and MASP in deuterostome invertebrates suggests that proteolytic complement cascades existed before the evolution of immunoglobulin antibodies, which are found exclusively in jawed vertebrates (gnathostomes). The only component of a lectin-mediated complement pathway that has not yet been identified among invertebrates is the lectin itself. Here, we report the identification of a collagenous, carbohydrate binding protein from the tunicate, *Styela plicata*, that is involved in the proteolytic activation of tunicate complement.

A COLLAGENOUS LECTIN FROM *S.PLICATA*

Purification of galactosyl-specific lectins

Lectin activity in the hemolymph of *S. plicata* was initially characterized using an ELISA-type assay. Biotinylated neoglycoproteins were used to identify hemolymph proteins bound to the surface of microtiter plates, or visa versa. Two distinct carbohydrate binding specificities, toward galactose and glucose-based neoglycoproteins, were identified in hemolymph. In both cases carbohydrate binding required Ca^{2+} and could be inhibited by soluble galactose or glucose. These characteristics are typical of C-type lectins (Drickamer and Taylor, 1993).

The galactosyl-binding components of hemolymph were purified on an affinity column of N-acetyl-D-galactosamine (galNAc)-conjugated agarose beads eluted with EDTA. Fractions from the galNAc affinity column retained their specificity for galactose-based neoglycoproteins and could also opsonise yeast for ingestion by tunicate phagocytes.

Physicochemical characterization of the *S. plicata* lectin

Reducing SDS-Page identified two proteins in galNAc fractions of hemolymph. A predominant protein appeared at 44 kDa with a less intense band at 65 kDa. Analysis under non-reducing conditions revealed a series of proteins with molecular weights equivalent to homo-oligomers of the major 44 kDa reduced polypeptide. Two-dimensional (reducing vs non-reducing) SDS-PAGE confirmed that the higher molecular weight bands seen under non-reducing conditions represented homodimers, trimers and higher order oligomers (up to nonomers) of the 44 kDa reduced polypeptide (Fig. 1). The minor 65 kDa band identified in reducing SDS-PAGE did not alter molecular weight after reduction with dithiothreitol, and was not a component of the high molecular weight non-reduced proteins.

Amino acid sequence and amino acid composition analyses

Both amino acid sequencing and amino acid composition analysis indicated that the 44 kDa reduced protein from galNAc affinity columns resembles vertebrate collectins, particularly MBP. The tunicate lectin is within the size range of vertebrate collectins (approx. 30-50 kDa; Lu, *et al.*, 1992), and amino acid composition analysis indicated that a substantial proportion of the tunicate protein comprised a collagenous domain similar to those of collectins. Amino acid sequencing identified classic collagen-like Gly-X-Y amino acid repeats (where X and Y are any amino acid, frequently proline or hydroxyproline) within the tunicate protein. Collagenase digestion indicated that approximately 30% of the 44 kDa molecule comprised a collagen-like domain (Fig. 2) (Davis and Lachmann, 1984).

The collagen domain of the tunicate lectin is preceded by a short (19 AA) tail region that bears structural similarity to the terminal domains of MBP and other collectins. In particular, a cysteine at residue 5 is conserved between the tunicate lectin and the collectins, MBP and bovine conglutinin. In MBP and conglutinin this cysteine contributes to the formation of collagenous trimers and higher order oligomers equivalent to those of the tunicate lectin (Crouch, *et al.*, 1994, Holmskov, *et al.*, 1995, Laursen, *et al.*, 1995, Lu, 1997).

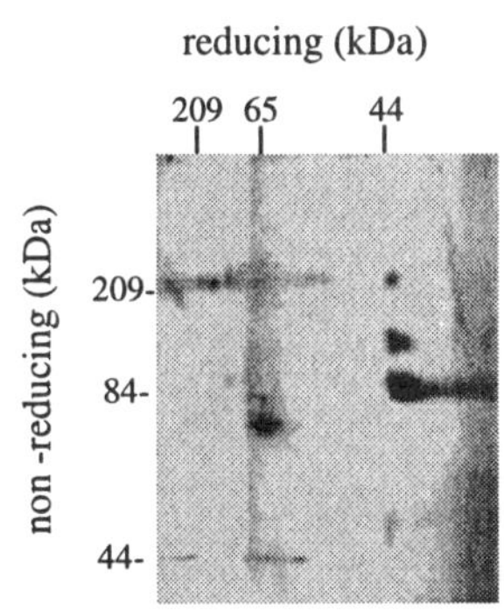

Figure 1. Two dimensional SDS-PAGE (non-reducing conditions - top to bottom; reducing conditions - left to right) of proteins isolated from hemolymph by galNAc affinity chromatography. The position of molecular weight markers (kDa) is shown at the top (\reducing) and left (non-reducing).

Evolutionary relationships of the tunicate lectin

Our analysis of the *S. plicata* lectin suggests that it has a similar secondary structure to vertebrate collectins of the type that can initiate complement activity. However, we do not yet have sufficient sequence to definitively assign homology to the tunicate protein. Our evidence cannot exclude the possibility that the tunicate lectin is related to complement component C1q. The domain associations and overall quaternary structures of vertebrate MBP and C1q are remarkably convergent (Sellar *et al.*, 1991; Reid, 1979; Reid *et al.*, 1982). Both molecules have collagenous domains that begin 18-30 amino acids from the N-terminus and both form oligomers based on collagen-like triple helixes.

At this stage the functional similarities between MBP and the tunicate protein (carbohydrate binding and opsonic activity) favor the idea that the tunicate lectin is related to MBP and not C1q. However, the functional significance of this distinction is essentially irrelevant. Both MBP and C1q are involved in the proteolytic activation of complement components via serine proteases (C1r/s, and MASP) that associate with their collagenous domains (Matsushita and Fujita, 1992, Tan, *et al.*, 1996, Thiel, *et al.*, 1997). Hence, the identification of a tunicate protein with obvious structural similarity to these highly convergent molecules (MBP and C1q) suggests that a proteolytic complement cascade also exists in tunicates.

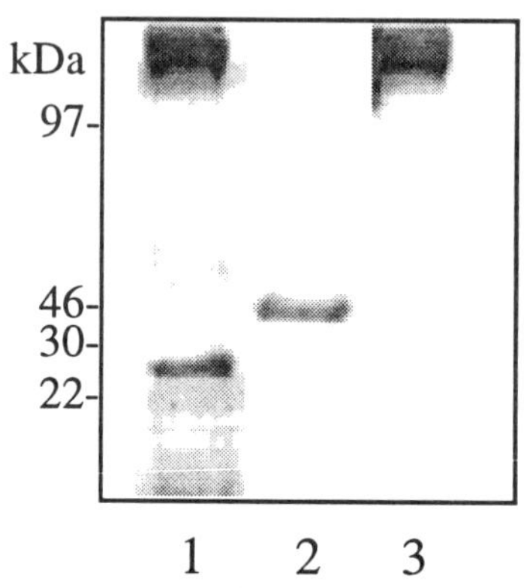

Figure 2. Reducing SDS-PAGE of proteins isolated by galNAc affinity chromatography and then digested with collagenase. Lane 1, collagenase digested galNAc fraction; Lane 2, non-digested galNAc fraction; Lane 3, collagenase alone. The position of molecular weight markers (kDa) is shown on the left. Fragments left after collagenase digestion were approximately 26kDa.

Possible MASPs in *S.plicata*

If the tunicate lectin does have role in the activation of complement, it should be associated with a protease equivalent to MASP or C1r/s. The minor 65 kDa band which co-precipitates with the 44 kDa collagenous lectin during galNAc affinity chromatography might represent such a protease. The 65 kDa protein does not have carbohydrate binding activity, even though it elutes with the 44 kDa lectin from galNAc affinity columns. This suggests that the 65 kDa protein is transiently associated with the collagenous lectin in a manner equivalent to the Ca^{2+} dependent interaction of MASP with MBP in mammals. The 65 kDa tunicate protein is also of similar molecular weight to vertebrate MASP, even though the tunicate protein is a monomer while MASP is a heterodimer (Matsushita and Fujita, 1992, Thiel, *et al.*, 1997).

COMPLEMENT ACTIVITY IN TUNICATES

Anaphylotoxin production

Complement-like proteins in *S. plicata* were originally identified by the formation of low molecular weight polypeptide fragments in hemolymph that resemble complement anaphylotoxins. In mammals, low molecular weight anaphylotoxins (approx. 7-9 kDa) are cleaved from the complement components C3, 4 & 5 as essential by-products of complement activation. We have found that an 8 kDa product is formed by activating *S. plicata* hemolymph with zymosan. The generation of this 8kDa peptide in hemolymph is inhibited by heating and serine protease inhibitors suggesting that it is the product of enzymatic cleavage analogous to vertebrate anaphyolotoxins.

Characterization of tunicate complement-like proteins

To confirm the existence of complement in *S. plicata*, we have probed hemolymph with polyclonal antibodies to vertebrate C3 (Figure 3). Those antibodies cross-react with known C3-like proteins from echinoderms and agnathans. In non-activated hemolymph, the cross-reactive tunicate proteins identified by anti-vertebrate C3 correspond precisely in size to intact mammalian C3 under non-reducing conditions. Under reducing conditions, the antibodies stained tunicate proteins of equivalent molecular weight to the two sub-units of mammalian C3, C3α and C3β. After activation of hemolymph with zymosan, anti-mammalian C3 antibodies also detected the 8kDa anaphylotoxin, and a corresponding decrease in size of the C3α component from which the anaphylotoxin is cleaved.

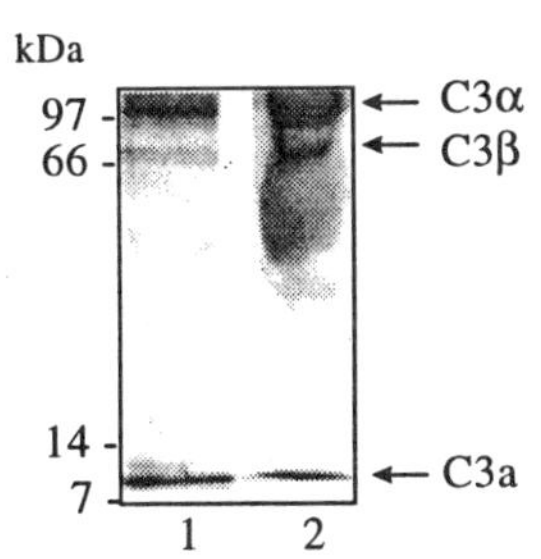

Figure 3. Reducing western blot of tunicate hemolymph (Lane 1) and human C3 (Lane 2) immunostained with a vertebrate C3 antiserum. The position of molecular weight markers is shown on the left. The position of the various C3 sub-units or fragments are shown with arrows on the right.

Zymosan binding activity and opsonisation is synergistically enhanced by tunicate lectin and complement-like proteins

The functional activity of the tunicate C3 was investigated using a zymosan binding assay in which activated tunicate complement components bound to zymosan and were detected with the anti-C3 antibodies. C3 cross reactive proteins in *S. plicata* hemolymph were shown to bind zymosan in a dose-dependent fashion with activities comparable to that of mammalian C3. This binding activity could be inhibited by serine protease inhibitors and by heating, suggesting that activation was proteolytic (Figure 4).

To test the binding activity of tunicate C3-like proteins in isolation, hemolymph was fractionated on an anti-vertebrate C3 affinity column. Under reducing conditions, fractions from the affinity column contained 3 predominant proteins, two of which corresponded precisely in size to mammalian C3α and β. The remaining protein probably represented a degradation product, and corresponded with a band evident in mammalian complement preparations. The zymosan binding activity of the anti-C3 affinity fractions was far lower than that of unfractionated hemolymph. However binding activity could be restored if zymosan was incubated simultaneously with the

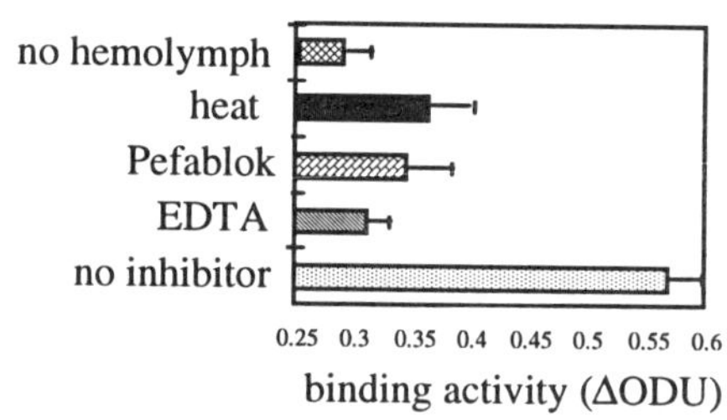

Figure 4. Zymosan binding activity of anti-C3 cross-reactive proteins from *S. plicata* hemolymph, and the effects of various potential inhibitors (heat, Pefablok and EDTA) on binding.

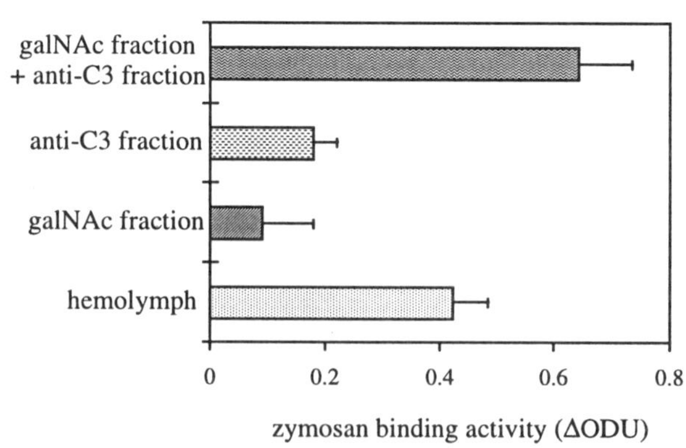

Figure 5. Zymosan binding activities of hemolymph, purified tunicate C3-like proteins (complement), galNAc fractions of hemolymph and a combination complement and galNAc fraction. Bars = standard errors, n>3.

collagenous lectin from galNAc affinity fractions and anti-C3 affinity purified proteins (Figure 5). The level of zymosan binding activity that could be achieved by co-incubation of the tunicate C3-like proteins with galNAc affinity fractions was not simply additive, but represented a synergistic enhancement of activity that could be suppressed by serine protease inhibitors.

The synergistic activity of collectin and complement was also evident in opsonisation. When incubated with target cells at low concentration, the independent opsonic activities of galNAc and anti-C3 affinity fractions amounted to only 17% of that evident when the two were combined (Figure 6). This synergistically enhanced opsonic activity was abrogated by serine protease inhibitors and EDTA.

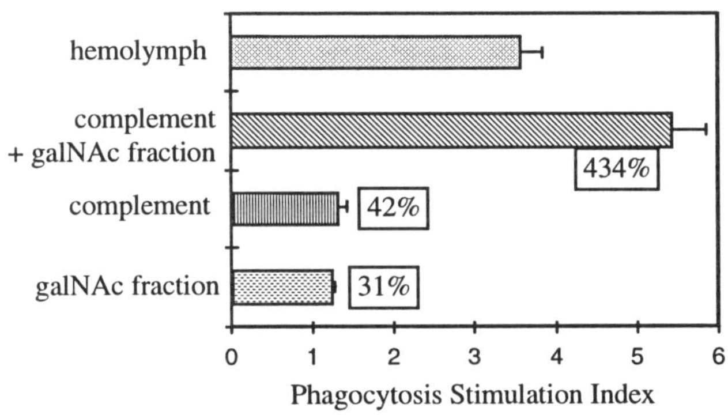

Figure 6. Opsonic activities (phagocytic stimulation index) of galNAc affinity fractions, affinity purified C3-like proteins, hemolymph and a combination of galNAc fractions with affinity purified C3-like proteins. Boxed numbers represent the percentage increase in phagocytic activity relative to non-opsonised controls. Bars = SEM, n>6.

CONCLUSIONS

The ability of tunicate galNAc fractions to synergistically activate C3-like proteins indicates for the first time that a proteolytic complement cascade exists in invertebrates. Given that invertebrates do not express soluble immunoglobulin antibodies, the identification of complement activity in tunicates has considerable ramifications. Most importantly it indicates that proteolytic complement activation exists in the absence of antibodies, and by corollary, that immunoglobulin antibodies co-opted a pre-existing complement system that may be relatively ubiquitous in the animal kingdom.

ACKNOWLEDGMENTS

This study was funded in part by grants from the Australian Research Council and the Macquarie University Internal Research Grants Scheme.

REFERENCES

Al-Sharif WZ, Sunyer JO., Lambris JD Smith, LC. Sea urchin coelomocytes specfically express a homologue of the complement component C3. J. Immunol. 1998;160:2983-2997

Azumi K, Satoh NYokosawa H. Functional and structural characterization of hemocytes of the solitary ascidian, *Halocynthia roretzi*. Journal of Experimental Biology . 1993;265:309-316.

Crouch E, Persson A, Chang D, Heuser J. Molecular structure of pulmanory surfactat protein D (SP-D). J. Biol. Chem. 1994;269:17311-17319

Davis AE, Lachmann PJ. Bovine conglutinin is a collagen-like protein. Biochemistry. 1984;23:2139-2144

Drickamer K, Taylor ME. Biology of animal lectins. Annu. Rev. Cell Biol. 1993;9:237-264

Epstein J, Eichbaum Q, Sheriff S, Ezekowitz RAB. The collectins in innate immunity. Current Opinions Immunol. 1996;8:29-35

Holmskov U, Laursen SB, Malhotra R, Wiedemann H, Timpl R, Stuart GR, Tornoe H, Madsen PS, Reid KBM, Jensenius JC. Comparative study of the structural and functional properties of a bovine plasma C-type lectin, collectin-43, with other collectins. Biochem. J. 1995;305:889-896

Holmskov U, Malhotra R, Sim RB, Jensenius JC. Collectins: collagenous C-type lectins of the innate immune defense system. Immunol. Today. 1994;15:67-74

Hoppe, HJReid, KBM. Collectins - Soluble proteins containing collagenous regions and lectin domains and their roles in innate immunty. Protein Science. 1994;3:1143-1158

Laursen SB, Hedemand JE, Thiel S, Willis AC, Skriver E, Madsen PS, Jensenius JC. Collectin in a non-mammalian species: isolation and charaterisation of mannan binding protein (MBP) from chicken serum. Glycobiology. 1995;5:553-561

Lu J, Willis AC, Reid KBM. Purification, characterisation and cDNA cloning of human lung surfactant protein D. Biochem. J. 1992;284:795-802

Lu JH. Collectins - Collectors of microorganisms for the innate immune system. Bioessays. 1997;19:509-518

Matsushita M, Fujita T. Activation of the classical complement pathway by mannose-binding protein in association with a novel C1s-like serine protease. J. Exp. Med. 1992;176:1497-1502

Nonaka M, Azumi K, Ji X, Namikawa-Yamada C, Sasaki M, Saiga H, Dodds AW, Sekine H, Homma MK, Matsushita M, Endo Y, Fujita T. Opsonic complement component C3 in the solitary ascidian, *Halocynthia roetzi*. J. Immunol. 1998;162:387-391

Reid, KBM. Complete amino acid sequences of the three collagen-like regions present in subcomponent C1q of the first component of human complement. Biochem. J. 1979; 179:367-371

Reid KBM, Gagnon J, Frampton J. Completion of the amino acid sequences of the A and B chains of subcomponent C1q of the first component of human complement. Biochem. J. 1982;203:559-569

Sellar GC, Blake DJ, Reid KBM. Characterization and organization of the genes encoding the A-, B- and C-chains of human complement subcomponent C1q. The complete derived amino acid sequence of human C1q. Biochem. J. 1991; 274:481-490

Smith LC, Chang L, Britten RJ, Davidson EH. Sea urchin genes expressed in activated coelomocytes are identified by expressed sequence tags. J. Immunol. 1996;156:593-602

Smith LC, Shih C-S Dachenhausen SG. Coelomocytes express SpBf, a homologue of Factor B, the second component in the sea urchin complement system. J. Immunol. 1998;161:6784-6793

Tan SM, Chung MCM, Kon OL, Thiel S, Lee SH, Lu J. Improvement on the purification of mannan-binding lectin and demonstration of its Ca^{2+} association with a C1s-like serine protease. Biochem. J. 1996;319:329-332

Thiel S, Vorup-Jensen T, Stover CM, Schwaeble W, Laursen SB, Poulsen K, Willis AC, Eggleton P, Hansen S, Holmskov U, Reid, KBM, Jensenius JC. A second serine protease associated with mannan-binding lectin that activates complement. Nature. 1997;386:506-510

IMMUNOLOGICAL PROPERTIES OF HEAT SHOCK PROTEINS ARE PHYLOGENETICALLY CONSERVED

Jacques Robert[1], Antoine Ménoret[2], Pramod K. Srivastava[2], and Nicholas Cohen[1]

[1]Department of Microbiology and Immunology University of Rochester School of Medicine and Dentistry, Rochester, NY 14642 and [2]Center for Immunotherapy of Cancer and Infectious Diseases, University of Connecticut School of Medicine, Farmington, CT 06030

INTRODUCTION

Heat shock proteins (HSPs) serve as molecular chaperones of polypeptides transported between cell organelles, and contribute to the folding of nascent and altered proteins. A rapidly growing body of data suggests that at least in mammals, the HSPs 70 and gp96, are involved in immunity (review in Srivastava *et al.,* 1998). Considering some analogy between MHC molecules and HSPs, it has been hypothesized that HSPs may have been forerunner presenting molecules, and that this primitive antigen presenting mechanism was integrated in the newly evolved MHC system (Srivastava and Heike,

Phylogenetic Perspectives on the Vertebrate Immune System.
Edited by Beck *et al.*, Kluwer Academic/Plenum Publishers, 2001.

1991). Structural similarities between hsp70 and major histocompatibility complex (MHC) class I peptide binding sites led Flajnik (1991) to a partially similar suggestion. Three characteristics of HSPs are reminiscent of MHC molecules (Table 1). First, these HSPs can physiologically associate with a large variety of antigenic peptides derived from cellular proteins. Second, after immunization with HSP-peptide complexes, T-cell immune responses are elicited against the chaperoned peptide. Finally, under certain circumstances (cell surface expression, cell death), these intracellular proteins end up in the extracellular compartment. Given the high degree of structural conservation of HSPs during evolution, their possible involvement in immune function, in general, and tumor immunity in particular, may be reflective of an ancient origin of HSPs that was contemporary with, or even antecedent to, the emergence of the vertebrate adaptive immune system that uses MHC presentation and restriction. In this review, we will examine the extent to which those immunological properties of HSPs found in mammals (peptide binding and chaperoning, adjuvanticity, immunogenicity) are phylogenetically conserved, and by extension, whether HSPs may be an evolutionary bridge between innate and adaptive immunity.

Table 1. HSP and MHC molecules have multiple features in common

HSP (gp96)	MHC
Very ancient (present in prokaryotes & eukaryotes)	Evolutionarily recent (arose in gnathostomes)
No polymorphism	Allelic polymorphism
Induced by IFN-γ, heat shock, glucose deprivation, viral infection	Expression is induced by IFN-γ
Binds peptides and unfolded proteins	Binds peptides
Mainly intracellular (cytosol or ER) but also at the surface of some cells	Expressed at the cell surface (most cells for class I, professional antigen-presenting cells for class II)
HSP-peptide complex can prime T-cells in solution (possibly also when expressed at the surface of some cells)	MHC-peptide complex can prime T-cell only when expressed at the surface of antigen-presenting cells

PHYLOGENETIC CONSERVATION OF STRUCTURE, PEPTIDE BINDING, AND PEPTIDE CHAPERONING OF GP96

Heat shock proteins (HSPs), constituents of a complex multigenic superfamily of phylogenetically highly conserved intracellular polypeptides, are produced by almost all types of prokaryotic and eukaryotic cells. As previously mentioned, HSPs perform essential biological functions under normal as well as stressful physiological conditions. These functions include: preventing protein aggregation under physical stress; serving as molecular chaperones of peptides and protein transport between cell organelles; and contributing to the folding of nascent and altered proteins (reviewed in Hendrik and Hartl, 1993; Morimoto *et al.*, 1994). HSPs are grouped in different families according to their primary structure; the degree of similarity of members within a family is greater than that of members of different families. In eukaryotes, different members of the HSP family are localized in specific cellular compartments. For example, members of the hsp70 subgroup are localized in the cytosol (hsc and hsp70), the mitochondria (mhsp70) and the endoplasmic reticulum endoplasmic reticulum (grp78 or BiP).

Our research has focused on gp96, the HSP that has been most extensively studied in the context of mammalian immunity. Gp96 is an ER-resident member of the hsp90 family that in vertebrates, can exist as two cytoplasmic isoforms known as hsp90a (or hsp 86) and hsp90b (or hsp 84) (Moore *et al.*, 1989; Krone and Sass, 1994). Gp96 constitutes the major protein in the ER lumen and contains the KDEL motif. In addition to mammals and chickens, genomic homologues of gp96 open reading frames have been identified in *C. elegans* and in the barley plant, but not in yeast (Gupta, 1995; Nicchita, 1998). As with HSPs in general, gp96 is phylogenetically highly conserved (*e.g.*, 65% amino acid sequence identity between mouse and *C. elegans*). Although limited, these data suggests that gp96 may be restricted to metazoans. To extend our information about the structural conservation of gp96 in the Metazoa, we used immunoblotting to determine the conservation of an epitope recognized by a rat anti-gp96 monoclonal antibody (clone 9G10) in different vertebrate and invertebrate species (Fig 1). We detected a single protein of identical apparent molecular weight in a representative teleost, urodele and anuran amphibian, reptile, bird and mammal. In addition, we detected gp96 homologues in the lamprey and hagfish, two primitive agnathan species. However, no specific signal was detected in any invertebrate species tested including snail, cricket, and tunicate. Interestingly, the epitope recognized by the anti-gp96 mAb encompasses 27 amino acid residues (position 287 to 314 on published sequences, accession number: J03297) that are highly conserved between chicken and mammalian homologues (more than 90% amino acid identity), but are poorly conserved in either *C. elegans* (26.3% identity, 42.1 % similarity) or barley (33.3 % identity, 44.4 % similarity). In addition, extensive screening of data banks for this epitope sequence did not reveal other significant matches. This dichotomy might be an example of a structural adaptation of this particular gp96 sequence during its evolution, to the emerging adaptive immune system of vertebrates.

In mammals, gp96 and hsp70 physiologically associate with a broad range of antigenic peptides derived from cellular proteins including viral antigens (Ciupitu *et al.*, 1998; Nieland *et al.*, 1996), tumor antigens (Tamura *et al.*, 1997; Srivastava *et al.*, 1998; Wells *et al.*, 1998), model antigens (Arnold *et al.*, 1995; Suto and Srivastava, 1995; Lammert *et al.*, 1997), minor histocompatibility (H) antigens (Arnold *et al.*, 1995) and other peptides (Ménoret *et al.* 1999). *In vitro*, various synthetic peptides can also be complexed with purified HSP (Blachere *et al.*, 1998). The peptide binding by HSP is non-covalent and stable at 37C (they remain bound when the complex is analyzed by SDS-PAGE without boiling the sample). The specificity of the interaction HSP-peptide is still unclear, and it remains to be determined whether there is a motif or length requirement as is the case for peptide presentation by MHC molecules.

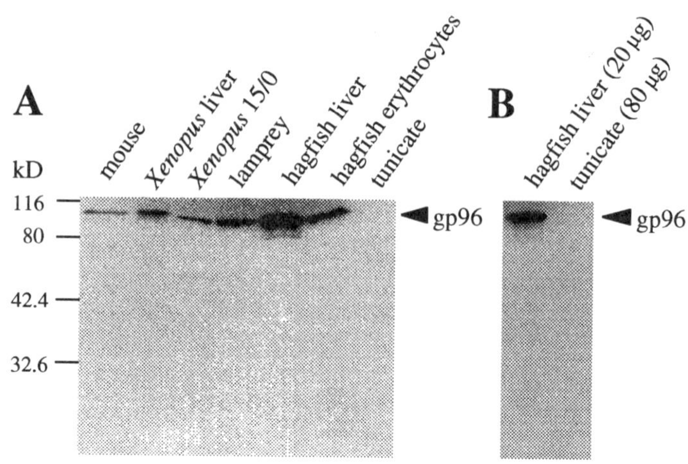

Figure 1. Conservation of a gp96 epitope in the whole vertebrate phylum including an Agnathan which lacks adaptive immunity. **A.** Western blot of protein lysate (20µg/lane) from mouse (*Mus musculus*) liver; frog (*Xenopus laevis*) liver; the *Xenopus* lymphoid tumor cell line, 15/0; lamprey (*Entosphenus tridentatus*) liver; hagfish (*Eptatretus stoutii*) liver; hagfish erythrocytes; and tunicate (*Clavelina picta*) whole tissue, probed with rat anti-gp96 mAb and rabbit anti-rat HRP conjugated Ab. No gp96 was detected in the tunicate sample, or in samples of snail, cricket, worm, or yeast (results not shown). **B.** Western blot of 20 µg of hagfish liver protein and 80 µg of tunicate protein, probed with the same Abs as in A, for confirmation that gp96 is not present in a lower concentration in the tunicate.

To determine if the high degree of structural conservation of gp96 in vertebrates reflects its functional conservation as an important molecule in immunity, we compared the peptide binding capacity of *Xenopus* gp96 *in vitro* to that of mammalian gp96 (Robert *et al.*, 1998; Robert *et al.*, 1999 submitted). *Xenopus* gp96 was purified from liver following the same protocol used in mammals (ammonium salt precipitation and Con-A sepharose followed by ion-exchange chromatography; Srivastava, 1997). A homogeneous protein of 96 kD was obtained that reacts with a rat anti-gp96 mAb and an antibody specific for the ER retention signal KDEL, and contains 5 of the 6 N-terminal residues found in mouse gp96 (mouse: XDEVDVD; *Xenopus*: XDEVEVD). This further substantiates the high degree of conservation of gp96 with respect to its molecular mass, glycosylated moiety, charge and amino-terminal sequence. Recently, homogeneous gp96 protein was obtained from the lamprey using exactly the same protocol (unpublished). This reveals the conservation of these biochemical properties in the whole vertebrate phylum.

Of direct relevance to the postulated conservation of immunological properties of gp96 is the observation that purified *Xenopus* gp96 can bind synthetic peptide (*e.g.*, VSV19, a precursor derived from the vesicular stomatitis virus nucleoprotein) *in vitro* as efficiently and as stably as human and mice gp96 (Robert *et al.*, 1998; Robert *et al.*, 1999 submitted). As shown for mouse gp96 (Blachere *et al.*, 1997), binding of exogenous peptide to *Xenopus* gp96 requires a short incubation at 50°C, and is dissociated by boiling. Even more interesting, is the observation that *in vitro*-

reconstituted *Xenopus* gp96 peptide complexes may be used, just like mouse gp96-peptide complexes, to pulse mouse peritoneal macrophages and specifically stimulate a CD8$^+$ CTL line that recognizes the K^b-restricted 8mer (RGYVYQGL) contained in the VSV19 peptide, to produce IFN-γ (Robert *et al.*, 1999 submitted). MHC class I molecules expressed on the macrophage surface cannot be directly loaded with the 19 mer VSV peptide molecule. This suggests that the amphibian HSP-peptide complex can be recognized by mouse macrophages, and that the peptides chaperoned by the amphibian gp96 are taken up and endogenously processed by these cells and then re-presented by the macrophage MHC class I in a manner identical to that in which a peptide chaperoned by the mouse gp96 is processed and re-presented. In mice, re-presentation of peptide chaperoned by gp96 is 200-400 times more efficient in stimulating class I-restricted CTLs than is the loading of free peptide (Suto and Srivastava, 1995), and it appears to be mediated by a cell surface receptor (Arnold *et al.*, 1999; Wassenberg *et al.*, 1999; Binder *et al.*, 1998). The ability to exchange *Xenopus* gp96 for mouse gp96 strongly suggests that this cell surface receptor is also conserved.

PHYLOGENETIC CONSERVATION OF IMMUNOGENICITY AND T-CELL ADJUVANTICITY OF GP96

A most remarkable feature of gp96 is its capacity, when purified from tumors, to elicit potent therapeutic as well as prophylactic cellular anti-tumor immunity that is dependent on the chaperoned peptides (reviewed in Srivastava *et al.*, 1998; Nicchita, 1998). Priming with small amount (10 μg) of HSP-peptide complexes purified from murine tumors completely protects syngeneic hosts against subsequent challenges with the tumor from which the HSP-peptide complexes were purified (Srivastava *et al.*, 1998; Udono and Srivastava, 1994). Such protective immunity is not elicited with HSPs derived from other sources such as normal tissues or a different tumor. More importantly, therapy for pre-existing murine tumors can be achieved by multiple injections of tumor-derived HSP-peptide complexes (Tamura *et al.*, 1997). Preliminary therapeutic clinical trials using gp96 purified from tumor tissues are encouraging (Amato *et al.*, 1999; Lewis *et al.*, 1999). Generation of specific T-cell responses by immunization with HSP-chaperoned peptides has also been shown for viral (Ciupitu *et al.*, 1998; Nieland *et al.*, 1996), and minor H-antigens (Arnold *et al.*, 1995). The *in vitro* demonstration of efficient re-presentation, by APCs, of peptides chaperoned by HSPs in the context of MHC class I (Suto and Srivastava, 1995), as well as the *in vivo* abrogation of anti-tumor protective immunity by depletion of macrophage and T-cells during priming (Udono et al., 1994), strongly suggest that HSP is a strong T-cell adjuvant for chaperoned peptides.

Our *Xenopus* model of transplantable tumors has provided an opportunity to investigate whether immunogenicity of gp96 is also evolutionarily conserved. The immune system of *Xenopus* is fundamentally similar to that of mammals (*e.g.*, rearranging TCR and Ig genes, MHC class I- and class II-restricted T-cell recognition; Du Pasquier *et al.*, 1989, Salter-Cid and Flajnik, 1995). NK-cell activity has also been recently characterized (Horton *et al.*, 1998). Relative to murine systems, however, the *Xenopus* model allows studying tumor immunity either in the presence or in the complete absence of MHC class I molecules (Robert and Cohen, 1998). Spontaneously arising thymic lymphoid tumors in *Xenopus* have been characterized, and transplantable cell lines developed (Du Pasquier and Robert, 1992; Robert *et al.*, 1994; 1995; Robert and Du Pasquier, 1996). Importantly, some of these lymphoid cell lines do not express any MHC class I or class II antigen mRNA or protein (Robert *et al.*, 1994). In previous studies using a MHC homozygous inbred strain, we have shown that the class I^+ ff-2 cell line is tumorigenic in larval but not in fully grown adult MHC identical hosts (Robert *et al.*, 1995), and that the ff-2 tumor rejection is mediated by a thymus-dependent immune response against tumor-specific antigens that develop during metamorphosis (Robert *et al.*, 1997). On the other hand, the MHC negative tumor cell line, 15/0, derived from the LG-15 clone, is more tumorigenic (Robert *et al.*, 1994; Robert and Cohen, 1998). It grows both in larvae and adults, even after transplanting as few as 100 cells intraperitoneally or subcutaneously. Priming with irradiated 15/0 cells results in a modest but reproducible delay in the first appearance of tumor after challenge (Robert and Cohen, 1998). This effect is not markedly improved by increasing the number of irradiated cells injected.

This *Xenopus* tumor system provided a unique opportunity to test the ability of gp96, presumably complexed to peptide, to generate an anti-tumor immune response against the MHC class I-negative 15/0 tumor cells. As seen in experiments involving priming with irradiated 15/0 tumor cells, but more dramatically, immunization with minute amount of gp96 purified from 15/0 tumors delayed the first sign of tumor appearance resulting from the challenge inoculum, and inhibited tumor growth (Robert *et al.*, 1998, 1999 submitted). Furthermore, in contrast to immunization with irradiated 15/0 cells, immunization with greater amounts of tumor-derived gp96 resulted in a more pronounced anti-tumor effect. The anti-tumor effect elicited by 15/0-derived gp96 is specific in that it is not elicited by immunization with intact LG-15 irradiated splenic lymphocytes, LG-15 total liver protein lysate, or gp96 purified from either LG-15 liver or mouse. As in mice, *Xenopus* gp96 generates an immune response against 15/0 tumor only when it has been purified from this tumor; this provides indirect evidence that *Xenopus* gp96 chaperones tumor antigen peptides.

Although the effector components of the immune response elicited by immunizing *Xenopus* with 15/0-derived gp96 remains to be characterized, the absence of detectable antibody- or complement-mediated killing suggests that it is primarily a cellular immune response. A typical $CD8^+$ cytotoxic T cell, however, seems improbable given the absence of class I expression by tumor cells. While it possible that NK cells (Horton *et al.*, 1998) are involved, two other hypotheses merit testing. First, although 15/0 cells do not express classical MHC class I, they do express mRNA for non-classical class Ib (Robert *et al.*, 1994). Class Ib genes are minimally polymorphic in *Xenopus* and share similar structural and genetic features with their mammalian counterparts (Flajnik *et al.*, 1993; Salter-Cid *et al.*, 1998). In mammals, most class Ib molecules are associated with β2m and bind peptides (reviewed in Parham, 1994). Given their similar sequence, *Xenopus* class Ib molecules expressed by 15/0 tumor cells could also present endogenous tumor-specific peptide and become targets for specific effector cells that were previously primed by re-presentation of the gp96-peptide complexes by professional APCs as has been previously demonstrated in mouse and in humans (Suto and Srivastava, 1995). Second, gp96 expressed on the tumor surface may

Table 2. Immune-related characteristics of MHC and HSP in vertebrates

		MHC Immune-related characteristics			**HSP (gp96)** Immune-related characteristics			
	Taxon	Gene present	Peptide binding and presentation	Surface expression	Gene present	Peptide-binding	Immuno-genicity	Surface expression
Adaptive and innate immunity	Mammals	+	+	+	+	+	+	+
	Birds	+	+	+	+	?	?	?
	Reptiles	+	+	+	+	?	?	?
	Amphibians	+	+	+	+	+	+	+
	Fish	+	+	+	+	?	?	+
Innate immunity	Agnathans	-	-	-	+	?	?	+
	Protochordates	-	-	-	+	?	?	?

directly present immunogenic peptides to specific effector cells, in a similar way to to that observed previously for hsp70 in a human situation (Kim *et al.*, 1995).

This hypothesis is supported by our observation that the immunogenicity of gp96 correlates with the level to which it is expressed on the surface of various *Xenopus* tumor cell lines (see next section and Table 2).

PHYLOGENETIC CONSERVATION OF GP96 SURFACE EXPRESSION

It has been reported that hsp70 expressed on the cell surface of some murine tumors may directly interact with $\gamma\delta$-T cell (Van Buskirk *et al.*, 1989; Tamura *et al.*, 1993; Kim *et al.*, 1995) and/or NK-cells (Multhof *et al.*, 1997). In fact, the surface expression of HSPs such as hsp70 (Ferrarini *et al.*, 1992; Multhoff *et al.*, 1995; Vanbuskirk *et al.*, 1989), BiP (Takemoto *et al.*, 1992), hsp90 (Ullrich *et al.*, 1986; Ferrarini *et al.*, 1992) and gp96 (Srivastava *et al.* 1986) has been observed for a variety of human and murine tumor cells, but not on the limited number of transformed or normal cells that have been examined (human EBV-transformed B-cells, peripheral blood leukocytes, and fibroblasts). Similarly, gp96 has also been shown to be expressed on the surface of some mouse tumor cells but not on normal embryonic mouse fibroblasts (Altmeyer *et al.*, 1996), despite the fact that the cell surface gp96 contains a carboxy terminal KDEL sequence (Lys-Asp-Glu-Leu) which is a feedback signal from the Golgi to the ER (Kock *et al.*, 1986). Recently, reports of surface expression of several ER-resident molecular chaperones, including gp96, on a subset of murine immature thymocytes (Wiest *et al.*, 1997) reveal that this phenomenon might be not restricted to tumor cells.

Given the remarkable phylogenetic conservation of gp96 structure, peptide binding capacity, and immunogenicity, we were interested in determining whether surface expression of this HSP is also a general phenomenon. Indeed, our data reveal that gp96, most probably containing the ER-retention motif, is expressed on the surface of all the different *Xenopus* lymphoid tumor cell lines tested (Robert *et al.*, 1999). As noted, previously, there is a correlation between the level of gp96 surface expression determined by flow cytometry with anti-gp96 mAb, and the immunogenicity of the tumor cell line. The least immunogenic tumor, 15/0, expresses the lowest level of surface gp96, whereas the highly immunogenic B3B7 tumor line expresses the most. The ff-2 cell line, which has an intermediate level of gp96 surface expression, is tumorigenic in larval (or less immunogenic), but not in fully grown adult MHC identical hosts (Robert *et al.*, 1994; 1995). Furthermore, tumor rejection of ff-2 cells is mediated by thymus-dependent immune response against tumor-specific antigens that develop during metamorphosis (Robert *et al.*, 1997). Another tumor clone (15/40, clone A1) that recently lost its tumorigenicity, displays levels of surface gp96 that are as high as the immunogenic B3B7 tumor (unpublished observations.). No obvious correlation between the immunogenicity of these three tumors and the level of surface MHC class I and class II molecules or other molecules (*i.e.*, CD8, CTX) was found. Surface gp96 was also detectable by immunoprecipitation of surface biotinylated tumor cells using either anti-gp96 or to lesser extent, anti-KDEL mAbs (Robert *et al.*, 1999).

Interestingly, while *Xenopus* gp96 cannot be detected by flow cytometry or immunoprecipitation of cell surface biotinylated lysate on the surface of most nontransformed cells including peritoneal and splenic macrophages, erythrocytes, as well the majority of splenocytes, a minor subset of splenic lymphocyte that appears to be mainly sIgM$^+$ B cells, express a high level of surface gp96 (Robert *et al.,* 1999). By flow cytometry, this level is 50-100 times greater than that on the different lymphoid tumor cells tested. Extensive control analysis has ruled out the possibility that gp96 present at the cell surface of tumor or lymphocyte results from a passive loading from free extracellular gp96. Rather, monitoring *in vitro* of gp96 re-expression after complete digestion of surface protein by pronase, indicates that gp96 surface expression on *Xenopus* tumor and normal lymphocytes (Robert *et al.*, 1999) is actively directed to, and retained on, the plasma membrane by a process that is dependent on translocation of protein from the ER to the Golgi (*i.e.*, it is brefeldin A-sensitive).

The report that different ER-resident molecular chaperones are expressed on the surface of immature murine thymocytes (Wiest *et al.*, 1997), together with our finding that surface gp96 is expressed on a subset of normal *Xenopus* lymphocytes implies that this phenomenon is not restricted to transformed cells, but is limited to some cell type or differentiation stage. To obtain further evidence of the biological relevance of this phenomenon, we took advantage of the conservation of the epitope recognized by the anit-gp96 monoclonal antibody to study other ectothermic vertebrates. A similar expression of gp96 homologues was found on the surface of some, but not all, T- and B-cell lines from the channel catfish (Robert et al., 1999). In several experiments, this pattern of expression was reproducibly detected (i.e., surface positive and negative clones always expressed the same phenotype). This clearly indicates that gp96 surface expression in this fish species is not merely the result of *in vitro* culture conditions but rather is restricted to certain cell types or stages of differentiation. Most interestingly, the restricted cell surface expression of gp96 homologues has been extended to the agnathan hagfish (Robert *et al.*, 1999). Agnathans, represented today by hagfish and lampreys, are the only known vertebrates where component of the adaptive immune system (*e.g.*, MHC, rearranging TCR, Ig) have not been found despite repeated molecular approaches to identifying them (Du Pasquier and Flajnik, 1999). Surface gp96 was detected on the surface of freshly harvested lymphocyte-like cells in the peripheral blood and pronephros cells, but not erythrocytes, of the hagfish. As in *Xenopus* the gp96 hagfish homologue is loaded at the cell surface by an active mechanism that is sensitive to brefeldin-A.

CONCLUSIONS AND SPECULATIONS

In mammals, a model has been proposed (Srivastava *et al.*, 1994) that links cytosolic and ER resident HSPs to MHC class I antigen presentation. Peptides, generated in the cytosol from protein degraded by proteasomes, are transported from the cytoplasm to the ER through an assembly of HSPs that bind and release them, in an ATP-dependent manner. In such a setting, HSPs would constitute a natural adjuvant facilitating the re-presentation of chaperoned peptides and modulating an adaptive cellular immune response (Srivastava *et al.*, 1998). On the other hand, based on upregulation of HSP expression by IFN-γ (Anderson *et al.*, 1994), and the fact that following "stressful" conditions (*e.g.*, glucose deprivation, viral infection), HSP may be released by dead cells, it has been postulated that HSPs may be "danger molecules" (Matzinger, 1994) that present an alarm signal to the immune system that does not depend on chaperoned antigenic peptides (Srivastava *et al.*, 1998). It has been further proposed that because of their very ancient origin and because of their capacity to stimulate macrophages (in a non-peptide-specific manner) to secrete low levels of cytokines (Basu *et al.*, 1998), HSPs may also play a key role in innate immunity as messengers of cell death. In such a scenario, HSPs would have been used in primitive multicellular organisms only as a danger signal. Over time, HSPs would have become integrated into an adaptive immune system as accessory agents in MHC-restricted antigen presentation.

Based on our data and on the many analogous features of MHC and gp96 molecules (Table 1 and 2), we hypothesize that during the emergence of the adaptive immune system, the role of gp96 (and perhaps other HSPs) has been more than that of a danger signal. Specifically, we propose that the surface expression of gp96 may have provided a primitive mechanism for presentation of cell surface antigens. The actual picture of the evolution of the vertebrate adaptive immune system derived from recent extensive molecular studies (review in Parham, 1998; Du Pasquier and Flajnik, 1999) is an abrupt and somehow simultaneous emergence of all the different gene elements in the ancestors of Gnathostomata (jawed fish). Indeed T and B lymphocyte receptors as well as MHC class I and class II genes have been identified in the living representatives of all the classes of gnathostomes but have yet to be found in agnathans and prochordates. Expressions as extreme as the "big bang" have been used (Marchalonis *et al.*, 1998) to describe such a scenario. However, according to the well-documented synthetic Darwinian theory (Mayr, 1980), evolution occurs at the level of populations of organisms, and sexual populations change essentially in a gradual manner without sudden saltation (*i.e.*, without the sudden origin of a new kind of individual who becomes the progenitor of a new kind of organism). This means that although the emergence of a functional adaptive immune system may have resulted from an accelerated evolutionary rate, it cannot have happened in a sudden and single step. Rather, a complex multi-step process involving natural selection was needed to obtain integration of the many different elements that constitute the contemporary adaptive immune system. In addition, from a strict Darwinian viewpoint, production of phenotypic diversity precedes natural selection. In agreement with this theory, Du Pasquier and Flajnik (1999) proposed a scenario in which a primitive repertoire of lymphocyte receptors, probably TCR-like, was first generated by somatic rearrangement, possibly, as originally proposed by Marchalonis *et al.* (1998), as a result of an horizontal transfer of the rearrangement machinery (a possibility propped up by the recent finding of the transposition activity of Rag1 and 2 genes; Agrawal *et al.*, 1998; Hiom *et al.*, 1998). During a second step, MHC genes evolved to effect a more efficient lymphocyte receptor repertoire and to avoid autoimmunity. It is possible that HSPs played an important role during the evolutionary time needed for the emergence of MHC genes and the functional and regulatory fine-tuning of this system. In other words, HSPs may have been used as a kind of surrogate for the MHC during that period required for the assembly of MHC genes to have evolved.

The expression of gp96 on the surface of lymphocyte-like cells of the primitive hagfish which lack a MHC, TCR, and Ig genes, suggests that such a system may be still operative in the Agnatha as well as in the immature *Xenopus* tadpole where class I is not expressed. It is also interesting to speculate that HSPs may still play a functional role in some fish (*e.g.*, elasmobranchs) and in urodeles where despite the identification of MHC genes (Sammut *et al.*, 1997; Bartl, 1998), the relative weakness of cytotoxic effector functions (Cohen, 1968) suggests that an optimal system of MHC-restricted antigen presentation may not yet have been realized at these phylogenetic levels.

ACKNOWLEDGMENTS

Research from these authors' laboratory (J.R. and N.C.) has been supported by NIH grants R37 HD-07901 and RO1 CA-76312; research from these authors' laboratory (A.M. and P.K.S.) has been supported by NIH grants CA-64394 and CA-44786 and by a research agreement with Antigenics, Inc.

REFERENCES

Agrawal, A., Eastamn, Q., and Schatz, D. (1998). Transposition mediated by Rag1 and Rag2 and its implication for the evolution of the immune system. Nature 394: 744-751.

Altmeyer, A., Maki, R.G., Feldweg, A.M., Heike, M., Protopopov, V.P., Masur, S.K. and Srivastava, P. (1996). Tumor-specific cell surface expression of KDEL-containing, endoplasmic reticular heat shock protein gp96. Int. J. Cancer 69: 340-349.

Amato. R., Murray L., Wood., B., Savary.C., Tomasovic S., Srivastava P.K,, and Reitsma, D. (1999). ASCO meeting

Anderson, S.L., Shen, T., Lou, J., Xing, L., Blachere, N.E., Srivastava, P. and Rubin B.Y. (1994). The endoplasmic reticular heat shock protein gp96 is transcriptionally upregulated in interferon-treated cells. J. Exp. Med. 180: 1565-1569.

Arnold, D., Faath, S., Rammensee, H. and Shild H. (1995). Cross-priming of minor histocompatibility antigen-specific cytotoxic T cells upon immunization with the heat shock protein gp96. J. Exp. Med. 182: 885-889.

Bartl, S. (1998). What sharks can tell us about the evolution of MHC genes. Immunol. Rev. 166: 317-331.

Basu, S., Suto, R., Binder, J. and Srivastava, P. (1998). Heat-shock proteins as novel mediators of cytokine secretion by macrophages. Cell Stress & Chaperones 3: (suppl. 1), 11.

Binder, J., Ménoret, A. and Srivastava, P. (1998). Receptor-dependent and receptor-independent re-presentation of heat-shock protein-chaperoned peptides. Cell Stress & Chaperones 3: (suppl. 1), 2.

Blachere, N.E., Li, Z., Chandawarkar, R.Y., Suto, R., Jaikaria, N.S., Basu, S., Udono, H and Srivastava, P. K. (1997). Heat shock protein-peptide complexes, reconstituted *in vitro*, elicit peptide-specific cytotoxic T lymphocyte response and tumor immunity. J. Exp. Med. 186: 1315-1322.

Ciupitu, A., Peterson, M., O'Donnell, C., Williams, K., Jindal, S., Kiessling, R., and Welsh, M. (1998). Immunization with a lymphocytic choriomeningitis virus peptide mixed with heat shock protein 70 results in protective antiviral immunity specific cytotoxic T lymphocytes. J. Exp. Med. 187: 685-691.

Clem, L.W., Bly, J. E., Wilson, M., Chinchar, V.G., Stuge, T., Barker, K., Luft, C., Rycyzyn, M., Hogan, R. J., van Lopik, T.and Miller, N. W. (1996). Fish immunology: The utility of immortalized lymphoid cells—a mini review. Vet. Immunol. Immunopath. 54: 137-144.

Cohen, N. 1968. Chronic skin graft rejection in the Urodela. I. A comparative study of the first- and second-set allograft reactions. J. Exp. Zool. 167:37-48.

Du Pasquier, L., and Flajnik, M. (1999). Origin and evolution of the vertebrate immune system. In Paul, W. E. ed., *Fundamental Immunology*, Fourth Edition p. 605-649. Raven Press New York.

Du Pasquier L. and Robert J. (1992). *In vitro* growth of thymic tumor cell lines from *Xenopus*. Devel. Immunol. 2: 295-307.

Du Pasquier, L., Schwager, J., and Flajnik, M. F. (1989). The immune system of *Xenopus*. Annu. Rev. Immunol. 7: 251-275.

Flajnik, M.F., Canel, C., Kramer, J., and Kasahara, M. (1991). Which came first, MHC class I or class II? Immunogenetics 33: 295-300.

Flajnik, M., Kasahara, M., Shum, P., Salter-Cid, L., Taylor, A., and Du Pasquier L. (1993). A novel type of class I gene organization in vertebrates: a large family of non-MHC-linked class I genes is expressed at the RNA level in the amphibian *Xenopus*. EMBO J 12: 4385-4495.

Gupta, R. (1995). Phylogenetic analysis of the 90kK heat shock family of protein sequences and ab examination of the relationship among animals, plants, and fungi species. Mol. Biol. Evol. 12: 1063-1073.

Hendrick, J.P. and Hartl, F.U. (1993). Molecular chaperone functions of heat-shock proteins. Annu. Rev. Biochem. 62: 349-375.

Hiom, K., Melek, M., and Gellert, M. (1998). DNA transposition by the RAG1 and RAG2 proteins: a possible source of oncogenic translocation. Cell 94: 463-470.

Horton, T. L., Ritchie, P., Watson, M., and Horton, J. D. (1998). Natural cytotoxicity towards allogeneic tumour targets in *Xenopus* mediated by diverse splenocyte populations. Dev. Comp. Immunol. 22: 217-230.

Kim, H.T, Nelson, E.L, Clayburger, C., Sanjanwala, M., Sklar, J. and Krensky, A.M. (1995). Gamma delta T cell recognition of tumor Ig peptide. J. Immunol. 154: 1614-1623.

Kobel, H.R. and Du Pasquier, L. (1977). Strains and species of *Xenopus* for immunological research. In: *Developmental Immunology*, Solomon, J.B. and Horton, J.D. Eds, (Amsterdam, Elsevier North-Holland), pp. 299-306.

Kock, G., Smith, M., Macer, D., Webster, P., and Mortara, R. (1986). Endoplasmic reticulum contains a common, abundant calcium-binding glycoprotein, endoplasmin.J. Cell. Sci. 86: 217-232.

Krone, P.H. and Sass, J. (1994). Hsp90a and hsp90b genes are present in the zebrafish and are differentially regulated in developing embryos. Biochem. Biophys. Res. Comm. 204: 746-752.

Lammert, E., Arnold, D., Nijenhuis, M., Momburg, F., Hammerliing, G., Brunner, J., Stevanovic, S., Rammensee, H. and Schild, H. (1997). The endoplasmic reticulum-resident stress protein gp96 binds peptides translocated by TAP. Eur. J. Immunol. 27: 923-927.

Lewis, J., Janestzki, S., Livingston, P., Desantis, D., Williams, L., Klitnstra, W., Reitman ,D., Hougaton A., Srivastava, P., and Bronnan, J. (1999). ASCO meeting.

Marchalonis, J. Schluter, S., Bernstein, R. And Hohman, V. (1998). Antibodies of sharks: revolution and evolutions. Immunol. Rev. 166: 103-122.

Mayr, E. (1997). This is biology. The science of the living world. The Belknap press of Harvard University press. Cambridge MS. 323 pp.

Matzinger, P. (1994). Tolerance, danger, and the extended family. Ann. Rev. Immunol. 12: 991-1045.

Ménoret, A., Peng, P. and Srivastava, P.K. (1999). Association of peptides with the heat shock protein gp96 occurs in vivo and not post cell lysis. Bioch. Biophys.Res. Com. (In press)

Moore, S.K., Kozak, C., Robinson, E.A., Ullrich, S.J. and Appella, E. (1989). Murine 86- and 84-kDa heat shock proteins, cDNA sequences, chromosome assignments, and evolutionary origins. J. Biol. Chem. 264: 5343-5351.

Morimoto, E., Tissiere, A., and Georgopoulos, eds (1994). The biology of heat shock proteins and molecular chaperones. Cold Spring Harbor Laboratory, Cold Sprong Harbor, NY.

Multhoff, G., Botzler, C., Wiesnet, M., Müller, E., Meier, T., Wilmann, W., and Issels, R. (1995). A stress-inducible 72-kDa heat-shock protein (hsp72) is expressed on the surface of human tumor cells, but not on normal cells. Int. J. Cancer 61: 272.

Multhoff, G., Botzler, C., Jennen, L., Schmidt, J., Ellwart, J., and Issels, R. (1997). Heat shock protein 72 on tumor cells a recognition structure for natural killer cells. J. Immunol. 158: 4341-4350.

Nicchitta CV. (1998). Biochemical, cell biological and immunological issues surrounding the endoplasmic reticulum chaperone GRP94/gp96. Current Op. Immunol; 10: 103-109.

Nieland, T.J.F., M.C.A.Tan, M. van Muijen, F. Koning, A. Kruisbeek, and G.M. van Bleek., (1996) Isolation of an immunodominant viral peptide that is enbdogenously bound to the stress protein gp96/grp94. Proc. Natl. Acad. Sci.(USA) 93: 6135-6139.

O'Brien, R., Happ, M., Dallas, A., Palmer, E., Kubo, R. and Born, W. (1989). Stimulation of a major subset of lymphocytes expressing T cell receptor gamma delta by an antigen derived from *Mycobacterium tuberculosis*. Cell 57: 667-674.

Parham, P. (1994). The rise and fall of great class I genes. Semin. Immunol. 6: 373-382

Parham, P. Ed. (1998). Immune system of ectothermic vertebrates. Immunol. Rev. 166. 384 pp.

Robert, J. and Du Pasquier L. (1996). *Xenopus* lymphoid tumor cell lines. In: *Manual of Immunological Methods* ed. I. Lefkovitz. Academic Press, pp. 2367-237.

Robert, J., Guiet, C. and Du Pasquier L. (1994). Lymphoid tumors of *Xenopus laevis* with different capacities for growth in larvae and adults. Devel. Immunol. 3: 297-307.

Robert, J., Guiet, C. and Du Pasquier L. (1995). Ontogeny of the alloimmune response against transplanted tumor in *Xenopus laevis*. Differentiation 59: 135-144.

Robert, J., Guiet C. , Cohen, N., and Du Pasquier L. (1997). Effects of thymectomy and tolerance induction on tumor immunity in adult *Xenopus laevis*. Internat. J. Canc. 70: 330-334.

Robert, J., Ménoret, A., Srivastava, P. and Cohen, N. 1998. The tumor-specific immunogenicity of the heat-shock protein gp96 is conserved during evolution. Cell Stress & Chaperones 3:suppl.1, 22.

Robert, J., and Cohen, N. (1998). Evolution of immune surveillance. Immunol. Rev. 166:231-243.

Robert, J., Ménoret, A., and Cohen, N. (1999). Cell surface expression of the endoplasmic reticular heat shock protein gp96 is phylogenetically conserved. J. Immunol. (In press).

Robert, J., Ménoret, A., Cohen N., and Srivastava, A. (1999). The immunological properties of the heat shock protein gp96 are conserved across phylogenetically distant species. (submitted).

Salter-Cid, L. and Flajnik, M. (1995). Evolution and developmental regulation of the major histocompatibility complex. Crit. Rev. Immunol. 15: 31-75.

Salter-Cid, L., Nonaka, M., and Flajnik, MF. (1998). Expression of MHC class Ia and class Ib during ontogeny: high expression in epithelia and coregulation of class Ia and lmp7 genes. J. Immunol. 160: 2853-61

Sammut, B., Laurens, V., and Tournefier, A. (1997). Isolation of Mhc class I cDNAs from the axolotl *Ambystoma mexicanum*. Immunogenetics 45: 285-294.

Srivastava, P.K., Deleo, A., and Old, L.J. (1986). Tumor rejection antigens of chemically induced sarcomas of inbred mice. Proc. Natl. Acad. Sci. USA 83: 3407-3411.

Srivastava, P.K and Heike, M. (1991). Tumor-specific immunogenicity of stress-induced proteins: convergence of two evolutionary pathways of antigen presentation? Sem. Immunol. 3: 57-64.

Srivastava,, P.K and Maki, R.G. (1991). Stress-induced proteins in immune response to Cancer. Curr. Top. Microbiol. Immunol. 167: 109-121.

Srivastava, P.K., Udono, H., Blachere, N.E. and Li, Z. (1994). Heat shock proteins transfer peptides during antigen processing and CTL priming. Immunogenetics 39: 93-98.

Srivastava, P.K. (1997). Purification of heat shock protein-peptide complexes for use in vaccination against cancers and intracellular pathogens. Immunol. Methods Manual 9.11: 737-747.

Srivastava PK. Ménoret A. Basu S. Binder RJ. McQuade KL. 1998. Heat shock proteins come of age: primitive functions acquire new roles in an adaptive world. Immunity. 8:657-65

Suto, R. and Srivastava, P.K. (1995). A mechanism for the specific immunogenicity of heat shock protein-chaperoned peptides. Science 269: 1585-1588.

Tamura, Y., Tsuboi, N., Sato, N. and Kibuchi, K. (1993). 70-kDa heat-shock cognate protein is a transformation-associated antigen and a possible target for host's anti-tumor immunity. J. Immunol. 151: 5516-5524.

Tamura, Y, Peng, P, Liu, K, Daou, M and Srivastava, P.K. (1997) Immunotherapy of tumors with autologous tumor-derived heat shock protein preparations. Science 278: 117-119.

Takemoto, H., Yoshimori, T., Yamamoto, A., Myiata, Y., Yahara, I., Inoue, K., and Tashiro, Y. (1992). Heavy chain binding protein (BiP/GRP78) and endoplasmin are exported from the endoplasmic reticulum in rat exocrine pancreatic cells, similar to protein disulfide-isomerase. Arch. Biochem. Biophysics 296: 129-134.

Udono, H. and Srivastava, P.K. (1993). Heat shock protein 70-associated peptides elicit specific cancer immunity. J. Exp. Med. 178: 1391-1396.

Udono, H. and Srivastava, P.K. (1994). Comparison of tumor-specific immunogenicities of stress-induced proteins gp96, hsp90 and hsp70. J. Immunol. 152: 5398-5403.

Udono, H., Levey, D.L. and Srivastava, P.K. (1994). Cellular requirements for tumor-specific immunity elicited by heat shock proteins: tumor rejection antigen gp96 primes CD8$^+$ T cells *in vivo*. Proc. Natl. Acad. Sci. USA 91: 3077-3081.

Ullrich, S., Robinson, A., Law, W., Willingham, M., and Appella E. (1986). A mouse tumor-specific transplantation antigen is a heat shock-related protein. Proc. Natl. Acad. Sci. USA 83: 3121-3128.

Van Buskirk, A., Crump, B., Margoliash, E. and Pierce, S. (1989). A peptide-binding protein having a role in antigen presentation is a member of the hsp70 family. J. Exp. Med. 170: 1799-1809.

Wassenberg, J., Dezfulian, C., and Nicchitta, C. (1999). Receptor mediated and fluid phase pathways for internalization of the ER Hsp90 chaperone GRP94 in murine macrophages. J. Cell Science 112: 2167-2175.

Wells, A., Rai, S., Salvato, M., Band, H., and Malkovsky, M. (1998). Hsp 72-mediated augmentation of MHC class I surface expression and endogeneous antigen presentation. Int. Immunol. 10: 609-617.

Wiest, D., Bhandoola, A., Punt, J., Kreibich, G., McKean, D. and Singer, A. (1997). Incomplete endoplasmic reticulum (ER) retention in immature thymocytes as revealed by surface expression of "ER-resident" molecular chaperones. Proc. Natl. Acad. Sci. USA 94: 1884-1889.

DROSOPHILA CELLULAR IMMUNE SYSTEM: *DOROTHY* ENCODES A UDP GLYCOSYLTRANSFERASE

Z. Zhou, A. Rodriguez, C.-Y. Wu and D. A. Kimbrell

Department of Biochemistry and Cell Biology
Rice University, 6100 Main Street
Houston, TX 77005-1892

INTRODUCTION

Immune defense in *Drosophila* has humoral and cellular components that act effectively to control infectious organisms through the production of antimicrobial proteins and the phagocytosis and encapsulation of invading organisms (reviewed in 9). The main tissue of humoral immune defense is the fat body, which corresponds to the mammalian liver. Cellular immune defense is mediated by the hematopoietic system, which consists of hemocytes and lymph glands.

In embryos, hemocytes derive from the head mesoderm and phagocytose embryonic cells that have undergone programmed cell death (40). In larvae, hemocytes originate from the hematopoietic organ, the lymph glands, which derive from the cardiogenic region in the lateral mesoderm (34, 36). When fully developed, the lymph glands appear as 5 to 7 paired lobes along the dorsal vessel next to the ring gland and brain,

Phylogenetic Perspectives on the Vertebrate Immune System.
Edited by Beck *et al.*, Kluwer Academic/Plenum Publishers, 2001.

with pericardial cells positioned between each lobe and continuing along the length of the dorsal vessel (29; Fig. 1). The lymph glands take on the role of hematopoietic organs during larval development, and give rise to two types of hemocytes (36). The primary hemocytes are plasmatocytes, and these serve the functions of defense against pathogens (31), including secretion of antimicrobial peptides (reviewed in 18). The plasmatocyte lineage clearly functions as phagocytic cells that can be compared to vertebrate macrophages and natural killer cells (1, 23, 32). The other major hemocytes produced by the lymph glands are the crystal cells. These cells contain crystalline inclusions, and function in wound healing and melanization (28. 29). In addition, hemocytes appear to be important for the extensive tissue remodeling that takes place during metamorphosis (27).

In many mutations in *Drosophila*, hemocyte and lymph gland function with regard to self appears to be affected (reviewed in 5, 6, 37, 41). The defining characteristic of these mutations is melanotic tumors, black masses of tissue that are aggregates of hemocytes or hemocytes surrounding other cell types. The hemocytes thus appear to form capsules and deposit melanin as a normal strain would during an immune response, but in this case the encapsulation is directed against the organism itself. These mutations can be viable or lethal, and may show an overgrowth of hemocytes and lymph glands as well. These overgrowth mutations are part of the growing list of mutations and genes that has established *Drosophila* as a model system for the study of cancer (reviewed in 6, 41). The genes that these overgrowth mutations identify have so far been found to be expressed in many tissues, even though the overgrowth is limited to the hematopoietic system. Other genes found to be expressed in lymph glands and/or hemocytes (*e.g. malvolio*, 32; *Ds47*, 12) are also expressed in additional tissues.

Lymph gland and hemocyte specific genes would be very useful for providing information about hematopoietic specific processes in immunity and development. We therefore conducted an enhancer detector screen for strains with reporter gene activity (ß-galactosidase) in lymph glands and/or hemocytes, and identified *Dorothy* (33). The *Dorothy* enhancer detector strain is characterized as having ß-galactosidase localized specifically in the lymph glands and pericardial cells. We have now determined the molecular identity of *Dorothy* and confirmed that *Dorothy* expression is specific to the hematopoietic system and pericardial cells. *Dorothy* is the first member found in an insect of the UDP glycosyltranferase gene superfamily. This type of transferase has been identified in many organisms, from baculoviruses to humans, and has been shown to be involved in regulation of steroid hormone levels and detoxification of xenobiotics and endogenous compounds (reviewed in 2, 3). In addition, our observations using the reporter gene expression of the *Dorothy* strain further the description of lymph gland development in mutant backgrounds, and indicate that *Dorothy* is also expressed in adult hemocytes.

MATERIALS AND METHODS

Drosophila stocks

The *P{lacW}80/12* strain (*P{lacW}Dot*) was identified by Rodriguez *et al.* (33). The stocks constructed are *yw; P{lacW}Dot* or *w: P{lacW}Dot* with the following third chromosomes: *bx34e Ubx1/TM3; Df(3R)Ubx109 e gl/Dp(3;3)P5 e; Df(3L)H99 ri1 pp/TM6 B; l(3)hem2P/TM6 B;* and *P{w+, hsp70: Ubx[1a+]}22/TM6 B* (Bloomington Stock Center; 15). *l(3)hem2P* is an enhancer trap mutation of *l(3)hem* (Elisabeth Gateff, personal communication).

Histology and tissue *in situ* hybridization

ß-galactosidase staining was performed as in ref. 33. Digoxigenin-labeled probes were prepared as described in the Boehringer-Mannheim DNA Labeling and Detection Kit and used for tissue *in situ* hybridization as described in ref. 39.

Heat shock

Embryos were heat shocked essentially as in ref. 8.

Isolation of *Dorothy* genomic DNA and cDNA

Genomic DNA flanking the P-element insertion site was recovered by the plasmid rescue technique (42) and analyzed by standard methods (35). The 1.5 kb *Cla*I/*Pst*I fragment was labeled by random priming and used as a hybridization probe to screen a lambda gt11 cDNA library (38) prepared from 9-12 hr embryos and a lambda ZAPII cDNA library prepared from 1-24 hr embryos (kindly provided by Carl Thummel). The ZAPII library yielded one positive clone, and two identical clones were recovered from the gt11 library; these cDNA inserts were subcloned into pBluescript (Stratagene). DNA sequencing and analysis was as in ref. 43.

RESULTS AND DISCUSSION

Molecular characterization of *Dorothy*

The *Dorothy* strain has a P-*lacW* insertion with reporter gene expression that is restricted to the lymph glands and pericardial cells (33; Fig. 1A). To investigate the target gene of the lymph gland/pericardial cell specific enhancer(s), the technique of plasmid rescue was used to isolate genomic DNA fragments flanking the P-*lacW* insertion (Fig. 2A). In order to determine the region that encodes the target gene, restriction enzyme fragments were used as probes in whole mount tissue *in situ* hybridization to embryos. The 1.5 kb *Cla*I/*Pst*I fragment (Fig. 2A) detected RNA localized to the lymph glands and pericardial cells. This fragment was then used as a probe to screen two embryonic cDNA libraries. Two different cDNAs were recovered, each approximately 1.7 kb. These cDNA also hybridize only to lymph glands and pericardial cells in whole mount tissue *in situ* hybridization to embryos and first instar larvae, as in Figure 1.

To determine the structure of *Dorothy*, 2.6 kb of genomic DNA flanking the 3' end of the P-element insertion was sequenced and compared to the cDNAs (Fig. 2). Analysis of these aligned sequences suggests that *Dorothy* is composed of three exons, two introns and extends over 2.3 kb of genomic DNA. The 3' end of the longest cDNA terminates approximately 10 bp after the second possible polyadenylation signal, but neither cDNA contains the 5' end. Since the 5' portion of *Dorothy* cDNA was not obtained, the possibility that there are more 5' exons cannot be excluded. However, the site of P-*lacW* insertion is approximately 500 bp from the longest methionine-initiated open reading frame of the first exon, and sequencing from the 5' end of the insertion shows several promoter elements in the first few hundred base pairs (data not shown).

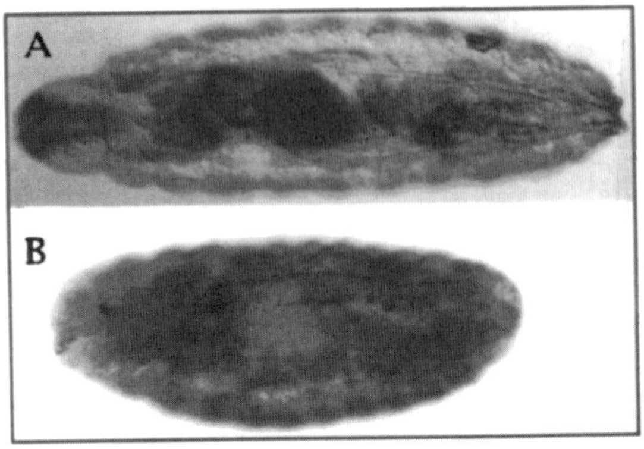

Figure 1. Expression pattern of *Dorothy.* (A) The *Dorothy* enhancer detector strain exhibits reporter gene expression in the lymph glands and pericardial cells of a first instar larva. (B) Expression of *Dorothy* in the lymph glands and pericardial cells is evident in a stage 16 embryo hybridized with *Dorothy* cDNA.

The inferred *Dorothy* protein is 537 amino acids, corresponding to an apparent molecular weight of 60 kD in the absence of cleavage and other post-translational modifications, and a pI of 7.6. The ALOM method (13) was used to detect the potential transmembrane domain which is underlined near the carboxy terminal in Figure 2B. The putative signal peptide is also underlined. Therefore it is likely that DOROTHY contains an extracellular domain, a membrane spanning domain, and an intracellular domain. Two putative N-glycosylation sites were also found. Several lines of evidence suggest that translation initiates at the methionine indicated in Fig. 2B. First, all three reading frames contain stop codons upstream of this methionine; second, the first 19 amino acids encoded by the open reading frame are hydrophobic and have the characteristic features of a cleaved signal peptide; third, the amino acid aligned sequence of DOROTHY beginning at this methionine matches well with the sequence similarities found for the remaining portion of the protein (Fig. 3).

A search of the GenBank protein sequence database using BLAST-P revealed that DOROTHY shares sequence similarity with a variety of ecdysteroid UDP-glucosyltransferases found in multinucleocapsid nuclear polyhedrosis baculoviruses (MNPV): *e.g. Lymantra dispar* MNPV ecdysteroid UDP-glucosyltransferase (LdMNPV), *Autographa californica* MNPV UDP-glucosyltransferase (AcMNPV), *Mamestra brassicae* MNPV UDP-glucosyltransferase (4, 19, 25) and also with the UDP-glucoronosyltranferases: *e.g.* rat UDP-glucoronosyltransferase 1F precursor (rUGT1F), human UDP-glucoronosyltransferase 1A (hUGT1A) (10, 26). As shown in Figure 3, the alignment of DOROTHY with three baculovirus ecdysteroid UDP-glucosyltransferases and two UDP-glucoronosyltransferases shows sequence similarity throughout the length of the DOROTHY sequence. DOROTHY shows the highest identity (18%) with MbMNPV (35% similarity), followed by LdMNPV and rUGT1F, both with 17% identity (37 and 34% similarity respectively). The regions between amino acids 26 and 43, 385 and 405, and 460 and 472 are highly conserved in all six proteins and exhibit 61, 45, and 61% identity respectively with DOROTHY.

Baculoviruses are known to use the ecdysteroid UDP-glucosyltransferase to alter the hormonal balance of the insect host, thereby causing a delay in molting and more

A

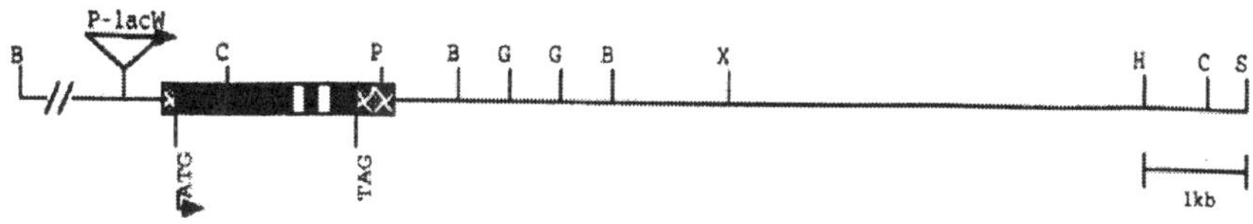

B

```
   1  AGCGATCCATTAGAATCTCACTACACAATGGGAGATAAGGTTATCGGATCAGAGAGTTGGTAACCGATTCAATCGGTGAAAAACTACGGAACGATGACG   100
                                                                                              M T   2

 101  ACGTGTTTGCGCAGCGGAATGTTAATAATTGTTTTGCTTGGGATTCTGGGAATTCCCGAACCGATATCTGCCGGTAATATTCTGGCCGTGTATCCACATT   200
   3  T C L R S G M L I I V L L G I L G P E P I S A G N I L A V Y P H F   36

 201  TCGGCTTCAGCCACTTCAAGGTGGTGATGCCCATCCTAAACGAGCTGGCGCATCGTGGTCATGACATCACGGTGATCTCATACGTGAAGAACCCCCAAGC   300
  37  G F S H F K V V M P I L N E L A H R G H D I T V I S Y V K N P Q A   69

 301  CGGAGCTTATCCCAATTACGAGGAGTTACTGATTTCGGCGCCCGGTGAGGATCAGTCGAGTACCACGATCAATTTGGTTCCCCTGACGGAGCACACACCC   400
  70  G A Y P N Y E E L L I S A P G E D Q S S T T I N L V P L T E H T P   102

 401  ACCAGATCCCTAGGAGTCCTCATCCGTGAGTATGTGGCTCTGCACGAAGAGGGCCAGAAGACTTGCGAGCACCTCTTCGCCAGTGGTCACATAGAACGGG   500
 103  T R S L G V L I R E Y V A L H E E G Q K T C E H L F A S G H I E R A   136

 501  CCATCGAACGACATCAAAATAGGCCCTATGATCTGCTACTCACGGAATACTTCAACTCCGATTGCCAACTGGCCTTGGCCAAGTTATTAAATCTGCCAAT   600
 137  I E R H Q N R P Y D L L L T E Y F N S D C Q L A L A K L L N L P I   169

 601  TATAGGCTTAAGTACCTGTGCACTGATGCCCTACTATTACGATCGCATCGATTTGCCCGACACACCTGCCTTCATACAGTCGGAATTCGTGGGATTCGCT   700
 170  I G L S T C A L M P Y Y Y D R I D L P D T P A F I Q S E F V G F A   202
             ⇒                           t(V->L)  t              c
 701  GGCCAACTGAACTGGCACGAACGGCTGCTCAACTTTGTTCAGGCCAAGTTAGTAAAGTTCCTGTACAAGTACCATTCCAATAGGGCTGACAACGAGTTGG   800
 203  G Q L N W H E R L L N F V Q A K L V K F L Y K Y H S N R A D N E L V   236

 801  TACGGAAGTATCTTGGAGTGGAAGTAGACGTAGAAGAAGTAGCTCGAACTCAAACTGCTTTTGTATTCGGAAACCAACATTACTCCCTGATGGGAAGTCG   900
 237  R K Y L G V E V D V E E V A R T Q T A F V F G N Q H Y S L M G S R   269

 901  TCCACAATCACTGCAATTTGTGGAAATCGGAGGAGTACATATCACCAAGAAAGCGGAGCAGGAGTTACCCCAGAATATCGCCAATTTCCTTAACCAATCA   1000
 270  P Q S L Q F V E I G G V H I T K K A E Q E L P Q N I A N F L N Q S   302
          c(V->L)                   t
1001  GCAGAGGGTGTGATCTTCATAAGCTGGGGTTCAATGGTGCGGGCCTCAAGTATTGACGAAGATAAACTGTCAGCCATACTAGAAGTTCTAAAAAGCCAGC   1100
 303  A E G V I F I S W G S M V R A S S I D E D K L S A I L E V L K S Q P   336
              g  t(E->D)                c                             ∇
1101  CCTTGAAGATCATCTGGAAATGGGAGGCAGAAGAGACTCCCGATACAGATGCCTCCAAGTTTCTCTTTGTCAAATGGGCTCCCCAATTGGCGCTTCTATG   1200
 337  L K I I W K W E A E E T P D T D A S K F L F V K W A P Q L A L L C   369

1201  TAAGTTTAAATTCACTAAGCAGGTTTCCAATTTATTAATGGACTTAAAACCAGTTATGCCGTGAAACGCAAACGAGAGCTATTTACATTTTCTAACTTAA   1300
              ∇                   c                           t
1301  GATCGCTTCTACTTTAGGTCACCCCAAAGTAAAGTTGTTTTGGTCACATGGCGGCTTACTTGGAACCACTGAATCGGTGCATTGTGGCAAACCCCTACTA   1400
 370                    H P K V K L F W S H G G L L G T T E S V H C G K P L L   396
                                                              c
1401  GTCACCCCAATCTACGGAGATCAGTTCCTGAATGCATTCTCGGTGCAAAATCGGGGAATGGGTCTGAAGTTGGACTACAAAGACATCACTGTACCAAACC   1500
 397  V T P I Y G D Q F L N A F S V Q N R G M G L K L D Y K D I T V P N L   430
      g                   ∇                                   ∇
1501  TAAAGAAAGCTCTTGCAGAGCTCAGCAAAAACAGGTGATTTTCTAATGCTTTTCAAATAAGCAAAGTTGTTTTAACCAATTTCTTTGCAGCTATGCCCAA   1600
 431  K K A L A E L S K N S                                               Y A Q   444
                                              t   c     t  a
1601  CGATCTCTTGAGGTCTCCAAGGTGTTCAATGAACGTCAGCAGACGCCTCTGGAATCAGCCATTTGGTCTGTGGAACACGTCATTAGCAATGGATTGATTG   1700
 445  R S L E V S K V F N E R Q Q T P L E S A I W S V E H V I S N G L I A   478
         t                             c
1701  CCGCCAGACTACTTCAATCTCCAGGCATAGAACTGAATGGATTTGTCTACCATTCACTGGATAGTGTAGCCCTGATCCTGTTACCTTTAATTTTACTGAT   1800
 479  A R L L Q S P G I E L N G F V Y H S L D V S A L I L L P L I L L I 511
                                 a
1801  CCTAGTCCTATGCTATCTGTGCAAACGGAATCCATCGAAGTTGGCCCACAAGAAAGTTGGCAAGAAGCCAAAGAGATCGTAGTCTACTAAGATTTACGCA   1900
 512  L V L C Y L C K R N P S K L A H K K V G K K P K R S *                      537

1901  TGACTTGACAATTTCTGTGACAACAAATATATATATTAAGCATCGTTAAAAGCGACTTCAGTAGCGCAGACGTTGCATGCAATCCCTGCGACACTCCTCG   2000
2001  AAGGTCAGGAAATTGTTCTCAGTGGAGGCACAGCCCGTGTAGATGAAGCTCTCGCAGTTTCCGGTCTTCTTGTTGTAGGCATATCGCAACTGGTGACCCC   2100
2101  GGCACAAACCAGGTGCCTTTGGCTGTCTGCAGGCTGAATGTAGAATAGAAAAGAGTTATTAAAATCTAGTTTCAATAAATCCACGACTCACCTA   2200
2201  TTATCTGTTCGATTTGTTCGTGTGTCTGGTTTTCCAAGCTCGGAAGACCCAACGATAGGCCCACAACCAATAGCAACACCAGCCAAAGTTGCCCCCAGCT   2300
                                                                 ⇐
2301  CGAAATCCAAACCATTTTGTTCAGTTCATTAACATATGGCGGCAATCTCTCTAAGTTAATAAATTATATATCCAATCGATCCTCTAGTTGCACTTT   2400
```

Figure 2. Molecular characterization of the *Dorothy* locus. (A) Genomic restriction map of the *Dorothy* locus. The boxed region indicates transcribed sequences starting from a putative CCATTAG transcription start site inferred from genomic DNA sequencing, to the second polyadenylation signal in the *Dorothy* cDNA. Gray boxes represent the 5' leader and 3' trailer sequence respectively. Black boxes represent translated sequences, and white boxes represent intron sequences. The site of the P-*lacW* insertion and direction of *lacZ* transcription are also shown. C-*Cla*I; B-*Bgl*II; X-*Xho*I; H-*Hind*III; S-*Sac*I; P-*Pst*I: only one *Pst*I site shown; G-*Bgl*II: only one is present; *Bgl*II left of *Dorothy* coding region is 5 kb from P-*lacW*. (B) Sequence of *Dorothy*. The sequence is a composite of genomic and cDNA sequences. The deduced amino acid sequence is shown beneath nucleotide sequence. The composite longest cDNA sequence obtained extends between the bases marked by open arrows. Hydrophobic stretches corresponding to the putative signal peptide and transmembrane domain are underlined beneath the amino acid sequence. The putative transcription start site and polyadenylation signals are also underlined. Splice sites are marked with triangles. The differences identified in the cDNAs are shown in lowercase letters above the genomic DNA sequence. One base change in the cDNAs results in amino acid change and the other two changes result in conserved substitutions. (GenBank # AF250328)

opportunity for the virus to replicate before the host is killed (19, 20, 22). *In vitro* studies show that the mechanism of action of ecdysteroid UDP-glucosyltransferase can be through both the glucose and galactose conjugation of ecdysteroids (20). Human UDP-glucoronosyltransferases have been extensively studied and found to be important components of detoxification of endogenous compounds and a variety of drugs, carcinogens and other xenobiotics (reviewed in 3). The UDP-glucoronosyltransferases catalyze the transfer of glucoronic acid to the various compounds, such as bilirubin and steroid hormones.

Dorothy is the first member of this gene family to be found in an insect. DOROTHY is predicted to be membrane bound, whereas the viral ecdysteroid UDP-glucosyltransferases are secreted (20). In many of the mammalian UDP-glucoronosyltransferases, however, this region of carboxy-terminal amino acids appears to form a transmembrane domain across the endoplasmic reticulum and the active site is on the lumen (reviewed in 3). Many related sequences have now been identified and a recommended nomenclature established under the general name UDP glycosyltransferase gene superfamily (16). A short amino acid signature sequence has been identified that characterizes the family (16), and this sequence occurs in DOROTHY amino acids 376-404. Of the 13 residues with 1-3 specific residues permitted in the signature sequence, DOROTHY matches in 12 of the 13.

Temporal and spatial patterns of *Dorothy* expression

Dorothy is expressed during all developmental stages and northern analysis shows a complexity of transcript sizes and expression pattern (data not shown). The major transcript is 2.3kb, corresponding to the estimated size from DNA analysis, but as many as seven transcripts ranging from 0.6kb to 4.6kb can be discerned. We have not resolved the issue as to whether this reflects differential splicing and/or related genes. It is important to note in this regard that multiple isoenzymes with UDP-glucosyltransferase activity have been proposed previously (see 24). Further sequence

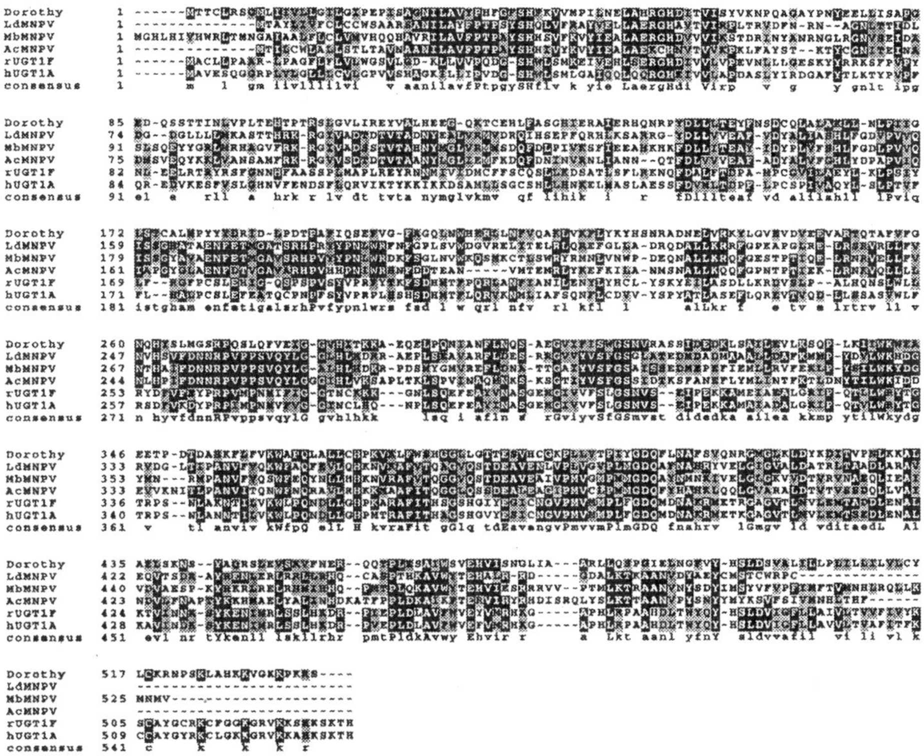

Figure 3. Similarity of the *Dorothy* sequence with several ecdysteroid UDP-glucosyltransferases and UDP-glucoronosyltransferases. Aligned amino acid sequences of *Dorothy*, LdMNPV (*Lymantria dispar* MNPV egt: GenBank accession number U04321), MbMNPV (*Mamestra brassicae* MNPV egt: GenBank U41999), AcMNPV (*Autographa californica* MNPV egt: GenBank P18569), rUGT1F (rat UDP-glucoronosyltransferase 1F precursor: GenBank P08430), hUGT1F(human UDP-glucoronosyltransferase 1A: GenBank P22309). Sequences were aligned using the Clustal W program. Residues conserved in half or more of the proteins are shown in inverted black print, conserved substitutions are shaded in gray. Dashes denote gaps in the alignments. The consensus sequence was determined with the Boxshade program.

analysis at the *Dorothy* locus has, however, revealed that the protein kinase gene *foraging* (21) is located immediately downstream of *Dorothy* and transcribed from the opposite strand. Also, a putative protease inhibitor may be transcribed from the opposite strand of a region that includes the *Dorothy* 3'UTR.

In contrast to the *Dorothy* transcript complexity, tissue specific localization of *Dorothy* transcripts in embryos and larvae is straightforward. *Dorothy* is first evident during embryonic stages 15 and 16 in the primordium of the lymph glands and pericardial cells as dorsal closure takes place. After dorsal closure *Dorothy* continues to be expressed in the lymph glands and pericardial cells into the larval stages. This localization matches that of β-galactosidase in *P{lacW}Dot* and suggests that the reporter gene is a reliable indicator of *Dorothy* expression (Fig. 1). Since this insertion is also homozygous viable and does not detectably effect the level of *Dorothy* transcripts, stocks can be constructed in which all progeny have the *Dorothy* enhancer trap and vary only for the segregation of selected mutations. We used these features to study lymph gland and pericardial cell specific differentiation in mutant backgrounds of the *Bithorax Complex* (*BXC*) and *lethal(3)hematopoietic organs missing* (*l(3)hem*).

The genes of the *Bithorax Complex* (*BXC*) regulate segmental identity in the abdomen and posterior thoracic segments (see 15), and thus were chosen as candidates to test for regulation of lymph gland development. We examined this initially by determining the β-galactosidase pattern in *P{lacW}Dot; bx Ubx* embryos and found that the lymph glands are duplicated (33). In more extreme mutation of the *BXC*, more tissue along the dorsal vessel develops lymph gland characteristics. In *P{lacW}Dot*; *Df(3R)Ubx109* (lacking Ubx and abd-A functions; 11) mutant embryos, most of the length of the dorsal vessel is associated with lymph gland morphology and the number of pericardial cells is reduced (Fig. 4A). We also examined the distribution of *Dorothy* transcripts in *Df(3R)Ubx109* embryos by whole mount *in situ* hybridization, and found a larger, more widespread patch of *Dorothy* expression along the dorsal vessel, the same as the pattern of ß-galactosidase in *P{lacW}Dot*; *Df(3R)Ubx109*. These embryos thus have an increase in transcript occurring in segments A2-A6 and this is consistent with homeotic transformation of the abdominal segments. The usual pattern of pericardial cells is not evident and suggests that the pericardial cells were replaced by lymph gland-like tissue. These results are consistent with the results of Mastick *et al.* (17), in which the lymph gland expression of the A7.1 gene in the T3/A1 apodeme is extended to segment A2 in *Ubx*$^-$ embryos. We also determined β-galactosidase localization in *P{lacW}Dot* embryos that were expressing Ubx ectopically. These embryos were found to lack lymph gland β-galactosidase in T3/A1 and to have an extension of pericardial type cells instead (Fig. 4B). Interestingly, ectopic levels of Ubx resulted in the enhancer trap also expressing the reporter gene such that β-galactosidase appears in a segmentally repeated manner, reminiscent of the pair rule genes *sloppy paired* and *odd skipped* which are located in the same polytene chromosome region as *Dorothy* (15).

The *lethal(3)hematopoietic organs missing* (*l(3)hem*) mutation causes homozygotes to die soon after formation of the puparium, and histological analysis shows the absence of mature hemocytes in the hemolymph and the absence of primordial blood cells in the lymph glands (7). In a stock that produces *P{lacW}Dot*; *l(3)hem2P* homozygotes, many embryos are found in which β-galactosidase does not occur at the T3/A1 apodeme, site of the lymph glands (Fig. 4C, D). Pericardial cells appear normal, as found by Gateff (7) for *l(3)hem* after puparium formation. The level of β-galactosidase in the pericardial cells does, however, appear diminished in both homozygous and heterozygous embryos (Fig. 4C, D). It is an interesting possibility that *l(3)hem* would have a dominant effect on regulation of *Dorothy*, but later in development, as third instar larvae, β-galactosidase in the pericardial cells appears the same for *l(3)hem2P* homozygotes, heterozygotes and *P{lacW}Dot* homozygotes (data not shown). Determining whether or not the absence of *Dorothy* expression in the lymph glands is a major factor in the lethality of *l(3)hem* will require mutations of *Dorothy*.

Mutations of *Dorothy* are necessary to dissect function at all levels. Initial experiments to produce *Dorothy* mutations have resulted in defining a mutation that deletes part of

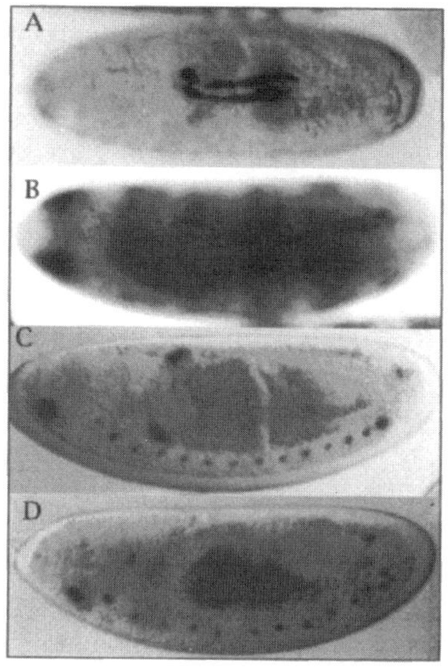

Figure 4. Expression of *Dorothy* in mutant backgrounds. (A) Stage 16 *P{lacW}Dot; Df(3R)Ubx109* embryo. Pericardial cells are transformed to a more anterior, lymph gland identity. (B) Stage 16 *P{lacW}Dot* embryo with ectopic expression of Ubx. Pericardial cell morphology only is seen along the dorsal vessel. (C) Stage 16 *P{lacW}Dot; l(3)hem2P/TM6B* embryo shows wild type dorsal vessel pattern with reduced level of β-galactosidase expression. Additional sites of staining result from the *l(3)hem2P* enhancer detector. (D) Stage 15 *P{lacW}Dot; l(3)hem2P* embryo lacks lymph glands and has reduced expression in pericardial cells.

Dorothy, missing 369 bases at the 5' end, producing transcripts without this region, and flies with this mutation are fully viable. We have yet to define a null mutation and it is thus an open question as to whether or not *Dorothy* is an essential gene.

We have not tested for *Dorothy* transcript localization in larval or adult hemocytes, but rather have used the *P{lacW}Dot* reporter gene as an indicator to study the hemocytes associated with the fat body in newly eclosed adults. As reviewed in Rizki (30), some larval fat body cells are carried over through metamorphosis and are present along with adult fat body cells at eclosion. We postulated that during this time after eclosion, when the fat body is still being restructured, that this would be an opportune developmental period to score for hemocyte specific expression. We found numerous hemocytes near larval fat body cells in recently eclosed adults, and β-galactosidase localization in *P{lacW}Dot* flies is easily recognizable in these hemocytes (Fig. 5). This result strongly suggests that *Dorothy* is expressed in these hemocytes of adults. Since β-galactosidase is stable, the actual time of transcription may be before and/or after eclosion. β-galactosidase activity is not detectable in older flies, suggesting that transcription does not continue or is greatly reduced. Alternatively, there is the possibility of different populations of hemocytes being involved.

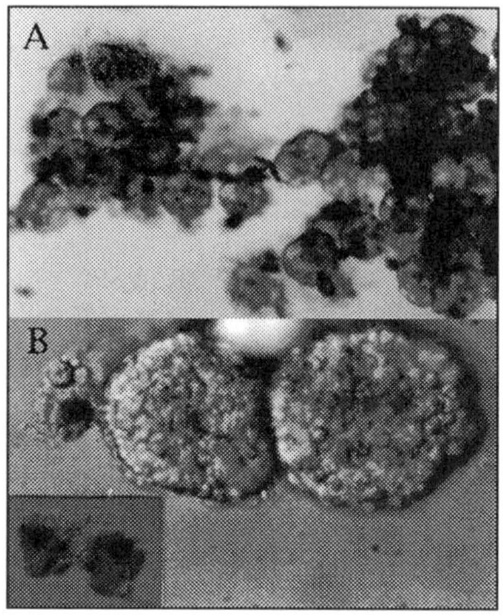

Figure 5. *Dorothy* expression in adult hemocytes of *P{lacW}Dot.* (A) Numerous hemocytes staining for ß-galactosidase are seen associated with the fat body cells of a recently eclosed adult. (B) Higher magnification of A shows nuclear β-galactosidase staining of a hemocyte associated with fat body cells. Inset shows two free floating hemocytes.

CONCLUSION

Dorothy encodes a protein with extensive similarity to a large family of transferases with members ranging from baculoviral ecdysteroid UDP-glucosyltransferases to mammalian UDP-glucoronosyltransferases. This is the first report of a UDP glycosyltransferase gene superfamily member in an insect. Although *Drosophila* is not a host for baculoviruses, *Dorothy* may be able to provide insight into the relationship between baculoviruses and their insect hosts. Baculoviruses are known to use the ecdysteroid UDP-glucosyltransferase to alter the hormonal balance and development of the insect as part of its repertoire during destruction of the host (20, 22). *Dorothy* may thus contribute to an understanding of the regulation of ecdysteroid hormone levels in *Drosophila* and also provide a model useful with respect to pest control. DOROTHY may be more similar to the mammalian UDP-glucoronosyltransferases and act against a variety of xenobiotics and endogenous compounds. Direct biochemical assays are required to determine the activity of DOROTHY.

The tissue specific sites of expression of *Dorothy* are interesting, and consistent in that hemocytes and pericardial cells are expected to have detoxifying activities (29). Lymph glands are less obvious candidates and may have this function and/or may be transcribing *Dorothy* as part of the differentiation of its cells toward hemocytes. A more obvious candidate for a UDP glycosyltransferase would be the fat body, but this just underscores the need to determine the precise function(s) of DOROTHY. For example,

DOROTHY may have a variety of roles in immune defense through detoxification, steroid regulation through metabolism, and general protection through inactivation of xenobiotics.

Human UDP-glucoronosyltransferase deficiencies are associated with Gilbert's syndrome, a mild, chronic unconjugated hyperbilirubinemia, and Crigler-Najjar syndrome, a lethal form of unconjugated hyperbilirubinemia (reviewed in 3). Various genetic lesions have been associated with these syndromes, and interestingly a chimeric oligonucleotide has been used to restore gene function using the Gunn rat model of Crigler-Najjar syndrome (14). The power of *Drosophila* genetics can now also be applied to this type of transferase gene through further production and analysis of wild type and mutant *Dorothy*. The *Dorothy* enhancer trap will continue to be useful as a marker of lymph gland, hemocyte and pericardial cell development, but also the *Dorothy* promoter can be used directly to drive lymph gland, hemocyte and/or pericardial cell specific expression of a variety of genes through production of transgenes. A transgenic system would allow study from many of the angles that are needed in order to acquire a better understanding of hematopoietic specific processes in immunity and development.

ACKNOWLEDGEMENTS

We thank Elisabeth Gateff for providing the *l(3))hem2P* stock, and Juan Botas for the *Df(3R)Ubx109* stock. We also thank Carl Thummel and John Tamkun for cDNA libraries. This work was supported by grants from the American Cancer Society (IM-697 and IM-697B) to DAK.

REFERENCES

1. Abrams, J.M., Lux, A., Steller, H. and Krieger, M. (1992) Macrophages in *Drosophila* embryos and L2 cells exhibit scavenger receptor-mediated endocytosis. Proc. Natl. Acad. Sci. USA 89, 10375-10379.
2. Barret, J.W., Krell, P.J. and Arif, B.M. (1995) Characterization, sequencing and phylogeny of the ecdysteroid UDP-glucosyltransferase gene from two distinct nuclear polyhedrosis viruses isolated from *C. fumiferna*. J. Gen. Virol. 76, 2447-2456.
3. Burchell, B. and Coughtrie, M.W.H. (1989) UDP-glucuronosyltransferases. Pharmac. Ther. 43, 261-289.
4. Clarke, E.E., Tristem, M., Cory, J. and O'Reilly, D.R. (1996) Characterization of the *Mamestra brassicae* multicapsid nuclear polyhedrosis virus ecdysteroid UDP-glucosyltransferase (egt) gene. J. Gen. Virol. 77, 2865-2871.
5. Gateff, E. (1978) Malignant and benign neoplasms of *Drosophila_melanogaster*. In Ashburner, M. and Wright, T.R.F. (eds.), Genetics and Biology of *Drosophila*. Academic Press, New York, Vol. 2B, pp. 181-275.
6. Gateff, E. (1994a) Tumor suppressor and overgrowth suppressor genes of *Drosophila melanogaster.* Developmental aspects. Int. J. Dev. Biol. 38, 565-590.
7. Gateff, E. (1994b) Tumor suppressor genes, hematopoietic malignancies and other disorders of *Drosophila melanogaster*. In Beck, G. et al. (eds.), Primordial Immunity: Foundations for the Vertebrate Immune System, The New York Academy of Science, New York, Ann. NY Acad. Sci. Vol. 712, pp. 260-279.
8. Gonzalez-Reyes, A. and Morata, G. (1990) The developmental effect of overexpressing a *Ubx* product in *Drosophila* embryos is dependent on its interactions with other homeotic products. Cell 61, 515-522.
9. Hoffmann, J. A. (1995) Innate immunity of insects. Curr. Opin. Immunology 7, 4-10.

10. Iyanagi, T., Haniu, M., Sogawa, K., Fujii-Kuriyama, Y., Watanabe, S., Shively, J.E. and Anan, K.F. (1986) Cloning and characterization of cDNA encoding 3-methylcholanthrene inducible rat mRNA for UDP-glucoronosyltransferase. J. Biol. Chem. 261, 15607-15614.
11. Karch, F., Weiffenbach, B., Peiffer, M., Bender, W., Duncan, I., Celniker, S., Crosby, M. and Lewis, E.B. (1985) The abdominal region of the bithorax complex. Cell 43, 81-96.
12. Kirkpatrick, R.B., Matico, R.E., McNulty, D.E., Strickler, J.E. and Rosenberg, M.R. (1995) An abundantly secreted glycoprotein from *Drosophila melanogaster* is related to mammalian secretory proteins produced in rheumatoid tissues and by activated macrophages. Gene 153, 147-154.
13. Klein, P., Kanehisa, M. and DeLisi, C. (1985) ALOM method to detect potential transmembrane segments. Biochem. Biophys. Acta 815, 468.
14. Kren, B.T., Parashar, B., Bandyopadhyay, P., Chowdhury, N.R., Chowdhury, J.R. and Steer C.J. (1999) Correction of the UDP-glucuronosyltransferase gene defect in the Gunn rat model of Crigler-Najjar syndrome type I with a chimeric oligonucleotide. Proc. Natl. Acad. Sci. USA 96, 10349-10354.
15. Lindsley, D.L. and Zimm, G.G. (1992) The Genome of *Drosophila melanogaster.* Academic Press, San Diego.
16. Mackenzie, P.I., Owens, I.S. Burchell, B., Bock, K.W., Bairoch, A. *et al.* (1997) The UDP glycosyltransferase gene superfamily: recommended nomenclature update based on evolutionary divergence. Pharmacogenetics 7, 255-269.
17. Mastick, G.S., McKay, R., Oligino, T., Donovan, K. and Lopez, A.J. (1995) Identification of target genes regulated by homeotic proteins in *D. melanogaster* through genetic selection of Ultrabithorax protein-binding sites in yeast. Genetics 139, 349-363.
18. Meister, M., Georgel, P., Lemaitre, B., Kappler, C., Lagueux, M., Reichhart, J.-M. and Hoffmann, J.A. (1994) Immune Gene Expression in *Drosophila*. In Hoffmann, J.A., Janeway, Jr., C.A. and Natori, S. (eds.), Phylogenetic Perspectives in Immunity: The Insect Host Defense, R. G. Landes Co., Austin, TX, pp. 167-181.
19. O'Reilly, D.R. and Miller, L.K. (1989) A baculovirus blocks insect molting by producing ecdysteroid UDP-glucosyl transferase. Science 245, 1110-1112.
20. O'Reilly, D.R., Brown, M.R. and Miller, L.K. (1992) Alteration of ecdysteroid metablism due to baculovirus infection of the fall armyworm *Spodoptera frugiperda*: host ecdysteroids are conjugated with galactose. Insect Biochem. Mole. Biol. 22, 313-320.
21. Osborne, K.A., Robichon, A., Burgess, E., Butland, S., Shaw, R.A., Coulthard, A., Pereira, H.S., Greenspan, R.J. and Sokolowski, M.B. (1997) Natural behavior polymorphism due to a cGMP-dependent protein kinase of *Drosophila*. Science 277, 834-836.
22. Park, E.J., Yin, C.M. and Burand, J.P. (1996) Baculovirus replication alters hormone-regulated host development. J. Gen. Virol. 77, 547-554.
23. Pearson, A., Lux, A. and Krieger, M. (1995) Expression cloning of dSR-CI, a class C macrophage-specific scavenger receptor from *Drosophila melanogaster*. Proc. Natl. Acad. Sci. USA 92, 4056-4060.
24. Rausell, C., Llorca, J. and Real, M.D. (1997) Separation by FPLC chromatofocusing of UDP-glucosyltransferases from three developmental stages of *Drosophila melanogaster*. Arch. Insect Biochem. Physiol. 34, 347-358.
25. Riegel, C.I., Lanner-Herrera, C. and Slavicek, J.M. (1994) Identification and characterization of the ecdysteroid UTP-glucosyltransferase gene of the *Lymantria dispar* multinucleocapsid nuclear polyhedrosis virus. J. Gen. Virol. 75, 829-838.
26. Ritter, J.K., Chen, F., Sheen, Y.Y., Tran, H.M., Kimura, S., Yeatman, M.T. and Owens, I.S. (1992) A novel complex locus UGT1 encodes human bilirubin, phenol, and other UDP-glucoronosyltransferase isozymes with identical carboxyl termini. J. Biol. Chem. 267, 3257-3261.
27. Rizki, M. T. M. (1957a) Tumor formation in relation to metamorphosis in *Drosophila melanogaster.* J. Morphology 100, 459-472.
28. Rizki, M.T.M. (1957b) Alterations in the haemocyte population of *Drosophila melanogaster*. J. Morphology 100, 437-458.
29. Rizki, T. M. (1978a) The circulatory system and associated cells and tissues. In Ashburner, M. and Wright, T.R.F. (eds.), Genetics and Biology of *Drosophila*. Academic Press, New York, Vol. 2B, pp. 397-452.
30. Rizki, T. M. (1978b) Fat body. In Ashburner, M. and Wright, T.R.F. (eds.), Genetics and Biology of *Drosophila*. Academic Press, New York, Vol. 2B, pp. 561-601.
31. Rizki, R.M. and Rizki, T.M. (1984) Selective destruction of a host blood cell type by a parasitoid wasp. Proc. Natl. Acad. Sci. USA 81, 6154-6158.

32. Rodrigues, V., Cheah, P.Y., Ray, K. and Chia, W. (1995) *malvolio*, the *Drosophila* homologue of mouse NRAMP-1 (*Bcg*), is expressed in macrophages and in the nervous system and is required for normal taste behaviour. EMBO J. 13, 3007-3020.
33. Rodriguez, A., Zhou, Z., Tang, M. L., Meller, S., Chen, J., Bellen, H. and Kimbrell, D.A. (1996) Identification of immune system and response genes, and novel mutations causing melanotic tumor formation in *Drosophila melanogaster*. Genetics 143, 929-940.
34. Rugendorff, A., Younossi-Hartenstein A. and Hartenstein V. (1993) Embryonic origin and differentiation of the *Drosophila* heart. Roux's Arch. Dev. Biol. 203, 266-280.
35. Sambrook, J., Fritsch, E.F. and Maniatis, T. (1989). Molecular Cloning: A Laboratory Manual. Cold Spring Harbor Laboratory Press, Cold Spring Harbor, NY.
36. Shrestha, R. and Gateff, E. (1982) Ultrastructure and cytochemistry of the cell types in the larval hematopoetic organs and hemolymph of *Drosophila melanogaster*. Dev. Growth Differ. 24, 65-82.
37. Sparrow, J.C. (1978) Melanotic tumors. In Ashburner, M. and Wright, T.R.F. (eds.), Genetics and Biology of *Drosophila*. Academic Press, New York, Vol. 2B, pp. 277-313.
38. Tamkun, J.W., Kahn, R.A., Kissinger, M., Brizuela, B.J., Rulka, C., Scott, M.P. and Kennison, J.A. (1991) The arflike gene encodes an essential GTP-binding protein in *Drosophila*. Proc. Natl. Acad. Sci. USA 88, 3120-3124.
39. Tautz, D. and Pfeifle, C. (1989) A non-radioactive *in situ* hybridization method for the localization of specific mRNAs in *Drosophila* embryos reveals translational control of the segmentation gene *hunchback*. Chromosoma 98, 81-85.
40. Tepass, U., Fessler, L., Aziz, A. and Hartenstein, V. (1994) Embryonic origin of hemocytes and their relationship to cell death in *Dros*. Development 120, 1829-1837.
41. Watson, K. L., Justice, R.W. and Bryant, P.J. (1994) *Drosophila* in cancer research: the first fifty tumor supressor genes. J. Cell Science 18, 19-33.
42. Wilson, D., Bellen, H.J., Pearson, R.K., O'Kane, C.J., Grossniklaus, U. and Gehring, W. (1989) P-element-mediated enhancer detection: Isolation and characterization of developmentally regulated genes in *Drosophila*. Genes Dev. 3, 1301-1333.
43. Zhou, X, Nguyen, T. and Kimbrell, D. A. (1997) Identification and characterization of the *Cecropin* antibacterial protein gene locus in *Drosophila virilis*. J. Mol. Evol. 44, 272-281.

EICOSANOIDS IN INSECT IMMUNE SIGNAL TRANSDUCTION

D.W. Stanley and Ralph W. Howard

Insect Biochemical Physiology Laboratory, University of Nebraska - Lincoln, Lincoln, NE 68583-0816 and USDA/ARS, GMPRC, 1515 College Ave., Manhanttan, KS 66502

INTRODUCTION

Eicosanoid is a collective term for all biologically active, oxygenated metabolites of arachidonic acid and two other C20 polyunsaturated fatty acids. There are three major groups of eicosanoids. The prostaglandins (PGs), so named because they were first discovered in seminal fluids originating in the prostate gland, are products of the cyclooxygenase pathways. The cytochrome P_{450} "epoxygenase pathways" yield the epoxyeicosatrienoic acids. Various lipoxygenase pathways are responsible for biosynthesizing a wide range of compounds, including leukotrienes and hydroxyeicosatetraenoic acids. The biosynthesis and structures of representatives of the major groups of eicosanoids are outlined in Figure 1, and described in detail elsewhere (Stanley, 1999). One or more members of the major eicosanoid groups have been detected in virtually all mammalian tissues and body fluids, where they serve as lipid mediators of cellular events. Among their important actions, eicosanoids modulate ion transport physiology, mediate contraction or relaxation of smooth muscles and exert tremendous influence in host defense mechanisms. It would be difficult to exaggerate the importance of eicosanoids in the vertebrate immune system. Sir John Vane shared in the 1982 Nobel Prize in Medicine or Physiology, which recognized his discovery that the analgesic effects of aspirin are due to inhibition of prostaglandin biosynthesis. This discovery launched a very large pharmaceutical research enterprise aimed at discovery of new non-steriodal anti-inflammatory drugs, all of which act through their inhibitory influence on prostaglandin biosynthesis.

Bengt Samuelsson and Sune Bergstrom also shared in the 1982 Prize, recognizing their work on the chemistry and biological significance of eicosanoids. Samuelsson, Berstrom and their colleagues were the first to determine the chemical structure of PGE_1 and a few other PGs (Oates, 1982); they inferred from the chemical structures that PGs must be synthesized by enzymatic oxygenation of arachidonic acid. We now know that PGs are among the key mediators of inflammatory and other host defense processes in vertebrates. Various PGs positively mediate inflammatory physiology, and they serve as negative modulators of other cellular defense reactions. Mast cell degranulation, for example, is inhibited by PGE_2. The authors went on to use radioactive arachidonic acid to trace the biosynthesis of PGs. In the course of this work, they investigated the biosynthesis of PGs from radioactive arachidonic acid in mammalian leukocytes. Arachidonic acid metabolism was not inhibited by the presence of aspirin in the leukocyte preparations. This observation led to the suggestion that arachidonic acid must be oxygenated by enzymatic pathways other than the cyclooxygenase pathway responsible for PG biosynthesis. Samuelsson and his colleagues finally determined that leukocytes yielded a suite of oxygenated triene metabolites of arachidonic acid. They named these compounds leukotrienes, and later showed that the Slow Reacting Substance of anaphylaxis is a mixture of leukotrienes (Samuelsson, 1983). Discovery of still other eicosanoids followed this work, although more of this interesting history lies outside the scope of this brief chapter. The point of our comments is that eicosanoids serve as signal transduction moieties in virtually every aspect of the vertebrate immune system.

EICOSANOIDS IN INVERTEBRATE IMMUNITY

Drawing on the background of vertebrate immunity, we considered the hypothesis that eicosanoids similarly serve as signal transduction moieties in invertebrate immunity. We selected a large, convenient insect, fifth instars of the tobacco hornworm, *Manduca sexta*, for our experimental work. We designed experiments to investigate the roles of eicosanoids in clearing artificial bacterial infections from hemolymph circulation. In our preliminary exploration, we first injected large dosages (26 μg) of the glucocorticoid dexamethasone into the hemocoel of tobacco hornworms, then injected 10^5 cells of a red-pigmented strain of the bacterium, *Serratia marcescens*, on the contralateral side of each insect. Control insects were injected with 2 ml of ethanol, then similarly infected. At 2.5 hour post infection (PI), we collected 100 μl hemolymph samples, which we diluted and spread on agar plates. The plates were incubated for 40 hours, after which we counted the number of red bacterial colonies on each plate.

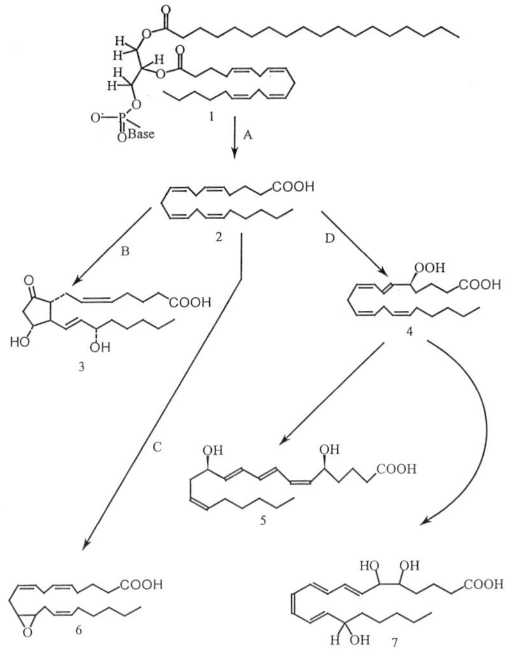

Figure 1. An overview of 20:4n-6 metabolism as understood from the mammalian background. Three polyunsaturated fatty acids, 20:3n-6, 20:4n-6 and 20:5n-3 are potential substrates for eicosanoid biosynthesis. Of these, metabolism of 20:4n-6 is most well studied. Chemical structures are denoted by numerals. **1** = a cellular phospholipid. **2** = hydrolyzed 20:4n-6. **3** = prostaglandin E_2. **4** = 5-hydroperoxyeicosatetraenoic acid. **5** = leukotriene B_4. **6** = 11,12-epoxyeicosatrienoic acid. **7** = lipoxin A. Capital letters indicate major enzyme systems responsible for eicosanoid biosynthesis. **A** = phospholipase A_2; **B** = cyclooxygenase and associated enzyme steps; **C** = cytochrome P_{450} epoxygenase; **D** = lipoxygenase.

We recovered very few bacterial colonies from control insects, from which we inferred that tobacco hornworms were competent to clear bacterial infections from hemolymph circulation. The results were different for the experimental insects, from nearly all of which we recovered large numbers of bacterial colonies. We interpreted these results to suggest that the dexamethasone treatments somehow impaired the insects' ability to clear bacterial infections from hemolymph circulation. Dexamethasone inhibits eicosanoid biosynthesis indirectly by inhibiting the synthesis and activity of phospholipase A_2, the enzyme responsible for initiating eicosanoid biosynthesis by releasing arachidonic acid from cellular PLs (Figure 1). In subsequent experiments we found that dexamethasone exerted its effects in a dose-dependent manner, and the effects were reversed by treating dexamethasone-injected hornworms with arachidonic acid, the substrate for eicosanoid biosynthesis immediately after they were infected with bacteria. Supported by these results, we set forth the hypothesis that eicosanoids mediate cellular immune reactions to bacterial infections in insects and possibly other invertebrates (Stanley-Samuelson *et al.*, 1991).

Perhaps the main advantage of working with tobacco hornworms is that Dunn and his colleagues had developed a detailed background on the immune reactions to bacterial infections in this animal. They showed that nodulation, a process in which infecting bacteria cells are cleared from circulation by entrapping them within matrices of the hornworm hemocytes, is responsible for clearing most bacterial cells from circulation within the first two hours PI (Horohov and Dunn, 1983).

On the basis of Dunn's work, we considered the possibility that eicosanoids mediate nodulation reactions to bacterial infections. We carried out a series of experiments similar to the work just described. In the first experiment, tobacco hornworms were injected with dexamethasone, then infected with a standard bacterial challenge. Control insects were treated with ethanol, then infected. At six hours PI, we assessed nodulation by counting the numbers of melanized nodules in the hemocoels of all hornworms. The dexamethasone treatments resulted in severely impaired nodulation in experimental insects, reducing nodulation by over 50%. Again, the dexamethasone effects were expressed in a dose-dependent manner and they were reversed by treating infected hornworms with arachidonic acid. Following this protocol, we used specific inhibitors to probe two specific pathways of eicosanoid biosynthesis, the cyclooxygenase pathway and a lipoxygenase pathway (Fig. 1). These treatments similarly resulted in impaired nodulation.

A couple of important conceptual issues attended these findings. First, we could not be sure, on the basis of counting melanized nodules, whether eicosanoids influenced the oxidative reactions involved in melanization, which would result in reduced numbers of visible nodules without actually impacting the nodulation process. We addressed this issue by determining the influence of dexamethasone on numbers of microaggregations in hemolymph circulation at one hour PI. Formation of microaggregations is an early phase of the nodulation process. This work revealed that dexamethasone treatments impaired the formation of microaggregations, from which we reasoned that eicosanoids are somehow involved in one or more cellular actions in nodulation. The second issue relates to our finding that inhibitors of both the cyclooxygenase and lipoxygenase pathways impaired nodule formation. How might members of two distinct groups of eicosanoids mediate nodulation in a physiological way? The formation of visible, melanized nodules is the outcome of many separate cellular actions. These might include recognition of bacterial cells, release of chemical signals to attract neighboring hemocytes, directed hemocyte movement, release of granules from granulocytes, adhesion between bacterial cells and hemocytes, hemocyte-hemocyte adhesion, attraction of plasmatocytes, investing the growing nodules with plasmatocytes, and prophenyloxidase activation. If we imagine that some of these steps are mediated by cyclooxygenase products, while others are mediated by lipoxygenase products, then the overall nodulation process would depend on products of both major pathways. Inhibition of either pathway would be registered in terms of impairing nodule formation.

Again, we selected the tobacco hornworm as a large, convenient model of insect immunity. The idea that eicosanoids mediate nodulation reactions to bacterial infections in this insect begs the question of whether eicosanoids similarly act in other species. We hypothesized that eicosanoids mediate nodulation in all insect species that form nodules in reaction to bacterial infections. Drawing on our protocols with tobacco hornworms (Miller and Stanley, 1998), we assessed the influence of eicosanoid biosynthesis inhibitors on nodulation in other insect species. So far, we found that eicosanoids mediate microaggregation and nodulation reactions in larvae of the tenebrionid beetle, *Zophobas atratus* (Miller *et al.*, 1996), three additional moth species, true armyworms, *Pseudaletia unipuncta*, and black cutworms, *Agrotis ipsilon* (Jurenka *et al.*, 1997), and silkworms, *Bombyx mori* (Stanley-Samuelson *et al.*, 1997), and larvae of the butterfly, *Colias eurytheme* (Stanley-Samuelson *et al.*, 1999). We have also cast the net into hemimetabolous insect orders. Eicosanoids also mediate nodulation reactions in adults of the cricket, *Gryllus assimilis* (Miller *et al.*, 1999) and adults of periodical 17-years cicadas, *Magicicada septendecim* and *M. cassini* (Tunaz *et al.*, 1999). Taken together, these findings support the hypothesis that eicosanoids mediate nodulation reactions to bacterial infections in insects. It is appropriate to temper this generalization. For one point, it is not certain that all insects react to bacterial challenge with nodules.

Downer and his colleagues advanced our understanding of specific immune actions mediated by eicosanoids (Mandato *et al.*, 1997). Using hemocytes from the great waxmoth, *Galleria mellonella*, Mandato *et al.* (1997) showed that two phases of nodule formation, specifically cell spreading and prophenyloxidase activation, depend on eicosanoid biosynthesis. Moving further, they also investigated the role of eicosanoids in phagocytosis, a distinct cellular defense reaction to bacterial infection. Their results indicated that phagocytosis also depends on eicosanoid biosynthesis. Looking at the broad area of cellular immunity in insects, phagocytosis and three aspects of nodulation microaggregation, cell spreading and prophenyloxidase activation are mediated by eicosanoids.

Aside from hemocytic reactions to bacterial infections, insects also elaborate a complex of humoral responses to infection. These take the form of induced biosynthesis of various anti-bacterial infections. As seen elsewhere in this volume, a great deal of research attention has been directed toward understanding the structures of anti-bacterial proteins and toward understanding regulation of the genes that code for these proteins. Morishima *et al.* (1997) contributed a substantial advance in this area with their work on induced biosynthesis of anti-bacterial proteins in fat body of the silkmoth, *B. mori*. They found that eicosanoids mediate fat body expression of two genes for anti-bacterial proteins, specifically cecropin and lysozyme. This is the first report of the role of eicosanoids in the signal transduction pathways responsible for inducing gene expression in insect tissues. As we will note just ahead, there is reason to suppose that eicosanoids are intimately involved in regulating fat body gene expression in insects.

We note two lines of work on eicosanoids in silkmoth immunity. In our work on nodulation, we learned that eicosanoids mediate nodulation reactions to bacterial infections in *B. mori* (Stanley-Samuelson *et al.*, 1997). Morishima *et al.* (1997) also found that eicosanoids also mediate induction of genes for anti-bacterial proteins in the same species. Looking back to the early days of research on the roles of prostaglandins in insect physiology, Yamaja Setty and Ramaiah (1982) suggested that prostaglandins influence the mobilization of lipids from fat body stores in silkmoths. If this is supported with additional work, we would recognize two independent eicosanoid actions in larval and adult fat body. We speculate that multiple independent eicosanoid actions in any given invertebrate species will become the norm as more information on eicosanoid actions comes to light.

The experimental results described so far support the hypothesis that eicosanoids mediate insect cellular and humoral immune reactions to bacterial infections. We note, however, that virtually all of the experimental work on this hypothesis has been designed to investigate the influence of various pharmaceutical eicosanoid biosynthesis inhibitors on one or another immune reaction. While very helpful, this approach is not without difficulty (Stanley-Samuelson, 1994). There are several potential problems. One, we know very little about the biochemical actions of the pharmaceutical products, which were developed for use in human and veterinary medicine, in invertebrates. To highlight the issue, indomethacin is a potent cyclooxygenase inhibitor in mammals and insects. Nonetheless, indomethacin did not inhibit PG biosynthesis in preparations from reproductive tracts of adult male crickets, *Acheta domesticus* (Destephano *et al.*, 1976). In general, we can not be sure the common products express their inhibitory actions in invertebrates. For another issue, there has been only one published report on the pharmacology of any eicosanoid biosynthesis inhibitor in an invertebrate (Miller and Stanley-Samuelson, 1996). In general, we can not be certain that any given inhibitor would be delivered to the target tissue at all, nor if it would be subjected to inactivating metabolism. These and other issues stress the value of additional experimental approaches to testing the eicosanoid hypothesis.

Let us turn to recent work on one such approach. The emerging wisdom suggests that most animals continually biosynthesize and degrade eicosanoids in the unbroken flow of homeostatic physiology (Stanley, 1999). This constant eicosanoid turnover is responsible for the basal amounts of eicosanoids that have been detected in virtually all animals taken under appropriate scrutiny. Aside, and in addition to this routine eicosanoid biosynthesis, particular tissues can be stimulated to greatly increase eicosanoid biosynthesis. This is a particularly active area of research in mammalian systems, in which PG biosynthesis is stimulated by a variety of challenges, including injury. The mechanism of increased PG biosynthesis involved induced synthesis of new enzymes responsible for biosynthesizing PGs (Otto and Smith, 1995). Among invertebrates, various events in reproductive biology also stimulate increased PG biosynthesis, although the mechanism remains unknown. Relative to our interest in immunity, the logic of most experiments with eicosanoid biosynthesis inhibitors is that the inhibitors would block stimulated eicosanoid biosynthesis within immunity-conferring tissues immediately after the experimental animals detect the presence of bacterial cells.

We reasoned that bacterial infections ought to stimulate increased biosynthesis of one or more eicosanoids in insects. The experimental approach to this idea is conceptually straightforward. An insect would be challenged with a bacterial infection, then a few minutes later quantities of at least some eicosanoids in circulating hemolymph would be determined. Relative to appropriate controls, increases in hemolymph eicosanoid concentrations would indicate that increased eicosanoid biosynthesis begins soon after a bacterial infection.

A serious technical problem stood athwart this research approach: it is difficult and expensive to determine quantities of eicosanoids. The most common technique is radioimmunoassay (RIA), and very satisfactory commercial RIA kits are available. Operationally, the kits are quite expensive and each eicosanoid would require a separate RIA procedure. The use of RIA to assess quantitative changes in any of an unknown number of eicosanoids would amount to a very expensive and possibly inconclusive fishing trip.

An effective alternative is to separate eicosanoids on high-performance liquid chromatography (HPLC) and detect the compounds using a flow-through detector. In the early 1980's we prepared p-bromophenacyl esters of extracted PGs, separated the PG derivatives on HPLC and estimated quantities using a flow-through UV detector (Stanley-Samuelson *et al.*, 1983). The p-bromophenacyl moiety absorbs UV very strongly, but sensitivity remains a problem. More recently, Osada et al. (1989) derivatized PGs with the fluorescent reagent 9-anthryldiazomethane (ADAM). The derivatives were separated on HPLC and detected using a flow-through fluorescence spectrophotometer. The fluorescent derivatives increased sensitivity, however, in our hands the derivatization chemistry yielded problematic results, with confounding side products, mainly because ADAM is very unstable at room temperatures. We designed a modified protocol, in which the amount of ADAM and the derivatization reaction times were reduced (Jurenka *et al.*, 1999). Our improved protocol facilitates relatively rapid and cost-effective determinations of PG quantities.

We applied the new technique to the biological issue of assessing the influence of bacterial infections on eicosanoid production in insects (Jurenka *et al.*, 1999). True armyworms were convenient for this work. Sixth instars were injected with heat-killed bacteria, or with saline for control treatments. After 30 min incubations, hemolymph was collected and eicosanoids were extracted. After passing the extract over a silicic acid column to clean up the extract, the samples were reacted with ADAM, then diluted to appropriate concentrations and analyzed on HPLC.

We detected a range of eicosanoids in the hemolymph of unchallenged, control true armyworms. One product exhibited chromatographic behavior identical with $PGF_{2\alpha}$, which was present at about 0.9 ng/ml hemolymph. The results with bacteria-challenged armyworms were substantially different. At 30 min post-infection, $PGF_{2\alpha}$ was present at 4.0 ng/ml hemolymph. $PGF_{2\alpha}$ also was the predominant PG in the hemolymph of challenged armyworms. In another experiment, we co-injected another set of larvae with heat-killed bacteria with 10 mg of arachidonic acid. At 30 min post-injection, we extracted eicosanoids from the hemolymph, then isolated $PGF_{2\alpha}$ by HPLC. We formed the trimethylsilyl derivative of $PGF_{2\alpha}$, which was analyzed by gas chromatography-mass spectrometry. We obtained a mass spectrum identical to published spectra of $PGF_{2\alpha}$, and with the diagnostic ions of $PGF_{2\alpha}$, confirming the identity of this PG in true armyworm hemolymph.

In addition to $PGF_{2\alpha}$, peaks representing four additional eicosanoids, which remain to be identified, were detected for hemolymph from control and experimental hornworms. These peaks were larger for the bacterial-challenged larvae than the control larvae. Some of these products may be lipoxygenase products, because they emerged from the HPLC column with retention times similar to hydroxyeicosatetraenoic acids. We considered the question of whether these products were eicosanoids with another experiment. One set of larvae were challenged with bacterial cells and a second set were co-injected with bacterial cells plus phenidone, a dual lipoxygenase and cyclooxygenase inhibitor. At 30 min post-injection, both groups of larvae were processed for eicosanoid analysis. Compared to the bacterial-challenged larvae, the peaks representing the putative eicosanoids were much reduced for the phenidone-treated armyworms. We take this finding to support the view that the unidentified products were, indeed, eicosanoids.

These experiments indicate that bacterial infections exert a powerful stimulatory influence on true armyworm eicosanoid biosynthesis. Taken with the earlier work showing that inhibition of eicosanoid biosynthesis pathways impairs the ability of insects to react to bacterial infections, these findings strongly support the hypothesis that eicosanoids mediate insect immune responses to bacterial infections.

To be sure, many questions attend the hypothesis that eicosanoids mediate invertebrate immune reactions to bacterial infections. For one, the identities of specific eicosanoids that mediate specific steps in cellular and humoral immune reactions largely remain to be revealed. For another, the biochemical modes of eicosanoid actions in invertebrate immunity are completely unknown. Perhaps the most prominent issue lies in the overall physiology of invertebrate immunity. We know other groups of compounds, including biogenic amines, serve mediating roles in immunity. How are the actions of all signal moieties involved in immunity coordinated to produce biologically meaningful immune reactions to infections?

REFERENCES

Destephano, D.B., Brady, U.E. and Woodall, L.B. (1976) Partial characterization of prostaglandin synthetase in the reproductive tract of the male house cricket, *Acheta domesticus*. Prostaglandins 11, 261-273.

Horohov, D.W. and Dunn, P.E. (1983) Phagocytosis and nodule formation by hemocytes of *Manduca sexta* larvae following injection of *Pseudomonas aeruginosa*. J. Invertebr. Pathol. 41, 203-213.

Jurenka, R.A., Miller, J.S., Pedibhotla, V.K., Rana, R.L. and Stanley-Samuelson, D.W. (1997) Eicosanoids mediate microaggregation and nodulation responses to bacterial infections in black cutworms, *Agrotis ipsilon*, and true armyworms, *Pseudaletia unipuncta*. J. Insect Physiol. 43, 125-133.

Jurenka, R.A., Pedibhotla, V.K. and Stanley, D.W. (1999) Prostaglandin production in response to bacterial infection in true armyworm larvae. Arch. Insect Biochem. Physiol. 45, 75-83.

Mandato, C.A., Diehl-Jones, W.L., Moore, S.J. and Downer, R.G.H. (1997) The effects of eicosanoid biosynthesis inhibitors on prophenoloxidase activation, phagocytosis and cell spreading in *Galleria mellonella*. J. Insect Physiol. 43, 1-8.

Miller, J.S. and Stanley-Samuelson, D.W. (1996) The pharmacology of indomethacin, a prostaglandin biosynthesis inhibitor, in larvae of the tobacco hornworm, *Manduca sexta*. J. Insect Physiol., 42, 893-901.

Miller, J.S. and Stanley, D.W. (1998) The nodule formation reaction to bacterial infection: Assessing the role of eicosanoids. In Techniques in Insect Immunology, ed. A. Wiesner, G.B. Dunphy, V.J. Marmaras, I. Morishima, M. Sugumaran, and M. Yamakawa, pp 265-270. SOS Publications, Fair Haven, NJ.

Miller, J.S., Nguyen, T. and Stanley-Samuelson, D.W. (1994) Eicosanoids mediate insect nodulation responses to bacterial infections. Proc. Natl. Acad. Sci.USA 91, 12418-12422.

Miller, J.S., Howard, R.W., Nguyen, T., Nguyen, A., Rosario, R.M.T., and Stanley-Samuelson, D.W. (1996) Eicosanoids mediate nodulation responses to bacterial infections in larvae of the tenebrionid beetle, *Zophobas atratus*. J. Insect Physiol. 42, 3-12.

Miller, J.S., Howard, R.W., Rana, R.L., Tunaz, H. and Stanley, D.W. (1999) Eicosanoids mediate nodulation reactions to bacterial infections in adults of the cricket, *Gryllus assimilis*. J. Insect Physiol. 45, 75-83.

Morishima, I., Yamano, Y., Inoue, K. and Matsuo, N. (1997) Eicosanoids mediate induction of immune genes in the fat body of the silkworm, *Bombyx mori*. FEBS Letters 419, 83-86.

Oates, J.A. (1982) The 1982 Nobel Prize in Physiology or Medicine. Science, 218, 765-768.

Osada, M., Nishikawa, M. and Nomura, T. (1989) Involvement of prostaglandins in the spawning of the scallop, *Patinopecten yessoensis*. Comp. Biochem. Physiol. 94C, 595-601.

Otto, J.C. and Smith, W.L. (1995) Prostaglandin endoperoxide synthases-1 and -2. J. Lipid Mediators Cell Signalling 12, 139-156.

Samuelsson, B. (1983) Leukotrienes: mediators of immediate hypersensitivity reactions and inflammation. Science 220, 568-575.

Stanley, D.W. (1999) Eicosanoids in Invertebrate Signal Transduction Systems. pp 272. Princeton University Press, Princeton, NJ.

Stanley, D.W., Hoback, W.W., Bedick, J.C., Tunaz, T., Rana, R.L., Nor Aliza, A.R. and Miller, J.S. (1999) Eicosanoids mediate nodulation reactions to bacterial infections in larvae of the butterfly, *Colias eurytheme*. Comp. Biochem. Physiol., in press.

Stanley-Samuelson, D.W. (1994) Assessing the significance of prostaglandins and other eicosanoids in insect physiology. J. Insect Physiol. 40, 3-11.

Stanley-Samuelson, D.W., Klocke, J.A., Kubo, I. and Loher, W. (1983) Prostaglandins and arachidonic acid in nervous and reproductive tissue from virgin and mated female cricket *Teleogryllus commodus*. Ent. exp. appl. 34, 35-39.

Stanley-Samuelson, D.W., Jensen, E., Nickerson, K.W., Tiebel, K., Ogg, C.L. and Howard, R.W. (1991) Insect immune response to bacterial infection is mediated by eicosanoids. Proc. Natl. Acad. Sci. USA 88, 1064-1068.

Stanley-Samuelson, D.W., Pedibhotla, V.K., Rana, R.L., Nor Aliza, A.R., Hoback, W.W. and Miller, J.S. (1997) Eicosanoids mediate nodulation responses to bacterial infections in larvae of the silkmoth, *Bombyx mori*. Comp. Biochem. Physiol. 118A, 93-100.

Tunaz, H., Bedick, J.C., Miller, J.S., Hoback, W.W., Rana, R.L. and Stanley, D.W. (1999) Eicosanoids mediate nodulation reactions to bacterial infections in adults of two 17-year periodical cicadas, *Magicicada septendecim* and *M. cassini*. J. Insect Physiol., in press.

Yamaja Setty, B.N. and Ramaiah, T.R. (1982) Effect of PGE_1 on lipid mobilization from pupal fat bodies of the silkworm, *Bombyx mori*. Indian J. Exp. Biol. 19, 115-118.

LECTINS FROM TUNICATES: STRUCTURE-FUNCTION RELATIONSHIPS IN INNATE IMMUNITY

Gerardo R. Vasta, Michael S. Quesenberry, Hafiz Ahmed, and Nuala O'Leary

Center of Marine Biotechnology, University of Maryland Biotechnology Institute, 701 E. Pratt Street, Baltimore, Maryland 21202, USA

INTRODUCTION: ANIMAL LECTINS. A HISTORICAL PERSPECTIVE

Since the discovery of lectins in the 1890s, it has remained a challenge to define the biological function(s). A basic premise of lectin study is that their physiological function(s) relate to their carbohydrate-binding properties. Thus, detailed and useful descriptions of tissue distribution, sugar specificity, and molecular structures are available for many lectins. From the earliest reports, possible roles in fertilization and immune responses have been investigated. Lectins secreted from the albumin gland of the snail *Helix pomatia* able to agglutinate erythrocytes (Camus, 1899) were later named "protectins" to suggest that these protect against microbial or fungal infection (Prokop, 1965). Studies on arthropod species, such as the horseshoe crab *Limulus*

Phylogenetic Perspectives on the Vertebrate Immune System.
Edited by Beck *et al.*, Kluwer Academic/Plenum Publishers, 2001.

polyphemus, the lobster *Homarus americanus*, the crab *Eupagurus prideauxii*, and the spider crab *Maia squinado*, extended the hemagglutination properties to hemolymph lectins, providing evidence that expression of these proteins is inducible by challenge with agglutinated bacteria, and furthermore that the induced lectins have opsonic properties (Cantacuzene, 1919).

The modern era of research on animal lectins has seen a vast expansion on these foundations. The number of lectins described and the variety of species in which these are known has increased rapidly. Ashwell's group described the first vertebrate lectins in the 1970's, and by the next decade sequence data became available for sufficient number such that these were classified into structurally similar groups with distinct properties, the C-type and S-type (later renamed galectins) lectins. Functional understanding of lectins from vertebrates revealed their participation in innate immune functions, as non-self recognition factors binding to LPS or bacterial surfaces, opsonizing bacteria, and activating complement. The most well-characterized molecule, now recognized as a participant in the innate immune system, is the serum mannose-binding lectin (MBL), a C-type lectin. As a component of the acute phase response, the lectin-dependent complement-activation pathway initiated by MBL may constitute the most ancient non-self recognition/defense mechanism (Vasta *et al*, 1994; Thiel, *et al*, 1997; Ji *et al*, 1997; Magor and Vasta *et al*, 1998). In vertebrates, C-type lectins in another subcategory known as selectins function to facilitate the adaptive immune response through lymphocyte, neutrophil and platelet homing or localization.

TYPES OF LECTINS PARTICIPATING IN IMMUNITY IN VERTEBRATES AND INVERTEBRATES

The role of C-type lectins in mammalian immunity has been recently well reviewed (Weis, 1998). Purification and cloning of an MBL-homolog from the chicken (Laursen *et al*, 1998) reflects the need to examine MBL function and evolution in non-mammalian species. The presence in vertebrate model systems of the adaptive immune response, i.e. that mediated by immunoglobulins and lymphoid cells, complicates studies of innate immune function. The role(s) of lectins may be more prominent in those species lacking immunoglobulins, or in which immunoglobulins are relatively less efficient in their recognition and effector functions. Thus, we have examined lectins in a nematode (*Caenorhabditis elegans*), protochordates (*Didemnum candidum* and *Clavelina picta*) and ectothermic vertebrates, such as teleost fish (*Morone saxatilis*) and amphibians (*Bufo arenarum*). Comparative studies on the amphibian *Bufo arenarum* and other species within this genus may prove especially instructive concerning their environmental adaptations and disease resistance due to the widespread distribution of this well conserved taxon throughout Africa, Asia and the Americas (Ahmed *et al.*, 1996).

Particularly advantageous for studies of innate immune mechanisms may be those extant species of taxa most closely related to vertebrates, yet lacking an adaptive immune response, such as the protochordates. Studies of hemolymph from ten tunicate species have indicated the presence of multiple lectin populations existing in each, all with considerable diversity in sugar specificity (Vasta and Marchalonis, 1984). This suggests that the expression of a diverse lectin repertoire improves the probability that any given microorganism may be recognized. Indeed, the colonial tunicate *C. picta* (Vasta *et al.*, 1996) expresses a number of lectins which all recognize the simple sugar L-fucose (Fuc; all other monosaccharides discussed here are in the D-configuration) but which interact differentially with more complex substrates such as glycoproteins, polysaccharides, blood cells or bacterial strains isolated from the tunicate's natural environment (Vasta, *et al.*, manuscript in preparation). One of the *C. picta* lectins is a C-type lectin with sequence similar to MBL. The tunicate *D. candidum* also contains a lectin repertoire, among which those with galactose (Gal) specificity have been characterized in detail (Vasta et al., 1986a; Vasta and Marchalonis, 1986b; Vasta et al., 1986c).

In addition to the MBL-like lectins, other lectin types participate in mammalian innate and adaptive immune functions. Galectins are a family of cation-independent, soluble β-galactoside-binding lectins which are ubiquitous in vertebrates invertebrates and fungi (Hirabayashi, 1997). A galectin has been found recently to bind, and therefore potentially to modulate, CR3, the macrophage receptor for iC3b, a product of the MBL-mediated complement activation pathway (Avni *et al.*, 1998). This possible connection in innate immune pathways may be better elucidated in invertebrate model systems.

LECTIN TYPES IN TUNICATE SPECIES

Among the lectins described in tunicates for which primary structure data are available, there exist some which are homologous to mammalian C-type lectins, some to the galectin family of Ca^{2+}-independent Gal-binding lectins, and there are those which are not assigned to either of these categories. Among the tunicate lectins for which amino acid or nucleotide sequence data is lacking, an LPS-binding lectin from *Halocynthia roretzi* (Azumi *et al.*, 1991) and a Gal-binding lectin from *Botryllus schlosseri* , shown to have opsonic properties (Ballarin *et al.*, 1999), have been characterized. Although it remains to be seen from its primary structure, due to the Ca^{2+}-independent nature of the latter, as well as apparent size (17-19 kDa by SDS-PAGE) and restriction of carbohydrate-specificity to Gal, this molecule was hypothesized to be galectin-like by these authors.

Lectins from the tunicate *Didemnum candidum* have been purified, and their properties described (Vasta et al., 1986a; Vasta and Marchalonis, 1986b; Vasta et al., 1986c). Partial sequence data from the N-terminus indicates that one of these lectins, designated DCL-I, is related to proteins in the pentraxin family (Vasta *et al.*, 1986c). A lectin from a related species, *Didemnum ternatanum*, has been recently described (Belogortseva *et al.*, 1998) but no primary structure information is available.

C-Type Lectins

Lectins with C-type CRD sequence homology, have been reported in the literature from three tunicate species, *Polyandrocarpa misakiensis*, *Botryllus schlosseri*, and *Clavelina picta*. Other proteins which may be included in the C-type lectin family have been clearly identified from primary structure data, but no sugar-binding activity has been reported. This is not uncommon, in fact, because the CRD appears to have been shuffled into association with various accessory domains resulting in multidomain proteins, and frequently a clone may be isolated on the basis of properties conferred by one of these ancillary domains for other purposes, and the C-type CRD is recognized only after the complete sequence is known. This is frequently and increasingly the case as the tools of molecular biology are applied to isolate cDNA clones, and a case in point is the C-type CRD found in the translation product of a cDNA clone isolated from *Botryllus schlosseri* (Pancer *et al.*, 1997).

For contextual purposes it is useful to consider other invertebrate sources of C-type lectins, especially when attempting to interpret the significance of primary structural features such as sequence motifs. Among the lectins to consider are echinoidin, from sea urchin *Anthocidaris crassispina* (Giga *et al.*, 199), bra-2 and bra-3, from the acorn barnacle *Megabalanus rosa* (Muramoto and Kamiya, 1990; Takamatsu *et al.*, 1993), and two insect lectins, granulocytin, from the flesh fly *Sarcophaga peregrina* (Fujita *et al.*, 1998), and the C-type lectin isolated from *Drosophila melanogaster* (Haq *et al.*, 1996).

Gal-Binding C-type Lectins

Besides CPL-III, the only C-type tunicate lectin for which sugar-binding properties have been described is from the *P. misakiensis*, which has Gal-binding activity, as do all other invertebrate C-type lectins characterized so far, with the single exception of the *C. picta* C-type lectin, which has Fuc preference. The crystal structure of *P. misakiensis* lectin was recently published, revealing some surprises as discussed below. Other (non-tunicate) invertebrate sources of C-type lectins include the sea cucumbers *Stichopus japonicus* and *Cucumaria echinata*, the acorn barnacle *Megabalanus rosa*, and the sea urchins *Lytechinus variegatus* and *Anthocidaris crassispina*. These C-type lectins all bind Gal.

A lectin originally reported with Ca^{2+}-dependent Gal-binding properties from the sea cucumber *S. japonicus* has amino acid sequence homology with C-type lectins (Himeshima et al., 1994) and is most closely related to the sea urchin lectin echinoidin isolated from *Anthocidaris crassispina*. Four Gal/GalNAc-specific lectins have been isolated from the sea cucumber, *Cucumaria echinata* and determined to have primary structures related to the C-type lectin family (Hatakeyama, 1994). Of these, one lectin was found to lyse red blood cells from humans and rabbits in a Ca^{2+}- and Gal-/GalNAc-dependent fashion.

Fuc-type

To date there is only one C-type tunicate lectin with specificity directed toward Fuc, the lectin described from *C. picta* sometimes designated as CPL-III to distinguish it from other *C. picta* Fuc-binding lectins which were similarly isolated. Here we will discuss CPL-III under the nomenclature CpMBL in recognition of the structural and functional relationship with the vertebrate serum MBL, from studies of which the recognition of the MBL-pathway stemmed. There are two C-type CRDs per polypeptide in CpMBL, which dimerize in solution. The sequence alignments of both CpMBL CRDs, designated according to either N- or C-terminal disposition in the polypeptide, shows the closest similarity to MBL of any invertebrate lectin reported (Vasta *et al.*,

1999). Although the overall sequence comparison is quite favorable, MBL CRD mutagenesis studies of Drickamer focussed our attention on the presence of the tripeptide EPN in both CpMBL CRDs, as well as the conspicuous absence of this tripeptide in all other invertebrate C-type CRDs. As discussed below, this sequence difference provides a possible rationale for the sugar-binding preferences among these lectins which distinguish CpMBL from all the others.

The recently-described *P. misakiensis* crystal structure modifies this interpretation for the invertebrate taxa, however, because the sequence EPSN occurs rather than QPD at this position. In fact, many invertebrate Gal-binding C-type lectins do not conserve the canonical QPD. It is possible that protochordates contain many lectin CRDs, of which only a select few were selected and preserved in the vertebrate lineages (although absence of proof does not constitute proof of absence, so we base these observations on the sequences available). We speculate that Gal-binding C-type lectins exist in tunicates and provide defensive, and we wonder why no Gal-specific collectin has been described. One may speculate that this absence, if true, may be due to the superior efficiency of Gal-binding immunoglobulins, or even the asialoglycoprotein receptor (ASGPR) in clearing Gal-displaying microbes from circulation. To our knowledge, no description of the potential for ASGPR-mediated clearance of Gal-bearing pathogens exists in the literature, so this remains speculation.

Galectins (S-Type Lectins)

Studies carried out mostly in mammals have suggested that galectins are directly or indirectly involved in immune functions, such as inflammation, apoptosis, and superoxide production, as discussed below. The role(s) of the few invertebrate galectins described so far, from the sponge *Geodia cydonium* (Hanisch, *et al.*, 1996), the nematode *Caenorhabditis elegans* (Arata *et al.*, 1997), and the tunicate *C. picta* (Ahmed *et al*, in preparation) require further study.

C. picta galectins isolated by sequential ion-exchange were next purified by lactosyl-Sepharose affinity chromatography. Observed on SDS-PAGE, the relative mobilities of the lactose eluates were 14.8 kDa (45%), 15.8 kDa (31%), and 37.5 kDa (24%), with a minor component at 33.5 kDa (Ahmed and Vasta, unpublished). These sizes are in good agreement with literature values (Barondes *et al.*, 1994; Vasta *et al.*, 1997). Western blot of the purified *C. picta* galectins using a rabbit anti-striped bass muscle galectin (Henrikson *et al.*, unpublished) showed that the 14.8 kDa band cross-reacts with the teleost 14.5 kDa galectin. Two approaches to obtain amino acid sequence were taken: PCR isolation of putative galectin cDNA clones using degenerate oligonucleotide primers on using galectin consensus sequences as a design basis, and peptide sequencing of protease-digested galectin. These sequence data in alignment with *bona fide* galectin family members indicate the relatedness of *C. picta* galectin to this family of proteins (O'Leary *et al.,* unpublished).

FUNCTIONS

Although a considerable body of evidence has been obtained with regards to the functional aspects of protein-carbohydrate interactions, there is still a great deal of speculation concerning the specific biological roles role(s) of C-type lectins and galectins in both vertebrates and invertebrates. This is illustrated in the following discussion, which focuses on those lectin function(s) related to innate immune mechanisms in tunicates.

Immobilization and opsonization

A classical definition of lectin activity includes agglutination of a target cell, reflecting cross-linking abilities due to the presence of two or more sugar-binding sites in the lectin, whereas the more recent definitions based on structural aspects rely on the identification of sequence motifs that define the known lectin families. Most invertebrate lectins described in the literature have been tested for their agglutination and opsonic properties upon purification, even prior to obtaining sequence information, and many display either or both activities, as noted above. Despite the possible merits of naming as a lectin the translation product of a cloned cDNA with sequence similarity to a family of proteins, to include the binding activities and biochemical properties of a native, or at least recombinant protein when making such a categorization may constitute a more reliable approach. This criterion reflects consideration of the biological role(s) of the protein, for which immobilization of a potential pathogen requires cross-linking ability. In combatting infection, an organism's first line of defense is to exclude the microbe from its tissues. Failing that, the host must limit the spread of the pathogen, which may potentially reach sites where it may avoid detection and thrive. Immobilization through lectin cross-linking is one way to slow or stop internal spread of infection. Following this accomplishment, the task of neutralization may proceed through phagocytosis , or if the target particle is too large, encapsulation followed by melanization (Sugumaran, 1991). The latter process is a consequence of activation of the prophenoloxidase cascade, which is triggered upon recognition of the pathogen by one or more of the various proteins identified so far (Soderhall), including lectins, as shown for some invertebrate models (Chen *et al*, 1999; Kanost, this volume). The role for lectins in the recognition, immobilization and elimination of pathogenic microbes through the humoral and cell-mediated pathways discussed below, is somewhat more clear at present time. Although significant effectors in this processes, antimicrobial peptides will not be discussed here.

Complement activation

Complement C3 activation and subsequent thiolester reaction with the target surface is required for coordination of molecular events leading to pathogen destruction. The humoral response pathway referred to as the complement cascade results in tagging the pathogen for enhanced phagocytosis or deposition of pore-forming protein complexes which compromise the integrity of the cell wall, leading to death of the microbial pathogen. To date there is no evidence for assembly of a membrane-attack complex in invertebrates. However, the initial steps leading to a covalent association of complement C3 with the targeted microbial surface leading to opsonization, phagocytosis and intracellular killing.

Vertebrate plasma contains higher levels of complement C3 than any other component in the pathway. A *tour de force* study from Lambris and coworkers comparing C3 from swine, rabbit, mouse, cobra, *Xenopus*, axolotl, and trout identified conserved structural and functional elements (Alsenz *et al.*, 1992), to which data from chicken and sea bream were recently added (128,129). There is absolute conservation of α/β chain structure and of the α thioester site. The ConA-binding site was missing in only one example. The thioester binding site, critical for covalently bonding to the "target", requires positionally-conserved Cys and His residues for reactivity towards hydroxyl-containing substrates, the absence of which distinguishes structurally-related molecules such as α-2 macroglobulin and C4 (Gadjeva *et al.*, 1998).

In order to establish the existence of key elements of the lectin-mediated complement activation pathway in tunicates, it was critical to establish the presence of a C3 homolog in *C. picta*. The approach taken to accomplish this goal was to isolate, clone and sequence PCR products derived from primers designed on the basis of conserved amino acid segments identified from sequence alignments of published C3 molecules, using *C. picta* cDNA as a template. The success of this approach, while reassuring, is not unexpected since the existence of complement-like molecules in tunicates should be predicted from the observation of C3 in the sea urchin (Al-Sharif *et al.*, 1998). Indeed, we have identified substantial portions of C3-like sequence from cDNA clones of *C. picta*, including the region of the predicted thioester site (O'Leary *et al.*, unpublished).

Phagocytic activation/Cell activation

Opsonization of surfaces may be achieved directly upon lectin binding to the target surface, as was discussed among the examples of invertebrate lectins, above. Alternatively, phagocytic cells may bind iC3b through CR3, potentially modulated by galectin as described above. Internalization of the phagocytized particle triggers cell responses such as superoxide production. Other cytolytic components, too many to elaborate here, are released into the phagocytic vacuole.

In the case of mammalian MBLs, which are collectins, opsonization is accomplished directly through the collagen-like domain and can be abolished in the mammalian systems by collagenase treatment. This leads either to direct lysis (Schweinle *et al.*, 1989) or to recognition by phagocytic cells (Valdimarsson *et al.*, 1998), and it has been shown that uptake of potentially pathogenic strains of bacteria can also be mediated by the Man-receptor of macrophages (Ezekowitz *et al.*, 1991). Despite this ability, it is now apparent that MBL-associated serine proteases (MASPs) act in the MBLectin pathway to activate complement (Thiel *et al.*, 1997).

Prophenoloxidase activation

A β1,3-glucan (laminarin) binding protein (βGBP) associated with prophenoloxidase activation has been identified in crayfish, cockroach, earthworm, and silkworm. The crayfish (*Pacifasticus leniusculus*) βGBP purified on laminarin has no peptidase or phenoloxidase activity, but does promote these hemocyte-derived activities specifically in the presence of laminarin (Duvic and Soderhall, 1990). Similar characteristics were reported for βGBPs from silkworm (*Bombyx mori*, Ochiai and Ashida, 1988), earthworm (*Eisenia foestida*, Beschin *et al*, 1998) and cockroach (*Blaberus discoidalis*, Chen *et al*, 1999). A protein with similar prophenoloxidase-enhancing activities but distinct in binding to peptidoglycan, with no measurable binding to β1,3-glucan, has also been described in silkworm (Yoshida *et al.*, 1996).

The role of lectins in the activation of the prophenoloxidase cascade has been demonstrated in insects. A lectin was purified from serum of the cockroach *Blaberus discoidalis* and sequenced at the N-terminus to show relatedness to some limulus C-reactive proteins (Chen *et al.*, 1999). This lectin, specific for β1,3 glucans, was characterized according to agglutination properties, amino acid composition, cellular distribution, electron microscopic structural analysis, and prophenoloxidase activation properties. The cockroach β1,3 glucan-binding lectin was shown to activate the prophenoloxidase pathway specifically in the presence of laminarin. Similarly, the role of lectins in prophenoloxidase cascade activation has been demonstrated in the tobacco horned worm *Manduca sexta* (Kanost, this conference). Tunicates also exhibit internal defence mechanisms of protection mediated by prophenoloxidase activation and melanization activities demonstrated in *Ciona intestinalis* and eight additional ascidians (Jackson *et al*, 1993). The tunicate prophenoloxidase activity shows an LPS dose-dependent response, and is likewise enhanced by laminarin, but not other sugars. Hemolymph drained from *C. picta* colonies will show progressive browning in storage, presumably due to melanization, a process that can be dramatically enhanced by the addition of carbohydrates for which the hemolymph lectins are specific (Vasta, unpublished observation).

INTEGRATED MODEL FOR IMMUNITY AND WOUND REPAIR

Although a full range of monosaccharide specificities may be found among the protochordate lectins, it is interesting to note that among the C-type lectins found in mammals the Gal-specific lectins are not typically involved in any known innate immune pathway. The asialoglycoprotein receptor, which is Gal-specific, has activity which appears to be directed toward senescent serum glycoproteins. The S.A.-binding C-type lectins, the selectins, mediate cell-cell regognition among "self" ligands to direct trafficking of lymphocytes. So far, it appears that Man-/Fuc- binding C-type CRDs are more typically involved in innate immunity. Whether this is the case for tunicates remains to be seen, but it may be expected that a variety of lectin properties are needed in protochordates, only some of which have been propogated by evolution into the vertebrates.

The evolutionary niche occupied by tunicates, bridging invertebrates and vertebrates, suggests these organisms may express extant examples of the most advanced defense mechanisms which preceded the advent of immunoglobulins, perhaps a homolog of the mammalian MBL. There are multiple lectins in plasma of the colonial tunicate *C. picta*, which includes CpMBL (Vasta *et al.*, 1996). The recent discovery of MBL-associated serine proteases in another tunicate species (Ji *et al.*, 1997), key molecular partners in the lectin-mediated complement pathway found in man (Matsushita *et al.*, 1992; Thiel *et al.*, 1997), is further proof that these model organisms harbor homologs of mammalian defense molecules. Study of the lectin-mediated defense pathways in these organisms is useful, since factors which would otherwise compete in a higher model system, such as antibodies, are absent in the tunicates. Together with galectins, we have now isolated and characterized additional key elements of an MBL-like defense pathway in *C. picta*.

The presence of complement in a sea urchin begs the question 'What is complement doing in a species unable to mount an antibody-mediated response?'. As indicated in Fig. 1, it is likely that CpMBL participates through only one of two tandem CRDs in

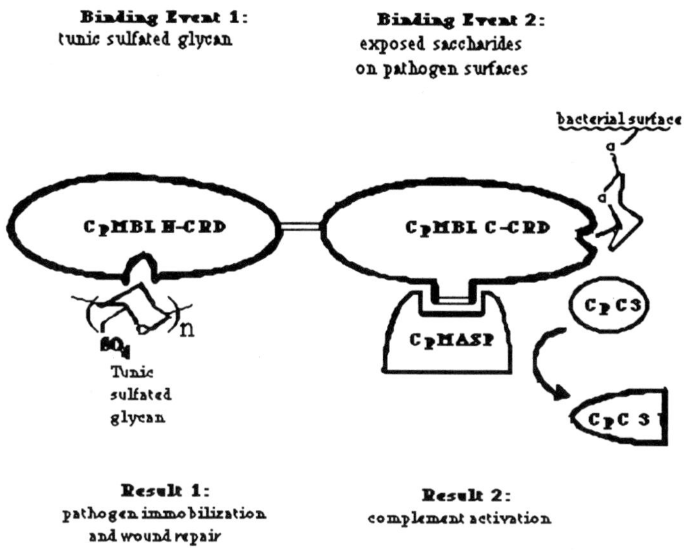

Figure 1. Differences in the tandem CRD sequences suggest their different functions: N-CRD similarity to selectin suggest this may bind tunic sulfated glycan, and C-CRD similarities to MBL suggests this domain may recognize pathogens.

recognition of non-self leading to CpC3 activation. There is other physical evidence supporting this aspect of the model. Fluorescence- quenching studies indicate only one of these two CRDs is probably binding only Fuc (Vasta *et al.*, unpublished). The rest of the pathway rests in part on the basis of cDNA evidence for molecular partners of CpMBL in a functional homolog of the vertebrate MBL-pathway, suggesting that the integral pathway has been evolutionarily conserved, not only some of the molecules. Thus, the role of complement in the MBL pathway should predate participation in the antibody-mediated pathways. As indicated in the model only one of the tandem CRDs functioning in self-recognition (Fig. 1), consistent with the data discussed above.

As discussed above, the relatedness suggested at the primary structure level between CpMBL N-CRD and selectin CRDs was supported by the 3D modeling studies originally focused on the homology with MBL, helping to explain the second function of CpMBL in binding tunic sulfated glycan and wound repair. In fact, CpMBL is structurally similar to both of these sub-families of C-type CRDs and most likely performs roles similar to both MBL and selectins. What is remarkable is the high degree of similarity between these molecules from such distant species, and the consistency between structure and function which allows us to rationalize the dual role of CpMBL. As indicated in the model (Fig. 1), it is likely that self-recognition is mediated by the other of the two tandem CRDs, the CRD which does not mediate recognition of non-self and which is more like selectins. The finding of binding activity in one molecule for two classes of ligand, both exogenous and endogenous, led to the 'dual-role hypothesis' for CpMBL which states that CpMBL participates in both wound repair and host defense. The efficiency of combining in one molecule two properties required in meeting the needs arising from a compromised tunic wall makes teleological sense, as both functions are required at the wound site, but further work is required to better understand CpMBL and the molecular partners with which it interacts. CpMBL has been discussed here in terms of properties common to both MBLs and selectins, but the unique nature of this molecule must be recognized since to our knowledge such dual properties have been described in no other lectin.

It is also critical to examine the role of galectins in innate and adaptive immune pathways. Recent studies in mammalian systems have shown T-lymphocytes triggered by L-selectin bind dendritic cells through galectin-3 (Vivette *et al.*, 1998). Also, recall that dendritic cells are among those which display CR3, which is modulated by galectin-1. With this many potential interactions, the simpler the model system without losing physiological relevance, the better. The protochordate *C. picta* offers the desired advantages as a model for such studies. It remains to be seen whether a connection exists between *C. picta* galectins and the hemocyte CpMBL-like molecules mentioned above, which may be putative selectins.

The complete cDNA sequence of CpMBL, and most of the polypeptide amino acid sequence as well (Quesenberry, *et al.*, manuscript in preparation) provide a foundation for searching *C. picta* for the presence of an 'integral' MBL pathway (Fig. 2). Although CpMBL does not exhibit a collagen-like tail structure similar to MBL and C1q, this does not appear to be an essential feature for complement activation because CRP, which lacks the collagen-like domain, also activates complement and is endowed with opsonic properties. The identification of MASP, C3, and galectin sequences confirm that the molecular partners are present in this tunicate species. This implies the ability of *C. picta* to execute each step in the pathway, recognition of the potential pathogen by CpMBL, activation of CpC3 by CpMASP, and modulation of cell-mediated responses through CpCR3 by galectin. The existence of a CR3-homolog in *C. picta*, CpCR3, remains to be established.

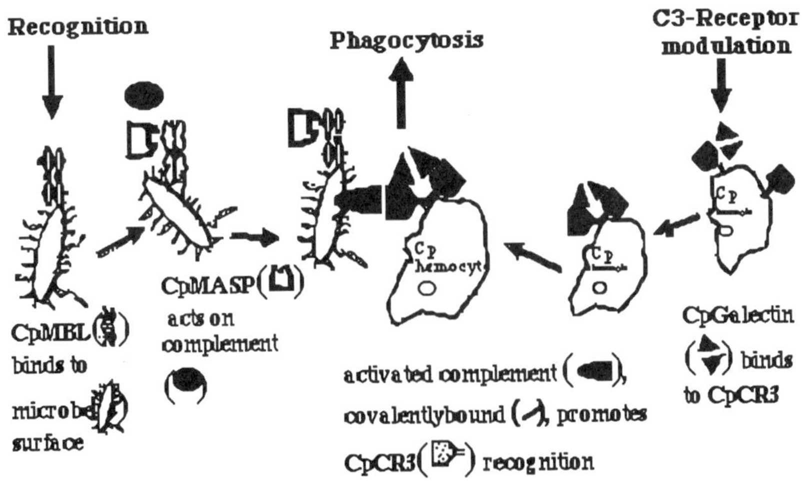

Figure 2. A Hypothetical CpMBL Pathway.

We have described a MBL-homolog in *C. picta* and propose that this molecule executes dual functions. CpMBL mediation of humoral defense is supported by structural relation to MBL, measured binding activities of the protein, and the detection of molecular cohorts in the MBL-pathway in *C. picta*. The evidence discussed here for coexistence of the key lectin-mediated complement activation pathway components in a single protochordate species suggests this pathway preceded the immunoglobulin pathway in evolution.

ACKNOWLEDGEMENTS

Supported by Grant MCB-94-06649 from the National Science Foundation and Grant No. 95-31 from the Lucille P. Markey Trust Fund to G.R.V.

REFERENCES

Ahmed, H., Pohl, J., Fink, N.E., Strobel, F., and G.R. Vasta The primary structure and carbohydrate specificity of a beta-galactosyl-binding lectin from toad (*Bufo arenarum* Hensel) ovary reveal closer similarities to the mammalian galectin-1 than to the galectin from the clawed frog *Xenopus laevis*, *J. Biol. Chem.*, 1996, 271, 33083-33094.

Al-Sharif, W.Z., J.O. Sunyer, J.D. Lambris, and L.C. Smith Sea urchin coelomocytes specifically express a homologue of the complement component C3. *J. Immunol.*, 1998, 160, 2983-2997.

Alsenz, J., D. Avila, H.P. Huemer, I. Esparza, J.D. Becherer, T. Kinoshita, Y. Wang, S. Oppermann, and J.D. Lambris Phylogeny of the third component of complement, C3: analysis of the conservation of human CR1, CR2, H, and B binding sies, concanavalin A binding sites and thiolester bond in the C3 from different species. *Dev. Comp. Immunol.*, 1992, 16, 63-76.

Avni, O., Z. Pur, E. Yefenof, and M. Baniyash Complement receptor 3 of macrophages is associated with galectin-1-like protein. *J. Immunol.*, 1998, 160, 6151-6158.

Azumi K., Ozeki S., Yokosawa H, and S. Ishii A novel lipopolysaccharide-binding hemagglutinin isolated from hemocytes of the solitary ascidian, *Halocynthia roretzi*: it can agglutinate bacteria. *Dev Comp Immunol.* 1991 Winter;15(1-2):9-16.

Baish, M.A., R.L. Lohr, and S. Bartl Molecular evidence for complement and alpha 2-macroglobulin family members in the colonial ascidian. *Botryllus schlosseri. Dev. Comp. Immunol.*, 1997, 21, 147.

Ballarin L., Tonello C., Guidolin L., and A. Sabbadin Purification and characterization of a humoral opsonin, with specificity for D-galactose, in the colonial ascidian *Botryllus schlosseri.. Comp Biochem Physiol B Biochem Mol Biol.* 1999 May;123(1): 115-23.

Barondes, S.H., D.N.W. Cooper, M.A. Gitt, and H. Leffler Galectins. Structure and function of a large family of animal lectins. *J. Biol. Chem.*, 1994, 294, 20807-20810.

Belogortseva N., Molchanova V., Glazunov V., Evtushenko E., and P. Luk'yanov N-Acetyl-D-glucosamine-specific lectin from the ascidian *Didemnum ternatanum. Biochim Biophys Acta.* 1998 Apr 10;1380(2):249-56.

Camus, M.L. Recherches experimentales sur une agglutinine produite par la glande de l'albumen chez l'*Helix pomatia. C.R. Acad. Sci.*, 1899, 129, 233.

Cantacuzene, J. Sur certains anticorps naturels observes chez *Eupagurus prideauxii. C.R. Soc. Biol.*, 1912, 73, 663.

Chen C, Rowley AF, Newton RP, and N.A. Ratcliffe Identification, purification and properties of a beta-1,3-glucan-specific lectin from the serum of the cockroach, *Blaberus discoidalis* which is implicated in immune defence reactions. *Comp Biochem Physiol B Biochem Mol Biol.* 1999 Mar;122(3):309-19.

Drickamer, K. Engineering galactose-binding activity into a C-type mannose-binding protein. *Nature*, 1992, 360, 183-186.

Ezekowitz RA, Williams DJ, Koziel H, Armstrong MY, Warner A, Richards FF, and RM Rose Uptake of Pneumocystis carinii mediated by the macrophage mannose receptor. Nature. 1991 May 9;351(6322):155-8.

Fujita Y., Kurata S., Homma K., and S. Natori A novel lectin from Sarcophaga. Its purification, characterization, and cDNA cloning. *J Biol Chem*. 1998 Apr 17;273(16):9667-72.

Gadjeva, M., A.W. Dodds, A. Taniguchi-Sidle, A.C. Willis, D.E. Isenman, and S.K.A. Law The covalent binding reaction of complement component C3. *J. Immunol.*, 1998, 161, 985-990.

Giga Y, Ikai A, and K.Takahashi The complete amino acid sequence of echinoidin, a lectin from the coelomic fluid of the sea urchin *Anthocidaris crassispina*. Homologies with mammalian and insect lectins. *J Biol Chem*. 1987 May 5;262(13):6197-203.

Hatakeyama T., Kohzaki H., Nagatomo H., Yamasaki N. Purification and characterization of four Ca^{2+}-dependent lectins from the marine invertebrate, *Cucumaria echinata*. *J. Biochem.*, 1994, 116, 209-214.

Hayat U, Reddy GP, Bush CA, Johnson JA, Wright AC, and JG, Jr, Morris Capsular types of *Vibrio vulnificus*: an analysis of strains from clinical and environmental sources. *J Infect Dis*. 1993 Sep;168(3):758-62.

Haq S Kubo T, Kurata S, Kobayashi A, and S Natori Purification, characterization, and cDNA cloning of a galactose-specific C-type lectin from *Drosophila melanogaster*. *J Biol Chem*. 1996 Aug 16;271(33):20213-8.

Himeshima T., Hatakeyama T., and Yamasaki N. Amino acid sequence of a lectin from the sea cucumber, *Stichopus japonicus*, and its structural relationship to the C-type animal lectin family. J. Biochem., 1994, 115, 689-692.

Hirabayashi, J., ed. Recent topics on galectins.*Trends Glycosci. Glycotechnol.* Vol. 9. 1997. 1-180.

Jackson AD, Smith VJ, and CM Peddie *In vitro* phenoloxidase activity in the blood of *Ciona intestinalis* and other ascidians. *Dev Comp Immunol*. 1993 Mar-Apr;17(2):97-108.

Ji, X., K. Azumi, M. Sasaki, and M. Nonaka Ancient origin of the complement lectin pathway revealed by molecular cloning of mannan binding protein-associated serine protease from a urochordate, the Japanese ascidian, *Halocynthia roretzi*. *Proc. Natl. Acad. Sci. USA*, 1997, 94, 6340-6345.

Laursen, S.B., T.S. Dalgaard, S. Thiel, B.L. Lim, T.V. Jensen, H.R. Juul-Madsen, A. Takahashi, T. Hamana, M. Kawakami, and J.C. Jensenius Cloning and sequencing of a cDNA encoding chicken mannan-binding lectin (MBL) and comparison with mammalian analogues. *Immunology*, 1998, 93, 421-430.

Magor KE, and G.R. Vasta Ancestral immunity comes of age. *Immunol Today*. 1998 Feb;19(2):54-6.

Matsushita, M. and T. Fujita Activation of the classical complement pathway by mannose-binding protein in association with a novel C1s-like serine protease. *J. Exp. Med.*, 1992, 176, 1497-1502.

Mavroidis, M., J.O. Sunyer, and J.D. Lambris Isolation, primary structure, and evolution of the third component of chicken complement and evidence for a new member of the α2-macroglobulin family. *J. Immunol.*, 1995, 154, 2164-2174.

Muramoto K, and H Kamiya The amino-acid sequence of multiple lectins of the acorn barnacle *Megabalanus rosa* and its homology with animal lectins. *Biochim Biophys Acta*. 1990 May 31;1039(1):42-51.

Prokop, O., G. Uhlenbruck, and W. Kohler A "new" human blood group receptor A_{hel} tested with saline extracts from *Helix hortensis* (garden snail). *J. Forensic Med.*, 1965, 12, 108.

Schweinle JE, Ezekowitz RA, Tenner AJ, Kuhlman M, and KA Joiner Human mannose-binding protein activates the alternative complement pathway and enhances serum bactericidal activity on a mannose-rich isolate of *Salmonella*. *J Clin Invest*. 1989 Dec;84(6):1821-9.

Sugumaran M. Molecular mechanisms for mammalian melanogenesis. Comparison with insect cuticular sclerotization. FEBS Lett. 1991 Dec 16;295(1-3):233-9.

Sumiya, M., M. Super, P. Tabona, R.J. Livinsky, T. Arai, M.W. Turner, and J.A. Summerfield Molecular basis of opsonic defect in immunodeficient children. *Lancet*, 1991, 337, 1569-1570.

Sunyer, J.O., L. Tort, and J.D. Lambris Diversity of the third form of complement, C3, in fish: functional characterization of five forms of C3 in the diploid fish *Sparus aurata*. *Biochem J.*, 1997, 326, 877-881.

Takamatsu N, Takeda T, Kojima M, Heishi M, Muramoto K, Kamiya H, T Shiba Acorn barnacle *Megabalanus rosa* lectin (BRA-3): cDNA cloning, gene structure and seasonal changes of mRNA and protein levels. Gene. 1993 Jun 30;128(2):251-5.

Thiel, S., T. Vorup-Jensen, C.M. Stover, W. Schwaeble, S.B. Laursen, K. Poulsen, A.C. Willis, P. Eggleton, S. Hansen, U. Holmskov, K.B.M. Reid, and J.C. Jensenius A second serine protease associated with mannan-binding lectin that activates complement. *Nature*, 1997, 386, 506-510.

Valdimarsson H, Stefansson M, Vikingsdottir T, Arason GJ, Koch C, Thiel S, and J.C. Jensenius Reconstitution of opsonizing activity by infusion of mannan-binding lectin (MBL) to MBL-deficient humans. *Scand J Immunol*. 1998 Aug;48(2):116-23.

Vasta GR, and J.J. Marchalonis Distribution, specificity and macromolecular properties of tunicate plasma lectins. *Prog Clin Biol Res.* 1984;157:125-41.

Vasta G.R,Marchalonis, J.J., Decker, J.M. Binding and mitogenic properties of a galactosyl-specific lectin from the tunicate Didemnum candidum for murine thymocytes and splenocytes. *J. Immunol.*, 1986a, 137, 3216-3223.

Vasta GR, and J.J. Marchalonis Galactosyl-binding lectins from the tunicate Didemnum candidum. Carbohydrate specificity and characterization of the combining site. *J Biol Chem.* 1986b Jul 15;261(20):9182-6.

Vasta GR, Hunt JC, Marchalonis JJ, and WW Fish Galactosyl-binding lectins from the tunicate Didemnum candidum. Purification and physicochemical characterization. *J Biol Chem.* 1986c Jul 15;261(20):9174-81.

Vasta, G.R., M.S. Quesenberry, and H. Ahmed, A tunicate fucose-binding lectin is a homologue of the mammalian binding proteins, In *New Directions in Invertebrate Immunology*, eds. Soderhall, K., Iwanaga, S., Vasta, G.R. SOS Publications, Fair Haven, N.J., 1996, pp. 189-227.

Vasta, G.R., H. Ahmed, L.M. Amzel, and M.A. Bianchet Galectins from amphibian species: carbohydrate specificity, molecular structure, and evolution. *Trends In Glycoscisciences Glycotechnology*, 1997, 9, 131-144.

Vasta, G.R., M.S. Quesenberry, H. Ahmed, and N. O'Leary, C-type lectins and galectins mediate innate and adaptive immune functions: their roles in the complement activation pathway, Dev. Comp. Immunol., 1996, 23, 401-420.

Weis, W.I., M.E. Taylor, and K. Drickamer The C-type lectin superfamily in the immune system. *Immunol. Rev.*, 1998, 163, 19-34.

CONTROL MECHANISMS OF THE PROPHENOLOXIDASE CASCADE

Manickam Sugumaran

Department of Biology, University of Massachusetts
Boston, MA 02125

INTRODUCTION

Insects do not possess immunoglobulins and other complicated immune proteins found in higher animals. However, they are equipped with a highly effective defense arsenal to protect themselves from invading parasites and other intruders. The hard armor like exoskeleton of insects affords the first line of defense by providing a physical barrier against the foreign organisms (Ashida & Brey, 1995; Sugumaran 1996,1998a). Once the cuticular barrier is surcumvented, the organisms entering the soft body of insects face a plethora of cellular and humoral host defense reactions. The response to the presence of non-self matter includes, phagocytocysis, nodule formation, encapsulation, melanization, and synthesis and secretion of antibacterial proteins (Hoffman *et al.*, 1999; Gillespie *et al.*, 1997; Soderhall *et al.*, 1990; Sugumaran 1996). Organisms too large to be phagocytosed are usually found encapsulated and often melanized in the hemolymph of insects. Therefore, phenoloxidase (PO), the enzyme responsible for the biosynthesis of melanin is considered as an important component of insects' immune system (Ashida & Yamazaki, 1990; Gillespie *et al.*, 1997; Sugumaran & Kanost 1993; Nappi & Sugumaran 1993; Soderhall et al., 1990). Interestingly, PO is not only involved

Phylogenetic Perspectives on the Vertebrate Immune System.
Edited by Beck *et al.*, Kluwer Academic/Plenum Publishers, 2001.

in defense reactions, but also in a few other physiologically important and relevant processes. Cuticular sclerotization which is essential to the survival of all insects is initiated by PO (Sugumaran 1998a). The quinonoid products formed, are acted upon by other cuticular enzymes such as quinone isomerase and quinone methide isomerase. These enzymes in association with PO, generate a number of reactive intermediates that crosslinking structural proteins and chitin. These reactions render the cuticular proteins insoluble and eventually cause the hardening of cuticle. In the last process, viz., wound healing, massive amounts of hemolymph loss is partly prevented by the action of PO which rapidly deposits melanin pigment at the wounding site (Ashida & Brey 1995; Lai-Fook, 1966; Sugumaran 1996). In addition, PO generated quinones being cytotoxic, may play a critical role in killing opportunistically invading microorganisms at the wound site (Nappi & Sugumaran 1993; Sugumaran 1996).

In order to serve in such important and diverse processes, the activity of PO must be precisely controlled. Therefore insects and other arthropods have developed a number of control mechanisms (Sugumaran & Kanost 1993). These include a) preservation of the enzyme in its inactive proform (prophenoloxidase, PPO) b) protease mediated activation of PPO, c) preservation of the activating enzyme in its inactive proform, d) Use of protease inhibitor to control activating proteases, e) compartmentalization of enzymes and substrates, f) protection of substrates against PO action, g) lipid mediated activation of PPO, h) self inactivation of PO, i) aggregation and stickiness of PO, j) complex formation with other proteins, and k) use of PO inhibitors. In this article, evidence for each of these control mechanisms are presented.

PRESERVATION OF THE ENZYME IN ITS INACTIVE PROFORM

Phenoloxidase forms the terminal component of a cascade of reactions associated with the immune response in arthropods (Ashida & Yamazaki 1990; Soderhall *et al.*, 1990; Sugumaran & Kanost 1993; Sugumaran 1996). It is a key enzyme causing a) melanization and encapsulation of parasites and other foreign objects in the hemolymph, b) wound healing and c) sclerotization of cuticle. To participate in these vital processes, PO is present through out the life stage of the insects in the hemolymph. Since the quinonoid products of PO are reactive and harmful, insects and other arthropods keep this enzyme in the inactive proform and activate it during a need (Ashida & Yamazaki 1990). This prevents unwanted melanization reactions in the hemolymph that might be harmful to the arthropods.

POs are synthesized in the hemocytes and stored as inactive proenzymes. The proenzyme from *Bombyx mori* (Ashida, 1971; Yasuhara *et al.*, 1995), *Manduca sexta* (Aso *et al.*, 1985; Hall *et al.*, 1995; Jiang 1997a), *Hyalophora cecropia* (Andersson *et al.*, 1989), *Galleria mellonella* (Kopacek *et al.*, 1995), *Holotrichia diomphalia* (Kwon, *et al.*, 1997), *Calliphora erythrocephala* (Pau & Eagles 1975), *Musca domestica* (Hara *et al.*, 1993), *Drosophila melanogaster* (Fujimoto *et al.*, 1993), *Sarcophaga bullata* (Chase *et al.*, 2000), *Blaberus discoidalis* (Durrant *et al.*, 1993), *Tenebrio molitor* (Heyneman, 1965), and *Locusta migratoria* (Cherqui *et al.*, 1996) have been purified and several of their properties have been characterized. As many as twenty cDNAs encoding for insect PPO has been sequenced. These include two from *Manduca sexta*

(Hall *et al.*, 1995; Jiang *et al.*, 1997a), two from *Bombyx mori* (Kawabata *et al.*, 1995), two from *Hyphantria cunea* (Park *et al.*, 1997) three from *Drosophila melanogaster* (Fujimoto *et al.*, 1995; Chase & Sugumaran 2000), six from *Anopheles gambiae* (Jiang *et al.*, 1997b; Lee *et al.*, 1998a; Muller *et al.*, 1999), three from *Sarcophaga bullata* (Chase *et al.*, 2000), one from *Tenebrio molitor* and one from *Armigeres subalbatus* (Cho *et al.*, 1998). All these studies confirm that PO is synthesized and stored in insects as inactive proenzyme form.

PROTEASE MEDIATED ACTIVATION OF PPO

Prophenoloxidase needs to be activated to exhibit its biological activity. This is achieved by two mechanisms. The first and well established mechanism calls for the action of an activating protease; the second mechanism utilizes lipids. A specific protease designated as the prophenoloxidase activating enzyme (PAE) activates PPO. PAE has been characterized from *Bombyx mori* (Ashida & Dohke 1980; Ashida & Yoshida 1988; Satoh *et al.*, 1999), *Manduca sexta* (Jiang *et al.*, 1998), *Holotrichia diomphalia* (Lee *et al.*, 1998b) and *Drosophila melanogaster* (Chosa *et al.*, 1997). Its presence has been demonstrated in other insects also (Saul & Sugumaran 1987, 1988). It is a serine protease that cleaves PPO at the conserved RF site present on all PPOs. As a result, an inactive propeptide of about 5 kDa is severed from PPO to generate the active PO. Both *Manduca* and *Holotrichia* PAEs require another protein component for activating PPO (Lee *et al.*, 1998b; Jiang *et al.*, 1998). *Bombyx* PAE seems to be present as two isozymes and is able to activate PPO on its own (Satoh *et al.*, 1999). The gene coding for PAE has been cloned and sequenced from all these three organisms (Jiang *et al.*, 1998; Lee *et al.*, 1998c; Satoh *et al.*, 1999). It codes for a serine protease which appears to be glycoprotein with a molecular weight of about 40 kDa. After transport across the membrane and processing by a protease, the molecular weight of PAE is reduced to about 28 kDa. The fact that septic injury causes elevation of its mRNA levels confirm that it is associated with the immune reaction in insects. Further studies are necessary to understand the mode of activation of PAE. It is possible that more than one PAE may be responsible for activating PPO, because a second potential cleavage site has been identified on PPO (Hall *et al.*, 1995). This site which usually consists of REE sequence is also conserved in most, if not all PPOs. But the enzyme cleaving at this site is yet to be characterized.

PRESERVATION OF THE ACTIVATING ENZYME

As stated in the earlier section, PAE is present in an inactive proenzyme form. Hence it needs to be activated. A proteolytic activation site (conserved RI or KI) has been identified on PAEs. Apparently another serine proteases is carrying out this reaction. Moreover, several *in vitro* activation studies also confirm the participation of other serine proteases in the PPO activation cascade. Ashida's group (Ashida & Yoshida 1988; Yoshida & Ashida 1986) has shown that a benzoyl arginine ethyl esterase activity is triggered prior to the appearance of PAE and PPO activity in the silkworm hemolymph. We have shown the appearance of this enzyme activity in *Sarcophaga* as well as *Manduca* hemolymph prior to the generation of PAE activity during protease mediated activation of PPO (Saul & Sugumaran 1987, 1988). But the detail nature of this protease and its exact role in PPO activation has not yet been established.

USE OF PROTEASE INHIBITOR TO CONTROL ACTIVATING PROTEASES

From the above discussion, it is evident that PPO activation is achieved by serine proteases. The activity of protease are usually controlled by proteinase inhibitors in most systems. Insects also possess large amounts of proteinase inhibitors in their hemolymph. Two types of protease inhibitors that control PPO activation been characterized so far. We first reported the control of PPO activation in the hemolymph of *Sarcophaga* and *Manduca* by a group of low molecular weight, endogenous serine protease inhibitors (Sugumaran *et al.*, 1995; Saul & Sugumaran 1996; Ramesh *et al.*, 1988). We have determined the primary amino acid sequence of these inhibitors. They belong to bovine pancreatic trypsin type inhibitors family and inhibit the activation of PPO in the hemolymph as well as partially purified preparations. A similar inhibitor has been reported from *Locusta migratoria* also (Boigegrain *et al.*, 1992; Brehelin *et al.*, 1991). It is not clear whether they are inhibiting the PAE directly or through other serine proteases. The second group of inhibitors belong to serpin family. As many as 12 different serpins have been isolated, characterized and cloned from *Manduca* (Jiang *et al.*, 1996; Jiang & Kanost 1997). Serpin 1J inhibits the protease mediated activation PPO (Jiang & Kanost 1997). Similarly other serine protease inhibitors isolated from *Bombyx* inhibits activating proteases in the PPO cascade (Aso *et al.*, 1994; Ashida & Sasaki 1994). More research is needed on these inhibitors also.

COMPARTMENTALIZATION OF ENZYMES AND SUBSTRATES

It is now well established that PPO is synthesized in the hemocytes. However, bulk of the PPO in the insect seems to be present in the plasma rather than cells (Saul *et al.*, 1987; Sugumaran & Kanost 1993). Examination of the primary structure of the PPO clearly reveals that the proenzyme lacks the signal peptide needed for its transport across the membrane. The cells synthesizing PPO seem to burst open and release the proenzyme into the plasma (Ashida & Yamazaki 1990; Sugumaran & Kanost 1993). In crustacean, PPO is nearly always confined to the granular cells and none or very little of this enzyme is found in the plasma (Soderhall *et al.*, 1990). Even in these organisms, the cells burst open and release the PPO during a need.

PROTECTION OF SUBSTRATES AGAINST PHENOLOXIDASE ACTION

Tyrosine and dopa, the precursors of melanin and sclerotin (and substrates for PO) are only sparingly soluble in aqueous medium. However, their sulfate, phosphate and glucose derivatives are highly soluble. Insects need large amounts of dopa derivatives for sclerotization and melanization reactions. Obviously more soluble sulfate, phosphate and sugar derivatives are better suited for storage than poorly soluble dopa and tyrosine. Storing in such derivatized form also has an added advantage. Unlike dopa and dopamine which are susceptible to rapid oxidation, the protected compounds are immune to PO attack as well as aerial oxidation. Therefore, not only PO, but also its substrates are protected (Kramer & Hopkins 1987). The enzymes causing the deprotection of these groups can be stored in separate cellular compartments to avoid prematured removal of the protecting groups. This separation allows the operation of addition control measures in the PPO cascade.

LIPID MEDIATED ACTIVATION OF PPO

Apart from proteases, a number of other reagents also activate PPO. Heat, acid, detergents, high concentration of fatty acids and organic solvents have been known to activate PPO (Sugumaran & Kanost 1993). But, their physiological significance is not clear. However, the fact that fatty acids could activate PPO prompted us to test other forms of lipids as mediators of activation. Our studies revealed that trace amounts of lipids especially phospholipids could activate PPO dramatically (Sugumaran & Nellaiappan 1990). Phospholipids turned out to be the most potent activators of PPO. Lysophospholipid seems to be a potent activator of lobster PPO (Sugumaran & Nellaiappan 1991). While some of the insect enzymes tested are activatable by trace amounts of phospholipids, the natural activator needs to be found out.

SELF INACTIVATION, AGGREGATION AND STICKINESS OF PO

PPO is reasonably stable; however, active PO becomes inactive quite rapidly like the enzymes associated with mammalian complement activation. Thus for example, Jiang and Kanost (1997) has witnessed the rapid activation of *Manduca* PPO by the activating protease (peak of activation at 40 min) and subsequent gradual decrease of activated PO activity. As much as 90% of PO activity was lost in about 4 hr. We have witnessed with both *Manduca* PPO and *Sarcophaga* PPO, detergent activation leads to rapid loss of PO activity (unpublished results). It is a desirable property for the protection of self matter because continuous production of quinonoid products in the hemolymph would lead to undesired side reactions such as toxicity, protein polymerization and destruction of self matter. Inactivation of PO limits the extent to which these reactions can occur in the hemolymph. The mechanism by which this inactivation occurs is not known and need to be investigated.

Phenoloxidase has the tendency to form molecular aggregates. Thus formation of dimers, and other polymeric structures of active PO has been documented on gels (Ashida & Dohke 1980; Jiang *et al.* 1997a). Activated PO has the tendency to bind to abiotic substances such as glass, plasticware, and gels used for the purification of enzymes, as well as biotic substances. This property is well suited for its role in defense reaction as an encapsulating and melanizing agent (but has caused a lot of grief and frustration to scientists working in this area). Interestingly the enzyme appears to be stable once attached to the foreign surface. The mechanism of attachment is not certain at this time. In the case of mammalian tyrosinases a cysteine rich EGF like motif seems to help the binding of melanogenic enzymes with one another. However, this motif is absent in invertebrate enzymes. Typical RGD sequence that may be associated with adhesion is also completely absent in all PPOs with the exception of one mosquito PPO (Muller *et al.*, 1999).

The short stretch of sequence, GCGWPQH first identified in *Manduca* PPO (Hall et al., 1995) resembled the thiolester motif present in α_2-macroglobulin and complement proteins C'3 and C'4 (GCGEQNM). In these proteins, the thiol group of cysteine displaces the amide of glutamine located three residues away. During proteolytic activation, the exposed thiol group reacts rapidly with nearby molecules ensuring attachment. Although initially we were excited to find a match in *Manduca* PPO (not an exact match), both biochemical analysis (failure to react with methylamine) and comparison with other known PPO sequences (lack of conservation of glutamine in several sequences) has discounted any such role for this conserved motif. Perhaps this motif may be responsible for the structural integrity of PO. Therefore, the mechanism of PO stickiness remains unknown and needs to be investigated.

COMPLEX FORMATION WITH OTHER PROTEINS

Enzymes associated with sequential metabolic reactions are known to form complexes called metabolons (Srere 1987). If a compound is solely formed to serve as the substrate for the next enzyme, complex formation between the two enzymes assures efficient metabolic transfer. If the product formed is a reactive intermediate (as in the case of PO reaction), complex formation guarantees safe handling of that metabolite. Metabolon formation also raises the local concentration of the substrates at each others active site and avoids dilution and/or loss of metabolite in the microenvironment of the cell. To efficiently perform its duty as an encapsulating enzyme, PO has to work with other related proteins in the cell. Therefore metabolon formation is highly advantageous to PO and related enzymes. Accordingly, we have characterized three different complexes of PO (Beck *et al.*, 1996; Sugumaran 1998b). The first complex designated as defense complex is formed between PO, PPO, and interleukin 1 type molecule (Beck *et al.*, 1996). This complex is formed in the hemolymph of *Manduca sexta* upon wounding and exposure to nonsterile conditions. Occasionally, dopachrome isomerase (the subsequent enzyme in the melanogenic pathway) is also found in this nonstoichiometric complex. The rest two complexes are melanogenic and sclerotinogenic complexes (Sugumaran *et al.,* 1995; Sugumaran 1998b). Melanogenic complex seems to contain PO and dopachrome isomerase while the sclerotinogenic complex contains PO, quinone isomerase and quinone methide isomerase (Sugumaran 1998b; Sugumaran *et al.*, 2000).

Complex formation also seems to attribute better control of PO activity. As stated earlier, bound PO seems to be more stable and does not loose its activity as the uncomplexed PO. The activity of PO is also dramatically inhibited by dopachrome isomerase (as well as quinone isomerase). Both these enzymes act on PO generated quinones. By complexation, they seem to control the activity of PO so as to establish a steady and controlled synthesis of melanin and/or sclerotin. Some of the questions need to be addressed with regard to complex formation are: How are the complexes formed? How are they stabilized? What is the nature of these complexes?

USE OF PHENOLOXIDASE INHIBITORS

Direct inhibition of PO is undoubtedly the best way to control the activity of PO. The first inhibitor of PO was identified by Tsukamoto's group (Tsukamoto *et al.*, 1992; Daquinag et al., 1995; 1999). They have isolated and characterized three low molecular weight proteinacious inhibitors (A: 2,850; B: 3,100 and C: 3,350) from the pupal extracts of house flies. These inhibitors which are both heat and pH stable, showed competitive inhibition of endogenous phenoloxidase activity. The primary amino acid sequence of one of the inhibitor is determined to be: AVTDNEIVPQCLADGSKSYSHDVCCTKRSHD-dopa-AKKCVT (Daquinag *et al.*, 1995). It has dopa as an unusual amino acid at position 32 and has a cystine knot motif commonly found in snail and spider toxins (Daquinag *et al.*, 1999). Given the facts that it is an inhibitor of phenoloxidase and that phenoloxidases are known to hydroxylate tyrosine to dopa, it is tempting to suggest that the tyrosine containing precursor is probably hydroxylated to form dopa peptide.

We have isolated and characterized a high molecular weight inhibitor from the larvae of *Manduca sexta* (Sugumaran and Nellaiappan, 2000). Our phenoloxidase inhibitor, which is a heat and pH labile protein, is a glycoprotein that is distinctly different from *Musca* inhibitor. It forms a tight complex with PO which resists dissociation by SDS as well as triton X-100. In addition, enzymes such as quinone isomerase and dopachrome isomerase which act on PO generated quinones seems to bind to PO (see section above) and inhibit its activity drastically. We have also characterized a nonprotein inhibitor of PO, which is oxalate (Sugumaran & Nellaiappan, 1990). The deposition of oxalate inside the ootheca of cockroaches has been known for some time, but its significance remained unidentified till we first demonstrated the potent inhibition of colleterial gland phenoloxidase activity by oxalate. Oxalate seems to inhibit PO responsible for the sclerotization of cockroach ootheca, thereby protecting the eggs. In spite of their obvious function as PO inhibitors, it is not clear at present whether these inhibitors are associated with immune system or not. Further studies are definitely needed to throw more light on this aspect.

CONCLUSION

Apart from being an important enzyme participating in a number of physiologically important processes, PO also generates reactive quinonoid compounds as products. This calls for the control of not only PO activity, but also quinonoid reactivity. Uncontrolled PO action is unquestionably deleterious to the self matter. Therefore, insects and other arthropods have developed a variety of control mechanisms to regulate the activity of PO. In this article, I have summarized some of the regulatory aspects of PPO cascade. Due to lack of space, a lot of the excellent work carried out in other arthropods could not be included in this review. As we learn more and more, I am sure additional regulatory mechanisms may be discovered. Thus a lot work on the PPO cascade still needs to be carried out. Only recently rapid strides are being made in the molecular biological aspects of PPO and related enzymes. The regulatory aspects of the cascade at the molecular biological level as well as hormonal level are yet to be studied. With the rapid advances in these areas, a clear picture of the control mechanisms of the cascade will soon emerge.

ACKNOWLEDGMENTS

Financial assistance was provided by grants from U. Mass. and N. I. H. (AI-14753).

REFERENCES

Andersson K, Sun SC, Boman HG, Steiner H. Purification of the prophenoloxidase from *Hyalophora cecropia* and four proteins involved in its activation. Insect Biochem. 1989; **19**: 629-637.

Ashida M. Purification and characterization of prephenoloxidase from hemolymph of the silkworm, *Bombyx mori.* Arch. Biochem. Biophys. 1971; **144:** 749-762.

Ashida M, Brey P. Role of the integument in insect defense: Prophenoloxidase cascade in the cuticular matrix. Proc. Natl. Acad. Sci. USA. 1995; **92:** 10698-10702.

Ashida M, Dohke K. Activation of pro-phenoloxidase by the activating enzyme of the silkworm, *Bombyx mori.* Insect. Biochem. 1980; **10:** 37-47.

Ashida M, Sasaki T. A target protease of serpins in insect hemolymph. Insect. Biochem. Mol. Biol. 1994; **24:** 1037-1041.

Ashida M, Yamazaki HI. "Biochemistry of the phenoloxidase system in insects: With special reference to its activation". In *Molting and metamorphosis*, Ohnishi. E, Ishizaki. H. eds. Tokyo. Jpn. Sci. Soc. Press. 1990. pp. 239-265.

Ashida M, Yoshida H. Limited proteolysis of prophenoloxidase during activation by microbial products in insect plasma and effect of phenoloxidase on electrophoretic mobilities of plasma proteins. Insect Biochem. 1988; **18**: 11-19.

Aso Y, Kramer KJ, Hopkins TL, Lookhart GL. Characterization of hemolymph protyrosinase and a cuticular activator from *Manduca sexta* (L). Insect Biochem. 1985; **15**: 9-17.

Aso Y, Yamashita T, Meno K, Murakami M. Inhibition of prophenoloxidase-activating enzyme from *Bombyx mori* by endogenous chymotyrpsin inhibitors. Biochem. Mol. Biol. Int. 1994; **33**: 751-758.

Beck G, Cardinale S, Wang L, Reiner M, Sugumaran M. Characterization of a defense complex consisting of interleukin 1 and phenoloxidase from the hemolymph of the tobacco hornworm, *Manduca sexta.* J. Biol. Chem. 1996;. **271:** 11035-11038.

Boigegrain RA, Mattras H, Brehelin M, Paroutaud P, Coletti-Previero MA. Insect immunity: Two proteinase inhibitors from hemolymph of *Locusta migratoria* Biochem. Biophys. Res. Commun. 1992: **189**, 790-793.

Brehelin M, Boigegrain RA, Drif L, Coletti-Previero MA. Purification of a protease inhibitor which controls prophenoloxidase activation in hemolymph of *Locusta migratoria* (insecta). Biochem. Biophys. Res. Commun. 1991: **179**, 841-846.

Chase MR, Sugumaran M. Genomic and cDNA sequences of prophenoloxidases from *Drosophila melanogaster*. This volume, page 349-362. (2000)

Chase MR, Raina K, Bruno J, Sugumaran M. Purification, characterization and molecular cloning of prophenoloxidases from *Sarcophaga bullata*. Insect Biochem. Mol. Biol. (*in press*, 2000)

Cherqui A, Duvic B, Brehelin M. Purification and characterization of prophenoloxidase from the hemolymph of *Locusta migratoria*. Arch. Insect Biochem. Physiol. 1996; **32**: 225-235.

Cho WL, Liu HS, Lee CH, Kuo CC, Chang TY, Liu CT, Chen CC. Molecular cloning, characterization and tissue expression of prophenoloxidase cDNA from the mosquito *Armigeres subalbatus* inoculated with *Dirofilaria immitis* microfilariae. Insect Mol. Biol. 1998; **7**: 31-40.

Chosa N, Fukumitsu T, Fujimoto K, Ohnishi E. Activation of pro-phenoloxidase A_1 by an activating enzyme in *Drosophila melanogaster*. Insect Biochem. Mol. Biol. 1997; **27**: 61-68.

Daquinag AC, Nakamura S, Takao T, Shimonishi Y, Tsukamoto T. Primary structure of a potent inhibitor of phenol oxidase from *Musca Domestica*. Proc. Natl. Acad. Sci. USA. 1995: **92,** 2964-2968.

Daquinag AC, Sato T. Kodo H, Takao T, Fukuda M, Shimonishi Y, Tsukamoto T. A novel endogenous inhibitor of phenol oxidase from *Musca Domestica* has a cystine motif commonly found in snail and spider toxins. Biochemistry. 1999: **38,** 2179-2188.

Durrant HJ, Ratcliffe NA, Hipkin CR, Aspan A, Soderhall K. Purification of the prophenoloxidase enzyme from hemocytes of the cockroach *Blaberus discoidalis*. Biochem. J. 1993; **289**: 87-91.

Fujimoto K, Masuda K, Asada N, Ohnishi E. Purification and characterization of prophenoloxidase from the pupae of *Drosophila melanogaster*. J. Biochem (Tokyo). 1993; **113**: 285-291.

Fujimoto K, Okino N, Kawabata SI, Iwanaga S, Ohnishi E. Nucleotide sequence of the cDNA encoding the proenzyme of phenoloxidase A_1 of *Drosophila melanogaster*. Proc. Natl. Acad. Sci USA. 1995; **92**: 7769-7773.

Gillespie JP, Kanost MR, Trenczek T. Biological mediators of insect immunity. Ann. Rev. Entomol. 1997; **42:** 611-643.

Hall M, Scott M, Sugumaran M, Soderhall K, Law JH. Proenzyme of *Manduca sexta* phenoloxidase: Purification, activation, substrate specificity of the active enzyme and molecular cloning. Proc. Natl. Acad. Sci USA. 1995; **92**: 7764-7768.

Hara T, Miyoshi T, Funatsu M. Comparative studies on larval and pupal phenoloxidases of the housefly, *Musca domestica*. Comp. Biochem. Physiol. 1993; **106B**: 287-292.

Heyneman RA. Final purification of a latent phenoloxidase with mono-and diphenoloxidase from *Tenebrio molitor*. Biochem. Biophys. Res. Commun. 1965; **21**: 162-169.

Hoffmann JA, Kafatos FC, Janeway CA, Ezekowitz RAB. Phylogenetic perspectives in innate immunity. Science 1999; **284**: 1313-1318.

Jiang H, Kanost MR. Characterization and functional analysis of twelve naturally occurring reactive site variants of serpins-1 from *Manduca sexta*. J. Biol. Chem. 1997; **272**: 1082-1087.

Jiang H, Wang Y, Huang Y, Mulnix AB, Kadel J, Cole K, Kanost MR. Organization of serpin gene 1 from *Manduca sexta*. Evolution of a family of alternate exons encoding the reactive site loop. J. Biol. Chem. 1996; **271**: 28017-28023.

Jiang H, Wang Y, Ma C, Kanost MR. Subunit composition of prophenoloxidase from *Manduca sexta*: Molecular cloning of subunit ProPo-P1. Insect Biochem. Mol. Biol. 1997a; **27**, 835-850.

Jiang H, Wang Y, Korochkina SE, Benes H, Kanost MR. Molecular cloning of cDNAs for two prophenoloxidases subunits from the Malaria vector, *Anopheles Gambiae*. Insect Biochem. Mol. Biol. 1997b; **27**: 693-699.

Jiang H, Wang Y, Kanost MR. Prophenoloxidase activating proteinase from an insect, *Manduca sexta*. A bacterial inducible protein similar to *Drosophila* ester. Proc. Natl. Acad. Sci USA. 1998; **95**: 12220-12225.

Kawabata T, Yasuhara Y, Ochiai M, Matsuura S, Ashida M. Molecular cloning of insect prophenoloxidase: A copper containing protein homologous to arthropod hemocyanin. Proc. Natl. Acad. Sci. USA. 1995; **92**: 7774-7778.

Kopácek P, Weise C, Gotz P. The prophenoloxidase from the wax moth *Galleria mellonella*: Purification and characterization of proenzyme. Insect Biochem. Mol. Biol. 1995; **25**: 1081-1091.

Kramer KJ, Hopkins TL. Tyrosine metabolism for insect cuticle tanning. Arch. Insect Biochem. Physiol. 1987; **6**: 279-301.

Kwon TH, Lee SY, Lee JH, Choi JS, Kawabata SI, Iwanaga S, Lee BL. Purification and characterization of prophenoloxidase from the hemolymph of coleopteran insect, *Holotrichia diomphlia*. Molecules & Cells 1997; **7**: 90-97.

Lai-Fook J. The repair of wounds in the integument of insects. J. Insect Physiol. 1966; **12:**195-226.

Lee WJ, Ahmed A, Torre AD, Kobayashi A, Ashida M, Brey PT. Molecular cloning and chromosomal localization of a prophenoloxidase cDNA from the malaria vector *Anapheles gambiae*. Insect Mol. Biol. 1998a; **7**: 41-50.

Lee SY, Kwon TH, Hyun JH, Choi JS, Kawabata SI, Iwanaga S, Lee BL. *In vitro* activation of prophenoloxidase by two kinds of prophenoloxidase activating factors isolated from hemolymph of coleopteran, *Holotrichia diomphlia* larvae. Eur. J. Biochem. 1998b; **254**: 50-57.

Lee SY, Cho MY, Hyun JH, Lee KM, Homma K, Natori S, Kawabata SI, Iwanaga S, Lee BL. Molecular cloning of cDNA for prophenoloxidase activating factor I, a serine protease is induced by lipopolysaccharide or b-1,3-glucan in coleopteran insect *Holotrichia diomphlia*. larvae. Eur. J. Biochem. 1998c; **257**: 615-621.

Muller HM, Dimopoulos G, Blass C, Kafatos FC. A hemocyte-like cell line established from malaria vector *Anopheles gambiae* expresses six prophenoloxidase genes. J. Biol. Chem. 1999; **274:** 11727-11735.

Nappi AJ, Sugumaran M. "Some biochemical aspects of eumelanin formation in insect immunity". In: *Insect Immunity.* Pathak J. P. N. ed. New Delhi, India. Oxford & IBH Publishing Co. 1993. pp. 131-148.

Park DS, Shin W, Kim MG, Park SS, Lee WJ, Brey PT, Park HY. Isolation and characterization of the cDNA encoding the prophenoloxidase of Fall webworm, *Hyphantria cunea*. Insect Biochem. Mol. Biol. 1997; **27**: 983-992.

Pau RN, Eagles PAM. The isolation of *o*-diphenoloxidase from the third instar larvae of blowfly, *Calliphora erythocephala*. Biochem. J. 1975; **149**: 707-712.

Ramesh N, Sugumaran M, Mole JE. Purification and characterization of two trypsin inhibitors from the hemolymph of *Manduca sexta* larvae. J. Biol. Chem. 1988; **263:** 11523-11527.

Satoh D, Horii A, Ochiai M, Ashida M. Prophenoloxidase activating enzyme of silkworm *Bombyx mori* - Purification, characterization and cDNA cloning. J. Biol. Chem. 1999: **274**: 7441-7453.

Saul SJ, Sugumaran M. Protease inhibitor controls prophenoloxidase activation in *Manduca sexta*. F. E. B. S. Lett. 1986; **208:** 113-116.

Saul SJ, Sugumaran M. Protease mediated prophenoloxidase activation in the hemolymph of the tobacco hornworm, *Manduca sexta*. Arch. Insect Biochem. Physiol. 1987; **5:** 1-11.

Saul SJ, Sugumaran M. Prophenoloxidase activation in the hemolymph of *Sarcophaga bullata* larvae. Arch. Insect Biochem. Physiol. 1988; **7:** 91-103.

Saul SJ, Bin L, Sugumaran M. The majority of prophenoloxidase in the hemolymph of *Manduca sexta* is present in the plasma and not in the hemocytes. Dev. Comp. Immunol. 1987; **11:** 479-485.

Söderhäll K, Aspán A, Duvic B. The ProPO system and associated proteins. Role in cellular communication in arthropods. Res. Immunol. 1990; **141:** 896-907.

Srere PA. Complexes of sequential metabolic enzymes. Ann. Rev. Biochem. (1987). **56**: 89-124.

Sugumaran M. "Role of insect cuticle in immunity". In *New Directions in Invertebrate Immunology*. Söderhäll, K., Iwanaga, S. and Vastha, G. eds. Fair Haven, NJ. SOS Publications. 1996. pp. 355-374.

Sugumaran M. Unified mechanism for sclerotization of insect cuticle. Adv. Insect Physiol. 1998a; **27**: 229-334.

Sugumaran M. "characterization of phenoloxidase complexes". In *Techniques in Insect Immunology*. Wiesner A, Dunphy GB, Marmaras VJ, Morishima I, Sugumaran M, Yamakawa M. eds. Fair Haven, NJ. SOS Publications. 1998. pp. 205-215.

Sugumaran M, Kanost M. "Regulation of insect hemolymph phenoloxidases". In *Parasites and pathogens* Beckage, N. E., Thompson, S. N., & Frederick, B. A. eds. San Diego, Academic Press. 1993; Vol. I. Parasites. pp. 317-342.

Sugumaran M, Nellaiappan K. On the latency and nature of phenoloxidase present in the left colleterial gland of the cockroach, *Periplaneta americana*. Arch. Insect Biochem. Physiol. 1990; **15:** 165-181.

Sugumaran M, Nellaiappan K. Characterization of a new phenoloxidase inhibitor from the cuticle of *Manduca sexta*. Biochem. Biophys. Res. Commun. 2000; **268:** 379-383.

Sugumaran M, Nellaiappan K. Lysolecithin - A potent activator of prophenoloxidase from the hemolymph of the lobster, *Homarus americanas*. Biochem. Biophys. Res. Commun. 1991; **176:** 1371-1376.

Sugumaran M, Saul SJ, Ramesh N. Endogenous protease inhibitors prevent undesired activation of prophenolase in insect hemolymph. Biochem. Biophys. Res. Commun. 1985; **132:** 1124-1129.

Sugumaran M, Nellaiappan K, Scott T, Amarathunga C. Complex formation between mushroom tyrosinase and *Manduca* dopachrome isomerase. Pigment Cell. Res. 1995; **8:** 180-186.

Sugumaran M, Nellaiappan K, Amarathunga C, Cardinale S, and Scott T. Insect Melanogenesis. III Metabolon formation in the melanogenic pathway–Regulation of phenoloxidase activity by endogenous dopachrome isomerase (Decarboxylating) from *Manduca sexta*. Arch. Biochem. Biophys.2000; **378:** 393-403.

Tsukamoto, T., Ichimaru, Y., Kanegae, N., Watanabe, K., Yamaura I., Katasura, Y. and Funatsu, M. Identification and isolation of endogenous insect phenoloxidase inhibitors. *Biochem. Biophys. Res. Commun.* 1992; **184,** 86-92.

Yasuhara Y, Koizumi Y, Katagiri C, Ashida M. Reexamination of properties of prophenoloxidase isolated from larval hemolymph of the silkworm, *Bombyx mori*. Arch. Biochem. Biophys. 1995; **320**: 14-23.

Yashida H, Ashida M. Microbial activation of two serine enzymes and prophenoloxidase in the plasma fraction of hemolymph of the silkworm, *Bombyx mori*. Insect Biochem. Mol. Biol. 1986; **16**: 539-545.

ACTIVATED PROPHENOLOXIDASE ENGAGED IN THE CELL CLUMP/CELL ADHESION OF COLEOPTERAN INSECT, *TENEBRIO MOLITOR* LARVAE

Hyun Seong Lee, Mi Young Cho and Bok Luel Lee

Pusan National University, College of Pharmacy
Jangjeong-Dong, Kum-jeong Ku
Pusan, 609-735 Korea

INTRODUCTION

The endogenous defense system is classified into two broad categories, cellular and humoral immune responses. The latter include the synthesis of potent antimicrobial proteins and the activation of the pro-phenoloxidase (pro-PO) cascade (1-5). Previously, we examined the humoral responses of larvae of the coleopteran insects, *Tenebrio molitor* and *Holotrichia diomphalia* (1-3) and reported the primary structures of antimicrobial proteins in their hemolymph and pro-PO activating factor (PPAF-I).

As cellular defense reactions, encapsulation response is well known and we have reported the factors relating to the encapsulation response recently (6-7). Another

Phylogenetic Perspectives on the Vertebrate Immune System.
Edited by Beck *et al.*, Kluwer Academic/Plenum Publishers, 2001.

cellular defensive response, cell clump/ cell adhesion has been reported which occurs in insects in response to a variety of foreign substances including microbes and foreign tissue implants (8). The first stage of recognition is followed by the adherence of hemocytes. However, the biochemical processes of cell clump/cell adhesion in insects are still obscure. We will presents evidence that activated phenoloxidase (PO) from pro-PO is engaged in the cell clump/cell adhesion reaction.

IDENTIFICATION OF 72kDa PROTEIN

For the study of the protein involved in the cell clump/cell adhesion reaction, we induced cell clump/cell adhesion with the hemolymph of *T. molitor in vitro* as shown in Figure 1(A). When the hemolymph was collected in the presence of the decoagulation buffer [Fig. 1 (B)] clumped cells were not detected. However, once the hemolymph was harvested in the presence of the insect saline [Fig.1 (A)], hemocytes clumped and aggregated onto the surface of the petri dish within 20min.

To examine the difference between clumped cells and non-clumped cells, the proteins clumped were regained from the surface of the petri dishes and analyzed on the 12% SDS-PAGE. As shown in Figure 2, one band (72kDa band shown by the arrow) was specifically enriched from the cell-clumped residues.

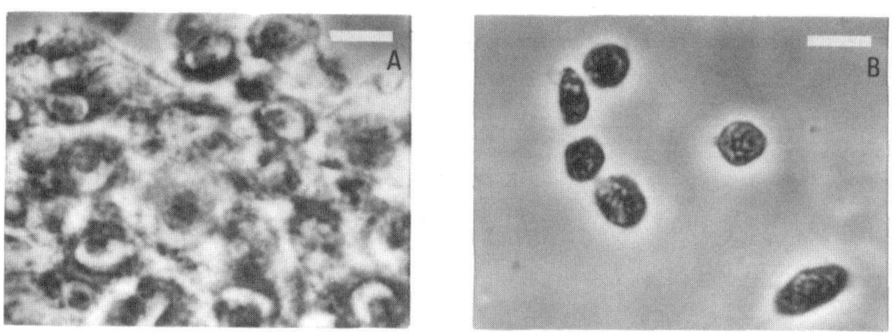

Figure 1. Morphology of *in vitro* induced cell clump/cell adhesion with the hemolymph of *T. molitor* larvae. Hemocytes in the presence of insect saline (A) and decoagulation buffer (B). A; Cell clump/cell adhesion was induced at room temperature for 20 min. B; Cell clump/cell adhesion was not observed in the decoagulation buffer at the same conditions. Bars indicate 10mm

cDNA CLONING OF 72kDa PROTEIN

To determine the whole amino acid sequence of the 72 kDa protein, cDNA library of *T. molitor* larvae was screened with an affinity-purified antibody against 72 kDa protein by an immunoscreening method. The positive clones for 72kDa protein contained an open reading frame of 2052 nucleotides corresponding to 684 amino acid residues. The chemically determined 15 amino acid residues of the amino terminal sequence of the 72 kDa protein were present in this sequence. To our surprise, the deduced amino acid sequence of 72 kDa protein coincided with known pro-POs and it had highly conserved copper binding sites with significant similarity to the known pro-POs, indicating that the 72 kDa protein is a pro-PO lacking 50 amino acid residues which are assumed to be a pro-sequence of *Tenebrio* pro-PO.

According to the study of other groups as well as ours, pro-PO was suggested to be activated by pro-PO-activating serine proteinase with a limited proteolysis at the sequence motif Asn-Arg-Phe-Gly. The amino terminal two residues of the 72 kDa protein were Phe and Gly, whereas the carboxy-terminal two residues of the prosegment were Asn and Arg. These results clearly indicate that the 72 kDa protein is an activated PO from *Tenebrio* pro-PO.

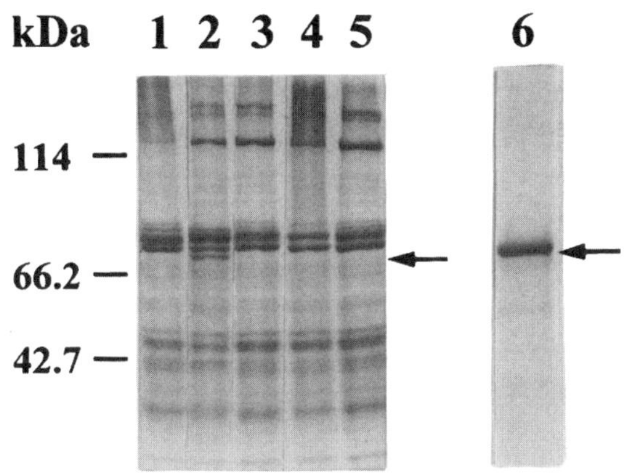

Figure 2. 12% SDS-PAGE analysis of cell clump/cell adhesion related protein. The arrow indicates the 72 kDa protein. Fresh hemolymph (10mg) of *T. molitor* larvae was loaded on the lane 1 after 10% TCA precipitation. Otherwise, hemolymph of *T. molitor* larvae was incubated in the presence of insect saline (lane 2, 3), or in the presence of decoagulation buffer (lane 4, 5). They were incubated for 20 min at room temperature, and their supernatant solution (lane 2, 4) was removed from cell clump residue (lane 3, 5). They also were loaded on 12% SDS-PAGE after 10% TCA precipitation. The purified 72 kDa protein was loaded on lane 6.

DISCUSSION

PO in arthropods catalyzes the oxidation of phenols to quinones, which are then polymerized non-enzymatically to melanin. It has been recognized that defense reactions in many insects are often accompanied by melanization. Activated PO has been suggested to be correlated with non-self recognition and a subsequent stimulation of celular responses, such as cell adhesion, degranulation , and hemocyte locomotion. Moreover, pro-POs from several invertebrate species have been cloned and sequenced as pro-PO. Nevertheless, little is known about the function of PO in the suggested mechanisms. Here, we have provided some evidences that activated PO is localized in cell clump/cell adhesion.

REFERENCES

1. Kwon, T. H., Lee, S. Y., Lee, J. H., Choi, J. S., Kawabata, S., Iwanaga S. Lee, B. L. Purification and characterization of pro-phenoloxidase from the hemolymph of coleopteran insect, *Holotrichia diomphalia* larvae. *Mol. Cells* 7, 90-97 (1997)
2. Lee, S. Y., Kwon, T. H., Hyun, J. H., Choi, J. S., Kawabata, S., Iwanaga, S., Lee, B. L., In vitro activation of pro-phenoloxidase by two kinds of pro-phenoloxidase activating factors isolated from hemolymph of coleopteran, *Holotrichia diomphalia* larvae. *Eur. J. Biochem.* **254**, 50-57 (1998)
3. Lee, S. Y., Cho, M. Y., Hyun. J. H., Lee K. M., Homma, K., Natori, S., Kwabata, S., Iwanaga, S., Lee, B. L. Molecular cloning of cDNA for pro-phenoloxidase activating factor I-, a serine protease that induced by lipopolysaccharide or 1,3-beta-glucan in coleopteran insect, *Holotrichia diomphalia* larvae.*Eur, J. Biochem.* **257**, 615-621 (1998)
4. Aspan, A. and Sderhll, K. Purification of pro-phenol-oxidase from crayfish blood cells and its activation by an endogenous serine protease. *Insect. Biochem.* **21**, 363-373 (1991)
5. Fujimoto, K., Okino, N., Kawabata, S., Iwanaga, S. and Ohnish, E. Nucleotide sequence of the cDNa encoding pro-phenoloxidase A1 of *Drosophila melanogaster. Poc. Natl Acad. Sci.* USA **92**, 7769-7773 (1995)
6. Cho, M. Y., Lee, H. S., Lee, K. M., Homma, K., Natori, S. and Lee, B. L. Molecular cloning and functional properties of two early-staged encapsulation-relating proteins from the coleopteran insect, *Tenebrio molitor* larvae. *Eur. J. Biochem.* 262, 737-744 (1999)
7. Cho, M. Y., Choi H. W., MoonG. Y., Kim, M. H., Kwon, T. H., Homma, K., Natori, S. and Lee, B. L. An 86kDa diapause protein 1-like protein is a component of early-staged encapsulation relating proteins in coleopteran insect, *Tenebrio molitor* larvae. *FEBS Letters* **451**, 303-307 (1999)
8. Tagahashi, H., Azumi, K. and Yokosawa, H. Hemocyte aggregation in the solitary ascidian *Halocynthia reretzi* : plasma factors, magnesium ion, and Met-Lys-bradykinin induce the aggregation. *Biol. Bull.* **186**, 247-253 (1994)

THREE KINDS OF ENCAPSULATION-RELATING FACTORS IN COLEOPTERAN INSECT, *TENEBRIO MOLITOR* LARVAE

Mi Young Cho, Hyun Seong Lee and Bok Luel Lee

Pusan National University, College of Pharmacy
Jangjeong-Dong, Kum-jeong Ku
Pusan, 609-735 Korea

INTRODUCTION

Insect's defensive roles have divided into two classes of mechanisms, humoral and cellular immune system (1, 2). Encapsulation is one of the major cellular immune system to protect themselves from macromolecular invaders (3-6). It has been observed in the relationship between the host insects and the parasites. Once the foreign invaders turn out to be non-self, the hemocytes of the host insect surround the foreign substances, harden and melanize the capsule not to let the parasites invade in their body. On these bases, we have been studying on the factors related to the early-staged encapsulation in the coleopteran insect, *Tenebrio molitor* larvae, to investigate non-self recognition mechanism (7-8).

Phylogenetic Perspectives on the Vertebrate Immune System.
Edited by Beck *et al.*, Kluwer Academic/Plenum Publishers, 2001.

INDUCTION OF ENCAPSULATION RESPONSE

To evoke encapsulation response, we injected DEAE-sepharose beads, CM-sepharose beads, Bio-gel and suture into the larvae. We could define the early-staged encapsulation as 10 min after the injection by the observation under the light microscope and scanning electron microscope. To examine what kind of proteins were bound to foreign surface, we regained the proteins from injected beads after 10min to find that at least four proteins are engaged in the early-staged encapsulation response. We named them early-staged Encapsulation-Relating Factors (ERPs) by DEAE-sepharose and they are shown in Figure 1.

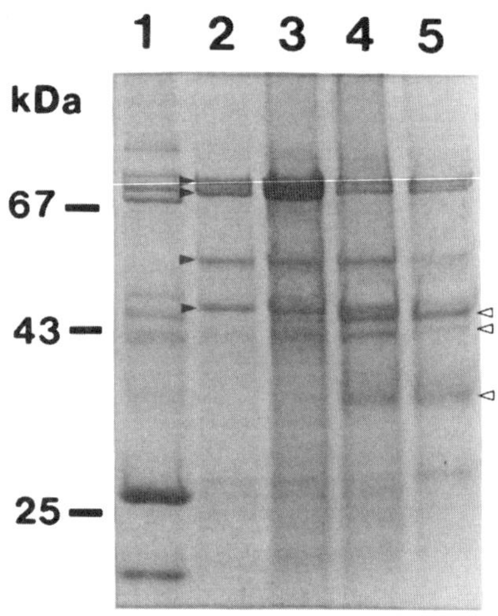

Figure 1. SDS-PAGE (12%) analysis of early-staged ERPs and three kinds of purified ERPs. Lane 1, 10 μg of the crude hemolymph; lane 2, the early-staged ERPs (10 μg) recovered from DEAE-sepharose beads after 10min incubation in the larvae; lane 3, the purified 56 kDa ERP (2 μg); lane 4, the purified 48 kDa ERP (2 μg); lane 5, the purified 86 kDa ERP (2 μg).

CHARACTERIZATION OF 56 AND 48kDa ERPS

We purified the ERPs to homogeneity by the method of electro-elution respectively. The antibodies against three of them were raised and analyzed through Western Blots. To our surprise, two of them, 56 kDa and 48 kDa antibodies cross-reacted against each other ERP. We carried out cDNA cloning of 56 kDa and 48 kDa ERP. 56 kDa ERP is novel glutamine-rich protein which consists of 579 amino acid residues and includes N-terminal amino acid sequence and the three partial amino acid sequences of the 48 kDa ERP. This indicates that the 48 kDa ERP is produced by the cleavage of Arg101-Gly102 of the 5 6kDa ERP by a limited proteolysis.

Western blotting analysis showed some basic information that the 56 kDa and 48 kDa ERPs are located in the hemocytic membrane fraction. In addition that, we ascertained that these two factors are crucial for forming capsule on the surface of non-self material by inhibition experiment using antibodies against these factors. When the early-staged encapsulated beads were coated with nonimmunized rabbit IgG or 56 kDa ERP antibody or 48 kDa ERP antibody and then re-injected into the body of the larvae, the early-staged capsule material coated beads were coated with many hemocytes forming thick capsule. However, as they were incubated with both the 56 kDa and 48 kDa ERP antibodies and re-injected into larvae, no further encapsulation reaction was observed (Fig. 2).

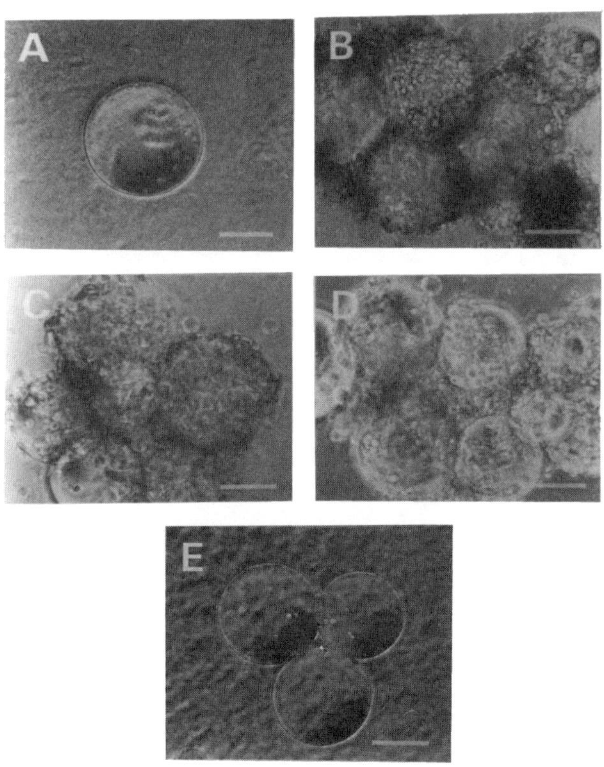

Figure 2. Effects of purified ERP antibodies on encapsulation response. (A) the recovered early-stage encapsulated beads; (B) beads in (A) incubated with nonimmunized rabbit IgG; (C) beads in (A) incubated with 56 kDa ERP antibody; (D) beads in (A) incubated with 48 kDa ERP antibody; (E) beads in (A) incubated with 56 kDa ERP and 48 kDa ERP antibody; (E) beads in (A) incubated with 56 kDa and 48 kDa ERP antibodies. The beads in (B-E) were re-injected into larvae. After 10 min, beads were recovered and observed by microscopy. Bar = 25 μm.

THE 86 kDa ERP WHICH RESEMBLES DIAPAUSE 1 PROTEIN

Another encapsulation-relating protein, 86 kDa ERP, was also purified homogeneously by electro-elution method and its antibody was raised. To determine its whole amino acid sequence, we screened the cDNA library of *T. molitor* larvae by the immunoscreening method. It was found that the 86 kDa ERP consists of 754 amino acid residues and it showed highly conserved sequence identity with diapause protein 1. The potential signal peptides are supposed to exist and the most likely cleavage site seems to be located between Ala15-Ser16. Moreover that, 86 kDa protein turned out to be exist in the late larval, prepupal with its maximum amount and pupal stages. These characters coincide with diapause protein 1, and we suggest that the diapause1-like protein play some roles in defense mechanism, especially early-staged encapsulation.

DISCUSSION

When foreign invaders are introduced into the hemocoel, insects react to them quickly and form capsules that consist of extracellular hemolymph components. But, the molecular mechanism of how to recognize intruding microorganisms and distinguish them from their own body component is not well known. We have searched the answer by studying the proteins that are related to the early step of encapsulation as a major cellular immune response.

We demonstrated that the 56 kDa and 48 kDa ERPs are novel glutamine-rich proteins, which are detected in the hemocytic membrane fraction and essential factors to inducing encapsulation response. Moreover, 86 kDa ERP was found out to be in good coincidence with well-known diapause protein 1.

We are under the study on another 78 kDa ERP and we expect to clarify the molecular mechanism of the early-staged encapsulation response and the non-self recognition factors.

REFERENCES

1. Jules A. Hoffmann, Fotis C. Kafatos, Charles A. Janeway Jr., R.A.B. Ezekowitz, Phylogenetic perspectives in innate immunity. *Science* **284**, 1313-1318 (1999)
2. Natori, S. Defense molecules in host defense and development of insects, in Phylogenetic Perspectives in immunity : the insect host defense (Hoffmann, J. A., Janeway, C. A. Jr. and Natori, S., eds) pp. 43-65, R. G. Landes Company, Austin TX. (1994)
3. Schmit, A. R. and Ratcliffe, N. A. The encapsulation of foreign tissue implants in *Galleria mellonella* larvae. *J. Insect Physiol. 23*, 175-184 (1977)
4. Pech, L. & Michael, R. S. Granular cells are required for encapsulation of foreign targets by insect haemocytes, *J. Cell, Sci., 109*, 2053-2060 (1996)
5. Asgari, S., Theopold, U., Wellby, C. & Schmidt, O. A protein with protective properties against the cellular defense reactions in insects, *Proc. Natl. Acad. Sci. USA. 95*, 3690-3695 (1998)
6. Hayakawa, Y. & Yazaki, K. Envelope protein of parasitic wasp symbiont virus, polydnavirus, protects the wasp eggs from cellular immune reactions by the host insect, *Eur. J. Biochem. 246*, 820-826 (1997)

7. Cho, M. Y., Lee, H. S., Lee, K. M., Homma, K., Natori, S. and Lee, B. L. Molecular cloning and functional properties of two early-staged encapsulation-relating proteins from the coleopteran insect, *Tenebrio molitor* larvae. *Eur. J. Biochem*. 262, 737-744 (1999)
8. Cho, M. Y., Choi H. W., MoonG. Y., Kim, M. H., Kwon, T. H., Homma, K., Natori, S. and Lee, B. L. An 86kDa diapause protein 1-like protein is a component of early-staged encapsulation relating proteins in coleopteran insect, *Tenebrio molitor* larvae. *FEBS Letters* **451**, 303-307 (1999)

A β-1,3-GLUCAN-BINDING PROTEIN FROM *Manduca sexta*

Congcong Ma and Michael R. Kanost

Department of Biochemistry, Kansas State University, Manhattan, KS, 66502

INTRODUCTION

Insects have pattern recognition molecules [1] which can recognize nonself molecules such as β-1,3-glucan, lipopolysaccharide, and peptidoglycan on the surface of invading pathogens. Upon binding to the foreign invaders, pattern recognition proteins further trigger defense pathways such as the prophenoloxidase pathway in insects. Insects thus can defend themselves efficiently against pathogens even though they lack the immunoglobulin system present in mammals.

Some pattern recognition molecules have been characterized from insects. A lipopolysaccharide-binding protein purified and cloned from the silkworm *Bombyx mori* [2] participates in hemocyte nodule formation and belongs to the C-type lectin super family. A peptidoglycan recognition protein purified and cloned from the silkworm [3, 4] can be induced by bacterial challenge and can trigger the prophenoloxidase pathway upon binding to peptidoglycan [3]. The silkworm peptidoglycan recognition protein has homologs in mice, nematodes, and bacteriophages. A peptidoglycan

Phylogenetic Perspectives on the Vertebrate Immune System.
Edited by Beck *et al.*, Kluwer Academic/Plenum Publishers, 2001.

recognition protein was also cloned from the moth *Trichoplusia ni*, and corresponding cDNA from mouse and human were also cloned [5]. Although β-1,3-glucan-binding protein has been purified from the silkworm [6] and the cockroach [7, 8], the primary structure has not been determined. Here, we describe a β-1,3-glucan-binding protein we have recently purified and cloned from the tobacco hornworm, *Manduca sexta*.

A cDNA sequence encoding a 53 kDa β-1,3-glucan-binding-protein (GBP) present in the plasma of *Manduca sexta*, was cloned from a fat body cDNA library. Western blot and Northern blot analysis showed that GBP is constitutively expressed in the fat body and secreted into plasma. A yeast challenge moderately induced GBP expression. The deduced amino acid sequence of GBP is similar to the gram negative bacteria binding protein from silkworm [9], and to β-1,3-glucanases from sea urchin [10] and bacteria [11].

The cDNA fragment coding for the mature β-1,3-glucan-binding protein was subcloned into expression vector H6-pQE60 [12]. The recombinant protein expressed in *E. coli* contains a six-histidine tag on the N-terminal and was purified by Ni-affinity chromatography. The expressed recombinant β-1,3-glucan-binding protein was not soluble but could be renatured after dissolving in 8 M urea. Figure 1 shows the gel filtration HPLC analysis of the renatured and purified *Manduca sexta* recombinant β-1,3-glucan-binding protein. The sample was injected to a Bio-Sil SEC250 column (300 mm x 7.8 mm, BioRad) and eluted with buffer containing 50 mM sodium phosphate, pH 6.8, and 150 mM sodium chloride. The flow rate was 1 mL/min. The elution time of β-1,3-glucan-binding protein was compared with a set of molecular weight standards (Bio-Rad) containing thyroglobulin (670 kDa), bovine gamma globulin (158 kDa), chicken ovalbumin (44 kDa), equine myoglobin (17 kDa), and vitamin B12 (1.35 kDa). The logarithms of the molecular weight of the standards

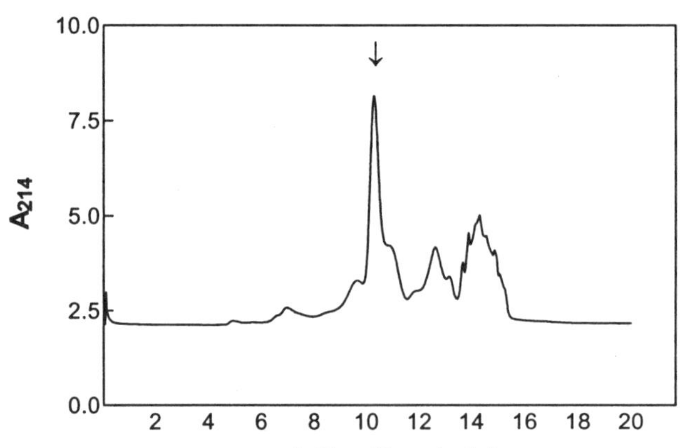

Figure 1. HPLC analysis of *Manduca sexta* recombinant β-1,3-glucan-binding protein. The arrow indicates the β-1,3-glucan-binding protein peak.

were plotted against the elution times and the standard curve was calculated by linear regression. The native molecular weight of the recombinant β-1,3-glucan-binding protein was calculated from the standard curve (Figure 2).

The calculated native molecular weight of the recombinant β-1,3-glucan-binding protein is 62 kDa which is very close to the calculated molecular weight according to the deduced amino acid sequence. This suggests that the recombinant protein exists as a monomer.

Functional study of the recombinant β-1,3-glucan-binding protein showed that it can cause the activation of the PPO pathway upon binding to β-1,3-glucan. It can also aggregate microorganisms. Thus, it is an important pattern recognition protein in insect defense system

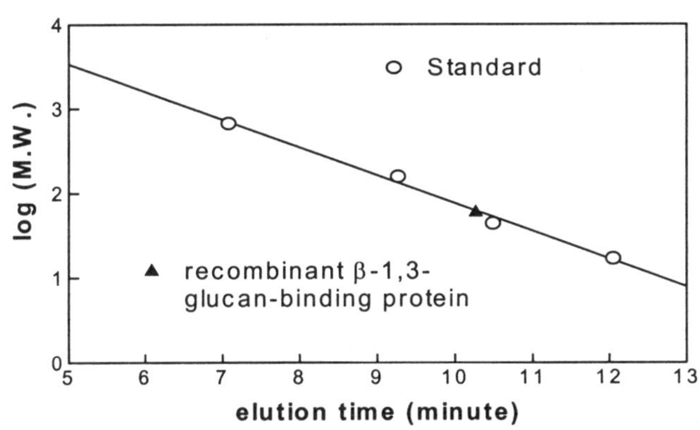

Figure 2. Calculation of the native molecular weight of the recombinant β-1,3-glucan-binding protein.

REFERENCES

1. Janeway CA Jr., Approaching the asymptote? Evolution and revolution in immunology. Cold Spring Harbor Symp Quant Biol 1989; 54:1-13
2. Koizumi N, Imamura M, Kadotani T, Yaoi K, Iwahana H, Sato R. The lipopolysaccharide-binding protein participating in hemocyte nodule formation in the silkworm *Bombyx mori* is a novel member of the C-type lectin superfamily with two different tandem carbohydrate-recognition domains. FEBS Letters 1999; 443:139-143
3. Yoshida H, Kinoshita K, Ashida M. Purification of a peptidoglycan recognition protein from hemolymph of the silkworm, *Bombyx mori*. J Biol Chem 1996; 271: 13854-13860
4. Ochiai M, Ahida M. A pattern recognition protein for peptidoglycan. J Biol Chem 1999; 274:11854-11858
5. Kang D, Liu G, Lundström A, Gelius E, Streiner H. A peptidoglycan recognition protein in innate immunity conserved from insects to humans. Proc Natl Acad Sci USA 1998; 95:10078-10082
6. Ochiai M, Ashida M. Purification of a $-1,3-glucan recognition protein in the prophenoloxidase activating system from hemolymph of the silkworm, *Bombyx mori*. J Biol Chem 1988; 263:12056-12062

7. Chen C, Durrant HJ, Newton RP, Ratcliffe NA. A study of novel lectins and their involvement in the activation of the prophenoloxidase system in *Blaberus discoidalis*. Biochem J 1995; 310:23-31
8. Söderhäll K, Rögener W, Söderhäll I, Newton RP, Ratcliffe NA. The properties and purification of a *Blaberus craniifer* plasma protein which enhances the activation of haemocyte prophenoloxidase by a β 1,3-glucan. Insect Biochem 1988; 18:323-330
9. Lee W-J, Lee J-D, Kravchenko VV, Ulevitch RJ, Brey PT. Purification and molecular cloning of an inducible Gram-negative bacteria-binding protein from the silkworm, *Bombyx mori*. Proc Natl Acad Sci USA 1996; 93:7888-7893
10. Bachman ES, McClay DR. Molecular cloning of the first metazoan β-1,3 glucanase from eggs of the sea urchin *Strongylocentrotus purpuratus*. Proc Natl Acad Sci USA 1996; 93:6808-6813
11. Yahata N, Watanabe T, Nakamura Y, Yamamoto Y, Kamimiya S, Tanaka H. Structure of the gene encoding β-1,3-glucanase A1 of *Bacillus circulans* WL-12. Gene 1990; 86:113-117
12. Lee E, Linder ME, Gilman AG. Expression of G-protein " subunits in *Escherichia coli*. Methods Enzymology 1994; 237:146-163

PROTEOLYTIC ACTIVATION OF PROPHENOLOXIDASE IN AN INSECT *Manduca sexta*

Haobo Jiang, Yang Wang, and Michael R. Kanost

Department of Biochemistry, Kansas State University, Manhattan, KS 66506

INTRODUCTION

Phenol oxidase (PO) (EC1.14.18.1) is important in cuticular sclerotization and in defense against pathogens and parasites (Söderhäll *et al.,* 1996; Sugumaran, 1996; Ashida and Brey, 1998). PO catalyzes hydroxylation of monophenols to *o*-diphenols and oxidation of *o*-diphenols to quinones. The quinones take part in sclerotization of cuticle and serve as precursors for synthesis of melanin. PO is synthesized in arthropods as an inactive zymogen, prophenoloxidase (proPO), which can be activated by proteolytic cleavage at a specific site in the protein. Wounding of an insect or exposure to certain microbial polysaccharides leads to the activation of a proteinase that specifically activates proPO, perhaps at the last step of a proteinase cascade (Ashida and Brey, 1998). The reactions catalyzed by PO then can result in melanization of foreign organisms trapped in capsules or hemocyte nodules (Gillespie *et al.*, 1997).

Phylogenetic Perspectives on the Vertebrate Immune System.
Edited by Beck *et al.*, Kluwer Academic/Plenum Publishers, 2001.

Melanization may be an important factor in determining the outcome of an infection. A strain of the mosquito, *Anopheles gambiae*, that is refractory to infection by *Plasmodium cynomolgi* encapsulates the parasite ookinetes in a melanin coat, whereas susceptible strains do not (Vernick, 1998). To understand how melanization is triggered and regulated, more information is required about the biochemistry and molecular biology of proPO activation.

To detect proPO activating proteinase (PAP) from the tobacco hornworm *Manduca sexta*, we selected a substrate similar to the amino-terminal side of the putative activation site in *M. sexta* proPO (Jiang, *et al.,* 1997). This substrate is useful for monitoring a proteinase activity in plasma, which appears upon addition of *Micrococcus lysodeikticus* (Fig. 1). The IEAR*p*Na hydrolyzing activity reaches its maximum while PO activity increases most rapidly, and then begins to decrease. Therefore, PAP may be responsible for the amidase activity, which is subject to regulation by a proteinase inhibitor present in the plasma.

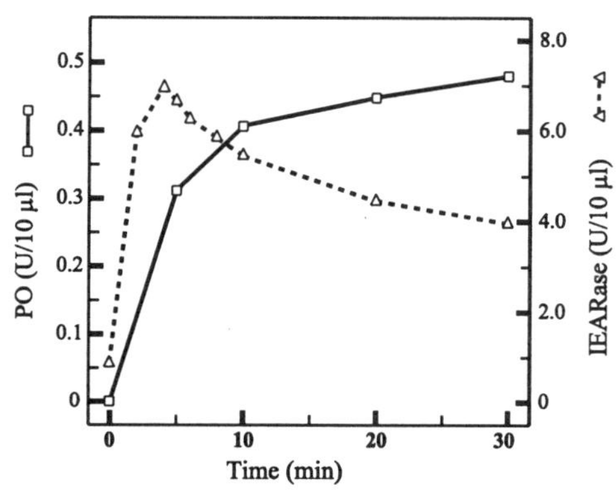

Figure 1. Activation of IEARase and PO by *Micrococcus lysodeikticus.* Elicitor (*M. lysodeiktius,* 1.0 µg/µl, 25 µl) was incubated with 250 µl of *M. sexta* plasma at room temperature. At different time intervals, IEARase and PO activities in 20 and 10 µl of the mixture were measured using acetyl- Ile-Glu-Ala-Arg-*p*-nitroanilide and dopamine as substrates, respectively (Jiang and Kanost, 1997). One unit of activity is defined as 0.001 ΔA/min.

Armed with the synthetic substrate and proPO, we have purified PAPs from hemolymph and cuticle of *M. sexta* pharate pupae. Purified proteins, designated hPAP and cPAP, are composed of a light chain and a heavy chain connected by a disulfide bond. The proteinase domain is located in the heavy chain, which can be affinity-labeled with ^{3}H-DFP. *M. sexta* hPAP and cPAP cleave proPO at the predicted activation site after Arg51. However, the cleavage products (~74 kDa) appear to be inactive. Complete activation of proPO, therefore, may require another protein to assist

PAP in a secondary proteolysis. The existence of such protein cofactor was confirmed during the purification of cPAP (Jiang, *et al.*, 1998).

Based on the amino-terminal sequence of their catalytic domains and a conserved sequence around the active site serine residue, we were able to design primers for polymerase chain reactions (PCRs). Using PCR-generated probes, we have isolated cDNAs for *M. sexta* hPAP and cPAP from induced fat body and hemocyte cDNA libraries, respectively. Genbank search and multiple sequence alignment indicated that the two proteinases belong to a family of arthropod serine proteinases with a clip domain. *M. sexta* hPAP is most similar in sequence to *Bombyx mori* proPO activating enzyme (PPAE, Satoh *et al.*, 1999), both containing two clip domins (Fig. 2). The phylogenetic analysis of all these sixteen proteinases indicated that *M. sexta* cPAP is closely related to prophenoloxidase activating factor I (PPAF-I, Lee et al., 1998) as well as *Drosophila* easter (Chasan and Anderson, 1989). Interestingly, serine

```
             1         2     3                        4         56
Ms hPAP-1    CTLPNNDKGTCKSLLQCDVASKIISKKPRTAQDEKFLRESACGFDGQTPKVCC
Bm PPAE-1    CRTPNGLNGNCVSVYECQALLAILNNQRRTQQDEKFLRDSQCGTKNSVPAVCC
Ms hPAP-2    CLTPDNKPGKCVNIKKCTHLAEIEEDP IGEDETTYLKNSVCAGPEDNS VCC
Bm PPAE-2    CNAADGQQGNCVNINSCPYVLQLLKNP  NEANLNYVRGSVCQGSEQQS ICC
Ms cPAP      CTTPQGADSNCISLYECPQLLSAFEQRPLPSPVVNYLRKSQCGFDGYTPRVCC
Hd PPAF-1   CRTPNGENARCVPINNCKILYDSVLTSDP   EVIRFLRASQCGYNG QPLVCC
Tt PCE      CSNRFTEEGTCKNVLDCRILLQKN         DYNLLKESICGFEGITPKVCC
```

Figure 2. Sequence alignment of clip domains in insect proPO activating proteinases and horseshoe crab proclotting enzyme. Clip domains of *M. sexta* hPAP (Ms hPAP-1 and -2), *Bombyx mori* proPO activating enzyme (Bm PPAE-1 and -2, Satoh *et al.* 1999), *M. sexta* cPAP (Ms cPAP, Jiang *et al.*, 1998), and *Holotrichia diomphalia* proPO activating factor I (Hd PPAF-1, Lee *et al.*, 1998) are aligned with that of *Tachypleus tridentatus* proclotting enzyme (Tt PCE, Muta *et al.*, 1990). In the horseshoe crab PCE, three disulfide bonds form between Cys1-Cys5, Cys2-Cys4, and Cys3-Cys6, which are conserved in all the sequences.

proteinases 14A, 14D1, as well as 14D2 from *Anopheles gambiae* are located in the same branch including the four PAPs, suggesting that they could actually be activating proteinases for proPO in the malaria vector.

Northern blot analysis indicates that *M. sexta* hPAP mRNA is present only in the fat body of larvae that have been challenged with *E. coli*. In contrast to the silkworm PPAE, *M. sexta* hPAP is not detected in epidermis. cPAP is synthesized in epidermis, as well as in naive hemocytes and fat body. Like hPAP, mRNA for cPAP is strongly induced in fat body upon injection of bacteria.

We have demonstrated previously that serpin-1J, at its estimated physiological concentration, blocks proPO activation (Jiang and Kanost, 1997). To find out if this is due to inhibition of the amidase, we added purified recombinant serpin-1J to serpin-depleted plasma, triggered the activation, and then measured IEAR*p*Na hydrolysis. The induced amidase activity was reduced in a concentration-dependent manner

(Fig. 3), and 50% inhibition occurred at an estimated hemolymph concentration for serpin-1J. Serpin-1J forms SDS-stable complexes with purified hPAP and cPAP. These results strongly suggest that serpin-1J is a physiological inhibitor of PAPs.

In order to further understand the mechanisms for proPO activation, we are now purifying and characterizing the unknown cofactor for PAPs in *M. sexta.* This will certainly contribute to our knowledge on protein-protein interactions among the proteinases, cofactor, and proPO. Moreover, inhibitory kinetic analysis will help to establish physiological regulation of PAPs by serpin-1J.

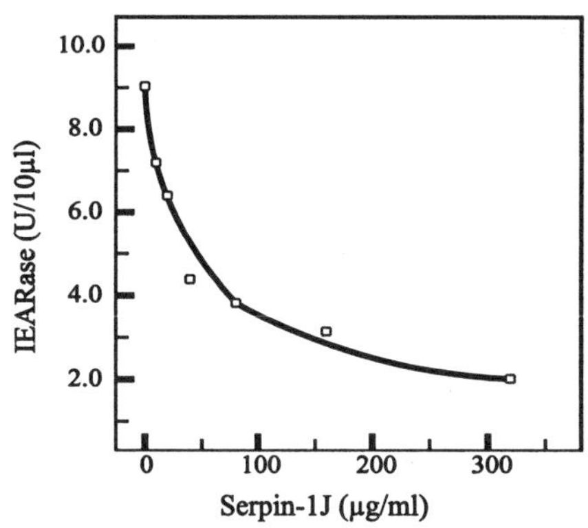

Figure 3. Concentration-dependent Inhibition of IEARase Activity by *M. sexta* Serpin-1J. Two microliters of recombinant serpin-1J at different concentrations were mixed with 20 μl aliquots of 0-50% ammonium sulfate fraction of *M. sexta* hemplymph (Jiang and Kanost, 1997). Then, *M. lysodeikticus* (1.0 μg/μl, 2 μl) was added to trigger the activation of IEARase and PO. Amidolytic activity was measured after 5 min using IEAR*p*Na as a substrate.

ACKNOWLEDGEMENTS

This work was supported by National Institutes of Health Grants GM58634 and GM41247.

REFERENCES

Ashida, M. and Brey P.T. (1998) Recent advances in research on the insect prophenoloxidase cascade. *Molecular Mechanisms of Immune Responses in Insects* (Brey, P.T. and Hultmark, D. eds), pp 135-172. Chapman & Hall, New York.

Chasan, R. and Anderson K. (1989) The role of easter, an apparent serine protease, in organizing dorsal-ventral pattern of the *Drosophila* embryo. *Cell,* **56**, 391-400.

Gillespie, J.P., Kanost, M.R., and Trenczek, T. (1997) Biological mediators of insect immunity. *Annu. Rev. Entomol.* **42**, 611-643.

Jiang, H. and Kanost, M.R. (1997) Characterization and functional analysis of 12 naturally occurring reactive site variants of serpin-1 from *Manduca sexta. J. Biol. Chem.* **272**, 1082-1087.

Jiang, H. Wang, Y., Ma, C., and Kanost, M.R. (1997) Subunit composition of pro-phenol oxidase from *Manduca sexta*: molecular cloning of subunit proPO-p1. *Insect Biochem. Mol. Biol.* **27**, 835-850.

Jiang, H., Wang, Y., and Kanost, M.R. (1998) Pro-phenol oxidase activating proteinase from an insect, *Manduca sexta*: a bacteria-inducible protein similar to *Drosophila* easter. *Proc.Natl. Acad. Sci. USA* **95**, 12220-12225.

Lee, S.Y., Cho, M.Y., Hyun, J.H., Lee, K.M. Homma, K., Natori, S., Kawabata, S., Iwanaga, S., and Lee, B.L. (1998) Molecular cloning of cDNA for pro-phenol-oxidase-activating factor I, a serine proteinase is induced by lipopolysaccharide or 1,3-ß-glucan in a coleopteran insect, *Holotrichia diamphalia* larvae. *Eur. J. Biochem.* **257**, 615-621.

Muta, T., Hashimoto, R., Miyata, T., Nishiway, H., Toh, Y., and Ichiwara, S. (1990) Proclotting enzyme from horseshoe crab hemocytes: cDNA cloning, disulfide locations, and subcellular localization. *J. Biol. Chem.* **265**, 22426-22433.

Satoh, D., Horii, A., Ochiai, M., and Ashida, M. (1999) Prophenoloxidase-activating enzyme of the silkworm, *Bombyx mori*: purification, characterization and cDNA cloning. *J. Biol. Chem.* **274**, 7441-7453.

Söderhäll, K. Cerenius, L., and Johansson, M.W. (1996) The prophenoloxidase activating system in invertebrates. *New Directions in Invertebrate Immunology* (Söderhäll, K., Iwanaga, S. and Vasta, G., eds), pp 229-253. SOS Publications, Fair Haven, NJ.

Sugumaran, M. (1996) Role of insect cuticle in immunity. *New Directions in Invertebrate Immunology* (Söderhäll, K., Iwanaga, S. and Vasta, G., eds), pp 355-374. SOS Publications, Fair Haven, NJ.

Vernick, K.D. (1998) Mechanisms of immunity and refractoriness in insect vectors of eukaryotic parasites. *Molecular Mechanisms of Immune Responses in Insects* (Brey, P.T. and Hultmark, D. eds), pp 261-309. Chapman & Hall, New York.

HEMOLYMPH PROTEINASES IN IMMUNE RESPONSES OF *Manduca sexta*

Michael R. Kanost, Haobo Jiang, Yang Wang, Xiao-Qiang Yu, Congcong Ma and Yifei Zhu

Department of Biochemistry, Kansas State University
Manhattan, KS 66506

INTRODUCTION

Extracellular signal transduction pathways in blood are often composed of cascades of serine proteinases to amplify rapidly a response to wounding or infection. The vertebrate blood coagulation and complement pathways are the most fully characterized of such proteinase cascades. Serine proteinases also play important roles in defensive responses in hemolymph of arthropods. The hemolymph coagulation system of horseshoe crabs has been very nicely elucidated to function as a complex and intricately regulated proteinase cascade for protection against microbial infection (Kawabata *et al.*, 1996).

We have been investigating serine proteinases in the hemolymph of a lepidopteran insect, the tobacco hornworm, *Manduca sexta*. We have identified and characterized a number of hemolymph proteinases by protein purification and by cDNA cloning methods and have begun to understand some of their functions. These proteinases may

Phylogenetic Perspectives on the Vertebrate Immune System.
Edited by Beck *et al.*, Kluwer Academic/Plenum Publishers, 2001.

participate in several defensive pathways, including activation of cytokine precursors and prophenol oxidase (Fig. 1).

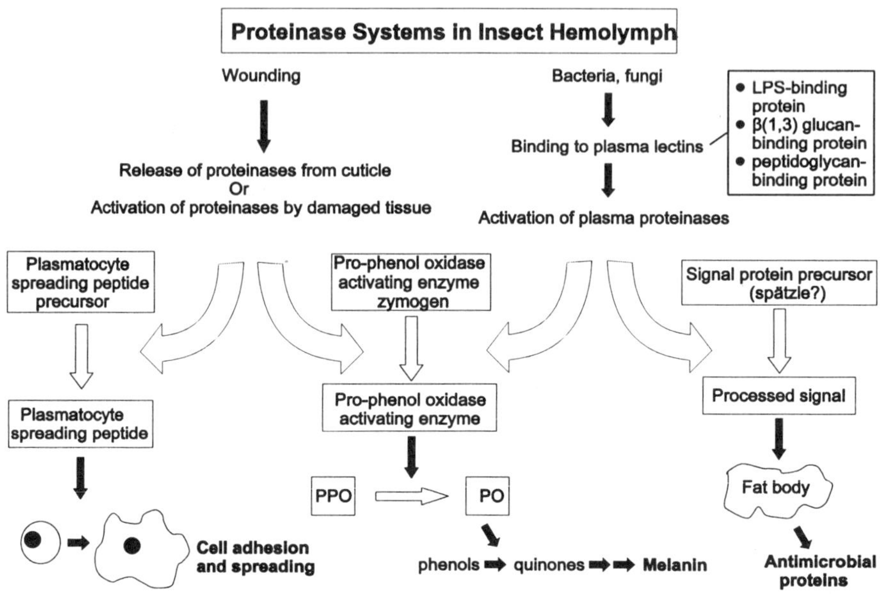

Figure 1. A model for potential serine proteinase cascade systems in hemolymph of *Manduca sexta* and other insects. Open arrows represent steps that involve proteolytic processing. The curved arrows represent proteinase cascades in which the number of steps is unknown and the proteinases involved have not been characterized. It is possible that some proteinases may participate in more than one pathway.

SERINE PROTEINASES IN *MANDUCA* HEMOLYMPH

Covalent labeling of the active site of serine proteinases (and perhaps serine esterases) in *Manduca* hemolymph with ^{3}H-diisopropylfluorophosphate detects five strongly labeled proteins with molecular masses from 31 to 100 kDa, visualized by SDS-PAGE and fluorography (Jiang *et al.* 1999). In addition to these, hemolymph may contain other proteinases that were present as zymogens or at concentrations below the limit of detection in this experiment. We have used PCR and cDNA library screening techniques to begin to characterize the group of *Manduca* serine proteinases synthesized by hemocytes or fat body and which are likely to be present in hemolymph. We have isolated cDNA clones for eight serine proteinases from hemocyte or fat body libraries, including two prophenol oxidase activating proteinases (PAPs) (Jiang *et al.*, 1998; Jiang, Wang and Kanost, unpublished results) and six proteinases whose functions are not yet known (Jiang *et al.* 1999; Jiang, Wang, Zhu, Scholz, Kanost, unpublished results).

The PAPs and four of the other proteinases (HP1, HP2, FP1, FP2) contain two domains, an amino-terminal clip domain followed by a carboxyl-terminal serine proteinase domain (Jiang *et al.*, 1998; 1999). This structural organization is similar to that of the horseshoe crab clotting enzyme (Kawabata *et al.*, 1996). The cleavage site for activation of these proteinases is in a region that links the two domains, such that proteolytic activation of the proenzymes produces two polypeptide chains, linked by a disulfide bond. In addition to the horseshoe crab clotting and insect prophenol oxidase activation pathways, proteinases with this clip domain-serine proteinase structure function in a cascade which regulates dorsal-ventral polarity in *Drosophila* embryonic development (Belvin and Anderson, 1996). Additional serine proteinases containing amino-terminal clip domains have recently been identified by cDNA cloning in several insect species and may be a common component of hemolymph proteins that participate in proteinase cascades, which are activated in response to wounding or infection (Jiang and Kanost, 1999).

Two other serine proteinase cDNAs (HP3 and HP4) isolated from a *Manduca* hemocyte library lack a clip domain-encoding sequence (Jiang *et al.*, 1999). These enzymes are most similar in sequence to mammalian serine proteinases expressed in neutrophils, mast cells, and natural killer cells, which are released in acute phase immune responses. Thus, HP3 and HP4 may be released from hemocytes during infections to participate in killing of microbial invaders. Another serine proteinase-like protein present in *Manduca* hemolymph is scolexin, which is synthesized in response to microbial infection, but whose function is not yet clear (Finnerty et al., 1999).

SERINE PROTEINASES IN BACTERIA-INDUCED HEMOLYMPH PROTEIN GENE EXPRESSION

Recent studies have suggested that a signal transduction system which regulates dorsal/ventral development in *Drosophila* embryos is also used in regulation of some genes for antibacterial and antifungal hemolymph proteins. In embryonic development this pathway depends on an extracellular serine proteinase cascade that eventually cleaves an inactive protein called spätzle, making it competent to bind to its receptor, Toll (reviewed in Belvin and Anderson, 1996). Binding of spätzle to Toll initiates a signal transduction pathway that leads to activation of a transcription factor named Dorsal. Toll is homologous in sequence to the mammalian interleukin-1 receptor. Dorsal is homologous to mammalian NF-κB, a member of the rel family, which is activated after interleukin-1 binds to the interleukin-1 receptor and stimulates expression of many mammalian acute phase proteins. This evolutionarily conserved pathway appears to also function in regulating immune protein genes in *Drosophila* (reviewed in Hoffmann and Reichhart, 1997). Dorsal and proteins related to Dorsal (Dif and Relish) are activated during expression of some inducible antimicrobial peptide genes, which contain NF-κB-like regulatory DNA sequences near their promoters. *Drosophila* mutants deficient in Toll or spätzle exhibit decreased induction of drosomycin, an antifungal peptide normally synthesized by the fat body after microbial challenge (Lemaitre *et al.*, 1996). These results suggest that infection may lead to proteolytic processing of spätzle present in hemolymph. The activated spätzle might bind to the Toll receptor in fat body membrane, leading to activation of the Dorsal pathway and expression of antimicrobial hemolymph proteins (Fig. 1). It is not yet known what proteinases in *Drosophila* participate in this pathway. Such signaling cascades are likely to be conserved in other insects and may include spätzle or other proteins as proteinase-activated ligands for fat body or hemocyte receptors.

In experiments to test the hypothesis that serine proteinases are involved in bacteria-induced synthesis of antimicrobial proteins by *Manduca* fat body, we have found that benzamidine, an inhibitor of enzymes with trypsin-like specificity, can partially block expression of two types of inducible genes. Injection of benzamidine into the hemocoel of larvae reproducibly decreases the induced expression of bactericidal (predominantly cecropin) activity and of a C-type lectin (immulectin) (Fig. 2). Benzamidine had no effect on lysozyme or hemolin induction (data not shown), indicating that there may be different signaling pathways for stimulation of different sets of antibacterial genes. These results suggest that hemolymph proteinases are involved in the humoral antibacterial response. Further research is needed to identify the proteinases, their substrates, and putative ligands for Toll-like receptors.

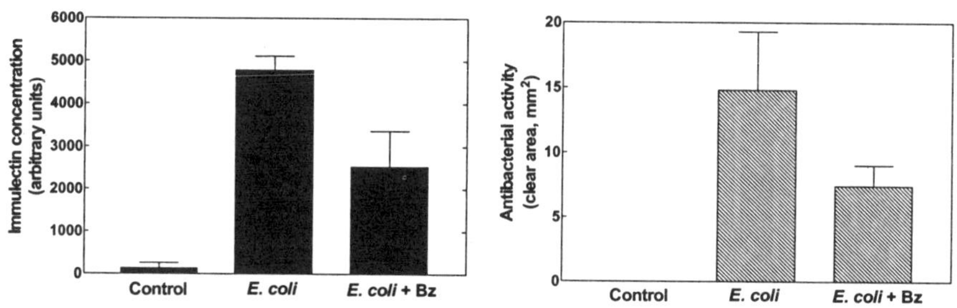

Figure 2. Decreased expression of bacteria-induced genes in insects injected with a proteinase inhibitor. Day 2 fifth instar larvae were injected with either 100 μL sterile 0.85%NaCl or 100 μL 1 M benzamidine (Bz). Ten minutes later, insects were injected with 10^8 formalin-killed *E. coli* in 10 μL saline or with 10 μL saline alone (Control). Injection of 50 μL of saline or 1 M benzamidine were repeated at one hour and three hours after injection of bacteria. Hemolymph was collected 24 hours after injection of bacteria, hemocytes were removed by centrifugation, and plasma samples from individual insects were assayed for lectin by Western blotting and for antibacterial activity by growth inhibition zone assay. Bars represent mean ±standard deviation. The *E. coli* + benzamidine treatment resulted in significantly reduced levels of induction of immulectin and antibacterial activity (t-test, $P<0.01$).

PROTEOLYTIC ACTIVATION OF A CYTOKINE PRECURSOR

Injection of bacteria into lepidopteran insects, including *Manduca,* leads to a rapid decrease in the number of plasmatocytes in circulation (Chain and Anderson, 1983; Geng and Dunn, 1989). A peptide that has the properties of a "plasmatocyte depletion factor," named plasmatocyte-spreading peptide, was isolated from plasma of *Pseudoplusia includens* (Clark *et al.*, 1997). This 23-residue peptide has a sequence very similar to peptides that were previously isolated from plasma of other lepidopterans, including *Pseudaletia separata* (Hayakawa, 1991), *Spodoptera exigua,*

Heliothis virescens, and Manduca sexta (Skinner *et al.*, 1991), which were reported to have different biological activities including paralysis (Skinner *et al.,* 1991), depression of juvenile hormone esterase level, and inhibition of growth and development (Hayakawa, 1991).

We have recently found that the *Manduca* paralytic peptide also acts to stimulate plasmatocyte spreading and adhesion and to inhibit bleeding (Wang *et al.*, 1999). Injection of the peptide causes depletion of plasmatocytes from circulation, and when hemocytes are treated with this peptide *in vitro*, the plasmatocytes are induced to spread on a surface and to aggregate (Clark *et al.*, 1997; Wang *et al.*, 1999). The paralytic and plasmatocyte adhesion peptides appear rapidly after hemolymph is collected. An explanation for the origin of the peptides is apparent in the amino acid sequences deduced from the cDNA clones for the peptides (Hayakawa *et al.*, 1995; Clark *et al.*, 1998; Wang *et al.*, 1999). The cDNAs encode polypeptides approximately 140 residues long, with an amino terminal secretion signal sequence. This precursor protein is synthesized by fat body and is normally present in plasma. The active form of the peptide is the carboxyl-terminal 23-residue fragment of the protein. An Arg residue directly preceding a conserved Glu at the amino terminus of the mature peptide is the site at which the precursor protein is processed by a proteinase with trypsin-like specificity. Upon injury the precursor is cleaved by a proteinase to liberate the C-terminal peptide, which can then act on a target such as plasmatocytes (Fig. 1). It is not yet known what proteinases are involved in this response or how they becomes activated.

The structures of three members of this growth blocking peptide/plasmatocyte spreading peptide (GBP/PSP) family, including *Manduca* paralytic peptide-1 (Fig. 3), have been solved by NMR spectroscopy (Aizawa *et al.*, 1999;Volkman *et al.*, 1999; Yu *et al.*, 1999a). All three members of the GBP/PSP family have a similar compact, ordered structure including a short, two strand antiparallel β-sheet and are stabilized by a single disulfide bond. Their folding pattern has some similarity with the carboxyl-terminal domain of epidermal growth factor. It will be intriguing to learn what type of receptor on plasmatocytes is stimulated by these peptides and to characterize the proteolytic processing required to generate the bioactive peptides.

ACTIVATION OF PROPHENOL OXIDASE

It has been appreciated for some time that melanization, a response to wounding and infection in insects and crustaceans, involves serine proteinases. Wounding or infection leads to rapid activation of a proteinase that in turn activates a phenol oxidase zymogen (prcphenol oxidase; PPO) (Fig. 1). Subsequent oxidation of phenols by phenol oxidase produce quinones that polymerize to form melanin. This process is discussed in greater detail in other chapters in this volume (Ashida; Jiang *et al.*; Söderhäll; Sugumaran). Experimental evidence has suggested that activation of prophenoloxidase involves a serine protease cascade (Söderhäll *et al.* 1996; Ashida and Brey, 1998). Like the horseshoe crab clotting system, PPO activation can be stimulated by LPS or β(1,3) glucans, and it can also be activated by peptidoglycan (Ashida and Brey, 1998). Recognition of these microbial polysaccharides triggers the pathway, leading to activation of a serine proteinase that cleaves PPO at a specific peptide bond near its amino terminus to produce active PO.

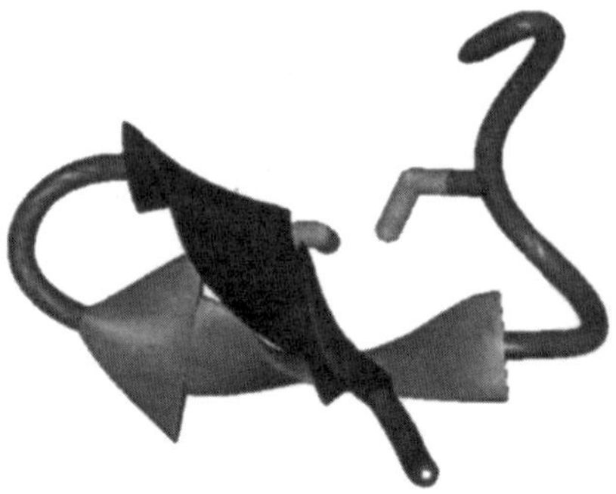

Figure 3. Solution structure of *Manduca* paralytic peptide-1. The figure shows a ribbon diagram of the structure of the 23-residue paralytic peptide, indicating the position of a disulfide bond between Cys^7 and Cys^{19} (Yu et al., 1999a).

We have recently identified two types of proteins from *Manduca* hemolymph that appear to function as non-self recognition molecules as they bind to microbial polysaccharides and activate the PPO pathway. We have characterized a family of C-type lectins, with different carbohydrate-binding specificity in *Manduca* plasma (Yu *et al.*, 1999b; Yu and Kanost, this volume). Two of these lectins bind to lipopolysacharides (LPS) from gram-negative bacteria, in a calcium dependent manner. When a complex of one of these lectins with LPS is added to plasma, the phenoloxidase cascade is rapidly activated (Fig. 4A). This response is reminiscent of effect of human mannose-binding lectin (another member of the C-type lectin family) on complement activation (Turner, 1996) and suggests that interaction of the *Manduca* lectin-LPS complex with a serine proteinase zymogen may lead to activation of the proteinase to trigger the pathway.

Another protein in *Manduca* hemolymph binds to a fungal cell wall polysaccharide, β(1,3)glucan (see Ma and Kanost, this volume) . This β-glucan recognition protein (GRP) is a constitutively expressed 53 kDa protein, which has sequence similarity to β(1,3) glucanases from bacteria and the sea urchin *Strongylocentrotus purpuratus*. However GRP lacks key catalytic residues such that it binds, but does not hydrolyze β(1,3) glucans. GRP agglutinates yeast, and a complex of GRP with the soluble β(1,3)glucan laminarin stimulates activation of the PPO system (Fig. 4B). As with the C-type lectin, we speculate that association of GRP with its corresponding microbial cell wall polysaccharide acts as a signal which leads to activation of a serine proteinase at the beginning of the PPO cascade.

Although the PPO cascade has not been fully defined, our laboratory and others have recently made progress in characterizing the terminal proteinase in the pathway, which specifically cleaves and activates PPO (see Jiang *et al.*, this volume). We have isolated two prophenol oxidase activating proteinases (PAP) from *Manduca*, one from cuticle (c-PAP; Jiang *et al.* 1998) and one from hemolymph (h-PAP; Jiang, Wang, and Kanost, unpublished results) of wandering stage fifth instar larvae and have cloned cDNAs which encode these enzymes. c-PAP contains an amino-terminal clip domain followed by a serine proteinase domain, whereas h-PAP contain two clip domains at its amino terminus. Similar prophenol oxidase activating proteinases have been characterized from a beetle, *Holotrichia diamphalia* (Lee *et al.*, 1998) and from the silkworm, *Bombyx mori* (Satoh *et al.*, 1999) by both protein purification and cDNA cloning. The *Holotrichia* proteinase contains a single clip domain, and the *Bombyx* proteinase contains two clip domains, like *Manduca* h-PAP. Future biochemical studies of these proteinases should lead to an understanding of how they interact specifically with their substrate, PPO, and of how they are themselves activated by a specific proteinase upstream in the cascade.

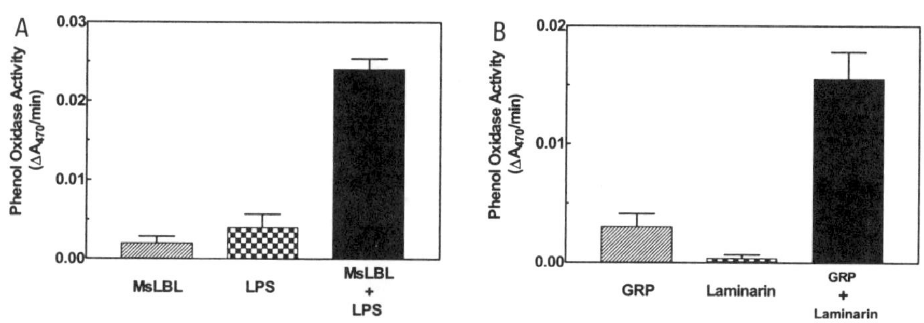

Figure 4. Activation of the prophenol oxidase pathway by microbial polysaccharide-binding proteins from *Manduca* plasma. **A**. Activation by a C-type lectin, *Manduca sexta* lipopolysaccharide binding lectin (MsLBL) + bacterial LPS. **B.** Activation by β(1,3)-glucan recognition protein + fungal laminarin.

REGULATION OF PROTEASES BY SERPINS IN HEMOLYMPH

Like blood clotting, phenol oxidase activation is normally regulated *in vivo* as a local reaction of brief duration. Also comparable to blood clotting, regulation of the proteinase cascade may be due in part to serine proteinase inhibitors in plasma (Kanost and Jiang, 1996; 1997). We have characterized a group of serpins from the serpin superfamily in *Manduca* plasma. These 46 kDa proteinase inhibitors are similar in amino acid sequence (Kanost *et al.*, 1989) and in three dimensional structure (Li *et al.*, 1999) to well-known mammalian serpins such as antithrombin-III, which regulates the activity of thrombin. Serpins with twelve different proteinase inhibitory selectivities are produced by alternative exon splicing from *Manduca* serpin gene-1 (Jiang *et al.*, 1996). In this gene there are twelve different copies of exon-9, which encodes the reactive site loop that interacts with a target proteinase. Use of different versions of exon-9 places different interchangeable inhibitory cassettes on the same protein framework, encoded by exons 1-8 (Fig. 5). The different amino acid sequences in the reactive site loops encoded by alternate versions of exon 9 inhibit an array of different serine proteinases (Jiang *et al.*, 1997). One of these serpin variants (serpin-1J) inhibits bacteria-stimulated prophenol oxidase activation (Jiang *et al.*, 1997) through its inhibition of PAP (Jiang *et al.*, this volume) and thus apparently functions to regulate the PPO activation pathway. The other serpin variants may regulate additional proteinases in the PPO cascade or may regulate hemolymph proteinases involved in other processes such as activation of cytokines. Further research is needed to discover the additional functions of these proteinase inhibitors in insect immune responses.

Figure 5. Ribbon structure of *Manduca* serpin-1K (Li *et al.*, 1999). The reactive site loop, which interacts with the target proteinase, is exposed at the surface of the molecule and shown at the top of the figure. The side chain of the P1 Tyr residue of serpin-1K is shown. Serine proteinases with chymotrypsin-like specificity attempt to cleave at this P1 residue, and then become trapped in a very stable serpin-proteinase complex. The carboxyl-terminal region of the proteinase, including the reactive site loop is encoded by exon-9 (the region colored the darkest shade), which can be substituted by alternative exon splicing for sequences with different proteinase selectivity.

CONCLUSION

It is apparent that hemolymph of *Manduca* and other insects contains a complex system of serine proteinases and inhibitors, which we do not yet understand very well. Future work in our laboratory will aim to identify the functions of these enzymes and to elucidate the molecular mechanisms by which proteinase cascades in hemolymph are triggered by pathogens or wounding and how they are regulated to prevent excessive and potentially harmful responses.

ACKNOWLEDGEMENTS

This research was supported by NIH grant GM14247. Contribution number 00-25-A from the Kansas Agricultural Experiment Station.

REFERENCES

Aizawa, T., Fujitani, N., Hayakawa, Y., Ohnishi, A., Ohkubo, T., Kumaki, Y., Kawano, K., Hikichi, K., and Nitta, K. (1999) Solution structure of an insect growth factor, growth-blocking peptide. *J. Biol. Chem.* **274**, 1887-1890.

Ashida, M., and Brey, P.T. (1998) Recent advances on the research of the insect prophenolxidase cascade. In *Molecular Mechanisms of Immune Responses in Insects* (Brey, P.T. and Hultmark, D. eds.), Chapman & Hall, London pp.135-172.

Belvin, M. and Anderson, K. (1996) A conserved signaling pathway: the *Drosophila* Toll-dorsal pathway. *Ann. Rev. Cell. Dev. Biol.* **12:** 393-416.

Chain, B.M. and Anderson, R.S. (1983) Inflammation in insects: the release of a plasmatocyte depletion factor following interaction between bacteria and haemocytes. *J. Insect Physiol.* **29**, 1-4.

Clark, K.D., Pech, L.L., and Strand, M.R. (1997) Isolation and identification of a plasmatocyte-spreading peptide from the hemolymph of the lepidopteran insect *Pseudoplusia includens. J. Biol. Chem.* **272**, 23440-23447.

Clark, K.D., Witherell, A., and Strand, M.R. (1998) Plasmatocyte spreading peptide is encoded by an mRNA differentially expressed in tissues of the moth *Pseudoplusia includens. Bioch. Biochem. Res. Commun.* **250**, 479-485.

Finnerty, C.M., Karplus, P.A., and Granados, R.R. (1999) The insect immune protein scolexin is a novel serine proteinase homolog. *Protein Sci.* **8**:242-248.

Geng, C. and Dunn, P.E. (1989) Plasmatocyte depletion in larvae of *Manduca sexta* following injection of bacteria. *Dev. Comp. Immunol.* **13**, 17-23.

Hayakawa, Y. (1991) Structure of a growth-blocking peptide present in parasitized insect hemolymph. *J.Biol.Chem.* **266**, 7982-7982.

Hayakawa, Y., Ohnishi, A., Yamanaka, A., Izumi, S., and Tomino, S. (1995) Molecular cloning and characterization of cDNA for insect biogenic peptide, growth blocking peptide. *FEBS Lett.* **376**, 185-189.

Hoffmann, J. and Reichhart, J. (1997) *Drosophila* immunity. *Trends Cell Biol.* **7**: 309-316.

Jiang, H. and Kanost, M.R. (1999) The clip domain family of serine proteinases from arthropod hemolymph. *Insect Biochem. Molec. Biol.* In press.

Jiang H., Wang, Y., Huang, Y., Mulnix, A.B., Kadel, J., Cole, K., and Kanost, M.R. (1996) Organization of serpin gene-1 from *Manduca sexta*: evolution of a family of alternate exons encoding the reactive site loop. *J. Biol. Chem.* **271:** 28,017-28,023

Jiang, H. and Kanost, M.R. (1997) Characterization and functional analysis of 12 naturally occurring reactive site variants of serpin-1 from *Manduca sexta. J. Biol. Chem.* **272:** 1082-1087.

Jiang, H., Wang, Y., and Kanost, M.R. (1998) Pro-phenol oxidase activating proteinase from an insect, *Manduca sexta*: a bacteria-inducible protein similar to *Drosophila* easter. *Proc. Natl. Acad. Sci. USA* **95**, 12220-12225.

Jiang, H., Wang, Y., and Kanost, M.R. (1999) Four serine proteinases expressed in *Manduca sexta* haemocytes. *Insect Mol. Biol.* **8**, 39-53.

Kanost, M.R. and Jiang, H. (1996) Proteinase inhibitors in invertebrate immunity. In "*New Directions in Invertebrate Immunology*", (K. Söderhäll, S. Iwanaga and G. Vasta, eds.) SOS Publications, Fair Haven, NJ. pp. 155-173.

Kanost, M.R. and Jiang, H. (1997) Serpins from an insect, *Manduca sexta.* In *"Chemistry and Biology of Serpins*", (F.C. Church, D.D. Cunningham, D. Ginsburg, M. Hoffman, S.R. Stone, and D.M. Tollefsen, eds.) Plenum, New York, pp. 155-161.

Kanost, M.R., Prasad, S.V. and Wells, M.A. (1989) Primary structure of a member of the serpin superfamily of proteinase inhibitors from an insect, *Manduca sexta. J. Biol. Chem.* **264,** 965-972.

Kawabata, S., Muta, T. and Iwanaga, S. (1996) The clotting cascade and defense molecules found in the hemolymph of the horseshoe crab. In *"New Directions in Invertebrate Immunology"*, (K. Söderhäll, S. Iwanaga and G. Vasta , eds.) SOS Publications, Fair Haven, NJ. pp. 255-283.

Lee, S.Y., Cho, M.Y., Hyun, H.H., Lee, K.M., Homma, K., Natori, S., Kawabata, S., Iwanaga, S., and Lee, B.L. (1988) Molecular cloning of cDNA for pro-phenol-oxidase activating factor I, a serine proteinase is induced by lipopolysaccharide or 1,3-ß-glucan in a coleopteran insect, *Holotrichia diamphalia* larvae. *Eur. J. Biochem.* **257,** 615-621.

Lemaitre, B., Nicolas, E., Michaut, L., Reichhart, J.M., and Hoffmann, J.A. (1996) The dorsoventral regulatory gene cassette spatzle/Toll/cactus controls the potent antifungal response in Drosophila adults. *Cell* **86:**973-983.

Li, J., Wang, Z., Canagarajah, B., Jiang, H., Kanost, M., and Goldsmith, E.J. (1999) The structure of active serpin 1K from *Manduca sexta. Structure* **7,** 103-109.

Satoh, D., Horii, A., Ochiai, M., and Ashida, M. (1999) Prophenoloxidase-activating enzyme of the silkworm, *Bombyx mori*: purification, characterization and cDNA cloning. *J. Biol. Chem.* **274,** 7441-7453.

Skinner, W.S., Dennis, P.A, Li, J.P., Summerfelt, R.M., Carney, R.L., and Quistad, G.B. (1991) Isolation and identification of paralytic peptides from hemolymph of the lepidopteran insects *Manduca sexta, Spodoptera exigua, and Heliothis virescens. J. Biol. Chem.* **266,** 12873-12877.

Söderhäll, K., Ceranius, L. Johansson, M.W. (1996) The prophenoloxidase activating system in invertebrates. In *"New Directions in Invertebrate Immunology"*, (K. Söderhäll, S. Iwanga and G. Vasta, eds.) SOS Publications, Fair Haven, NJ. pp. 229-253.

Turner, M. W. (1996) Mannose-binding lectin: the pluripotent molecule of the innate immune system. *Immunol. Today* **17,** 532-540.

Volkman, B.F., Anderson, M.E., Clark, K.D., Hayakawa, Y., Strand, M.R., and Markley, J.L. (1999) Structure of the insect cytokine peptide plasmatocyte-spreading peptide 1 from *Pseudoplusia includens. J. Biol. Chem.* **274,** 4493-4496.

Wang, Y., Jiang, H. and Kanost, MR. (1999) Biological activity of *Manduca sexta* paralytic and plasmatocyte spreading peptide and primary structure of its hemolymph precursor. *Insect Biochem. Molec. Biol.* In press.

Yu, X.-Q., Prakash, O., and Kanost, M.R. (1999a) Structure of a paralytic peptide from an insect, *Manduca sexta. J. Peptide Res.* In press.

Yu, X.-Q., Gan, H., and Kanost, M.R. (1999b) An inducible C-type lectin from an insect, *Manduca sexta,* stimulates activation of plasma prophenol oxidase. *Insect Biochem. Molec. Biol.* **29:** 585-597.

CYTOTOXIC REACTIONS ASSOCIATED WITH INSECT IMMUNITY

A. J. Nappi and E. Vass

Department of Biology, Loyola University Chicago
Chicago, IL 60626

INSECT INNATE IMMUNE EFFECTOR MECHANISMS

Insects and other invertebrates lack the immunoglobulins and adaptive responses that characterize vertebrates yet possess efficient innate immune systems comprised of both cellular and humoral elements (Gillespie et al., 1997; Vilmos and Kurucz, 1998). The evolutionary origins and molecular basis for the various recognitive mechanisms remain among the most intriguing of immunological puzzles at all levels of biological organization (Ratcliffe, 1993; Ottaviani and Franceschi, 1998). Frequently, the first line of defense against potentially invasive organisms are integumental or midgut defenses, which may involve various cytotoxic proteins and antimicrobial peptides synthesized by epidermal cells and transported to the sites of wounding (Brey et al., 1993; Furukawa et al., 1999). Foreign organisms that breach integumental barriers or instead pass through the gut wall to invade the host's hemocoel encounter reactive blood cells, and an array of both non-specific and specific inducible cytotoxic molecules.

The predominant cellular immune responses of insects are phagocytosis and encapsulation. The latter response is directed against eukaryotic organisms and is characterized in part by the adhesion of numerous hemocytes to foreign surfaces to form a multilayered capsule (Marmaras et al., 1996; Carton and Nappi, 1997). Humoral immunity against intrahemocoelic invasions involves the induction of proteolytic

Phylogenetic Perspectives on the Vertebrate Immune System.
Edited by Beck *et al.*, Kluwer Academic/Plenum Publishers, 2001.

cascades that cause hemolymph coagulation (Brehelin, 1977; Rowley, 1977; Minnick et al., 1986; Muta and Iwanaga, 1996; Duvic and Brehelin, 1998), localized melanization reactions that generate cytotoxic quinoids and reactive intermediates of oxygen (ROI) (Nappi and Vass, 1993; Carton and Nappi, 1997), and the synthesis of antimicrobial peptides (Hoffmann et al., 1996; Meister et al., 1997; Dimarcq et al., 1999). Several studies have demonstrated striking parallels between the innate immune responses of insects and the innate immunity of mammals that have led some investigators to propose a common ancestry for these immune systems (Cooper, 1992; Ottaviani and Franceschi, 1998; Dimarcq et al., 1999).

Initial host cellular responses to pathogens or parasites occur mainly at the cell surface and are rapid and transient. Some of these early responses are based predominantly on the activation of pre-existing components, and involve little or no biosynthetic activity by the responding cells (*e.g.*, changes in extra- and intra-cellular pH, ion fluxes, and cross-linking and oxidative immobilization of membrane proteins). The cell type most likely to have played a key role in the evolution of animal defense systems was the phagocyte. When activated, phagocytic cells manifest a rapid and coordinated series of reactions that typically include cellular proliferation, chemotaxis, degranulation, and cell lysis. The molecules synthesized by activated phagocytes can be categorized as being derived from oxygen-dependent or oxygen-independent mechanisms (Table 1). Included in this arsenal of defenses are lysozyme, antimicrobial peptides, ROI and reactive intermediates of nitrogen (RNI). In immune reactive larvae of *Drosophila*, two ROI have been identified, superoxide anion ($O_2^{\cdot-}$) (Nappi et al., 1995) and hydrogen peroxide (H_2O_2) (Nappi and Vass, 1998).

Production of ROI by phagocytic cells results from ligand-receptor interactions and is mediated in part by NADPH oxidase, a plasma membrane-localized enzyme complex that facilitates the transfer of electrons from cytosolic NADPH to molecular oxygen via an intermediary flavoprotein and cytochrome (Eq 1). The initial ROI formed by this reaction is $O_2^{\cdot-}$, which is converted to H_2O_2 through action of superoxide dismutase (SOD) (Eq 2). Ensuing reactions involving these two molecules generate other reactive molecules, including the hydroxyl radical ($^{\cdot}OH$) which is considered to be the most reactive of the oxygen species (Eqs 3-6). Hydrogen peroxide can produce $^{\cdot}OH$ through interactions with transition metals (Fenton reaction) or nitric oxide (NO) (Nappi and Vass, 1998). The involvement of H_2O_2 with the myeloperoxidase (MPO)-halide system generates cytotoxic hypochlorous acid (OCl^-). The latter oxidant is mainly responsible for the intracellular killing of phagocytosed bacteria, as well as the oxidative modifications of tissue macromolecules, the deamination and decarboxylation of proteins with the formation of aldhydes, sulfhydryl oxidation, and the covalent cross-linking of proteins, such as IgG. The large burst in oxygen (O_2) consumption that accompanies the NADPH oxidase activity of phagocytes is associated with an increase in metabolism through the hexose monophosphate shunt. Although phagocytic cells are the most efficient producers of ROI, virtually all eukaryotic cells generate these molecules, primilarly as products of electron transfer reactions. Enzymes other than NADPH oxidase that mediate the production of ROI by activated phagocytes include xanthine oxidase, glucose oxidase, enzymes involved in arachidonic acid metabolism and mitochondrial electron transport.

$$NADPH + 2O_2 \longrightarrow NADP^+ + H^+ + 2O_2^{\cdot-} \qquad \text{Eq 1}$$

$$2O_2^{\cdot-} + 2H^+ \xrightarrow{SOD} H_2O_2 + O_2 \qquad \text{Eq 2}$$

$$H_2O_2 + O_2^{\cdot-} \longrightarrow {}^{\cdot}OH + HO^- + O_2 \qquad \text{Eq 3}$$

$$H_2O_2 + Fe^{2+} \longrightarrow {}^{\cdot}OH + HO^- + Fe^{3+} \qquad \text{Eq 4}$$

$$H_2O_2 + NO \longrightarrow {}^{\cdot}OH + HNO_2 \qquad \text{Eq 5}$$

$$H_2O_2 + NO_2 \longrightarrow {}^{\cdot}OH + HNO_3 \qquad \text{Eq 6}$$

Reactive intermediates of nitrogen are derived from NO, a free radical that is synthesized from L-arginine by NOS (Fig. 1). Three isoforms of nitric oxide synthase (NOS) have been described, each the product of a distinct gene. Two isoforms are constitutive (cNOS) and require NADPH, an elevation in intracellular Ca^{2+}, and attendant activation of calmodulin to produce NO. The third isoform is inducible (iNOS) upon stimulation by LPS or cytokines, and functions independently of calcium/ calmodulin, but does require NADPH. The NADPH dependant oxidation of a guanidino nitrogen of L-arginine by iNOS requires iron protoporphyrin IX (heme), flavin adenine dinucleotide (FAD), flavin adenine mononucleotide (FMN), and tetrahydrobiopterin (BH_4). Nitric oxide activates heme-containing proteins such as soluble guanylyl cyclase, oxyhemoglobin, and cytochrome p450. Nitric oxide also reacts readily with itself, H_2O, O_2, $O_2^{\cdot-}$ and H_2O_2 to generate an array of reactive molecules including nitrous anhydride (N_2O_3), peroxynitrite anion ($ONOO^-$), nitrite (NO_2^-), nitrate (NO_3^-) and ${}^{\cdot}OH$ (Fig. 2). When mediated by the MPO-halide system, the interactions of phagocyte-derived NO with H_2O_2 may enhance the covalent cross-linking of proteins (MacMicking et al., 1997). Injurious RNI-mediated nitrosative chemistry involving nitrosonium cation (NO^+) donation generates nitrosamines and S-nitrosothiols, causes the deamination of bases in DNA, and inhibits the activity of DNA repair enzymes. Peroxynitrite anion, its acid form (ONOOH), and ${}^{\cdot}OH$ are strong oxidizing agents that can initiate cytotoxic membrane lipid peroxidations. Peroxynitrite anion can also react with thiols and ascorbate to generate thiyl and ascorbyl radicals, respectively. It is important to note that NO can inhibit the TCA cycle and mitochondrial respiration by binding to the iron redox centers of aconitase, complex I, complex II, and complex IV.

COAGULATION

One mechanism of coagulation, which has been studied extensively in the horseshoe crab (*Lymulus polyphemus*), involves hemocyte serine protease zymogens, which trigger a coagulation cascade in response to lipopolysaccharide (LPS) and β-1, 3-glucan (Muta et al., 1993). Some components of this reaction have been found to contain complement-related domains, a C-type lectin domain, an epidermal growth factor-like domain, and structural similarity to nerve growth factor. In *Manduca sexta*, an inducible glucose-specific lectin has been identified that also serves to initiate hemocyte-coagulating activity (Minnick et al., 1986). Scolexin is a coagulation-provoking plasma protein that is induced in larvae of *M. sexta* in response to bacterial or viral infection (Kyriakides et al., 1995). Hemagglutinating or lectin activity also has been attributed to scolexin, and when this function is blocked the coagulation-initiating function is likewise inhibited. Possible activation mechanisms of the zymogens include proteolysis and the formation of at least one salt bridge, or via substrate or cofactor binding to form a complex. The latter may be a type of lectin/serine proteinase complex similar to that formed by human mannose-binding protein and associated serine proteinases that activate human complement (Sato et al., 1994; Thiel et al., 1997). The serine proteinases of these complexes become activated by the lectin component after a sugar binding event

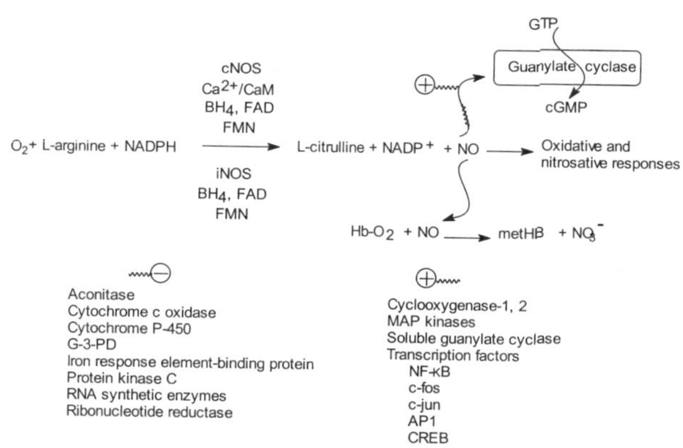

Figure 1. Nitric oxide is synthesized from L-arginine by NO synthase (NOS). Typically, NOS isoforms are either Ca^{2+}/calmodulin-dependent and constitutively expressed (cNOS), or Ca^{2+} -independent and inducible (iNOS). Inducible NOS is manifested by phagocytic cells in response to diverse stimuli, such as inflammatory cytokines (*e.g.*, IL-1, TNF-α) and LPS. Transforming growth factor B, IL-4 and IL-10 inhibit iNOS expression.

MELANIZATION

Melanization is a common manifestation in the hemocyte-mediated encapsulation reactions made against intrahemocoelic infections, and in cuticular wound healing responses that are generated at the site of integumental assault (Brey et al., 1993; Ashida and Brey, 1995). Little is known of the chemistry of the capsules that sequester intrahemocoelic parasites, but cross-linking processes similar to those involved in cuticular sclerotization may be involved in the formation of these melanotic protein-polyphenolic complexes (Nappi et al., 1992). Knowledge of the cellular and humoral activation processes that localize cytotoxic elements of the melanin proteolytic cascade is crucial to a better understanding of the non-self recognition mechanism of organisms that employ catecholamine-derived pigment intermediates in their defense against infectious agents.

In *Drosophila*, a melanogenic cascade is initiated early during parasite encapsulation, at about the time when crystal cells containing the requisite pigment precursors lyse at or near the foreign surface (Nappi, 1984; Nappi and Carton, 1986). Concomitantly, large numbers of phagocytic plasmatocytes precociously differentiate into larger hemocytes termed lamellocytes, which collaborate with plasmatocytes to form some of the capsule elements (Rizki and Rizki, 1990; Russo et al., 1996; Carton and Nappi, 1997). Studies involving mutants of *Drosophila* have demonstrated the important roles played by melanogenic proteins and specific hemocytes in the immune response of this insect. Parasitized mutant flies that either lack or are deficient in crystal cells are incapable of forming melanized capsules around eggs of the larval wasp parasitoid, *Leptopilina* (Rizki and Rizki, 1984). Similarly, mutant larvae with reduced dopa decarboxylase activity manifest a reduced melanotic encapsulation response against the parasitoid (Nappi et al., 1992). In the *domino* mutant, third-instar larvae lack circulating hemocytes, a situation that diminishes their ability to melanize, but does not alter their ability to maintain a wild-type antimicrobial response up to the time they die as prepupae (Braun et al., 1998). Instead of being released into the hemocoel, the differentiating blood cells are retained and destroyed within the hemocytopoetic (*i.e.*, lymph) glands, which exhibit varying degrees of melanization (Nappi and Carton, 1986). Similarly, melanization-deficient *Black cell* mutants exhibit a wild-type antimicrobial response, but fail to mount a wild-type melanization response (Braun et al., 1998). These observations establish the relevance of the proteolytic cascades that form antimicrobial peptides and produce melanization reactions in immune competent *Drosophila*.

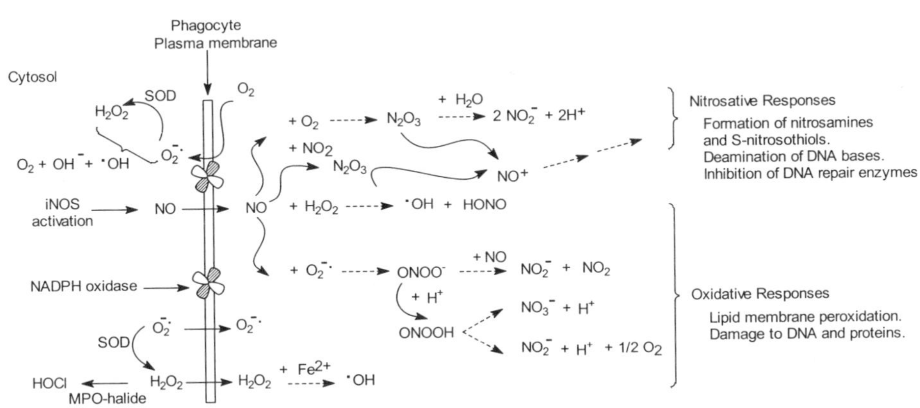

Figure 2. When activated, phagocytes generate reactive intermediates of oxygen (ROI) and nitrogen (RNI) which interact at sites of inflammation. An increase in O_2 uptake accompanies activation of a membrane-bound NADPH-oxidase and the production of $O_2^{-\cdot}$. Superoxide dismutase (SOD) converts $O_2^{-\cdot}$ to H_2O_2. Reactions involving H_2O_2 with $O_2^{-\cdot}$, transition metals, or NO generate $^{\cdot}OH$. Reactions of H_2O_2 with the myeloperoxidase (MPO)-halide system generate hypochlorous acid (HOCl). Activated

phagocytes also generate NO which reacts with H_2O_2 to generate $\cdot OH$. Nitric oxide also reacts with O_2 to form nitrous anhydride (N_2O_3) and nitrite (NO_2^-), and with $O_2^{\cdot-}$ to produce peroxynitrite (ONOO-), the acid form of which gives rise to NO_2^- or nitrate (NO_3^-). RNI-mediated nitrosative chemistry involving nitrosonium cation (NO^+) donation generates nitrosamines and S-nitrosothiols, causes the deamination of bases in DNA, and inhibits the activity of DNA repair enzymes.

The mechanism by which injury or pathogen invasion activates melanogenic enzyme cascades remain virtually unexplored. Two melanin biosynthetic pathways are known in animals, one producing brown to black eumelanins, the second forming yellow to reddish-brown pheomelanins (Fig. 3). Detailed molecular studies of melanogenesis have shown the process to be very complex, involving multiple enzymes (Hearing and Tsukamoto, 1991). In mammals, pigment biosynthesis typically occurs within melanocytes. These highly specialized cells contain at least three membrane-bound melanogenic oxidizing enzymes; tyrosinase (monophenol monooxygenase), tyrosinase-related protein 1 (TRP1, or DHICA oxidase), and tyrosinase-related protein 2 (TRP2, dopachrome tautomerase). Tyrosinase hydroxylates tyrosine (monophenol oxidase activity) to 3,4-dihydroxyphenylalanine (L-DOPA), and then oxidizes the latter (diphenoloxidase activity) to form DOPAquinone, the common precursor for both types of melanin (Prota, 1992; Ito et al., 1993). These reactions are the initial steps in melanin biosynthesis, making tyrosinase a key enzyme in melanin production. However, peroxidase can serve the same functions as tyrosinase in catalyzing the oxygenation of monophenols to o-diphenols, and the ensuing oxidation of o-diphenols to the corresponding quinones necessary for melanin synthesis (Fig. 4) (Prota, 1992). In the absence of cysteine or related thiol compounds, DOPAquinone is spontaneously converted into L-DOPAchrome. This semiquinone evolves to melanin via 5,6-dihydroxyindole (DHI), a highly unstable indolic o-diphenol intermediate derived from the decarboxylation of DOPAchrome. However, in the presence of TRP2, L-DOPAchrome undergoes a non-decarboxylative rearrangement leading to 5,6-dihydroxyindole-2-carboxylic acid (DHICA). This indolic o-diphenol is oxidized by TRP1 to an unstable o-quinone that further polymerizes to form an heterogeneous eumelanin biopolymer. Thus, the combined action of TRP2 and TRP1 accounts for incorporation of carboxylated indolic units into eumelanins. The three enzymes are believed to occur *in vivo* as a complex that may not only enhance pigment synthesis, but also mediate the release of potentially toxic melanogenic intermediates (Jimenez-Cervantes et al., 1994; Chakraborty et al., 1996; Schallreuter et al., 1998). Pheomelanins consist of benzothiazine, benzothiazole, and isoquinoline units derived from the oxidative polymerization of cysteinyl or glutathional conjugates of L-DOPA, or its decarboxylated derivative dopamine (Ito, 1986; Ito and Wakamatsu, 1989; Winder et al., 1994; Prota et al., 1995).

The melanization reactions associated with cuticular sclerotization, wound healing and defense reactions in insects may involve pathways similar to those outlined above (Hopkins et al., 1984; Wright, 1987; Pentz et al., 1990), but in many cases the aromatic substrate(s) and specific melanogenic oxidizing enzyme(s) responsible for pigment production in infected insects have not been identified. Insect phenol oxidase (PO; monophenol, dihydroxyphenylalanine:oxygen oxidoreductase; EC 1.14.18.1), also known as tyrosinase, is a copper-binding molecule that plays a role in sclerotization of cuticle and encapsulation of foreign particles. The components of insect PO comprise a serine proteinase cascade, and show a close phylogenetic relationship with the arthropod hemocyanins, the copper-binding proteins involved in oxygen transport (Hughes, 1999). In the case of cuticular melanization, it has been suggested that pro-phenol oxidase (proPO), a zymogen of PO, originates from circulating hemocytes and is actively transported across the epidermis and into the cuticular matrix. Recent studies demonstrate that the cuticular proPO of *Bombyx mori* is activated through a limited proteolysis by a serine proteinase, termed prophenol oxidase-activating enzyme (PPAE), which itself exists as a zymogen (Ashida and Brey, 1995). In *Drosophila*, proPO occurs in multiple forms (Al, A3 and Dox-A2) that appear to function as components in an activating cascade system (Pentz and Wright, 1991). Phenol oxidase activity in immune reactive larvae *of D. melanogaster* was shown to be markedly augmented during the early stages of melanotic encapsulation of the eggs of the parasitic wasp *Leptopilina boulardi*, but the mechanism of parasite-mediated activation was not examined (Nappi et al., 1991; Vass et al., 1993). The activities of other catecholamine-metabolizing enzymes (*e.g.*, tyrosine aminotransferase, N-acetyltransferase) were also found to be altered during parasite encapsulation in *Drosophila* (Vass et al., 1993). The type of pigment found in the hemocyte-derived capsules formed by larvae of *D. melanogaster* around eggs of the wasp parasitoid was identified as eumelanin based on the detection DHI (Nappi et al., 1992).

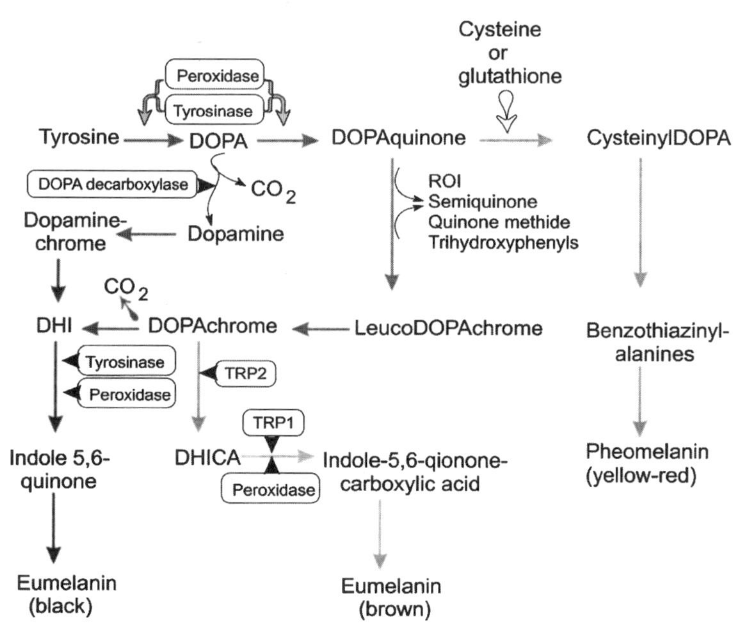

Figure 3. Diagram of the two melanogenic pathways in animals. Potentially cytotoxic molecules generated during melanogenesis include ROI, semiquinones, quinone methides, and trihydroxyphenyls. Trihydroxyphenyls, which are formed by nucleophilic addition of water to the o-quinone, can generate p-quinones, p-semiquinones, and p-quinone methides.

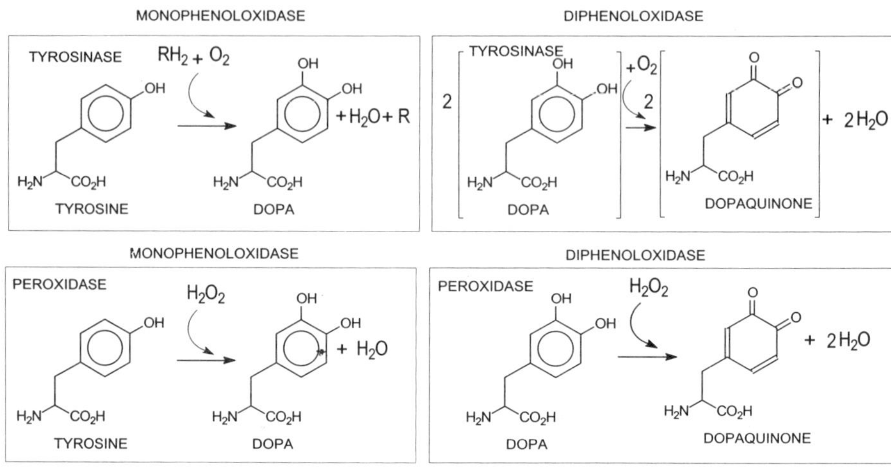

Figure 4. Diagram illustrating the conversion of tyrosine to DOPA, and DOPA to DOPAquinone via tyrosinase and peroxidase.

5,6-Dihydroxyindole and other quinoid intermediates of eumelanin (quinones, semiquinones, and quinone methides) are potentially cytotoxic, since they can bind to nucleophiles of foreign surfaces forming protein cross-linking complexes that immobilize endoparasites (Nappi and Vass, 1993). 5,6-Dihydroxyindole-2-carboxylic acid causes a marked increase in the levels of NO produced by LPS-induced iNOS, suggesting a messenger role for DHICA (D' Acquisto et al., 1995). Melanogenesis and melanin intermediates can switch aerobic glycolysis to anaerobic metabolism, stimulate the pentose phosphate pathway, and inhibit the phosphorylation of glycoproteins (Slominski et al., 1998). Because of the free radical scavenging properties of melanin (Riley, 1997; Rozanowska et al., 1999), its presence on the surface of a parasitoid's egg during capsule formation may facilitate the targeting of cytotoxic radicals, localizing immune reactions at foreign surfaces and preventing cytotoxic substances from disseminating throughout the open circulatory system of the host (Carton and Nappi, 1997). Other cytotoxic molecules that are generated in association with melanotic encapsulation reactions in larvae *of D. melanogaster* include two ROI, $O_2^{\cdot-}$ (Nappi et al., 1995) and H_2O_2 (Nappi and Vass, 1998) (Fig. 5).

ANTIMICROBIAL PEPTIDES

The humoral immune responses of insects also include the synthesis of a diverse set of broad-spectrum antimicrobial peptides and proteins (Meister et al., 1997), some of which may serve as cell adhesion molecules, or in regulating the activity of cell surface receptors or signal transduction pathways leading to altered gene expression (Lanz-

Mendoza et al., 1996). The immune peptides of insects are synthesized primarily by the fat body (a functional equivalent of the mammalian liver) (Hoffmann et al., 1996), and to a lesser extent by the hemocytes (Cociancich et al., 1994), cuticular cells (Brey et al., 1993), midgut and salivary glands (Kylsten et al., 1992), and reproductive structures (Samakovlis et al., 1991; Rosetto et al., 1996). Although considerable structural differences exist among the more than 150 antibacterial and antifungal peptides/polypeptides that have been described in insects, certain common structural patterns allow for the incorporation of many of these molecules into one of the following four groups; cecropins, cysteine-rich peptides, proline-rich peptides, and glycine-rich peptides/polypeptides (Hoffmann et al., 1996; Boman, 1998). Cecropins are linear (two linked a-helices) 4-kDa cationic peptides, frequently C-terminally amidated, and devoid of cysteine residues. They are produced in response to septic injury with Gram-positive and Gram-negative bacteria. The group of cysteine-rich peptides contains members with 1-4 disulfide bridges and molecular masses ranging from 2 to 6 kDa. Insect defensins, which belong to this family are active against Gram-positive bacteria. Cysteine-containing peptides exhibiting antifungal activity also have been described, drosomycin from *Drosophila,* thanatin from the hemipteran *Podisus maculiventris* (Fehlbaum et al., 1996), and heliomicin from the lepidopteran *Heliothis virescens* (Lamberty et al., 1999). Drosomycin and heliomicin are inactive against bacteria, whereas thanatin exhibits both fungicidal and bactericidal activities. The proline-rich group is comprised of molecules (e.g., drosocin, metchnikowin) that are active against Gram-negative bacteria and fungi (Levashina et al., 1995; Levashina et al., 1998). The glycine-rich family (attacins, sarcotoxins II, and diptericins) range from 8–24 kDa and are induced by Gram-negative bacteria. Presumably, these substances all share a common ability to permeabilize microbial membranes (Christensen et al., 1988; Cociancich et al., 1993; Morishima et al., 1995; Lockey and Ourth, 1996; Rosetto et al., 1996; Carlsson et al., 1998; Dimarcq et al., 1999; Hoffmann et al., 1999), a view that can be more critically assessed and strengthened with additional experimental approaches involving natural infections.

Substances known to stimulate the expression of immune peptide genes both *in vivo* and *in vitro* include arachidonic acid, phorbol esters, phytohemagglutinin, lipopolysaccharides, ROI, and RNI. Despite the usefulness of injecting bacteria or bacterial products into insects to monitor the production of polypeptides, it has yet to be established whether many of these substnaces are in fact produced during natural infections.

IMMUNE RECOGNITION

The ability of insects and other invertebrates to detect and combat infectious agents and abnormally developing endogenous cells necessitates the involvement of receptor-mediated responses and various endocytic and exocytic processes that facilitate cell to cell communication and the synthesis of cytotoxic molecules (Ottaviani et al., 1995). Experimental approaches designed to elucidate the molecular basis for discrimination between self and non-self in insects must address the concept that anticipatory non-self recognition responses are receptor-mediated, and that sufficient numbers of receptors be available to deal with all possible epitopes.

The identity of non-self recognition mechanism(s) that activate immune effector responses in insects and other invertebrates has not been irrefutably established, primarily because of the difficulty in identifying cause and effect relationships among temporally correlated responses. Non-self recognition and immune effector responses are viewed as separately regulated processes. Numerous substances have been implicated in the non-self recognition systems of invertebrates including lectins (Vasta et al., 1994; Parrinello, 1995), eicosinoids (Stanley-Samuelson et al., 1991; Uscian and Stanley-Samuelson, 1993; Morishima et al., 1997; Stanley, 1997), cytokines (Ottaviani et al., 1995), the proPO activating system and related melanogenic enzymes (*i.e.*, dopa decarboxylase, peroxidase) (Ratcliffe, 1991; Soderhall and Aspan, 1993; Soderhall et al., 1994; Ashida and Brey, 1995; Johansson and Soderhall, 1996; Marmaras et al., 1996), melanogenic intermediates (quinones, semiquinones, quinone methides) (Nappi and Vass, 1993), and a wide range of peptides and proteins that have been shown to modify hemocyte behavior during immune challenge (*e.g.*, cytokine- and complement-like molecules, hemomucin (Theopold et al., 1996), hemocytin (Kotani et al., 1995), hemolin (Sun et al., 1990; Bettencourt et al., 1997; Daffre and Faye, 1997; Faye and Kanost, 1997; Hughes, 1998), lipophorin (Mandato et al., 1996), peroxidasin (Nelson et al., 1994), and eicosanoids (Stanley-Samuelson et al., 1991; Stanley, 1997)), to exhibit antimicrobial activity, and molecules with structural and functional properties similar to some vertebrate complement components (Marchalonis and Schluter, 1989; Marchalonis and Schluter, 1990).

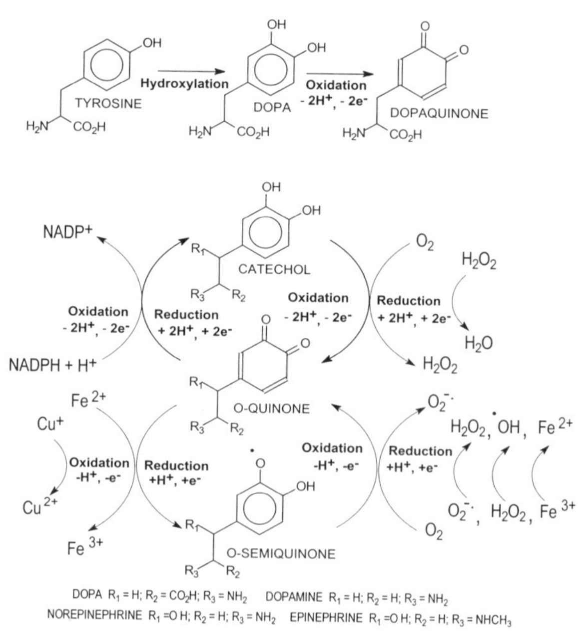

Figure 5. Diagram illustrating some possible redox reactions involving catechols, quinones, semiquinones, and certain metal ions, which can generate potentially cytotoxic ROI during melanotic encapsulation reactions.

PATTERN RECOGNITION PROTEINS

In insects, several proteins have been identified as putative pattern recognition proteins because of their ability to bind specific molecules, including lipopolysaccharide-binding protein, Gram-negative bacteria-binding protein, peptidoglycan binding protein, and β-1,3-glucan recognition protein (Trenczek, 1998). Pattern recognition proteins may be present in the plasma as free-floating molecules, or as plasma membrane receptors, where they are called pattern recognition receptors (Medzhitov et al., 1997). A protein that specifically binds to peptidoglycan (PGBP) in the silkworm *B. mori* is an essential component for exogenous peptidoglycan to trigger the proPO casacade that results in melanin synthesis (Ochiai and Ashida, 1999). Apparently, the binding of peptidoglycan and PGBP forms a complex that activates a protease zymogen of the proPO cascade. The PGBP gene is constitutively expressed in the fat body, epithelial cells, and hemocytes of naive silkworms. Upon bacterial challenge, the expression of the PGBP gene intensifies.

OTHER PROTEINS WITH PUTATIVE ROLES IN INSECT DEFENSE

In *Drosophila*, a macrophage receptor for apoptotic cells has been identified and termed croquemort (catcher of death) (Franc et al., 1996). This receptor is a member of the CD36 superfamily of scavenger receptors and is implicated in recognition. Croquemort-mediated phagocytosis suggests that phagocytosis evolved primarily as a cellular process for the removal of effete cells.

Another protein expressed by phagocytosing cells in *Drosophila* is peroxidasin (Nelson et al., 1994). This heme protein is synthesized by the fat body, and by the hemocytes that phagocytose apoptotic cells. Peroxidasin catalyzes certain hydrogen peroxide (H_2O_2)-mediated oxidations, and the formation of dityrosine. Peroxidasin is believed to function in phagocytosis and in extracellular matrix consolidation.

Hemolin is a molecule with a C2-type Ig-domain that appears to be involved in insect defense responses (Faye and Kanost, 1997). The protein is induced by bacterial infections in lepidopteran insects. It has been proposed that hemolin functions in recognition, phagocytosis, and hemocyte adhesion (Sun et al., 1990; Ladendorff and Kanost, 1991; Schmidt et al., 1993; Kanost et al., 1994; Lanz-Mendoza et al., 1996; Russell and Dunn, 1996; Zhao and Kanost, 1996; Shin et al., 1998; Su et al., 1998).

It has also been suggested that lipophorins may play a role in insect immune reactions by interacting with lipopolysaccharides in bacterial cell walls (Iimura et al., 1998). The formation of a serum lipoprotein-complex may represent a common pathway employed by insects and mammals to inactivate lipopolysaccharide and kill bacteria (Kato et al., 1994).

Lectins

Lectins are carbohydrate binding, non-enzymatic proteins or glycoproteins of ubiquitous occurrence. Because of their opsonizing properties and ability to agglutinate cells, invertebrate lectins have long been viewed as primitive recognition molecules (Lackie and Vasta, 1988; Vasta et al., 1994; Parrinello, 1995; Vasta and Ahmed, 1996), despite their heterogeneous and non-immunoglobulin composition. Several studies have demonstrated the ability of invertebrate lectins to agglutinate and immobilize microbial pathogens or metazoan parasites, reactions that may mediate subsequent phagocytosis or encapsulation of these organisms. Some endogenous invertebrate lectins have been shown to enhance the activation of the proPo system (Ratcliffe et al., 1991), with ensuing activation of the melanogenic cascade. Lectins have also been found accompanying melanin deposits on microfilariae of *Brugia malayi* infected *Anopheles quadrimaculatus* mosquitos (Nayar et al., 1995).

Melanin components

Part of the rationale for considering the insect proPO complex as a non-self recognition system is that a melanization reaction invariably accompanies hemocytic encapsulation responses made against endoparasites. However, in many cases the aromatic substrate(s) and specific melanogenic oxidizing enzyme(s) responsible for a melanization response occurring in an infected insect have not been identified. Various microbial products and exogenous substances (*e.g.*, fatty acids, detergents, organic solvents, transition metals) can activate the proPO system generating a cascade of chemical reactions that form melanin. When activated, the PO complex enhances the phagocytic ability of certain hemocytes (Ratcliffe et al., 1991). Interestingly, the proPO of some arthropods have been shown to contain a short sequence of amino acids that exhibits some sequence similarity to the thiol ester region of a-macroglobulins and complement proteins C3 and C4 (Hall et al., 1995).

It should be noted, however, that the use of various substances to activate proPO and to produce melanin merely establishes these compounds as melanogenic elicitors. Enhanced phagocytic activity following activation of the proPO system establishes the opsonic activity of melanin intermediates, but provides no evidence for a specific receptor-mediated non-self recognition process. Moreover, melanin-like pigments can be formed by the action of tyrosinase on substrates other than tyrosine or L-DOPA, including opioid peptides (Rosei et al., 1994; Rosei et al., 1995; Rosei, 1996), tetra-hydroisoquinolines (Rosei and Mosca, 1995; Mosca et al., 1998), and tryptophan (Hegedus et al., 1986; Aroca et al., 1991; Chakraborty and Chakraborty, 1993; Bosetto et al., 1995; Chakraborty et al., 1996). Also, melanogenic enzymes other than PO may be activated during melanogenesis, and products derived from such activation may play essential roles during immune responses. Of importance is the fact that other oxidative systems are known to produce both eumelanin and pheomelanin pigments from catecholamines and 5-S-cysteinyl-catecholamines, respectively, including peroxidase (d'Ischia et al., 1991), lipoxygenase/H_2O_2 (Rosei et al., 1994; Mosca et al., 1996), xanthine oxidase (Foppoli et al., 1997), cytochrome c (Rosei et al., 1998), and prostaglandin H synthase (*i.e.*, cyclooxygenase or COX) (Stokes et al., 1999). Thus, a variety of enzymes are capable of oxidizing various catecholamines and related substrates to produce melanin. Unless the specific reactions producing melanotic encapsulation reactions in insects are identified, the role proposed for melanogenic enzymes in insect immunity can not be established with any confidence.

Cytokines

Cytokines have long been studied with the idea of obtaining a better understanding of the evolution of the complex cellular, humoral and molecular interactions that regulate vertebrate host defenses (Beck et al., 1989; Beck and Habicht, 1991; Ottaviani et al., 1995). A prerequisite for the evolutionary acquisition of non-self recognition processes is an effective system of communication among cells that potentiates their collaborative interactions with foreign entities. Although invertebrates lack an acquired immune response based upon immunoglobulins or major histocompatibility complex antigens, these organisms produce various cytokines which function in non-specific immunity. Cytokines, which are released by a variety of activated immune cells, exert profound effects on cells of the immune system and the host defense system in general and are critical in the induction of effector mechanisms mediating antimicrobial defenses. Cytokines include the interferons (IFN), the interleukins (IL), tumor necrosis factor-alpha (TNF-α) and lymphotoxin. Since cytokines initiate and regulate essential host defense responses, these molecules are likely to be highly conserved during evolution and found throughout the animal kingdom. Some functional analogs of cytokines, such as IL-1, IL-2, IL-6 and TNF-a have been identified in different groups of invertebrates (*e.g.* annelids, tunicates, molluscs, echinoderms) (Raftos and Cooper, 1991; Raftos et al., 1991; Raftos et al., 1992; Ottaviani et al., 1995; Ottaviani et al., 1995; Hanusova et al., 1999). These invertebrate-derived cytokine-like molecules have been shown to share many biological activities with vertebrate equivalents, to cross-react with antibodies elicited against vertebrate cytokines, and to exhibit mitogenic activity (Beck et al., 1989; Beck and Habicht, 1991; Beck et al., 1996).

Are there receptor molecules associated with invertebrate hemocytes that are structurally or functionally similar to the surface Ig or T-cell antigen receptors on vertebrate lymphocytes? There is some serological evidence and partial amino acid sequence data suggesting the existence in invertebrates of molecules of the Ig superfamily (limulin, didemnum, lectin, fasciclin II, amalgam, Thy-1, β_2-microglobulin, echinoidin, Ig-related proteins, hemolin) (Ratcliffe, 1993), and perhaps these or some other hemolymph components could be involved in the non-self recognition process of invertebrates. It has been proposed that the Thy-1 molecule has been conserved throughout metazoan evolution and that the Thy-1 gene is the most conserved among the immune related genes (immunoglobulins, β_2-microglobulin and major histocompatibility complex), which share extensive amino acid homology (Shalev et al., 1985).

Complement Components

Components of the vertebrate complement system play critical roles in immunoregulation, initiation and enhancement of phagocytosis, and in inflammatory responses. Each of the three known complement activation pathways (e.g., classical, alternative, and lectin) produces C5 convertase, which cleaves complement component (C5) to initiate the self-assembly of a membrane attack complex. The membrane attack complex creates site-specific transmembrane channels into the lipid bilayer causing the lysis of target cells. It is of interest to note that a complement system with components similar to those found in vertebrates has been reported in deuterostome invertebrates (echinoderms), where it appears to be involved mainly in initiating or enhancing phagocytosis (Marchalonis and Schluter, 1989; Vilmos and Kurucz, 1998).

Perhaps these invertebrate complement-like molecules represent an early phylogenetic emergence of a mechanism to enhance both pathogen destruction and the non-self recognition capabilities of these organisms. However, experimental evidence is needed to clearly establish whether these substances indeed function in specific cell surface recognition interactions, or if they serve as secondary or effector molecules.

CONCLUSIONS

Invertebrate host-pathogen relations are integrating adaptations of great complexity involving the host's immune ability to both recognize and destroy foreign organisms. Differing strategies of immune recognition have evolved in response to selection pressures from specific parasites, pathogens, and other environmental factors. The types of non-self recognition mechanisms manifested by organisms appear to correlate well with their life span and morphological and physiological complexity, with typically short-lived invertebrate species exhibiting mostly innate mechanisms, and longer-lived vertebrate species possessing adaptive immune responses. Adaptive immune mechanisms characteristically involve cell membrane receptors that engage non-self molecules, collaborative cell interactions involving cell signaling, the activation of complement proteins, and synthesis of specific immunoglobulins. The humoral and cellular immune systems of invertebrates may interact with each other to some degree in order to successfully eliminate various pathogens. At present, little is known of the specific molecular mechanisms associated with immune surveillance and non-self recognition in insects and other invertebrates, but this important topic is the primary focus of numerous investigations.

ACKNOWLEDGEMENTS

The authors acknowledge the assistance of the National Institutes of Health (GM59774) and Research Services at Loyola University Chicago.

REFERENCES

Aroca P, Solano F, Garcia-Borron JC, Lozano JA. Specificity of dopachrome tautomerase and inhibition by carboxylated indoles. Considerations on the enzyme active site. Biochemical Journal 1991; 277(Pt 2):393-397.

Ashida M, Brey PT. Role of the integument in insect defense - pro-phenol oxidase cascade in the cuticular matrix. Proc.Natl. Acad.Sci. USA 1995; 92(23):10698-10702.

Beck G, Cardinale S, Wang L, Reiner M, Sugumaran M. Characterization of a defense complex consisting of interleukin 1 and phenol oxidase from the hemolymph of the tobacco hornworm, *Manduca sexta*. Journal of Biological Chemistry 1996; 271(19):11035-11038.

Beck G, Habicht GS. Primitive cytokines: Harbingers of vertebrate defense. Immunology Today 1991; 12:180-183.

Beck G, RF OB, Habicht GS. Invertebrate cytokines: the phylogenetic emergence of interleukin-1. Bioessays 1989; 11(2-3):62-7.

Bettencourt R, Lanz-Mendoza H, Lindquist KR, Faye I. Cell adhesion properties of hemolin, an insect immune protein in the Ig superfamily. European Journal of Biochemistry 1997; 250(3):630-7.

Boman HG. Gene-encoded peptide antibiotics and the concept of innate immunity: an update review. Scandinavian Journal of Immunology 1998; 48(1):15-25.

Bosetto M, Arfaioli P, Ristori GG, Fusi P. Formation of melanin-yype compounds from L-tryptophan on Ca-saturated and Al-saturated clays. Fresenius Environmental Bulletin 1995; 4(6):369-374.

Braun A, Hoffmann JA, Meister M. Analysis of the *Drosophila* host defense in domino mutant larvae, which are devoid of hemocytes. Proceedings of the National Academy of Sciences of the United States of America 1998; 95(24):14337-14342.

Brehelin M. Hemolymph coagulation in *Locusta migratoria*. Ann Parasitol Hum Comp 1977; 52(1):98-99.

Brey PT, Lee WJ, Yamakawa M, Koizumi Y, Perrot S, Francois M, Ashida M. Role of the integument in insect immunity: Epicuticular abrasion and induction of cecropin synthesis in cuticular epithelial cells. Proceedings of the National Academy of Sciences of the United States of America 1993; 90(13):6275-6279.

Carlsson A, Nystrom T, de Cock H, Bennich H. Attacin—an insect immune protein—binds LPS and triggers the specific inhibition of bacterial outer-membrane protein synthesis. Microbiology 1998; 144(Pt 8):2179-2188.

Carton Y, Nappi AJ. *Drosophila* cellular immunity against parasitoids. Parasitology Today 1997; 13(6):218-227.

Chakraborty AK, Chakraborty DP. The effect of tryptophan on dopa-oxidation by melanosomal tyrosinase. International Journal of Biochemistry 1993; 25(9):1277-1280.

Chakraborty AK, Platt JT, Kim KK, Kwon BS, Bennett DC, Pawelek JM. Polymerization of 5,6-dihydroxyindole-2-carboxylic acid to melanin by the pmel 17/silver locus protein. European Journal of Biochemistry 1996; 236(1):180-188.

Chakraborty DP, Roy S, Chakraborty AK. Vitiligo, psoralen, and melanogenesis: some observations and understanding. Pigment Cell Research 1996; 9(3):107-116.

Christensen B, Fink J, Merrifield RB, Mauzerall D. Channel-forming properties of cecropins and related model compounds incorporated into planar lipid membranes. Proceedings of the National Academy of Sciences of the United States of America 1988; 85(14):5072-5076.

Cociancich S, Bulet C, Hetru C, Hoffmann JA. The inducible antibacterial peptides in insects. Parasitology Today 1994; 10:131-139.

Cociancich S, Ghazi A, Hetru C, Hoffmann JA, Letellier L. Insect defensin, an inducible antibacterial peptide, forms voltage-dependent channels in *Micrococcus luteus*. Journal of Biological Chemistry 1993; 268(26):19239-19245.

Cooper EL. Overview of immunoevolution. Boll. Zool. 1992; 59:119-129.

D'Acquisto F, Carnuccio R, d'Ischia M, Misuraca G. 5,6-Dihydroxyindole-2-carboxylic acid, a diffusible melanin precursor, is a potent stimulator of lipopolysaccharide-induced production of nitric oxide by J774 macrophages. Life Sci 1995; 57(26):PL401-6.

Daffre S, Faye I. Lipopolysaccharide interaction with hemolin, an insect member of the Ig-superfamily. FEBS Lett 1997; 408(2):127-30.

Dimarcq JL, Bulet P, Hetru C, Hoffmann J. Cysteine-rich antimicrobial peptides in invertebrates. Biopolymers 1999; 47:465-477.

d'Ischia M, Napolitano A, Prota G. Peroxidase as an alternative to tyrosinase in the oxidative polymerization of 5,6-dihydroxyindoles to melanin(s). Biochim Biophys Acta 1991; 1073(2):423-30.

Duvic B, Brehelin M. Two major proteins from locust plasma are involved in coagulation and are specifically precipitated by laminarin, a beta-1,3-glucan. Insect Biochemistry & Molecular Biology 1998; 28(12):959-967.

Faye I, Kanost M, editors. Function and regulation of hemolin. London: Chapman & Hall; 1997. pp. 173-188.

Fehlbaum P, Bulet P, Chernysh S, Briand JP, Roussel JP, Letellier L, Hetru C, Hoffmann JA. Structure-activity analysis of thanatin, a 21-residue inducible insect defense peptide with sequence homology to frog skin antimicrobial peptides. Proc Natl Acad Sci U S A 1996; 93(3):1221-5.

Foppoli C, Coccia R, Cini C, Rosei MA. Catecholamines oxidation by xanthine oxidase. Biochim Biophys Acta 1997; 1334(2-3):200-6.

Franc NC, Dimarcq JL, Lagueux M, Hoffmann J, Ezekowitz RA. Croquemort, a novel *Drosophila* hemocyte/macrophage receptor that recognizes apoptotic cells. Immunity 1996; 4(5):431-43.

Furukawa S, Taniai K, Yang J, Shono T, Yamakawa M. Induction of gene expression of antibacterial proteins by chitin oligomers in the silkworm, *Bombyx mori*. Insect Molecular Biology 1999; 8(1):145-148.

Gillespie JP, Kanost MR, Trenczek T. Biological mediators of insect immunity. Annual Review of Entomology 1997; 42:611-643.

Hall M, Scott T, Sugumaran M, Soderhall K, Law JH. Proenzyme of Manduca sexta phenol oxidase: purification, activation, substrate specificity of the active enzyme, and molecular cloning. Proceedings of the National Academy of Sciences of the United States of America 1995; 92(17):7764-7768.

Hanusova R, Bilej M, Brys L, De-Baestselier P, Beschin A. Identification of a coelomic mitogenic factor in *Eisenia foetida* earthworm. Immunological Letters 1999; 65:203-211.

Hearing VJ, Tsukamoto K. Enzymatic control of pigmentation in mammals. FASEB J 1991; 5(14):2902-9.

Hegedus ZL, Frank HA, Altschule MD, Nayak U. Human plasma lipofuscin melanins formed from tryptophan metabolites. Archives of International Physiology & Biochemistry 1986; 94(5):339-48.

Hoffmann JA, Kafatos FC, Janeway CA, Ezekowitz RAB. Phylogenetic perspectives in innate immunity. Science 1999; 284:1313-1318.

Hoffmann JA, Reichhart JM, Hetru C. Innate immunity in higher insects. Current Opinion in Immunology 1996; 8(1):8-13.

Hopkins TL, Morgan TD, Kramer KJ. Catecholamines in haemolymph and cuticle during larval, pupal and adult development of *Manduca sexta*. Insect Biochemistry 1984; 14:533-540.

Hughes AL. Protein phylogenies provide evidence of a radical discontinuity between arthropod and vertebrate immune systems. Immunogenetics 1998; 47(4):283-96.

Hughes AL. Evolution of the arthropod prophenoloxidase/hexamerin protein family. Immunogenetics 1999; 49(2):106-14.

Iimura Y, Ishikawa H, Yamamoto K, Sehnal F. Hemagglutinating properties of apolipophorin III from the hemolymph of *Galleria mellonella* larvae. Archives of Insect Biochemistry & Physiology 1998; 38(3):119-25.

Ito S. Reexamination of the structure of eumelanin. Biochimica et Biophysica Acta 1986; 883(1):155-161.

Ito S, Wakamatsu K. Melanin chemistry and melanin precursors in melanoma. Journal of Investigative Dermatology 1989; 92(5 Suppl):261S-265S.

Ito S, Wakamatsu K, Ozeki H. Spectrophotometric assay of eumelanin in tissue samples. Anal Biochem 1993; 215(2):273-7.

Jimenez-Cervantes C, Solano F, Lozano JA, Garcia-Borron JC. The DHICA oxidase activity of the melanosomal tyrosinases LEMT and HEMT. Pigment Cell Research 1994; 7(5):298-304.

Johansson MW, Soderhall K. The prophenoloxidase activating system and associated proteins in invertebrates. Progress in Molecular Subcellular Biology 1996; 15:46-66.

Kanost MR, Zepp MK, Ladendorff NE, Andersson LA. Isolation and characterization of a hemocyte aggregation inhibitor from hemolymph of *Manduca sexta* larvae. Arch Insect Biochem Physiol 1994; 27(2):123-36.

Kato Y, Motoi Y, Taniai K, Kadono-Okuda K, Yamamoto M, Higashino Y, Shimabukuro M, Chowdhury S, Xu J, Sugiyama M and others. Lipopolysaccharide-lipophorin complex formation in insect hemolymph: a common pathway of lipopolysaccharide detoxification both in insects and in mammals. Insect Biochem Mol Biol 1994; 24(6):547-55.

Kotani E, Yamakawa M, Iwamoto S, Tashiro M, Mori H, Sumida M, Matsubara F, Taniai K, Kadono-Okuda K, Kato Y and others. Cloning and expression of the gene of hemocytin, an insect humoral lectin which is homologous with the mammalian von Willebrand factor. Biochim Biophys Acta 1995; 1260(3):245-58.

Kylsten P, Kimbrell DA, Daffre S, Samakovlis C, Hultmark D. The lysozyme locus in *Drosophila melanogaster*: different genes are expressed in midgut and salivary glands. Mol Gen Genet 1992; 232(3):335-43.

Kyriakides TR, McKillip JL, Spence KD. Biochemical characterization, developmental expression, and induction of the immune protein scolexin from *Manduca sexta*. Arch Insect Biochem Physiol 1995; 29(3):269-80.

Lackie AM, Vasta GR. The role of galactosyl-binding lectin in the cellular immune response of the cockroach *Periplaneta americana* (Dictyoptera). Immunology 1988; 64(2):353-7.

Ladendorff NE, Kanost MR. Bacteria-induced protein P4 (hemolin) from *Manduca sexta*: a member of the immunoglobulin superfamily which can inhibit hemocyte aggregation. Arch Insect Biochem Physiol 1991; 18(4):285-300.

Lamberty M, Ades S, Uttenweiler-Joseph S, Brookhart G, Bushey D, Hoffmann JA, Bulet P. Insect immunity. Isolation from the lepidopteran *Heliothis virescens* of a novel insect defensin with potent antifungal activity. Journal of Biological Chemistry 1999; 274(14):9320-6.

Lanz-Mendoza H, Bettencourt R, Fabbri M, Faye I. Regulation of the insect immune response: the effect of hemolin on cellular immune mechanisms. Cell Immunol 1996; 169(1):47-54.

Levashina EA, Ohresser S, Bulet P, Reichhart JM, Hetru C, Hoffmann JA. Metchnikowin, a novel immune-inducible proline-rich peptide from *Drosophila* with antibacterial and antifungal properties. Eur J Biochem 1995; 233(2):694-700.

Levashina EA, Ohresser S, Lemaitre B, Imler JL. Two distinct pathways can control expression of the gene encoding the *Drosophila* antimicrobial peptide metchnikowin. Journal of Molecular Biology 1998; 278(3):515-27.

Lockey TD, Ourth DD. Formation of pores in *Escherichia coli* cell membranes by a cecropin isolated from hemolymph of *Heliothis virescens* larvae. European Journal of Biochemistry 1996; 236(1):263-71.

MacMicking J, Xie Q, Nathan C. Nitric oxide and macrophage function. Annu. Rev. Immunol. 1997; 15(323-350.).

Mandato CA, Diehljones WL, Downer RGH. Insect hemocyte adhesion *in vitro*: Inhibition by apoliphorin I and an artificial substrate. Journal of Insect Physiology 1996; 42(2):143-148.

Marchalonis JJ, Schluter SF. Immunoproteins in evolution. Dev Comp Immunol 1989; 13(4):285-301.

Marchalonis JJ, Schluter SF. On the relevance of invertebrate recognition and defence mechanisms to the emergence of the immune response of vertebrates. Scand J Immunol 1990; 32(1):13-20.

Marmaras VJ, Charalambidis ND, Zervas CG. Immune response in insects: the role of phenoloxidase in defense reactions in relation to melanization and sclerotization. Arch Insect Biochem Physiol 1996; 31(2):119-33.

Marmaras VJ, Charalambidis ND, Zervas CG. Immune response in insects: The role of phenoloxidase in defense reactions in relation to melanization and sclerotization. Arch Insect Biochem. Physiol. 1996; 31:119-133.

Medzhitov R, Preston-Hurlburt P, Janeway CA, Jr. A human homologue of the *Drosophila* Toll protein signals activation of adaptive immunity. Nature 1997; 388(6640):394-7.

Meister M, Lemaitre B, Hoffmann JA. Antimicrobial peptide defense in *Drosophila*. BioEssays 1997; 19(11):1019-26.

Minnick MF, Rupp RA, Spence KD. A bacterial-induced lectin which triggers hemocyte coagulation in *Manduca sexta*. Biochem Biophys Res Commun 1986; 137(2):729-35.

Morishima I, Horiba T, Iketani M, Nishioka E, Yamano Y. Parallel induction if cecropin and lysozyme in larvae of the silkworm, *Bombyx mori*. Developmental & Comparative Immunology 1995; 19(5):357-363.

Morishima I, Yamano Y, Inoue K, Matsuo N. Eicosanoids mediate induction of immune genes in the fat body of the silkworm, *Bombyx mori*. FEBS Letters 1997; 419(1):83-6.

Mosca L, Blarzino C, Coccia R, Foppoli C, Rosei MA. Melanins from tetrahydroisoquinolines: spectroscopic characteristics, scavenging activity and redox transfer properties. Free Radical Biology & Medicine 1998; 24(1):161-7.

Mosca L, Foppoli C, Coccia R, Rosei MA. Pheomelanin production by the lipoxygenase-catalyzed oxidation of 5-S-cysteinyldopa and 5-S-cysteinyldopamine. Pigment Cell Research 1996; 9(3):117-25.

Muta T, Iwanaga S. The role of hemolymph coagulation in innate immunity. Current Opinion in Immunology 1996; 8(1):41-7.

Muta T, Oda T, Iwanaga S. Horseshoe crab coagulation factor B. A unique serine protease zymogen activated by cleavage of an Ile-Ile bond. J Biol Chem 1993; 268(28):21384-8.

Nappi AJ. Hemocyte reactions and early cellular changes during melanotic tumor formation in *Drosophila melanogaster*. J. Invertebr. Pathol 1984; 43:395-406.

Nappi AJ, Carton Y. Cellular immune responses and their genetic aspects in *Drosophila*. In: Brehelin M, editor. Immunity in Invertebrates. Berlin: Springer-Verlag; 1986. p 171-187.

Nappi AJ, Carton Y, Frey F. Parasite-induced enhancement of hemolymph tyrosinase activity in a selected immune reactive strain of *Drosophila melanogaster*. Arch Insect Biochem Physiol 1991; 18(3):159-68.

Nappi AJ, Carton Y, Li J, Vass E. Reduced cellular immune competence of a temperature-sensitive dopa decarboxylase mutant strain of *Drosophila melanogaster* against the parasite *Leptopilina boulardi*. Comp. Biochem. Physiol. 1992; 101B:453-460.

Nappi AJ, Vass E. Melanogenesis and the generation of cytotoxic molecules during insect cellular immune reactions. Pigment Cell Res 1993; 6(3):117-26.

Nappi AJ, Vass E. Hydrogen peroxide production in immune-reactive *Drosophila melanogaster*. Journal of Parasitology 1998; 84(6):1150-7.

Nappi AJ, Vass E. Hydroxyl radical formation resulting from the interaction of nitric oxide and hydrogen peroxide. Biochimica et Biophysica Acta 1998; 1380:55-63.

Nappi AJ, Vass E, Carton Y, Frey F. Identification of 3,4-dihydroxyphenylalanine, 5,6-dihydroxyindole and N-acetylarterenone during eumelanin formation in immune reactive larvae of *Drosphila melanogaster*. Arch. Insect Biochem Physiol 1992; 20: 181-191.

Nappi AJ, Vass E, Frey F, Carton Y. Superoxide anion generation in *Drosophila* during melanotic encapsulation of parasites. Eur J Cell Biol 1995; 68(4):450-6.

Nayar JK, Mikarts LL, Chikilian ML, Knight JW, Bradley TJ. Lectin binding to extracellularly melanized microfilariae of *Brugia malayi* from the hemocoel of *Anopheles quadrimaculatus*. J Invertebr Pathol 1995; 66(3):277-86.

Nelson RE, Fessler LI, Takagi Y, Blumberg B, Keene DR, Olson PF, Parker CG, Fessler JH. Peroxidasin: a novel enzyme-matrix protein of *Drosophila* development. Embo J 1994; 13(15):3438-47.

Ochiai M, Ashida M. A pattern recognition protein for peptidoglycan. Cloning the cDNA and the gene of the silkworm, *Bombyx mori*. Journal of Biological Chemistry 1999; 274(17):11854-8.

Ottaviani E, Capriglione T, Franceschi C. Invertebrate and vertebrate immune cells express pro-opiomelanocortin (POMC) mRNA. Brain Behav Immun 1995; 9(1):1-8.

Ottaviani E, Caselgrandi E, Franceschi C. Cytokines and evolution: *in vitro* effects of IL-1 alpha, IL-1 beta, TNF-alpha and TNF-beta on an ancestral type of stress response. Biochem Biophys Res Commun 1995; 207(1):288-92.

Ottaviani E, Franceschi C. A new theory on the common evolutionary origin of natural immunity, inflammation and stress response: the invertebrate phagocytic immunocyte as an eye-witness. Domestic Animal Endocrinology 1998; 15(5):291-6.

Ottaviani E, Franchini A, Cassanelli S, Genedani S. Cytokines and invertebrate immune responses. Biology of the Cell 1995; 85(1):87-91.

Parrinello N. Humoral and cellular lectins of ascidians. J. Mar. Biotechnol. 1995; 3:29-34.

Pentz ES, Black BC, Wright TRF. Mutation affecting phenol oxidase activity in *Drosophila*: quicksilver and tyrosinase-1. Biochem. Genetics 1990; 28:151-171.

Pentz ES, Wright TR. *Drosophila melanogaster* diphenol oxidase A2: gene structure and homology with the mouse mast-cell tum- transplantation antigen, P91A. Gene 1991; 103(2):239-42.

Prota G. The role of peroxidase in melanogenesis revisited. Pigment Cell Res 1992; Suppl(2):25-31.

Prota G, Lamoreux ML, Muller J, Kobayashi T, Napolitano A, Vincensi MR, Sakai C, Hearing VJ. Comparative analysis of melanins and melanosomes produced by various coat color mutants. Pigment Cell Research 1995; 8(3):153-163.

Raftos DA, Cooper EL. Proliferation of lymphocyte-like cells from the solitary tunicate, *Styela clava*, in response to allogeneic stimuli. J Exp Zool 1991; 260(3):391-400.

Raftos DA, Cooper EL, Habicht GS, Beck G. Invertebrate cytokines: tunicate cell proliferation stimulated by an interleukin 1-like molecule. Proc Natl Acad Sci U S A 1991; 88(21):9518-22.

Raftos DA, Cooper EL, Stillman DL, Habicht GS, Beck G. Invertebrate cytokines II: release of interleukin-1-like molecules from tunicate hemocytes stimulated with zymosan. Lymphokine Cytokine Res 1992; 11(4):235-40.

Ratcliffe NA. The prophenoloxidase system and its role in arthropod immunity. In: Warr GW, Cohen N, editors. Phylogenesis of Immune Functions. Boca Raton, FL: CRC Press; 1991. p 45-71.

Ratcliffe NA. Cellular defense responses of insects: unresolved problems. In: Beckage NE, Thompson SN, Federici BA, editors. Parasites and Pathogens of Insects. San Diego: Academic Press; 1993. p 267-304.

Ratcliffe NA, Brookman JL, Rowley AF. Activation of the prophenoloxidase cascade and initiation of nodule formation in locusts by bacterial lipopolysaccharides. Dev Comp Immunol 1991; 15(1-2):33-9.

Riley PA. Melanin. International Journal of Biochemistry & Cell Biology 1997; 29(11):1235-9.

Rizki TM, Rizki RM. The cellular defense system of *Drosophila melanogaster*. In: King RC, H. Akai H, editors. Insect ultrastructure. Volume 2. New York: Plenum Press; 1984. p 579-604.

Rizki TM, Rizki RM. Encapsulation of parasitoid eggs in phenoloxidase-deficient mutants of *Drosophila melanogaster*. *J. Insect Physiol.* 1990; 36:523-529.

Rosei MA. Melanins from opioid peptides. Pigment Cell Res 1996; 9(6):273-80.

Rosei MA, Blarzino C, Coccia R, Foppoli C, Mosca L, Cini C. Production of melanin pigments by cytochrome c/H2O2 system. International Journal of Biochemistry & Cell Biology 1998; 30(4):457-63.

Rosei MA, Blarzino C, Foppoli C, Mosca L, Coccia R. Lipoxygenase-catalyzed oxidation of catecholamines. Biochem Biophys Res Commun 1994; 200(1):344-50.

Rosei MA, Mosca L. Production of melanin pigments by chemical and enzymatic oxidation of tetrahydroisoquinolines. Biochemistry & Molecular Biology International 1995; 35(6):1253-9.

Rosei MA, Mosca L, Coccia R, Blarzino C, Musci G, De Marco C. Some biochemical properties of melanins from opioid peptides. Biochim Biophys Acta 1994; 1199(2):123-9.

Rosei MA, Mosca L, De Marco C. Spectroscopic features of native and bleached opio-melanins. Biochim Biophys Acta 1995; 1243(1):71-7.

Rosetto M, Manetti AG, Giordano PC, Marri L, Amons R, Baldari CT, Marchini D, Dallai R. Molecular characterization of ceratotoxin C, a novel antibacterial female-specific peptide of the ceratotoxin family from the medfly *Ceratitis capitata*. European Journal of Biochemistry 1996; 241(2):330-7.

Rowley AF. The role of the haemocytes of *Clitumnus extradentatus* in haemolymph coagulation. Cell Tissue Res 1977; 182(4):513-24.

Rozanowska M, Sarna T, Land EJ, Truscott TG. Free radical scavenging properties of melanin interaction of eu- and pheo-melanin models with reducing and oxidising radicals. Free Rad. Biol. Med. 1999; 26:518-525.

Russell V, Dunn PE. Antibacterial proteins in the midgut of *Manduca sexta* during metamorphosis. Journal of Insect Physiology 1996; 42(1):65-71.

Russo J, Dupas S, Frey F, Carton Y, Brehelin M. Insect immunity: early events in the encapsulation process of parasitoid (*Leptopilina boulardi*) eggs in resistant and susceptible strains of *Drosophila*. Parasitology 1996; 112(Pt 1):135-42.

Samakovlis C, Kylsten P, Kimbrell DA, Engstrom A, Hultmark D. The andropin gene and its product, a male-specific antibacterial peptide in *Drosophila melanogaster*. Embo J 1991; 10(1):163-9.

Sato T, Endo Y, Matsushita M, Fujita T. Molecular characterization of a novel serine protease involved in activation of the complement system by mannose-binding protein. International Immunology 1994; 6(4):665-669.

Schallreuter K, Slominski A, Pawelek JM, Jimbow K, Gilchrest BA. What controls melanogenesis? Experimental Dermatology 1998; 7(4):143-50.

Schmidt O, Faye I, Lindstrom-Dinnetz I, Sun SC. Specific immune recognition of insect hemolin. Dev Comp Immunol 1993; 17(3):195-200.

Shalev A, Segal S, Eli MB. Evolutionary conservation of brain Thy-1 glycoprotein in vertebrates and invertebrates. Dev Comp Immunol 1985; 9(3):497-509.

Shin SW, Park SS, Park DS, Kim MG, Kim SC, Brey PT, Park HY. Isolation and characterization of immune-related genes from the fall webworm, Hyphantria cunea, using PCR-based differential display and subtractive cloning. Insect Biochemistry & Molecular Biology 1998; 28(11):827-37.

Slominski A, Paus R, Mihm MC. Inhibition of melanogenesis as an adjuvant strategy in the treatment of melanotic melanomas: selective review and hypothesis. Anticancer Research 1998; 18(5B):3709-15.

Soderhall K, Aspan A. Prophenoloxidase activating system and its role in cellular communication. In: Pathak JPN, editor. Insect immunity. New Delhi: Oxford & IBH Publishing Co. Pvt. Ltd.; 1993. p 113-129.

Soderhall K, Cerenius L, Johansson MW. The prophenoloxidase activating system and its role in invertebrate defence. Ann N Y Acad Sci 1994; 712:155-61.

Stanley DW. Eicosanoids mediate insect cellular immune reactions to bacterial infections. Advances in Experimental Medicine & Biology 1997; 433:359-62.

Stanley-Samuelson DW, Jensen E, Nickerson KW, Tiebel K, Ogg CL, Howard RW. Insect immune response to bacterial infection is mediated by eicosanoids. Proc Natl Acad Sci U S A 1991; 88(3):1064-8.

Stokes AH, Hastings TG, Vrana KE. Cytotoxic and genotoxic potential of dopamine. J, Neurosci. Res. 1999; 55:659-665.

Su XD, Gastinel LN, Vaughn DE, Faye I, Poon P, Bjorkman PJ. Crystal structure of hemolin: a horseshoe shape with implications for homophilic adhesion. Science 1998; 281(5379):991-5.

Sun SC, Lindstrom I, Boman HG, Faye I, Schmidt O. Hemolin: an insect-immune protein belonging to the immunoglobulin superfamily. Science 1990; 250(4988):1729-32.

Theopold U, Samakovlis C, Erdjument-Bromage H, Dillon N, Axelsson B, Schmidt O, Tempst P, Hultmark D. *Helix pomatia* lectin, an inducer of *Drosophila* immune response, binds to hemomucin, a novel surface mucin. J Biol Chem 1996; 271(22):12708-15.

Thiel S, Vorupjensen T, Stover CM, Schwaeble W, Laursen SB, Poulsen K, Willis AC, Eggleton P, Hansen S, Holmskov U and others. A second serine protease associated with mannan-binding lectin that activates complement. Nature 1997; 386(6624):506-510.

Trenczek T. Endogenous defense mechanisms of insects. Zoology 1998; 101:298-315.

Uscian JM, Stanley-Samuelson DW. Phospholipase A2 activity in the fat body of the tobacco hornworm *Manduca sexta*. Arch Insect Biochem Physiol 1993; 24(4):187-201.

Vass E, Nappi AJ, Carton Y. Alterations in the activities of tyrosinase, N-acetyltransferase, and tyrosine aminotransferase in immune reactive larvae of *Drosophila* melanogaster. Dev Comp Immunol 1993; 17(2):109-18.

Vasta GR, Ahmed H. Animal lectins as cell surface receptors: current status for invertebrate species. Progress in Molecular & Subcellular Biology 1996; 17:158-182.

Vasta GR, Ahmed H, E. FN, T. EM, G. MA, Snowden A, Odom EW. Animal lectins as self/non-self recognition molecules: Biochemical and genetic approaches to understanding their biological roles and evolution. Annals of the New York Academy of Sciences 1994; 712:55-73.

Vilmos P, Kurucz E. Insect immunity: evolutionary roots of the mammalian innate immune system. Immunology Letters 1998; 62(2):59-66.

Winder A, Kobayashi T, Tsukamoto K, Urabe K, Aroca P, Kameyama K, Hearing VJ. The tyrosinase gene family—interactions of melanogenic proteins to regulate melanogenesis. Cell Mol Biol Res 1994; 40(7-8):613-26.

Wright TRF. The genetic of biogenic amine metabolism, sclerotization and melanization in *Drosophila melanogaster*. Adv Genet 1987; 24:127-222.

Zhao L, Kanost MR. In search of a function for hemolin, a hemolymph protein from the immunoglobulin superfamily. Journal of Insect Physiology 1996; 42(1):73-79.

GENOMIC AND cDNA SEQUENCE OF PROPHENOLOXIDASES FROM *DROSOPHILA MELANOGASTER*

Michael R. Chase and Manickam Sugumaran

Department of Biology
University of Massachusetts - Boston
Boston, MA 02125

INTRODUCTION

Phenoloxidase (PO) is an important enzyme associated with the biochemistry of three physiologically important processes in insects. The first process, *viz.*, cuticular sclerotization, ensures the protection of the soft bodied insects from their predators and preys as well as from desiccation by hardening the cuticle (Andersen *et al.*, 1996; Sugumaran 1991; 1998). During sclerotization, PO generated quinones serve both as sclerotizing agents and as substrates for quinone isomerases that produce quinone methides (Sugumaran, 1998; Saul & Sugumaran 1988, 1989 a,b, 1990, Ricketts & Sugumaran 1994). The reactions of quinones and quinone methides with cuticular structural components *viz.*, chitin and proteins, lead to the hardening of cuticle (Andersen *et al.*, 1996; Sugumaran 1991, 1998). In the second process, organisms establishing successful entry into the insect hemocel face among other host defense reactions, the dreaded action of PO, which causes deposition of melanin pigment around the intruder (Ashida & Brey 1995; Gillespie *et al.*, 1997; Nappi & Sugumaran 1993;

Phylogenetic Perspectives on the Vertebrate Immune System.
Edited by Beck *et al.*, Kluwer Academic/Plenum Publishers, 2001.

Soderhall, *et al.*, 1990; Sugumaran, 1996; Sugumaran & Kanost, 1993). While the intruder may still live in side the melanotic encapsule, its ability to multiply and damage the host is dramatically hindered by the melanotic capsule. Finally during wounding, loss of insect hemolymph is arrested by the rapid deposition melanin at the wounding site (Lai-Fook, 1966; Sugumaran, 1996). Apart from sealing the wound, PO action may also provide cytotoxic quinones that could harm the opportunistically invading microorganism (Nappi & Sugumaran 1993; Sugumaran 1996).

To perform such important physiological processes, PO is present in the hemolymph through out the life cycle of insects. Active PO is deleterious to the host. Hence, it is present as an inactive proform (prophenoloxidase, PPO) and is activated during a need (Ashida & Yamazaki, 1990). In recent years, rapid strides have been made towards our understanding of the biochemical and molecular biological aspects of this protein. PPOs from *Bombyx mori* (Ashida 1971; Yasuhara *et al.*, 1995), *Manduca sexta* (Aso *et al.*, 1985; Hall *et al.*, 1995; Jiang *et al.*, 1997a), *Hyalophora cecropia* (Andersson *et al.*, 1989), *Galleria mellonella* (Kopecek *et al.*, 1995), *Holotrichia diompalia* (Kwon *et al.*, 1997), *Drosophila melanogaster* (Seybold *et al.*, 1975; Fujimoto *et al.*, 1993), *Calliphora erythocephala* (Pau & Eagles 1975), *Musca domestica* (Hara *et al.*, 1993), *Blaberus discoidalis* (Durrant *et al.*, 1993), *Tenebrio molitor* (Heyneman 1965), and *Locusta migratoria* (Cherqui *et al.*, 1996) have been purified and characterized. The cDNA encoding PPOs from the following organism have been sequenced and characterized: two of *Manduca sexta* (Hall *et al.*, 1995; Jiang *et al.*, 1997a), two of *Bombyx mori* (Kawabata *et al.*, 1995), two of *Hyphantria cunea* (Park *et al.*, 1997), one from *Drosophila melanogaster* (Fujimoto *et al.*, 1995), six from *Anopheles gambiae* (Jiang *et al.*, 1997b; Lee *et al.*, 1998; Muller *et al.*, 1999), one of *Armigeres subalbatus* (Cho *et al.*, 1998), and two from *Sarcophaga bullata* (Chase *et al.*, 2000). Since a wealth of molecular biological resources are available on *Drosophila melanogaster*, we decided to investigate the presence of PPO genes in this organism. By surveying public data base, we identified three genomic sequences of PPO and present some of our preliminary findings in this brief report.

MATERIAL AND METHODS

Expressed Sequence Tags Data

The Berkeley Drosophila Genome Project database (http://www.fruitfly.org/blast/) was searched for Expressed Sequence Tags (ESTs) with homology to D*rosophila melanogaster* A1 PPO sequence previously characterized by Fujimoto *et al.*, (1995) using a tblastx search. Multiple ESTs were identified, downloaded, sorted by similarity with Sequencher and then compared by alignment with CLUSTAL W (Thompson *et al.*, 1994).

A clone for each distinct PPOs (Genbank accession number AI257599 and AA941592) we identified was obtained from Research Genetics, Inc. (Huntsville, Alabama), purified using standard procedures and sequenced on an ABI 377 (Applied Biosystems) using dye terminator chemistry. Internal primers were designed and used to sequence the remainder of each insert. Sequence data was assembled and edited with the program Sequencher (Gene Corps). Sequences were then conceptually translated, aligned with CLUSTAL W and imported into MACLADE (Maddison & Maddison, 1993).

Genomic Sequence Data

For each PPO we identified, a blastn search was used to query High Throughput Genomic Sequence (HTGS). Sequence homologous to these ESTs were analyzed with the program GENSCAN (Burge, 1998; Burge and Karlin, 1997) using the MIT server (http://CCR081. mit.edu/GENSCAN.html). Splice sites were identified with GENSCAN and by directly comparing the complete cDNA sequences to the genomic sequence.

RESULTS

EST Database Search

In our database searches, we identified three distinct PPO genes in *Drosophila melanogaster*. One homologous to the previously reported PPO, PPOA1 (Fujimoto *et al.* 1995) and two novel genes, designated here as DmPPOA2 (AI257599) and DmPPOA3 (AA941592). Since DmPPOA1 (D45835) was already sequenced (Fujimoto *et al.*, 1995), we did not sequence a homologous EST for this gene. The complete nucleotide sequence and deduced amino acid sequence for DmPPOA2 and DmPPOA3 are presented in Figure 1 and 2 respectively. All *D. melanogaster* PPOs are closely related to the two PPO genes characterized in *Sarcophaga bullata* (Chase *et al.*, 2000). Identities among these and all reported PPO amino acid sequences range from 79% to 38% (Data not shown). DmPPOA2 and DmPPOA3 are 69% identical. DmPPOA1 is most identical to SbPPO2 (62%), but less identical to DmPPOA2 (49%) and DmPPOA3 (48%). The finding of two divergent PPO lineages within a species is consistent with what others have found (Jiang *et al.*, 1997a,b; Muller *et al.*, 1999).

Figure 3 shows the alignment of all three PPOs from *Drosophila*. The potential proteolytic cleavage sites are indicated by arrows. The cleavage site (RF site) usually found in most PPOs is conserved only in DmPPOA1. Both DmPPOA2 and DmPPOA3 have RV at this site. On the other hand, the REE site found in most PPOs is present in DmPPOA2 and DmPPOA3. DmPPOA1 has RQE at this site. Other notable conserved motifs are: copper binding sites (underlined), putative thiol ester motif (double underlined) and Conserved C-terminal site (marked by asterisks).

Genomic Sequence

Genomic sequence for DmPPOA1, DmPPOA2 and DmPPOA3 was identified and analyzed with GENSCAN. Splice sites were confirmed by directly comparing the entire sequence of cDNA to the genomic sequence. They corresponded exactly to the GENSCAN predictions, except for DmPPOA3 which produced a hybrid with a gene immediately downstream of its C-terminal end. DmPPOA2 was localized to accession AC006074 on chromosome 2R, cytological region 44F12-45A4, DmPPOA1 to accession AC004640 on chromosome 2R, cytological region 54F1-55A1 and DmPPOA3 to accession AC008352, cytological region 59C-59D. The genomic sequence of DmPPOA1 contains five short introns, while DmPPOA2 and DmPPOA3 contain only three introns. All intron positions are identical between DmPPOA2 and DmPPOA3, but none are in common with DmPPOA1 (Figure 4-6).

DISCUSSION

In this survey, we identified two novel PPOs, DmPPOA2 and DmPPOA3 in *Drosophila melanogaster* and characterized the genomic sequence for DmPPOA2, DmPPOA3 and a previously identified DmPPOA1, by searching a public DNA sequence database. All these sequences share considerable homology with other reported PPO and are consistent with the observation of two distinct genetic lineages within a species. This data, along with the data reported by Muller *et al*, (1999) clearly demonstrate that PPOs have undergone multiple duplications within dipteran taxa.

The relative position and number of introns between these divergent genes also distinguish the independence between PPO lineages. Since none of the intron positions are conserved between DmPPOA1 and DmPPOA3, they are likely to have separated for a considerable period of time. Though intron position has not been characterized for other PPOs, they tend to be conserved (Stoltzfus *et al.*, 1997) and may useful to infer the evolutionary relationship between and within lineages.

Genetic markers have been studied that correlate with DmPPOA1, DmPPOA2 and DmPPOA3 in *Drosophila melanogaster* and are being reported. DmPPOA1 has already been linked to the genetic marker Black Cell (Bc) 55A1—55A4 (Fujimoto *et al.*, 1995). We associate DmPPOA2 with the mutant phenotype Tyrosine-2 (Tyr-2) 43E18—51A8 (Lewis and Lewis, 1963). DmPPOA3 correlates with the mutant phenotype Freckled (Frd) 59C-59C (Perotti *et al.*, 1968).

Since most PPO genes have been isolated from reverse transcribed larval mRNA, it is likely that additional surveys will uncover new genes encoding for PPOs. Several mutant phenotypes identified in Drosophila (FLYBASE) indicate that several additional PPO genes may be present in this taxa, but nomenclature and differences in biochemical assays used to characterize the gene products, complicate a clear association with PPO activity. For example, the gene isolated as DOX-A2 (Pentz *et al.*, 1991) is likely a regulator of PO, rather than a PO, since the cDNA from this region encodes a product that is highly homologous to proteosomes. However, we cannot eliminate the possibility that more PPO genes exist in *D. melanogaster*. The soon to be completed sequencing of the *Drosophila* genome will provide conclusive data.

Additional studies involving spatial and temporal localization with gene specific probes (antibody and antisense) will be required to understand adaptive significance of multiple PPO genes. Since, PPOs are involved in diverse processes such as sclerotization, melanization and encapsulation; localization studies should include several developmental stages and responses to hormonal, immune and wounding challenges. The role of a particular gene within a specific mechanism will allow further dissection of the cascade of factors that ultimately activate phenoloxidase and how these factors coevolve with recently derived genes.

```
    1 TTTGTATAAA ATGTGCAGTT CTTTCGGTGA ATAGTCACAG TCGTGAAGTT CCCAGACTTC CAGTCACATT CCCCATTTAG TTTGCTCCGC GATCCAGCAG
   +2                            M  A   D  K  K   N  L  L  L   L  F  D   H  P  T   E  P  V  F   M  D  K   G  K  R
  101 GTCCTCCCTG ACATCCCATT GAAAATGGCC GACAAGAAGA ATCTCCTCCT GCTTTTCGAC CATCCCACCG AGCCAGTGTT CATGGACAAG GGCAAGAGGG
   +2 V  T  V  F   D  V  P   D  S  F   L  T  D  R   Y  R  P   I  S  N   E  V  Q  S   R  V  G   D  K  V   E  Q  R  V
  201 TGACCGTGTT CGATGTGCCC GACTCCTTCC TGACCGACCG CTACCGACCC ATCAGCAACG AGGTGCAGAG CCGCGTCGGC GACAAGGTGG AGCAGCGCGT
   +2   P  V  R   E  I  S   I  P  D  L   R  I  P   M  S  L   G  R  D  E   Q  F  S   L  F  L   P  K  H  R   R  I  A
  301 TCCAGTCCGG GAGATCTCCA TTCCCGATCT GCGGATCCCC ATGTCCCTGG GCCGGGACGA GCAGTTCTCG CTGTTCCTGC CCAAGCACCG CAGGATCGCC
   +2  G  R  L   I  D  I  F   M  N  M   R  S  V   D  D  L  Q   S  V  A   V  Y  A   R  D  R  V   N  P  V   L  F  N
  401 GGCCGCCTGA TCGATATATT CATGAACATG CGCTCCGTGG ACGACCTGCA GAGCGTGGCG GTGTACGCCA GGGATCGGGT CAACCCGGTG CTCTTCAACT
   +2 Y  A  L  S   V  A  L   L  H  R   P  D  T  Q   G  L  D   L  P  S   F  S  Q  T   F  P  D   R  F  I   D  S  Q  V
  501 ACGCCTTGTC GGTGGCGCTG CTCCATCGTC CGGACACCCA GGGCCTGGAT CTGCCATCCT TCTCGCAGAC TTTCCCCGAC CGCTTCATCG ACTCGCAGGT
   +2   I  R  K   M  R  E   E  S  F  V   V  Q  P   G  S  R   M  P  I  T   I  P  R   D  Y  T   A  S  D  L   D  P  E
  601 GATCCGCAAG ATGCGGGAGG AGTCTTTTGT GGTGCAGCCC GGTTCCCGGA TGCCCATCAC CATTCCCAGG GACTACACCG CCTCCGATCT GGATCCCGAG
   +2  H  R  L   W  Y  F  R   E  D  L   G  I  N   L  H  H  W   H  W  H   L  V  Y   P  F  E  A   S  D  R   S  I  V
  701 CACCGGCTGT GGTACTTCCG CGAGGATCTG GGCATCAACC TTCACCACTG GCATTGGCAT CTGGTCTACC CCTTCGAGGC CTCCGACCGC AGCATAGTGG
   +2 A  K  D  R   R  G  E   L  F  Y   Y  M  H  Q   Q  V  I   A  R  Y   N  A  E  R   F  S  N   N  L  A   R  V  L  P
  801 CCAAGGATCG TCGTGGCGAG CTCTTCTACT ACATGCACCA GCAGGTGATC GCCCGCTACA ATGCGGAGCG GTTCAGCAAC AACCTGGCCA GGGTTCTGCC
   +2   F  N  N   L  R  D   P  I  A  E   G  Y  F   P  K  M   D  S  L  V   A  S  R   A  W  P   P  R  F  E   S  T  R
  901 CTTCAACAAC CTGAGGGATC CCATTGCCGA GGGCTACTTC CCCAAGATGG ACTCTCTGGT AGCCAGCCGC GCGTGGCCAC CGCGCTTCGA GAGCACCAGG
   +2  L  S  D   L  N  R  E   S  D  Q   L  N  V   E  I  G  D   L  E  R   W  R  D   R  I  Y  E   A  I  H   Q  G  F
 1001 TTGAGCGATC TCAACCGGGA GTCCGACCAG CTGAATGTGG AGATTGGAGA TCTGGAGCGT TGGCGTGATC GCATCTACGA GGCCATTCAC CAAGGATTTG
   +2 V  M  D  E   R  G  N   R  V  P   L  D  E  A   T  G  I   D  T  L   G  N  M  I   E  S  S   I  L  S   P  N  R  V
 1101 TCATGGATGA GCGTGGCAAC CGGGTGCCAT TGGATGAGGC GACGGGCATT GATACCCTCG GCAACATGAT TGAATCATCC ATTCTGTCGC CCAACCGAGT
   +2   L  Y  G   D  L  H   N  N  G  H   T  F  I   S  Y  A   H  D  P  T   S  K  H   L  E  S   F  G  V  M   G  D  V
 1201 GCTGTATGGT GACCTGCACA ACAACGGACA CACCTTCATC TCGTACGCCC ACGATCCGAC CAGCAAGCAC CTGGAGTCCT TTGGCGTGAT GGGTGATGTG
   +2  S  T  A   M  R  D  P   V  F  Y   K  W  H   S  Y  I  D   R  I  F   Q  E  H   K  S  R  L   P  A  Y   T  E  N
 1301 TCCACGGCCA TGCGCGACCC GGTCTTCTAC AAGTGGCACT CGTACATCGA CCGCATCTTC CAGGAGCACA AGTCGCGCCT GCCCGCCTAT ACCGAGAACC
   +2 Q  L  N  Y   P  G  V   S  I  A   G  I  Q  V   D  T  N   G  G  R   P  N  N  L   T  T  F   W  Q  Q   S  D  V  D
 1401 AACTAAACTA CCCGGGCGTC TCGATCGCGG GCATCCAGGT GGACACCAAT GGAGGTCGTC CCAATAATCT GACCACCTTC TGGCAGCAGT CGGACGTGGA
   +2   M  S  R   G  F  D   F  L  P  R   G  N  V   F  A  R   F  T  H  L   Q  H  L   P  F  T   Y  T  I  S   L  N  N
 1501 CATGTCCCGT GGCTTCGACT TCCTGCCGCG AGGTAATGTC TTCGCCAGGT TCACCCACCT GCAGCATCTG CCCTTCACCT ATACCATMAG TTTGAACAAC
   +2  D  S  G   A  Q  R  F   G  Y  V   R  I  F   M  A  P  K   N  D  E   R  G  Q   P  M  L  M   R  D  Q   R  S  M
 1601 GATAGTGGTG CCCAGCGTTT CGGCTACGTG CGCATCTTCA TGGCGCCGAA GAACGACGAG CGTGGCCAGC CCATGCTGAT GCGAGATCAG CGCTCCATGA
   +2 M  I  E  L   D  K  F   V  T  S   L  N  P  G   P  N  T   I  R  R   R  S  T  E   S  S  V   T  I  P   F  E  R  T
 1701 TGATCGAGCT GGACAAGTTC GTGACTTCGC TGAACCCGGG ACCCAACACC ATCCGCCGCC GCTCCACGGA GTCGAGCGTG ACCATTCCGT TCGAGCGCAC
   +2   F  R  N   L  D  A   N  R  P  A   A  G  T   P  E  E   L  E  F  N   F  C  G   C  G  W   P  N  H  M   L  V  P
 1801 CTTCCGCAAC CTGGATGCCA ATCGTCCGGC GGCCGGCACT CCGGAGGAGC TGGAGTTCAA CTTCTGCGGC TGCGGCTGGC CCAACCACAT GCTGGTGCCA
   +2  K  G  L   P  E  G  L   Q  C  V   L  F  I   M  V  S  N   Y  E  N   D  R  I   D  Q  Q  L   V  G  R   C  S  D
 1901 AAGGGTCTGC CCGAGGGACT GCAGTGCGTG CTGTTCATCA TGGTGTCGAA CTATGAGAAC GACAGGATTG ACCAGCAATT GGTGGGTCGG TGCAGCGATG
   +2 A  A  S  Y   C  G  V   R  D  R   L  Y  P  D   R  Q  S   M  G  F   P  F  D  R   L  P  R   S  G  V   D  R  L  V
 2001 CCGCTTCCTA CTGTGGCGTC CGTGACCGTC TTTACCCGGA TCGCCAGTCC ATGGGATTCC CCTTCGACAG GTTGCCCCGC AGCGGAGTCG ACCGCCTGGT
   +2   N  F  L   T  P  N   M  S  I  V   D  V  N   I  R  H   E  N  R  T   V  Q  R   P  N
 2101 TAACTTCCTC ACGCCCAACA TGAGCATCGT GGACGTGAAC ATCCGCCACG AGAACCGCAC CGTGCAGCGC CCAAACTAGA CGATCCGCAC CTGATTCGCT
 2201 TTCCCGATTC GCCCTGACCC GATTCCATGA CCTCGCCCAC TTCTAGATAT ACAACTCTTA GCAATCTATT GCATCTTTAT TGAGCCGCTT AATAAACAAG
 2301 ACATCATGGC ATTCCA
```

Figure 1. cDNA sequence and deduced amino acid sequence of PPOA2

```
  +3                                          M A    D K K N   L L L    L F D    H P T E   P V F    M D K
   1 TTAAAGGTCC TCCTAATCGC GAATCCATCG AAAATGGCCG ACAAGAAGAA TCTGCTCCTT CTGTTCGATC ATCCCACGGA GCCCGTGTTC ATGGACAAGG
  +3 G G N G    T V F    D V P    D S Y V    T D R    Y N Q    M C K K   V Q R    R V S    S A S E
 101 GTGGCAATGG CACCGTCTTC GACGTGCCCG ACTCCTACGT GACCGATCGC TACAATCAGA TGTGCAAGAA GGTCCAGCGG CGCGTGAGCA GCGCGTCCGA
  +3  K N V    Q V K    E I A I    P D L    S C S    M R L G    R S E    Q F S    I F L K    S H R
 201 GAAGAACGTT CAAGTCAAGG AGATCGCCAT ACCGGATCTC AGTTGCTCCA TGAGACTGGG CCGCTCGGAG CAGTTCTCGA TCTTCCTGAA GTCCCACCGC
  +3  K M A    S H L I    E I F    T K M    Q T V D    E L Q    S V A    V Y A R    D R V    N P V
 301 AAGATGGCCA GCCACCTGAT CGAGATCTTC ACCAAAATGC AGACCGTGGA TGAGCTGCAG AGCGTGGCGG TGTACGCCAG GGATCGCGTC AATCCGGTGC
  +3 L F N Y    A L S    V A L    L H R P    D T K    D L E    L P A F    A Q T    F P D    R F I D
 401 TCTTCAACTA CGCCTTGTCG GTGGCCCTGC TCCACCGTCC GGACACCAAG GACCTCGAAC TGCCCGCCTT TGCGCAGACC TTCCCGGATC GCTTTATCGA
  +3  S K M    L R S    M R E E    S F V    V E R    S A A R    L P V    V S S    V K Y T    A S D
 501 CTCGAAGATG CTTCGCAGCA TGCGCGAGGA GTCCTTTGTG GTGGAGCGCT CCGCCGCCCG CCTGCCCGTC GTCAGTTCGG TCAAGTACAC CGCCTCCGAT
  +3  L D V    E H R L    W Y F    R E D    L G V N    L H H    W H W    H L V Y    P I E    A P D
 601 CTCGATGTGG AGCACCGGCT GTGGTACTTC CGCGAGGACC TGGGCGTCAA CCTGCACCAC TGGCACTGGC ATCTGGTCTA TCCCATAGAG GCGCCCGACC
  +3 R S I V    D K D    R R G    E L F Y    Y M H    Q Q I    I A R Y    N A E    R L S    N H M A
 701 GCAGCATCGT GGACAAGGAC CGTCGTGGCG AGCTGTTCTA CTATATGCAC CAGCAGATCA TTGCCCGCTA CAATGCCGAG CGGCTGAGCA ACCACATGGC
  +3  R V Q    P F N    N L D E    P I A    E G Y    F P K M    D S L    V A S    R A Y P    P R F
 801 CCGGGTGCAG CCGTTCAACA ACCTGGACGA ACCAATTGCC GAGGGCTACT TCCCCAAGAT GGACTCGCTG GTGGCCAGTA GGGCGTATCC TCCGCGCTTC
  +3  D N T    R L S D    V D R    P N N    Q L R V    G I D    D M K    R W R E    R I Y    E A I
 901 GATAACACCC GGCTGAGCGA TGTGGACCGG CCTAACAACC AGTTGAGGGT TGGCATCGAT GATATGAAGC GCTGGCGCGA ACGCATCTAC GAGGCCATCC
  +3 H Q G Y    V L D    T N N    E K I V    L D D    A K G    I D I L    G N I    I E A    S D L T
1001 ATCAGGGCTA CGTTTTGGAT ACGAACAATG AAAAGATCGT GCTGGATGAC GCTAAGGGCA TCGACATCCT GGGCAACATC ATCGAGGCTT CCGATCTGAC
  +3  P N S    T L Y    G D F H    N M G    H I L    I A Y S    H D P    T N K    H L E Y    A G V
1101 GCCCAACAGT ACGCTGTACG GCGACTTCCA CAACATGGGG CACATACTGA TTGCCTACTC ACACGATCCG ACCAACAAGC ACCTGGAGTA CGCGGGCGTG
  +3  M G D    A S T A    M R D    P I F    Y K W H    A F I    D N M    F Q E H    K R L    L S P
1201 ATGGGCGACG CCTCCACCGC GATGCGCGAT CCGATCTTCT ACAAGTGGCA CGCCTTCATC GACAATATGT TCCAGGAGCA CAAACGTCTG CTTTCTCCCT
  +3 Y E K Q    E L S    F P D    V R V E    S I Q    V E S    Q G Q V    N R L    T T F    W Q E S
1301 ATGAGAAGCA AGAGCTCAGC TTTCCGGACG TTCGCGTCGA GAGCATCCAG GTGGAGAGCC AGGGTCAGGT CAACAGGCTG ACCACCTTCT GGCAGGAGTC
  +3  D V D    M S R    G L D F    V P R    G H V    L A R F    T H L    Q H H    E F S Y    T I K
1401 CGATGTGGAC ATGTCCCGTG GCCTTGATTT TGTGCCGCGC GGCCATGTCC TCGCTCGGTT CACCCATCTG CAGCACCACG AATTCAGCTA CACCATCAAG
  +3  V E N    S S E A    T R Y    G Y V    R I F L    A P K    L D D    R N A P    M L L    E Q Q
1501 GTGGAGAACT CCAGCGAGGC CACGCGGTAT GGCTACGTGC GCATCTTCCT GGCGCCCAAG CTGGATGATC GCAATGCACC CATGCTGCTG GAGCAACAGC
  +3 R L M M    V E L    D K F    V V T M    P P G    S H T    I T R N    S T E    S S V    T I P F
1601 GCCTGATGAT GGTGGAGCTG GACAAGTTCG TGGTTACCAT GCCTCCCGGC AGCCATACGA TCACCCGGAA CTCCACGGAG TCGAGTGTCA CCATTCCGTT
  +3  E R T    F R N    L D K L    E E L    Q N F    L C G C    G W P    Q H M    L I P K    G R P
1701 CGAGCGCACC TTCCGCAACC TGGACAAGCT GGAGGAGCTG CAGAACTTCT TGTGCGGCTG CGGTTGGCCC CAGCACATGC TGATTCCGAA GGGTCGGCCC
  +3  E G L    R F E L    F V M    V S N    Y E E D    K V D    Q T V    A D C G    C S I    A A S
1801 GAGGGTCTGC GCTTCGAGCT CTTCGTCATG GTTTCCAACT ACGAAGAGGA TAAGGTTGAC CAGACGGTCG CAGATTGCGG CTGCAGCATC GCCGCTTCCT
  +3 Y C G L    R D R    L Y P    D R K S    M G F    P F D    R K P R    R G S    E I L    E N F L
1901 ACTGCGGCCT GCGCGATCGC CTCTATCCGG ATCGCAAGTC CATGGGCTTC CCCTTCGATC GCAAGCCGCG TCGAGGCTCC GAGATCCTGG AGAACTTCCT
  +3  T P N    M C A    V E V I    I T H    E D R    T E K L    R E V    P A R    S
2001 GACGCCCAAC ATGTGCGCCG TGGAGGTCAT CATTACCCAC GAGGACCGCA CCGAGAAGCT TCGCGAGGTA CCAGCTCGCT CCTGAATAAG AGAAGCATCT
2101 TGTTATGGCC AGTTTCATTT TTTGGCTTAA AGAGAAGCCA CTAATAACGG TCTATAGCTT TCTCGCTGTG TAGAAAATAC CAAATAAATT AATGAAATAC
2201 TAAAAAAAAA AAAAAAAAAA
```

Figure 2. cDNA sequence and deduced amino acid sequence of PPOA3

```
          10        20        30        40        50 ⇓      60        70        80        90       100
A1   MT-NTDLKALELMFQRPLEPAFTTRDSGKTVLELPDSFYTDRYRNDTEEVGNRFSKDVDLKIPIQELSNVPSLEFTKKIGLKNQFSLFNNRHREIASELI [99]
A2   -MAD-K.N.L.L.DH.T..V.MDKGKRV..FDV....L.....PISN..QS.VGDK.EQRV.VR.I.-I.D.RIPMSL.RDE.....LPK..R..GR.. [96]
A3   -MAD-K.N.L.L.DH.T..V.MDKGGNG..FDV...YV....NQMCKK.QR.V.SASEKNVQVK.IA-I.D.SCSMRL.RSE...I.LKS..KM..H.. [96]

          110       120       130       140       150       160      ⇓170       180       190       200
A1   TLFMSAPNLRQFVSLSVYTKDRVNPVLFQYAYAVAVAHRPDTREVPITNISQIFPSNFVEPSAFRDARQEASVIGESGARVHVDIPQNYTASDREDEQRL [199]
A2   DI..NMRSVDDLQ.VA..AR........N..LS..LL.....QGLDLPSF..T..DR.IDSQVI.KM.E.SF.VQ-P.S.MPIT..RD.....LDP.H.. [195]
A3   EI.TKMQTVDELQ.VA..AR........N..LS..LL.....KDLELPAFA.T..DR.IDSKML.SM.E.SF.VER.A..LP.VSSVK.....LDV.H.. [196]

          210       220       230       240       250       260       270       280       290       300
A1   AYFREDIGVNSHHWHWHLVYPTT-GPTEVVNKDRRGELFYYMHHQILARYNVERFCNNLKKVQPLNNLRVEVPEGYFPKILSSTNNRTYPARVTNQKLRD [298]
A2   W.....L.I.L.........FEASDRSI.A............Q.VI....A...S...AR.L.F....DPIA......MD.LVAS.AW.P.FESTR.S. [295]
A3   W.....L...L.........IEAPDRSI.D............Q..I.....A..LS.HMAR...F...DEPIA......MD.LVAS.A..P.FD.TR.S. [296]

          310       320       330       340       350       360       370       380       390       400
A1   VDRHDGR-VEISDVERWRDRVLAAIDQGYVEDSSGNRIPLDEVRGIDILGNMIEASPVLSINYNFYGNLHNEGHNIISFAHDPDYRHLEDFGVMGDVTT [396]
A2   LN.ESDQLN...G.L......IYE..H..F.M.ER...V....AT...T.......S.-I..P.RVL..D...N..TF..Y....TSK...S.......S. [394]
A3   ...PNNQLR.G.D.MK...E.IYE..H....L.TNNEK.V..DAK.......I....-D.TP.STL..DF..M..IL.AYS...TNK...YA.....AS. [395]

          410       420       430       440       450       460       470       480       490       500
A1   AMRDPIFYRWHGFIDTVFNKFKTRLNPYNAGELNFDGITVDYIEAKIGKSNTKANTLLTYWQKSSADLAAGLDFGPTTDRNIFASFTHLQNAPFTYTFNV [496]
A2   .....V..K..SY..RI.QEH.S..PA.TENQ..YP.VSIAG.QVDTN—GGRP.N.T.F..Q.DV.MSR.F..L.-RG.V..R.....HL.....ISL [490]
A3   ........K..A...NM.QEH.RL.S..EKQ..S.PDVR.ES.QVESQ—G-QV.R.T.F..E.DV.MSR....V.-RGHVL.R.....HHE.S..IK. [490]

          510       520       530       540       550       560       570       580       590       600
A1   TNNGAR-RTGTCRIFICPKVDERNQALNLEEQRLLAIEMDKFTVDLVPGENTIRRQSTESSVAIPFERSFRPVGADYQPKAADELARFKFCGCGWPQHLL [595]
A2   N.DSGAQ.F.YV...MA..N...G.PMLMRD..SMM..L...VTS.N..P.....R......T.....T..NLD.NRPAAGTP.ELE.N.......N.M. [590]
A3   E.SSEAT.Y.YV...LA..L.D..APML..Q...MMV.L...V.TMP..SH..T.N......T.....T..NLDK——L-.ELQNFL........M. [583]

          610       620       630       640       650       660       670       680       690       700
A1   LPKGNAQGMLFDLFVMISDYSQDSVEQPK-TPNDACSTAYSFCGLKDKLYPDRRTMGYPFDRRLP—NANLTELVGAFGNMAKTDLRIVFNDRVIDK- [689]
A2   V...LPE.LQCV..I.V.N.EN.-RIDQQLVGR-..D.A.Y..VR.R.....QS..F....LPRSGVDR.VNFLTP-..SIV.VN.RHEN.TVQR- [682]
A3   I...RPE.LR.E....V.N.EE.-KVDQTVADCG..I.A.Y...R.R.....KS..F....KPRRGSEI.ENFLTP-..CAVEVI.THE..TEKLREV [679]
                                                       **************

A1   A-    [690]
A2   PN-    [684]
A3   PARS    [683]
```

Figure 3. Alignment of all three PPO sequences from *Drosophila melanogaster*.

```
   1 TTGGATTTGT GAGAAAAAAG GCTCGACCAT TGCAGTTTTG CGGTTGGGAC TTGATAATGA ATATTTTTAA AAATATTATT TTTATATTCG GACTGCTAGT
 101 TAGCTATAAA ATGAGTTCGG CTTCCTGGGA AAATCACAGT TCTGCAACTG GCTTCGTTGA GTGAAGTTAA AAGTGTAGTT GAAAGTGTAC ATCCATAGAG
  +3 EXON 1 M  T  N  T  D  L  K  A  L  E  L  M  F  Q  R  P  L  E  P  A  F  T  T  R  D  S  G  K  T  V  L
 201 AAAACCATGA CTAACACGGA TCTGAAAGCC TTGGAACTTA TGTTCCAGCG ACCCCTGGAA CCGGCATTCA CCACTCGTGA CTCCGGAAAA ACTGTATTGG
  +3 E  L  P  D  S  F  Y  T  D  R  Y  R  N  D  T  E  E  V  G  N  R  F  S  K  D  V  D  L  K  I  P  I  Q  E
 301 AACTGCCCGA TTCCTTCTAT ACGGATCGCT ATCGCAATGA CACCGAGGAG GTGGGCAATC GATTCTCTAA GGACGTGGAT CTGAAGATCC CAATTCAGGA
  +3  L  S  N  V  P  S  L  E  F  T  K  K  I  G  L  K  N  Q  F  S  L  F  N  N  R  H  R  E  I  A  S  E  L
 401 GTTGAGCAAC GTTCCCAGTC TGGAGTTCAC CAAAAAGATT GGTCTCAAGA ATCAATTTTC GCTGTTTAAC AATCGCCATC GTGAGATCGC CAGCGAGCTG
  +3  I  T  L  F  M  S  A  P  N  L  R  Q  F  V  S  L  S  V  Y  T  K
 501 ATCACCCTCT TCATGAGCGC ACCAAATCTC AGACAATTCG TGTCGCTTTC CGTTTACACC AAGTAAGTAC TTGTACTAAA ATAAATATAT ATGTATTGTA
  +3             EXON 2     D  R  V  N  P  V  L  F  Q  Y  A  Y  A  V  A  V  A  H  R  P  D  T  R  E  V  P  I
 601 TAATGTATTA ATTTTTTAGG GATCGTGTGA ATCCTGTGTT GTTCCAATAC GCCTATGCCG TCGCCGTTGC CCATCGCCCG GATACGCGTG AAGTGCCCAT
  +3  T  N  I  S  Q  I  F  P  S  N  F  V  E  P  S  A  F  R
 701 TACCAACATT TCCCAGATCT TTCCGAGCAA CTTTGTAGAA CCTTCCGCCT TCCGGGTTCG TTTAAGTTCT TCTTTATTTG ACCAATCAAA GTCTTTAATG
  +3            EXON 3         D  A  R  Q  E  A  S  V  I  G  E  S  G  A  R  V  H  V  D  I  P  Q  N  Y  T
 801 GTGTATTTAT AAATTTGACC ACAGGATGCC CGTCAGGAAG CTTCTGTTAT TGGAGAGAGC GGCGCTCGTG TCCATGTGGA CATTCCGCAA AACTACACGG
  +3 A  S  D  R  E  D  E  Q  R  L  A  Y  F  R  E  D  I  G  V  N  S  H  H  W  H  W  H  L  V  Y  P  T  T  G
 901 CCTCGGATCG CGAGGATGAA CAGCGCCTGG CGTACTTCCG TGAGGACATC GGTGTTAATA GCCATCATTG GCACTGGCAC TTGGTCTACC CGACCACCGG
  +3  P  T  E  V  V  N  K  D  R  R  G  E  L  F  Y  Y  M  H  H  Q  I  L  A  R  Y  N  V  E  R  F  C  N  N
1001 ACCCACGGAG GTGGTCAACA AGGATCGTCG TGGCGAGCTC TTCTACTACA TGCACCACCA GATACTCGCG CGCTACAATG TGGAGCGCTT CTGCAACAAT
  +3  L  K  K  V  Q  P  L  N  N  L  R  V  E  V  P  E  G  Y  F  P  K  I  L  S  S  T  N  N  R  T  Y  P  A
1101 CTGAAGAAGG TTCAGCCGCT GAACAATCTG CGTGTGGAGG TTCCCGAGGG CTACTTCCCC AAGATCCTGT CCAGCACGAA TAACCGCACC TATCCCGCCC
  +3 R  V  T  N  Q  K  L  R  D  V  D  R  H  D  G  R  V  E  I  S  D  V  E  R  W  R  D  R  V  L  A  A  I  D
1201 GAGTGACCAA CCAAAAGCTG AGGGACGTGG ACCGCCATGA TGGGCGTGTG GAGATCTCCG ATGTGGAGCG CTGGCGTGAT CGCGTGCTGG CTGCCATAGA
  +3  Q  G  Y  V  E  D                                                                  EXON 4   +2  S  S
1301 TCAAGGATAC GTTGAAGATG TGAGTTGGGC CCGTTTGGAA TATTGAGCAT AGGTCATAGA AAGCACTAGA AATTTTTATT TGCTAATCCA TAGTCTTCTG
  +2 G  N  R  I  P  L  D  E  V  R  G  I  D  I  L  G  N  M  I  E  A  S  P  V  L  S  I  N  Y  N  F  Y  G  N
1401 GCAATCGTAT ACCATTGGAC GAAGTTCGTG GCATCGATAT CCTGGGCAAC ATGATTGAGG CATCGCCAGT CCTATCCATC AACTATAATT TCTATGGCAA
  +2  L  H  N  E  G  H  N  I  I  S  F  A  H  D  P  D  Y  R  H  L  E  E  F  G  V  M  G  D  V  T  T  A  M
1501 TCTACACAAC GAGGGGCACA ACATAATCTC CTTTGCCCAC GATCCCGATT ACCGTCACTT GGAGGAATTC GGTGTGATGG GCGATGTGAC CACGGCCATG
  +2  R  D  P  I  F  Y  R  W  H  G  F  I  D  T  V  F  N  K  F  K  T  R  L  N  P  Y  N  A  G  E  L  N  F
1601 AGGGATCCAA TCTTCTACAG GTGGCACGGA TTCATCGACA CGGTGTTCAA CAAATTCAAG ACTCGCCTGA ACCCGTACAA TGCCGGTGAG CTTAACTTCG
  +2 D  G  I  T  V  D  Y  I  E  A  K  I  G  K  S  N  T  K  A  N  T  L  L  T  Y  W  Q  K  S  S  A  D  L  A
1701 ATGGTATTAC CGTGGACTAC ATTGAGGCCA AGATTGGCAA GTCCAATACC AAGGCCAACA CCCTGTTGAC CTACTGGCAG AAGTCGAGTG CCGACTTGGC
```

```
  +2  A G L  D F G  P T T D  R N I  F A S  F T H L  Q N A  P F T  Y T F N  V T N
1801 AGCTGGCCTG GACTTTGGAC CAACGACCGA TCGCAACATA TTCGCCTCGT TCACCCATCT CCAGAATGCC CCCTTCACTT ATACATTCAA TGTGACCAAT
  +2 N G A  R R T G  T C R  I F I  C P K V  D E R  N Q A  L N L E  E Q R  L L A
1901 AATGGAGCCA GACGAACGGG TACTTGCCGT ATCTTCATCT GCCCCAAGGT GGATGAGCGT AATCAGGCTT TGAATCTGGA GGAGCAGCGA CTGTTGGCCA
  +2 I E M D  K F T  V D
2001 TCGAAATGGA CAAGTTCACA GTTGATTGTG AGTAATAACT GTATTTCTTA GTATATATAT TTATTTAATA TTTTGGTATT TAAAAAGTGT TTCTTTCGTA
  +3                                                                           EXON 5      L V  P G E N
2101 CAAAGCTAGC CAGTGATTTT AGCTTATACT TTTCCCTACT GACAACCTCC TACGATCTAT TAACATACTT TAATTAATCC CCAGTGGTGC CTGGTGAGAA

  +3  T I R  R Q S  T E S S  V A I  P F E  R S F R  P V G  A D Y  Q P K A  A D E
2201 CACCATACGC CGCCAGTCGA CAGAGTCCTC GGTGGCCATT CCCTTCGAGC GCTCATTCCG CCCGGTTGGA GCAGACTATC AGCCCAAGGC CGCCGGATGAG
  +3 L A R  F K F C  G C G  W P Q  H L L L  P K G  N A Q  G M V F  D L F  V M I
2301 TTGGCACGCT TCAAGTTTTG CGGATGTGGG TGGCCACAGC ATCTCCTGCT GCCCAAGGGC AATGCCCAGG GAATGGTCTT CGACCTGTTC GTCATGATTT
  +2 S D Y S  Q D S  V E Q P  K T                                                                    EXON 6    P
2401 CCGATTACTC GCAGGACTCC GTTGAGCAGC CCAAAACGTG AGTGCTATAA ACCCAATTTC CCATCATCAG CTGACATTCT CCTTGTCGCC CATTAGACCC
  +2 N D A  C S T A  Y S F  C G L  K D K L  Y P D  R R T  M G Y P  F D R  R L P
2501 AACGATGCCT GCAGCACGGC GTACTCGTTC TGTGGTCTGA AGGACAAATT GTATCCTGAC AGACGCACCA TGGGCTATCC CTTCGATAGG CGCCTGCCCA
  +2 N A N L  T E L  V G A  F G N M  A K T  D L R  I V F N  D R V  I D K  A @
2601 ATGCCAATCT CACCGAATTG GTCGGTGCCT TTGGCAACAT GGCCAAGACC GATTTAAGGA TTGTCTTCAA TGATCGTGTG ATTGACAAGG CCTAGATGAT
2701 TGCTCCAACT AACAGAGATG ATTTTAAGAA TCAATGGAAA CGGACTTGAA AACGGAAAAC TCTTGGAATC ATATACACTT TTTGCTATTA ACTTATAACT
2801 AACCAAACTT GTGCGCGTTT ATGTGGTGTT TTATCGTTTC TAATTACACT CTACAAAAAT AAATG
```

Figure. 4. Genomic sequence of *Drosophila melanogaster* PPO A1

```
   1 TTTTGTATAA AATGTGCAGT TCTTTCGGTG AATAGTCACA GTCGTGAAGT TCCCAGACTT CCAGTCACAT TCCCCATTTA GTTTGCTCCG CGATCCAGCA
  +3            EXON 1                 M  A  D  K  K  N  L  L  L  L  F  D  H  P  T  E  P  V  F  M  D  K  G  K  R
 101 GGTCCTCCCT GACATCCCAT TGAAA ATGGC CGACAAGAAG AATCTCCTCC TGCTTTTCGA CCATCCCACC GAGCCAGTGT TCATGGACAA GGGCAAGAGG
  +3  V  T  V  F  D  V  P  D  S  F  L  T  D  R  Y  R  P  I  S  N  E  V  Q  S  R  V  G  D  K  V  E  Q  R
 201 GTGACCGTGT TCGATGTGCC CGACTCCTTC CTGACCGACC GCTACCGACC CATCAGCAAC GAGGTGCAGA GCCGCGTCGG CGACAAGGTG GAGCAGCGCG
  +3 V  P  V  R  E  I  S  I  P  D  L  R  I  P  M  S  L  G  R  D  E  Q  F  S  L  F  L  P  K  H  R  R  I  A
 301 TTCCAGTCCG GGAGATCTCC ATTCCCGATC TGCGGATCCC CATGTCCCTG GGCCGGGACG AGCAGTTCTC GCTGTTCCTG CCCAAGCACC GCAGGATCGC
  +3  G  R  L  I  D  I  F  M  N  M  R  S  V  D  D  L  Q  S  V  A  V  Y  A  R  D  R  V  N  P  V  L  F  N
 401 CGGCCGCCTG ATCGATATAT TCATGAACAT GCGCTCCGTG GACGACCTGC AGAGCGTGGC GGTGTACGCC AGGGATCGGG TCAACCCGGT GCTCTTCAAC
  +3  Y  A  L  S  V  A  L  L  H  R  P  D  T  Q  G  L  D  L  P  S  F  S  Q  T  F  P  D  R  F  I  D  S  Q
 501 TACGCCTTGT CGGTGGCGCT GCTCCATCGT CCGGACACCC AGGGCCTGGA TCTGCCATCC TTCTCGCAGA CTTTCCCCGA CCGCTTCATC GACTCGCAGG
  +3 V  I  R  K  M  R  E  E  S  F  V  V  Q  P  G  S  R  M  P  I  T  I  P  R  D  Y  T  A  S  D  L  D  P  E
 601 TGATCCGCAA GATGCGGGAG GAGTCTTTTG TGGTGCAGCC CGGTTCCCGG ATGCCCATCA CCATTCCCAG GGACTACACC GCCTCCGATC TGGATCCCGA
  +3  H  R  L  W  Y  F  R  E  D  L  G  I  N  L  H  H  W  H  W  H  L  V  Y  P  F  E  A  S  D  R  S  I  V
 701 GCACCGGCTG TGGTACTTCC GCGAGGATCT GGGCATCAAC CTTCACCACT GGCATTGGCA TCTGGTCTAC CCCTTCGAGG CCTCCGACCG CAGCATAGTG
  +3  A  K  D  R  R  G  E  L  F  Y  Y  M  H  Q  Q  V  I  A  R  Y  N  A  E  R  F  S  N  N  L  A  R  V  L
 801 GCCAAGGATC GTCGTGGCGA GCTCTTCTAC TACATGCACC AGCAGGTGAT CGCCCGCTAC AATGCGGAGC GGTTCAGCAA CAACCTGGCC AGGGTTCTGC
  +3 P  F  N  N  L  R  D  P  I  A  E  G  Y  F  P  K  M  D  S  L  V  A  S  R  A  W  P  P  R  F  E  S  T  R
 901 CCTTCAACAA CCTGAGGGAT CCCATTGCCG AGGGCTACTT CCCCAAGATG GACTCTCTGG TAGCCAGCCG CGCGTGGCCA CCGCGCTTCG AGAGCACCAG
  +3  L  S  D  L  N  R  E  S  D  Q  L  N  V  E  I  G  D  L  E  R  W  R  D  R  I  Y  E  A  I  H  Q  G  F
1001 GTTGAGCGAT CTCAACCGGG AGTCCGACCA GCTGAATGTG GAGATTGGAG ATCTGGAGCG TTGGCGTGAT CGCATCTACG AGGCCATTCA CCAAGGATTT
  +3  V  M  D                                                 EXON 2        +1 E  R  G  N  R  V  P  L  D  E  A  T
1101 GTCATGGATG TGAGTTGTTG GTGGGTTTTC CATTAACTAA ATAACATGCC ATATTCCCTT TCAGGAGCGT GGCAACCGGG TGCCATTGGA TGAGGCGACG
  +1  G  I  D  T  L  G  N  M  I  E  S  S  I  L  S  P  N  R  V  L
1201 GGCATTGATA CCCTCGGCAA CATGATTGAA TCATCCATTC TGTCGCCCAA CCGAGTGCTG GTAATAAGTC CTTACCCTTA GTAAATTATT CTAGGGATTA
  +2 EXON 3                Y  G  D  L  H  N  N  G  H  T  F  I  S  Y  A  H  D  P  T  S  K  H  L  E  S  F  G
1301 ATCCTATTGC TCCCCTAGTA TGGTGACCTG CACAACAACG GACACACCTT CATCTCGTAC GCCCACGATC CGACCAGCAA GCACCTGGAG TCCTTTGGCG
  +2 V  M  G  D  V  S  T  A  M  R  D  P  V  F  Y  K  W  H  S  Y  I  D  R  I  F  Q  E  H  K  S  R  L  P  A
1401 TGATGGGTGA TGTGTCCACG GCCATGCGCG ACCCGGTCTT CTACAAGTGG CACTCGTACA TCGACCGCAT CTTCCAGGAG CACAAGTCGC GCCTGCCCGC
  +2  Y  T  E  N  Q  L  N  Y  P  G  V  S  I  A  G  I  Q  V  D  T  N  G  G  R  P  N  N  L  T  T  F  W  Q
1501 CTATACCGAG AACCAACTAA ACTACCCGGG CGTCTCGATC GCGGGCATCC AGGTGGACAC CAATGGAGGT CGTCCCAATA ATCTGACCAC CTTCTGGCAG
  +2  Q  S  D  V  D  M  S  R  G  F  D  F  L  P  R  G  N  V  F  A  R  F  T  H  L  Q  H  L  P  F  T  Y  T
1601 CAGTCGGACG TGGACATGTC CCGTGGCTTC GACTTCCTGC CGCGAGGTAA TGTCTTCGCC AGGTTCACCC ACCTGCAGCA TCTGCCCTTC ACCTATACCA
```

```
  +2 I  S  L  N   N  D  S   G  A  Q   R  F  G  Y   V  R  I   F  M  A   P  K  N  D   E  R  G   Q  P  M   L  M  R  D
1701 TCAGTTTGAA CAACGATAGT GGTGCCCAGC GTTTCGGCTA CGTGCGCATC TTCATGGCGC CGAAGAACGA CGAGCGTGGC CAGCCCATGC TGATGCGAGA
  +2  Q  R  S   M  M  I   E  L  D  K   F  V  T   S  L  N   P  G  P  N   T  I  R   R  R  S   T  E  S  S   V  T  I
1801 TCAGCGCTCC ATGATGATCG AGCTGGACAA GTTCGTGACT TCGCTGAACC CGGGACCCAA CACCATCCGC CGCCGCTCCA CGGAGTCGAG CGTGACCATT
  +2  P  F  E   R  T  F  R   N  L  D   A  N  R   P  A  A  G   T  P  E   E  L  E   F  N  F  C   G  C  G   W  P  N
1901 CCGTTCGAGC GCACCTTCCG CAACCTGGAT GCCAATCGTC CGGCGGCCGG CACTCCGGAG GAGCTGGAGT TCAACTTCTG CGGCTGCGGC TGGCCCAACC
  +2 H  M  L  V   P  K  G   L  P  E   G  L  Q  C   V  L  F   I  M  V   S  N  Y  E   N  D  R
2001 ACATGCTGGT GCCAAAGGGT CTGCCCGAGG GACTGCAGTG CGTGCTGTTC ATCATGGTGT CGAACTATGA GAACGACAGG GTATGTTAGC TGGCTAGTAA

  +1                        EXON 4                I   D  Q  Q  L   V  G  R   C  S  D   A  A  S  Y   C  G  V   R  D  R
2101 ATCAAGGATA TTTTTGACAA TTATCTCCAC CTCCAG ATTG ACCAGCAATT GGTGGGTCGG TGCAGCGATG CCGCTTCCTA CTGTGGCGTC CGTGACCGTC
  +1 L  Y  P  D   R  Q  S   M  G  F   P  F  D  R   L  P  R   S  G  V   D  R  L  V   N  F  L   T  P  N   M  S  I  V
2201 TTTACCCGGA TCGCCAGTCC ATGGGATTCC CCTTCGACAG GTTGCCCCGC AGCGGAGTCG ACCGCCTGGT TAACTTCCTC ACGCCCAACA TGAGCATCGT
  +1  D  V  N   I  R  H   E  N  R  T   V  Q  R   P  N  @
2301 GGACGTGAAC ATCCGCCACG AGAACCGCAC CGTGCAGCGC CCAAACTAGA CGATCCGCAC CTGATTCGCT TTCCCGATTC GCCCTGACCC GATTCCATGA
2401 CCTCGCCCAC TTCTAGATAT ACAACTCTTA GCAATCTATT GCATCTTTAT TGAGCCGCTT AATAA
```

Figure 5. Genomic sequence of *Drosophila melanogaster* PPO-A2

```
   1 ATGTAACCAT AGGGCATTTC GGCTGCTGGG TATAAAAGGA GCTGATAGTT GGGTGGGGAC GTGCAGTTGA GAATCAGCAA CTGCTCGCTA AGGTCCTCC
  +1            EXON1       M  A  D   K  K  N   L  L  L   F  D  H   P  T  E  P   V  F  M   D  K  G   G  N  G  T
 101 TAATCGCGAA TCCATCGAAA ATGGCCGACA AGAAGAATCT GCTCCTTCTG TTCGATCATC CCACGGAGCC CGTGTTCATG GACAAGGGTG GCAATGGCAC
  +1  V  F  D   V  P  D   S  Y  V  T   D  R  Y   N  Q  M   C  K  K  V   Q  R  R   V  S  S   A  S  E  K   N  V  Q
 201 CGTCTTCGAC GTGCCCGACT CCTACGTGAC CGATCGCTAC AATCAGATGT GCAAGAAGGT CCAGCGGCGC GTGAGCAGCG CGTCCGAGAA GAACGTTCAA
  +1  V  K  E   I  A  I  P   D  L  S   C  S  M   R  L  G  R   S  E  Q   F  S  I   F  L  K  S   H  R  K   M  A  S
 301 GTCAAGGAGA TCGCCATACC GGATCTCAGT TGCTCCATGA GACTGGGCCG CTCGGAGCAG TTCTCGATCT TCCTGAAGTC CCACCGCAAG ATGGCCAGCC
  +1 H  L  I  E   I  F  T   K  M  Q   T  V  D  E   L  Q  S   V  A  V   Y  A  R  D   R  V  N   P  V  L   F  N  Y  A
 401 ACCTGATCGA GATCTTCACC AAAATGCAGA CCGTGGATGA GCTGCAGAGC GTGGCGGTGT ACGCCAGGGA TCGCGTCAAT CCGGTGCTCT TCAACTACGC
  +1  L  S  V   A  L  L   H  R  P  D   T  K  D   L  E  L   P  A  F  A   Q  T  F   P  D  R   F  I  D  S   K  M  L
 501 CTTGTCGGTG GCCCTGCTCC ACCGTCCGGA CACCAAGGAC CTCGAACTGC CCGCCTTTGC GCAGACCTTC CCGGATCGCT TTATCGACTC GAAGATGCTT
  +1  R  S  M   R  E  E  S   F  V  V   E  R  S   A  A  R  L   P  V  V   S  S  V   K  Y  T  A   S  D  L   D  V  E
 601 CGCAGCATGC GCGAGGAGTC CTTTGTGGTG GAGCGCTCCG CCGCCCGCCT GCCCGTCGTC AGTTCGGTCA AGTACACCGC CTCCGATCTC GATGTGGAGC
  +1 H  R  L  W   Y  F  R   E  D  L   G  V  N  L   H  H  W   H  W  H   L  V  Y  P   I  E  A   P  D  R   S  I  V  D
 701 ACCGGCTGTG GTACTTCCGC GAGGACCTGG GCGTCAACCT GCACCACTGG CACTGGCATC TGGTCTATCC CATAGAGGCG CCCGACCGCA GCATCGTGGA
  +1  K  D  R   R  G  E   L  F  Y  Y   M  H  Q   Q  I  I   A  R  Y  N   A  E  R   L  S  N   H  M  A  R   V  Q  P
 801 CAAGGACCGT CGTGGCGAGC TGTTCTACTA TATGCACCAG CAGATCATTG CCCGCTACAA TGCCGAGCGG CTGAGCAACC ACATGGCCCG GGTGCAGCCG
  +1  F  N  N   L  D  E  P   I  A  E   G  Y  F   P  K  M  D   S  L  V   A  S  R   A  Y  P  P   R  F  D   N  T  R
 901 TTCAACAACC TGGACGAACC AATTGCCGAG GGCTACTTCC CCAAGATGGA CTCGCTGGTG GCCAGTAGGG CGTATCCTCC GCGCTTCGAT AACACCCGGC
  +1 L  S  D  V   D  R  P   N  N  Q   L  R  V  G   I  D  D   M  K  R   W  R  E  R   I  Y  E   A  I  H   Q  G  Y  V
1001 TGAGCGATGT GGACCGGCCT AACAACCAGT TGAGGGTTGG CATCGATGAT ATGAAGCGCT GGCGCGAACG CATCTACGAG GCCATCCATC AGGGCTACGT
  +2                                                                   EXON2      T  N  N   E  K  I   V  L  D  D
  +1  L  D
1101 TTTGGATGTG AGTGGGGAGC TCTCCGATTC CAGGAGTACA TGCATCCAAC AACTAATTGG TTATTTTTTA GACGAACAAT GAAAAGATCG TGCTGGATGA
  +2  A  K  G   I  D  I   L  G  N  I   I  E  A   S  D  L   T  P  N  S   T  L
1201 CGCTAAGGGC ATCGACATCC TGGGCAACAT CATCGAGGCT TCCGATCTGA CGCCCAACAG TACGCTGGTA AGGAATCTAA GTGATAATCT GTAGATCACC
  +3                       EXON3  Y   G  D  F   H  N  M   G  H  I  L   I  A  Y   S  H  D   P  T  N  K   H  L  E
1301 TTTCCATGTA TATTATTTCC ATCTGCAGTA CGGCGACTTC CACAACATGG GGCACATACT GATTGCCTAC TCACACGATC CGACCAACAA GCACCTGGAG
  +3  Y  A  G   V  M  G  D   A  S  T   A  M  R   D  P  I  F   Y  K  W   H  A  F   I  D  N  M   F  Q  E   H  K  R
1401 TACGCGGGCG TGATGGGCGA CGCCTCCACC GCGATGCGCG ATCCGATCTT CTACAAGTGG CACGCCTTCA TCGACAATAT GTTCCAGGAG CACAAACGTC
  +3 L  L  S  P   Y  E  K   Q  E  L   S  F  P  D   V  R  V   E  S  I   Q  V  E  S   Q  G  Q   V  N  R   L  T  T  F
1501 TGCTTTCTCC CTATGAGAAG CAAGAGCTCA GCTTTCCGGA CGTTCGCGTC GAGAGCATCC AGGTGGAGAG CCAGGGTCAG GTCAACAGGC TGACCACCTT
  +3  W  Q  E   S  D  V   D  M  S  R   G  L  D   F  V  P   R  G  H  V   L  A  R   F  T  H   L  Q  H  H   E  F  S
1601 CTGGCAGGAG TCCGATGTGG ACATGTCCCG TGGCCTTGAT TTTGTGCCGC GCGGCCATGT CCTCGCTCGG TTCACCCATC TGCAGCACCA CGAATTCAGC
  +3  Y  T  I   K  V  E  N   S  S  E   A  T  R   Y  G  Y  V   R  I  F   L  A  P   K  L  D  D   R  N  A   P  M  L
1701 TACACCATCA AGGTGGAGAA CTCCAGCGAG GCCACGCGGT ATGGCTACGT GCGCATCTTC CTGGCGCCCA AGCTGGATGA TCGCAATGCA CCCATGCTGC
  +3 L  E  Q  Q   R  L  M   M  V  E   L  D  K  F   V  V  T   M  P  P   G  S  H  T   I  T  R   N  S  T   E  S  S  V
1801 TGGAGCAACA GCGCCTGATG ATGGTGGAGC TGGACAAGTT CGTGGTTACC ATGCCTCCCG GCAGCCATAC GATCACCCGG AACTCCACGG AGTCGAGTGT
  +3  T  I  P   F  E  R   T  F  R  N   L  D  K   L  E  E   L  Q  N  F   L  C  G   C  G  W   P  Q  H  M   L  I  P
1901 CACCATTCCG TTCGAGCGCA CCTTCCGCAA CCTGGACAAG CTGGAGGAGC TGCAGAACTT CTTGTGCGGC TGCGGTTGGC CCCAGCACAT GCTGATTCCG
  +3 K  G  R   P  E  G  L   R  F  E   L  F  V   M  V  S  N   Y  E  E   D  K  V
2001 AAGGGTCGGC CCGAGGGTCT GCGCTTCGAG CTCTTCGTCA TGGTTTCCAA CTACGAAGAG GATAAGGTAA GGTGTCTTTA GAGTATACTG CGAATGCCTC
  +2                       EXON4      D  Q  T   V  A  D  C   G  C  S   I  A  A   S  Y  C   G  L  R  D   R  L  Y
2101 CTGATCCCAA TCCTCTTTCC TTGTGCAGGT TGACCAGACG GTCGCAGATT GCGGCTGCAG CATCGCCGCT TCCTACTGCG GCCTGCGCGA TCGCCTCTAT
  +2  P  D  R   K  S  M  G   F  P  F   D  R  K   P  R  R  G   S  E  I   L  E  N   F  L  T  P   N  M  C   A  V  E
2201 CCGGATCGCA AGTCCATGGG CTTCCCCTTC GATCGCAAGC CGCGTCGAGG CTCCGAGATC CTGGAGAACT TCCTGACGCC CAACATGTGC GCCGTGGAGG
  +2 V  I  I  T   H  E  D   R  T  E   K  L  R  E   V  P  A   R  S
2301 TCATCATTAC CCACGAGGAC CGCACCGAGA AGCTTCGCGA GGTACCAGCT CGCTCCTGAA TAAGAGAAGC ATCTTGTTAT GGCCAGTTTC ATTTTTTGGC
2401 TTAAAGAGAA GCCACTAATA ACGGTCTATA GCTTTCTCGC TGTGTAGAAA ATACCAAATA AATTAATGAA ATACT
```

Figure 6. Genomic sequence of *Drosophila melanogaster* PPOA3

ACKNOWLEDGEMENTS

This work was supported in part by grants from U. Mass/Boston and N.I.H. (AI 14753).

REFERENCES

Andersen SO, Peter MG, Roepstorff P. Cuticular sclerotization in insects. Comp Biochem. Physiol. 1996; **113B**: 689-705.

Andersson K, Sun SC, Boman HG, Steiner H. Purification of the prophenoloxidase from *Hyalophora cecropia* and four proteins involved in its activation. Insect Biochem. 1989; **19**: 629-637.

Ashida M. Purification and characterization of prephenoloxidase from hemolymph of the silkworm, *Bombyx mori.* Arch. Biochem. Biophys. 1971; **144:** 749-762.

Ashida M, Brey P. Role of the integument in insect defense: Prophenoloxidase cascade in the cuticular matrix. Proc. Natl. Acad. Sci. USA. 1995; **92:** 10698-10702.

Ashida M, Yamazaki HI. "Biochemistry of the phenoloxidase system in insects: With special reference to its activation". In Molting and metamorphosis, Ohnishi E, Ishizaki H. eds. Tokyo. Jpn. Sci. Soc. Press. 1990. pp. 239-265.

Aso Y, Kramer KJ, Hopkins TL, Lookhart GL. Characterization of hemolymph protyrosinase and a cuticular activator from *Manduca sexta* (L). Insect Biochem. 1985; **15**: 9-17.

Burge CB. "Modeling dependencies in pre-mRNA splicing signals". In Computational Methods in Molecular Biology, Salzberg, S., Searls, D. and Kasif, S. eds. Amsterdam, Elsevier Science, 1998. pp. 127-163.

Burge C, Karlin S. Prediction of complete gene structures in human genomic DNA. J. Mol. Biol. 1997; **268:** 78-94.

Chase MR, Raina K, Bruno J, Sugumaran M. Purification, characterization and molecular cloning of prophenoloxidases from *Sarcophaga bullata.* Insect Biochem. Mol. Biol. (*in press*, 2000)

Cherqui A, Duvic B, Brehelin M. Purification and characterization of prophenoloxidase from the hemolymph of *Locusta migratoria*. Arch. Insect Biochem. Physiol. 1996; **32**: 225-235.

Cho WL, Liu HS, Lee CH, Kuo CC, Chang TY, Liu CT, Chen CC. Molecular cloning, characterization and tissue expression of prophenoloxidase cDNA from the mosquito *Armigeres subalbatus* inoculated with *Dirofilaria immitis* microfilariae. Insect Mol. Biol. 1998; **7**: 31-40.

Durrant HJ, Ratcliffe NA, Hipkin CR, Aspan A, Soderhall K. Purification of the prophenoloxidase enzyme from hemocytes of the cockroach *Blaberus discoidalis*. Biochem. J. 1993; **289**: 87-91.

Fujimoto K, Masuda K, Asada N, Ohnishi E. Purification and characterization of prophenoloxidase from the pupae of *Drosophila melanogaster*. J. Biochem (Tokyo). 1993; **113**: 285-291.

Fujimoto K, Okino N, Kawabata SI, Iwanaga S, Ohnishi E. Nucleotide sequence of the cDNA encoding the proenzyme of phenoloxidase A_1 of *Drosophila melanogaster*. Proc. Natl. Acad. Sci USA. 1995; **92**: 7769-7773.

Gillespie JP, Kanost MR, Trenczek T. Biological mediators of insect immunity. Ann. Rev. Entomol. 1997; **42:** 611-643.

Hall M, Scott M, Sugumaran M, Soderhall K, Law JH. Proenzyme of *Manduca sexta* phenoloxidase: Purification, activation, substrate specificity of the active enzyme and molecular cloning. Proc. Natl. Acad. Sci USA. 1995; **92**: 7764-7768.

Hara T, Miyoshi T, Funatsu M. Comparative studies on larval and pupal phenoloxidases of the housefly, *Musca domestica*. Comp. Biochem. Physiol. 1993; **106B**: 287-292.

Heyneman RA. Final purification of a latent phenoloxidase with mono-and diphenoloxidase from *Tenebrio molitor*. Biochem. Biophys. Res. Commun. 1965; **21**: 162-169.

Jiang H, Wang Y, Ma C, Kanost MR. Subunit composition of prophenoloxidase from *Manduca sexta*: Molecular cloning of subunit ProPo-P1. Insect Biochem. Mol. Biol. 1997a; **27**, 835-850.

Jiang H, Wang Y, Korochkina SE, Benes H, Kanost MR. Molecular cloning of cDNAs for two prophenoloxidases subunits from the Malaria vector, *Anopheles Gambiae*. Insect Biochem. Mol. Biol. 1997b; **27**: 693-699.

Kawabata T, Yasuhara Y, Ochiai M, Matsuura S, Ashida M. Molecular cloning of insect prophenoloxidase: A copper containing protein homologous to arthropod hemocyanin. Proc. Natl. Acad. Sci. USA. 1995; **92**: 7774-7778.

Kopácek P, Weise C, Gotz P. The prophenoloxidase from the wax moth *Galleria mellonella*: Purification and characterization of proenzyme. Insect Biochem. Mol. Biol. 1995; **25**: 1081-1091.

Kwon TH, Lee SY, Lee JH, Choi JS, Kawabata SI, Iwanaga S, Lee BL. Purification and characterization of prophenoloxidase from the hemolymph of coleopteran insect, *Holotrichia diomphlia*. Molecules & Cells 1997; **7**: 90-97.

Lai-Fook J. The repair of wounds in the integument of insects. J. Insect Physiol. 1966; **12**:195-226.

Lee WJ, Ahmed A, Torre AD, Kobayashi A, Ashida M, Brey PT. Molecular cloning and chromosomal localization of a prophenoloxidase cDNA from the malaria vector *Anapheles gambiae*. Insect Mol. Biol. 1998; **7**: 41-50.

Lewis HW, Lewis HS. Genetic regulation of dopa oxidase activity in *Drosophila*. Ann. N.Y. Acad. Sci. 1963; **100**: 827-839.

Maddison WP, Maddison DR. MACLADE, Version 3.04. Sunderland Mass, USA. Sinauer Associates, Inc. Publishers. 1993.

Muller HM, Dimopoulos G, Blass C, Kafatos FC. A hemocyte-like cell line established from malaria vector *Anopheles gambiae* expresses six prophenoloxidase genes. J. Biol. Chem. 1999; **274**: 11727-11735.

Nappi AJ, Sugumaran M. "Some biochemical aspects of eumelanin formation in insect immunity". In: Insect Immunity. Pathak J. P. N. ed. New Delhi, India. Oxford & IBH Publishing Co. 1993. pp. 131-148.

Pau RN, Eagles PAM. The isolation of o-diphenoloxidase from the third instar larvae of blowfly, *Calliphora erythocephala*. Biochem. J. 1975; **149**: 707-712.

Park DS, Shin W, Kim MG, Park SS, Lee WJ, Brey PT, Park HY. Isolation and characterization of the cDNA encoding the prophenoloxidase of Fall webworm, *Hyphantria cunea*. Insect Biochem. Mol. Biol. 1997; **27**: 983-992.

Pentz ES, Wright TR. *Drosophila melanogaster* diphenoloxidase A2: Gene structure and homology with the mouse mast-cell tum-transplantation antigen, P91A. Gene 1991; **22**: 239-242.

Perotti ME, Bairati A. Ultrastructure of the melanotic masses in two tumorous strains of *Drosophila melanogaster* (tuB3 and Freckled). J. Invert. Pathol. 1968; **10**: 122-138.

Ricketts D, Sugumaran M. 1,2-dehydro-N-b-alanyldopamine as a new intermediate in insect cuticular sclerotization. J. Biol. Chem. 1994; **269**: 22217-22221.

Saul SJ, Sugumaran M. A novel quinone: quinone methide isomerase generates quinone methides in insect cuticle. F. E. B. S. Lett. 1988; **237**: 155-158.

Saul SJ, Sugumaran M. Characterization of a new enzyme system that desaturates the side chain of N-acetyldopamine. F.E.B.S. Lett. 1989a; **251**: 69-73.

Saul SJ, Sugumaran M. N-Acetyldopamine quinone methide/1,2-dehydro-N-acetyldopamine tautomerase - A new enzyme involved in sclerotization of insect cuticle. F.E.B.S. Lett. 1989b; **255**: 340-344.

Saul SJ, Sugumaran M. 4-alkyl-*o*-quinone/2-hydroxy-*p*-quinone methide isomerase from the larvae hemolymph of *Sarcophaga bullata*. I. Purification and characterization of enzyme catalyzed reaction. J. Biol. Chem. 1990; **265**: 16992-16999.

Seybold WD, Meltzer PS, Mitchell HK. Phenoloxidase activation in *Drosophila*: A cascade of reactions. Biochem. Genetics. 1975; **13**: 85-108.

Söderhäll K, Aspán A, Duvic B. The ProPO system and associated proteins. Role in cellular communication in arthropods. Res. Immunol. 1990; **141**: 896-907.

Sugumaran M. Molecular mechanisms of mammalian melanogenesis - comparison with insect cuticular sclerotization. F. E. B. S. Lett. 1991; **293**: 4-10.

Sugumaran M. "Role of insect cuticle in immunity". In New Directions in Invertebrate Immunology. Söderhäll, K., Iwanaga, S. and Vastha, G. eds. Fair Haven, NJ. SOS Publications. 1996. pp. 355-374.

Sugumaran M. Unified mechanism for sclerotization of insect cuticle. Adv. Insect Physiol. 1998; **27**: 229-334.

Sugumaran M, Kanost M. "Regulation of insect hemolymph phenoloxidases". In Parasites and pathogens Beckage, N. E., Thompson, S. N., & Frederick, B. A. eds. San Diego, Academic Press. 1993; Vol. I. Parasites. pp. 317-342.

Stoltzfus A, Logsdon JM, Palmer JD, Doolittle WF. Intron "sliding" and the diversity of intron positions. Proc. Natl. Acad. Sci. USA 1997; **94**: 10739-10744.

Thompson JD, Higgins DG, Gibson TJ. CLUSTAL W: Improving the sensitivity of progressive multiple sequence alignment through sequence weighting, positions-specific gap penalties and weight matrix choice. Nucleic Acid Res. 1994; **22**: 4673-4680.

Yasuhara Y, Koizumi Y, Katagiri C, Ashida M. Reexamination of properties of prophenoloxidase isolated from larval hemolymph of the silkworm, *Bombyx mori*. Arch. Insect Biochem. Physiol. 1995; **32**: 14-23.

THE COMPLEMENT SYSTEM IN SEA URCHINS

L. Courtney Smith

Department of Biological Sciences and the Graduate Program in Genetics, The George Washington University, Washington DC

HISTORICAL BACKGROUND ON ECHINODERM IMMUNE RESPONSIVENESS

Comparative immunology had its origin in the studies of inflammation by Elie Metchnikoff who proposed that the primary effectors of the immune response were circulating, amoeboid, phagocytic cells (Metchnikoff, 1891). In order to support this proposal, Metchnikoff reported that the bipinnaria larvae of a sea star would respond to physical injury (the introduction of rose prickles or glass rods into the larval blastocoel), to a normal infection by marine algae or to the artificial introduction of bacteria into the blastocoel with the encapsulation and phagocytosis of these foreign materials by accumulated amoeboid, phagocytic mesodermal cells. These studies were the first to demonstrate that echinoderms, and by inference according to Metchnikoff, all animals were capable of responding to inert substances, to injury and to foreign or pathogenic cells.

Phylogenetic Perspectives on the Vertebrate Immune System.
Edited by Beck *et al.*, Kluwer Academic/Plenum Publishers, 2001.

Many years later, Bill Hildemann and his colleagues investigated echinoderm immunity from the approach of using allograft rejection of transplanted skin to assess immune capabilities in these organisms. Using two species of sea stars and one sea cucumber species, they demonstrated that echinoderms could reject allogeneic tissues and hence differentiate between self and non-self (Hildemann & Dix, 1972; Karp & Hildemann, 1976). To define the capabilities and the specificity of non-self recognition in echinoderms, a large scale allograft rejection analysis was performed on the sea urchin, *Lytechinus pictus* (Coffaro, 1979). Results showed that second set grafts were rejected twice as fast as first set grafts, however, kinetics of third party rejections occurred at the same rate as that for second set allografts (Smith & Davidson, 1992). Similar results were observed for clearance rates of foreign particles that were introduced into the coelomic cavity since reintroduced particles were cleared at the same rate as that recorded for the initial challenge (reviewed in Smith & Davidson, 1994; Gross *et al.*, 1999). These results demonstrated that the sea urchin immune system functioned on non-adaptive, non-specific or innate mechanisms. Yet, as Metchnikoff had first demonstrated, the innate system of the echinoderm was very efficient at recognizing and removing or encapsulating foreign tissue.

MOLECULAR ASSESSMENT OF THE SEA URCHIN IMMUNE RESPONSE

The fluid filled coelomic cavity of the sea urchin contains a variety of circulating cells types known collectively as coelomocytes. These cells mediate host responses to immune challenges through activities that include increased chemotaxis and phagocytosis, and the formation of cellular clots (reviewed in Smith & Davidson, 1992; Gross *et al.*, 1999). Activities of the amoeboid, phagocytic coelomocytes require significant changes in cell shape which are mediated by changes in the actin cytoskeleton (Edds, 1985). In initial searches for immune relevant genes specifically expressed in coelomocytes, the sea urchin profilin gene, SpCoel1, was cloned and sequenced (Smith *et al.*, 1992). Profilin transcripts were found to increase significantly in coelomocytes responding to injury or to lipopolysaccharide (LPS) (Smith *et al.*, 1992; 1995). Profilin is a major actin binding protein involved in either inhibiting or promoting the polymerization of actin monomers, depending on slight, localized changes in profilin concentration (Sohn & Goldschmidt-Clermont, 1994) thereby modulating the shape of the cytoskeleton and the shape of the cell (Finke et al., 1994). Since a significant proportion of coelomocytes have been characterized as amoeboid phagocytes (Johnson, 1969), it was assumed that increases in profilin transcripts was directly related to alterations in the cytoskeleton required for increased amoeboid movement, phagocytosis, encapsulation, clot formation and perhaps secretion or degranulation in response to injury and/or challenge with LPS.

COELOMOCYTE ESTs

Using profilin transcript content as a marker for general immune activation in individual sea urchins responding to LPS, RNA from coelomocytes with the highest level of profilin transcripts per cell was used to generate a cDNA library. To identify genes that were specifically expressed in coelomocytes and that were involved in the immune functions mediated by the coelomocytes, the approach was to generate expressed sequence tags (ESTs) (Smith *et al.*, 1996). Of the 405 clones randomly selected for EST analysis, 307 that had inserts larger than 0.5 kb, were partially sequenced. The deduced amino acids from these clones were compared to known proteins on GenBank and about 30% of the clones (89) matched to 55 different proteins (see Table I in Smith *et al.*, 1996). These matches fell into several categories that included a) putative immune effector proteins, b) cell surface proteins and receptors, c) proteins involved in signaling systems, d) secreted proteins and chaperones in the endoplasmic reticulum involved in the secretory machinery, e) cytoskeletal and cytoskeletal modifying proteins, f) general cell function proteins, and g) proteins of unknown function. Of these categories, the immune relevant proteins were the most interesting.

SEA URCHIN COMPLEMENT COMPONENTS

The complement system in vertebrates is an important aspect of innate immunity and is composed of three initiating pathways, alternative, lectin and classical, that converge on the terminal pathway. Two homologues of components of the vertebrate alternative pathway were identified in the EST study. The sequence obtained from EST064 is part of a 9 kB message from a single copy gene (*Sp064*) that encodes a homologue of complement component C3, called SpC3 (Al-Sharif *et al.*, 1998). The sequence from EST152 is part of a 5.5 kB message from the single copy gene (*Sp152*) that encodes a homologue of factor B (Bf), called SpBf (Smith *et al.*, 1998). Both sea urchin complement genes are expressed specifically in coelomocytes.

The deduced amino acid sequences from SpC3 and SpBf were aligned with other members of the thioester and Bf/C2 protein families respectively, which enabled the identification of domains and locations of conserved sequences and revealed structural similarities and predicted conserved function within these two sea urchin proteins. SpC3 shows a number of sequence and structural similarities to the thioester complement component protein family. It has a conserved $\beta\alpha$ junction that, when cleaved, results in two chains, α and β, of sizes similar to those for vertebrate C3 proteins (Fig. 1A; Al-Sharif *et al.*, 1998). SpC3 has a conserved thioester site with several conserved amino acids surrounding the thioester that, in higher vertebrates, are involved in shielding and stabilizing it prior to forming covalent bonds with acceptor molecules (Lambris *et al.*, 1993, Isaac & Isenman, 1992). Furthermore, an associated and conserved histidine is located about 100 amino acids towards the C terminus which may also be important for thioester function (Dodds *et al.*, 1996). SpC3 also shows putative cleavage sites in conserved positions for C3-convertase-like and factor I-like protease activity (Fig. 1A, see also Al-Sharif *et al.*, 1998) which may be important in controlling the activity of the protein. Unprocessed SpC3 has a hydrophobic leader that is typical of a secreted protein and it is localized in both coelomocytes in the coelomic fluid (Al-Sharif *et al.*, 1998; Gross *et al.*, 1999). There are five locations (recognition sequences) within the SpC3 sequence that are putative N-linked glycosylation sites; four of which are located in the α chain. Protein alignments also show that many of the cysteines are found in conserved positions, including the two that are putatively involved in interchain disulfide bond formation. Phylogenetic analysis of the complement thioester family proteins, using the alpha 2 macroglobulin (α_2M) family as the out group, shows that SpC3 clusters at the base of the clade, indicating that it is the first diverging member of the family, or is most similar to the ancestral sequence (Al-Sharif *et al.*, 1998; Smith *et al.*, 1999).

Alignments between SpBf, the second complement component to be identified in the sea urchin (Smith *et al.*, 1998), and the other members of the Bf/C2 family reveal the domain structure and the positions of certain amino acids that are conserved in all members of this protein family including SpBf. Domains identified in SpBf include a serine protease domain, a von Willibrand factor domain, and short consensus repeats (SCRs) that are also referred to a complement control protein (CCP) modules (Smith *et al.*, 1998). Within the domains, certain conserved amino acids were identified may be important for a) cleavage of SpBf by a putative factor D-like protease, b) binding Mg^{++}, and c) serine protease activity (Fig. 1B). Within the Bf/C2 proteins in higher vertebrates, all of these regions are important for function in the alternative pathway of complement.

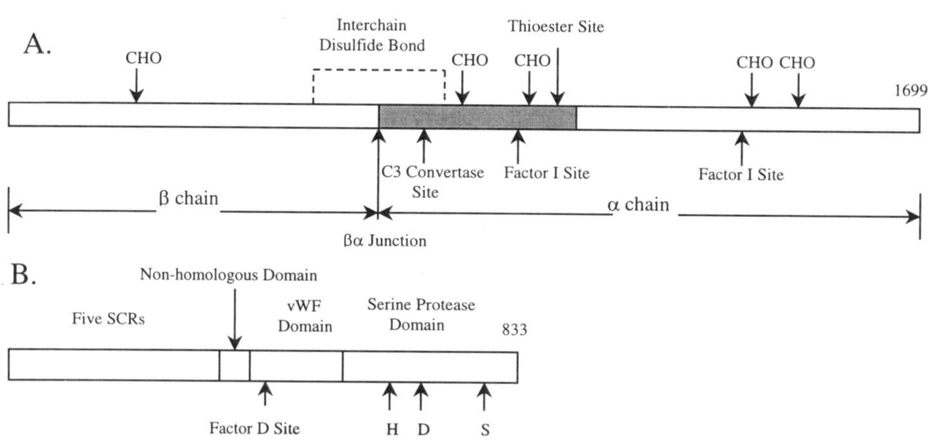

Figure 1. Diagram of SpC3 and SpBf showing domains and positions of putative cleavage sites. A. SpC3. Conserved N-linked glycosylation sites and interchain disulfide bonds are shown. Shaded area indicates the region of the a chain that was expressed as a 6 histidine fusion protein and used to produce an antiserum (anti SpC3-6-His) in rabbits. Autolytic cleavage at the thioester generates fragments of 50 kD (N terminal fragment) and 75 kD (C terminal fragment) from the α chain (see fig. 3). B. SpBf. Conserved amino acids in conserved positions that are putatively involved in the activity of the serine protease domain are indicated. The total number of amino acids in each protein are shown to the right of each diagram. Diagrams are drawn approximately to scale. Figure modified from Smith *et al* (1999).

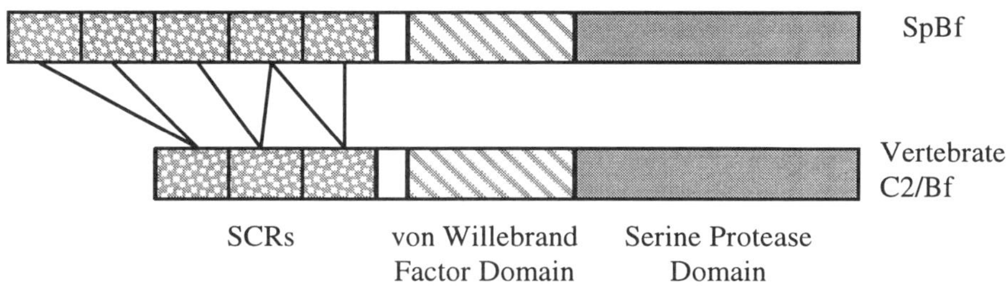

Figure 2. Similarities between SCR domains in SpBf and vertebrate C2/Bf proteins. Similarities and differences were identified among SCRs from SpBf and the SCRs from the Bf/C2 protein family using the Clustal W program for amino acid alignments followed by phylogenic analysis using the PAUP program (see Smith et al, 1998). Domains with similar sequence (implying similar function) are in the same order in the echinoderm and vertebrate proteins. Domain deletions in the vertebrate sequence has been assumed from the basal positions of the SpBf SCRs within the phylogenic clades.

Although protein alignments demonstrated the similarities between SpBf and other members of the Bf/C2 protein family, SpBf has five SCRs whereas all vertebrate members of the Bf/C2 family have only three SCRs. These small domains, that have been described as "beads on a string" are involved in Bf binding to other complement components (DiScipio, 1992; Hourcade *et al.*, 1995). To understand whether the sea urchin protein had duplicated some of its SCR domains, or whether the ancestral vertebrate protein had undergone one or more domain deletion events, a detailed sequence comparison of the SCRs from all the Bf/C2 proteins was undertaken (Smith *et al.*, 1998). To test whether this method was useful in identifying similarities and differences between SCRs, which are only 60 amino acids in length, individual SCRs were used as independent sequences for amino acid alignments and phylogenetic analysis. Clustering of sequences into clades was interpreted to indicate sequence similarities and to infer functional similarities. We found that the three SCRs from the Bf/C2 proteins (exclusive of those in SpBf) tended to cluster independently or to form independent clades (Smith *et al.*, 1998). This indicated that our methods were useful not only for identifying similarities and differences in SCR sequences, but that these differences corresponded to functional differences that have been identified for SCRs in the vertebrate Bf/C2 proteins (Horiuchi *et al.*, 1993; Hourcade *et al.*, 1995; Xu & Volanakis, 1997). When the SCRs from SpBf were added into the phylogenetic analysis, certain SCRs from SpBf tended to cluster with certain of the vertebrate SCRs indicating that the SpBf SCRs were in the same relative order as that for vertebrate SCRs (Fig. 2). In addition, upon inspection of the phylogenic trees generated by this analysis, the SCRs from SpBf tended to fall at the base of the clades with which they clustered indicating that they were the first diverging sequences. Together, this suggested that 1)

SpBf is ancestral, 2) the ancestral Bf/C2 protein had five SCRs, and 3) two SCRs were deleted from the Bf/C2 protein prior to the appearance of the vertebrates (Fig. 2).

SpC3 FUNCTION

The thioester group on vertebrate complement components C3 and C4 in addition to α_2M, in their activated form, covalently attaches to acceptor molecules with hydroxyl or amine groups, which is the basis for opsonization. The reactivity of this site is partially protected from deactivation by interaction with water molecules by the conformation of the protein surrounding the thioester site (for review, see Sim & Sim, 1983). Since the amino acids involved in thioester protection are important for thioester function, they have been well conserved across species (see fig. 3 in Al-Sharif *et al.*, 1998). Disruption of the protective protein conformation by heating under denaturing conditions with high pH results in the autolytic cleavage of the a chain at the thioester site, and has been well characterized for vertebrate C3, C4 and α_2M (Sim & Sim, 1981; 1983). This reaction does not occur if the thioester has become deactivated, either by being covalently bound, or if the a chain has been cleaved at the C3 convertase site. Although this reactivity is not physiologically relevant, autolysis only occurs in C3 molecules with active thioesters and it can be used as an assay to determine the subset of proteins with active thioester sites (Sim & Sim, 1983). Alternatively, decreased autolysis also be used to identify reagents to which the thioester will bind.

Autolysis of the SpC3 α chain

The conditions for autolysis, as described by Sim & Sim (1981; 1983), were duplicated for coelomic fluid and the effects on the a chain of SpC3 were analyzed by Western blot with a rabbit antiserum directed the SpC3 α chain. The antiserum, anti-SpC3-6His, was raised against a 50kD fragment that included the N terminal end of the SpC3 α chain (see Fig. 1A) and incorporated six histidines used for Ni affinity column purification before injection into rabbits. Treating coelomic fluid to induce autolytic reactions resulted in the appearance of a 50 kD band in addition to the presence of the full length α chain (Fig. 3, lane 6). Since anti-SpC3-6His only identifies the N terminal portion of the α chain, the 50 kD fragment must include the N terminus and would be cleaved at the thioester site after accounting for N-linked glycosylation. The remainder of the α chain, a 75 kD C terminal fragment, does not appear in Figure 3 because we have determined that anti-SpC3-6His does not recognize sequences to the C terminal side of the thioester (data not shown). This result suggests that the autolytic reaction occurs for SpC3 as it has been shown previously for mammalian C3 (Sim & Sim, 1981; 1983). Furthermore, this indicates that protein and thioester function in SpC3 is similar to that for other thioester containing proteins (C3, C4 and α_2M) from higher vertebrates. It should also be noted that the full length α chain, a 130 kD band, remains after autolysis takes place indicating that some of the SpC3 molecules are inactive. This is similar to what has been found for the thioester proteins from higher vertebrates (Sim & Sim, 1983).

	1	2	3	4	5	6
100mM MeNH$_4$	-	-	-	25°	37°	-
30mM EDTA	-	+	-	-	-	-
30mM EGTA	+	-	-	-	-	-
2mM Pefabloc	-	-	+	-	-	-

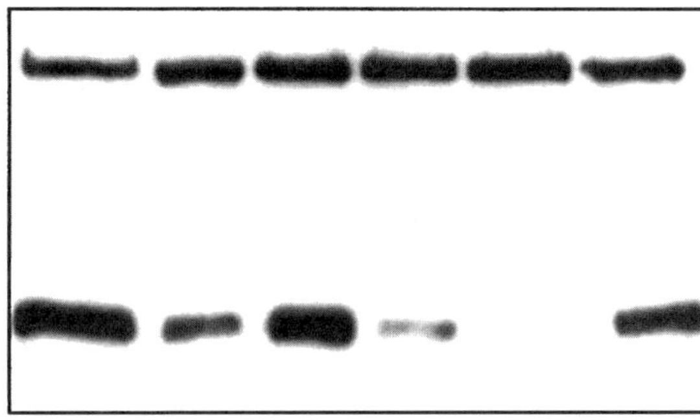

Figure 3. Autolytic activity of SpC3. Autolytic cleavage of SpC3 occurs in molecules with active thioester sites. Pre-incubation with methylamine or EDTA reduces the amount of 50 kD fragment generated, whereas EGTA and pefabloc, a serine protease inhibitor, have no effect. Materials: All materials were obtained from either Fisher Scientific Products or Sigma Chemical Co. unless stated otherwise. Methods: Coelomic fluid was withdrawn and incubated for 3 hrs at 25°C (or 37°C, sample #5) either alone (lane 6) or with EGTA (lane 1), EDTA (lane 2) or with methylamine ($MeNH_3$, lane 4 and 5). After incubation, autolysis sample buffer was added (4M Urea, 1% SDS, 100mM Tris pH 10, 20mM DTT) followed by heating to 95°C for 5 minutes. Fragments were separated by SDS-PAGE and electroblotted onto nitrocellulose in blotting buffer (150mM glycine, 20% methanol, 20mM Tris pH 8.8, 0.05% SDS) at 300 mA for 2 hr at 8°C. The filter was blocked over night in blotto (200 mM Tris, pH 7.4, 138 mM NaCl, 0.1% Tween 20, 5% milk proteins) incubated with rabbit antiserum to SpC3-6-Histidine fusion protein (1:15,000 dilution in blotto) which is specific for the SpC3 α chain (see Fig. 1A). After washing, the filter was incubated in goat antiserum to rabbit immunoglobulins labeled with horse radish peroxidase (Pierce; 1:80,000 dilution in blotto), washed, incubated in ECL reagent (Pierce) and exposed to film.

Since SpC3 α chain autolysis indicated that some of the molecules in the coelomic fluid were in an active state, it was of interest to determine whether this reaction could be blocked; *i.e.*, could be used to demonstrate thioester function. Coelomic fluid was pre-incubated with methylamine (reagent of choice to demonstrate thioester binding function), EDTA and EGTA (divalent ion chelators), and pefabloc (a serine protease inhibitor to ensure that the reaction was not due to proteases in the coelomic fluid), heated under denaturing conditions and analyzed by Western blot with anti-SpC3-6His (Fig. 3). Pre-treatment with methylamine (lanes 4 and 5) and EDTA (lane 2) decreased or blocked the formation of the 50 kD fragment while EGTA (lane 1) and pefabloc (lane 3) were ineffective at blocking autolysis. These results indirectly demonstrate that the binding capabilities of the SpC3 thioester functions in a comparable manner to other proteins containing thioester sites and is capable of binding methylamine. Results also show that EDTA (which chelates both Mg^{++} and Ca^{++}) does block autolysis while EGTA (which chelates Mg^{++}) does not. This indicates Ca^{++} is required for peptide bond cleavage. Since pefabloc does not decrease or block autolysis, this indicates that the reaction is not due to a serine protease present in the coelomic fluid.

Analyses of complement C3 homologues, AsC3 from a tunicate (Nonaka *et al.*, 1999; Nonaka & Azumi, 1999) and the complement-like protein from hagfish (Hanley *et al.*, 1992; Raftos *et al.*, 1992) have shown previously that these proteins directly bind methylamine and mediate opsonization and phagocytosis. Preliminary data on SpC3 indicates that it also functions in opsonization which augments phagocytosis (unpublished). Together, these data show that the primitive complement proteins in the lower deuterostomes are important in identification and clearance of pathogens and other foreign particles.

EVOLUTION OF THE COMPLEMENT SYSTEM IN DEUTEROSTOMES

The complement cascades in higher vertebrates appear to have evolved from a few primordial genes through gene duplication and subsequent divergence of function (Bentley, 1988). This hypothesis is based on similarities in protein sequence and function within complement protein families, the clustered organization of some complement genes, and parallels in organization and function of the lectin, alternative and classical pathways (Campbell *et al.*, 1988; Reid & Campbell, 1993). Complement systems have been identified in all vertebrate classes (Lambris *et al.*, 1994) and now, with the identification of complement components in echinoderms and ascidians (Smith *et al.*, 1996; 1998; Al-Sharif *et al.*, 1998; Nonaka *et al.*, 1999; Nonaka & Azumi, 1999), complement function has been identified within the two major phyla of the deuterostomes. Complement homologues have not been identified in insects, and none are present in the genome of *Caenorhabditis elegans* (see Science vol 282 no. 5396, 1999 for several analyses of the worm genome). This emphasizes the significant value of investigating the simpler deuterostomes for understanding the evolutionary origins and appearances of the several subsystems that function together in the vertebrate immune response.

The complement systems in the simpler deuterostomes is an important aspect of the innate defense system in these animals. Complement function in the vertebrates has been characterized as being of vital importance in defense against microbial pathogens and in enabling communications between the innate system and the adaptive system (Fearon & Locksley, 1996; Fearon, 1998). A central question regarding the evolution of immunity in higher vertebrates has been one of origins; was the occurrence of a simple, complement-based opsonin system involved? Duplications of complement genes and complement pathways seem to have occurred within the evolutionary timeframe during when the ancestral, retroviral recombinase genes were fortuitously incorporated into or near (probably) an immunoglobulin (Ig) encoding gene (Agrawal *et al.*, 1998; Hiom *et al.*, 1998) which eventually resulted in the invention of the rearranging Ig genes and the adaptive immune system of the higher vertebrates. Significant expansion of the complement system has been identified in the elasmobranchs (see fig. 3, Gross *et al.*, 1999) in which the rearranging Ig genes are first identified. The cyclostomes, which do not have rearranging Igs, do not have a typical higher vertebrate complement system (Nonaka *et al.*, 1984; Hanley *et al.*, 1992, Ishiguro *et al.*, 1992, Nonaka & Takahashi, 1992; Nonaka *et al.*, 1994). Perhaps the presence of an effective opsonization system in the form of the primitive complement components was a conducive "platform" onto which the nascent adaptive immune system could be advantageously exploited by the vertebrate ancestor which resulted in the expansion of both systems in the higher vertebrates. Continued analysis of the innate immune system in the purple sea urchin will further our understanding of the mechanisms by which the echinoderm immune response works and may provide keys for understanding how the multiple subsystems in the mammalian immune response came to function together.

ACKNOWLEDGEMENTS

The author would like to thank P.S. Gross, L.A. Clow, and D.P. Terwilliger for providing information on preliminary data presented in this review. P.S. Gross critically read and significantly improved the final manuscript

REFERENCES

Agrawal, A., Q.M. Eastman and D.G. Schats. 1998. Transposition mediated by RAG1 and RAG2 and it implications for the evolution of the immune system. *Nature* 394:744

Al-Sharif, W.Z., J.O. Sunyer, J.D. Lambris and L.C. Smith. 1998. Sea urchin coelomocytes specifically express a homologue of complement component C3. *J Immunol* 160:2983.

Bentley, D. R. 1988. Structural superfamilies of the complement system. *Expl Clin Immunogenet* 5:69.

Campbell, R. D., S. K. Law, K. B. M. Reid and R. B. Sim. 1988. Structure, organization, and regulation of the complement genes. *Ann Rev Immunol* 6:161.

Coffaro, K.A. 1979. Transplantation immunity in the sea urchin. Doctoral dissertation, University of California, Santa Cruz, CA.

DiScipio, R.G. 1992. Ultrastructure and interactions of complement factors H and I.. *J Immunol* 149:2592.

Dodds, A.W., X.-D. Ren, A.C. Willis and S.K.A. Law. 1996. The reaction mechanism of the internal thioester in the human complement component C4. *Nature* 379:177.

Edds, K.T. "Morphological and cytoskeletal transformation in sea urchin coelomocytes." In *Blood Cells of Marine Invertebrates: Experimental Systems in Cell Biology and Comparative Physiology*, W. D. Cohen, ed. A. R. Liss, New York, NY, 1985, 53.

Fearon, D.T. 1998. The complement system and adaptive immunity. *Semin Immunol* 10:355.

Fearon, D.T. and R.M. Locksley. 1996. The instructive role of innate immunity in the acquired immune response. *Science* 272:50

Finke, T., J.A. Theriot, K.R. Dise, G.F. Tomaselli and P.J. Goldschmidt-Clermont. 1994. Dynamic actin structures stabilized by profilin. *PNAS USA* 91:1510.

Gross, P.S., W.Z. Al-Sharif, L.A. Clow and L.C. Smith. 1999. Echinoderm immunity and the evolution of the complement system. *Dev Comp Immunol* in press

Hanley, P. J., J.W. Hook, D.A. Raftos, A.A. Gooley, R. Trent and R.L. Raison. 1992. Hagfish humoral defense protein exhibits structural and functional homology with mammalian complement components. *PNAS USA* 89:7910.

Hildemann, W.H. and Dix, T.G. 1972. Transplantation reactions of tropical Australian echinoderms. *Transplantation* 15:624.

Hiom, K., M. Melek and M. Gellert. 1998. DNA transposition by the RAG1 and RAG2 proteins: a possible source of oncogenic translocations. *Cell* 94:463.

Horiuchi, T., S. Kim, M. Matsumoto, I. Watanabe, S. Fujita and J.E. Volanakis. 1993. Human complement factor B: cDNA cloning, nucleotide sequencing, phenotype conversion by site-directed mutagenesis and expression. *Mol Immunol* 30:1587.

Hourcade, D.E., L.M. Wagner and T.J. Oglesby. 1995. Analysis of the short consensus repeats of human complement factor B by site-directed mutagenesis. *J Biol Chem* 270:19716.

Isaac, L. and D.E. Isenman. 1992. Structural requirements for thioester bond formation in human complement component C3. Reassessment of the role of thioester bond integrity on the conformation of C3. *J Biol Chem* 267:10062.

Ishiguro, H., D. Kobayashi, M. Suzuki, K. Titani, S. Tomonaga and Y. Kurosawa. 1992. Isolation of a hagfish gene that encodes a complement component. *EMBO J* 11:892.

Johnson, P.T. 1969. The coelomic elements of sea urchins (*Strongylocentrotus*). III. *In vitro* reaction to bacteria. *J Invert Pathol* 13:42.

Karp, R.D. and Hildemann, W.H. 1976. Specific allograft reactivity in the sea star *Dermasterias imbricata*. *Transplantation* 22:434.

Lambris, J. D., Z. Lao, J. Pang and J. Alsenz. 1993. Third component of trout complement. cDNA cloning and conservation of functional sites. *J Immunol.* 151:6123.

Lambris, J.D., M. Mavroidis and J.O. Sunyer. Phylogeny of third component of complement, C3. In *New aspects of complement structure and function.* A. Erdei, ed. R. G. Landes Co., Austin, Texas. 1994, 34.

Metchnikoff, E. *Lectures on the Comparative Pathology of Inflammation: Delivered at the Pasteur Institute in 1891.* trans. F.A. Starling and E.H. Starling, M.D. Kegan Paul, Trench, Trubner & Co., Ltd., London 1893.

Nonaka, M., R. Fugii, T. Kaidoh, S. Natsuume-Sadai, N. Yamaguchi and M. Takahashi. 1984. Purification of a lamprey complement protein homologous to the third complement of the mammalian complement system. *J Immunol* 133:3242.

Nonaka, M. and M. Takahashi. 1992 Complete complementary DNA sequence of the 3rd component of complement of Lamprey: implication for the evolution of thioester containing proteins. *J Immunol* 148:3290.

Nonaka, M., M. Takahashi and M. Sasaki. 1994. Molecular cloning of a lamprey homologue of the mammalian MHC class III gene, complement factor B. *J Immunol* 152:2263.

Nonaka, M., K. Azumi, X. Ji, C. Namikawa-Yamada, M. Sasaki, H. Saiga, A.W. Dodds, H. Sekine, M.K. Homma, J. Matsushita, Y Endo and T. Fujita. 1999. Opsonic complement component C3 in the solitary ascidian, *Halocynthia roretzi*. *J Immunol* 162:387.

Nonaka, M. and K. Azumi. 1999. Opsonic complement system of the solitary ascidian, *Halocynthia roretzi*. *Dev Comp Immunol* in press.

Raftos, D.A., J.W. Hook and R.L. Raison. 1992. Complement-like protein from the phylogenetically primitive vertebrate, *Eptatretus stouti*, is a humoral opsonin. *Comp Biochem Physiol* 103B:379.

Reid, K. M. B. and Campbell, R. D. 1993. Structure and organization of complement genes. *Immunol Med* 20:89.

Sim, R.B. and E. Sim. 1981. Autolytic fragmentation of complement components C3 and C4 under denaturing conditions, a property shared with a_2-macroglobulin. *Biochem J* 193:129.

Sim, R.B. and E. Sim. 1983. Autolytic fragmentation of complement components C3 and C4 and its relationship to covalent binding activity. *Ann NY Acad Sci.* 421:259.

Smith, L.C., R.J. Britten and E.H. Davidson. 1992. SpCoel1, a sea urchin profilin gene expressed specifically in coelomocytes in response to injury. *Mol Bio Cell* 3:403.

Smith, L.C. and E.H. Davidson. 1992. The echinoid immune system and the phylogenetic occurrence of immune mechanisms in deuterostomes. *Immunol Today* 13:356.

Smith, L.C. and E.H. Davidson. 1994. The echinoderm immune system: characters shared with vertebrate immune systems, and characters arising later in deuterostome phylogeny. In: *Primordial Immunity: Foundations for the Vertebrate Immune System,* G. Beck, E. L. Cooper, G. S. Habicht and J. J. Marchalonis, eds. *Ann NY Acad Sci* 712:213.

Smith, L.C., R.J. Britten and E.H. Davidson. 1995. Lipopolysaccharide activates the sea urchin immune system. *Dev. Comp. Immunol.* 19:217.

Smith, L.C., L. Chang, R.J. Britten and E.H. Davidson. 1996. Sea urchin genes expressed in activated coelomocytes are identified by expressed sequence tags (ESTs). Complement homologues and other putative immune response genes suggest immune system homology within the deuterostomes. *J Immunol* 156:593.

Smith, L.C., C.-S. Shih and S.G. Dachenhausen. 1998. Coelomocytes specifically express SpBf, a homologue of factor B, the second component in the sea urchin complement system. *J Immunol* 161:6784.

Smith, L.C., K. Azumi and M. Nonaka. 1999. Complement systems in invertebrates. The ancient alternative and lectin pathways. *Immunopharm* in press.

Sohn, R.H. and P. J. Goldschmidt-Clermont. 1994. Profilin: at the crossroads of signal transduction and the actin cytoskeleton. *Bioessays* 16:465.

Xu, Y. and J.E. Volanakis. 1997. Contribution of the complement control protein modules of C2 in C4b binding assessed by analysis of C2/Bf chimeras. *J Immunol* 158:5958.

PARALYTIC AND PLASMATOCYTE SPREADING PEPTIDE FROM *Manduca sexta*

Yang Wang, Haobo Jiang, and Michael R. Kanost

Department of Biochemistry, Kansas State University, Manhattan, KS, 66506, USA

INTRODUCTION

A family of hemolymph peptides with diverse biological activities has been identified in several Lepidopteran species. These peptides can cause rapid paralysis (Skinner *et al.*, 1991), block growth and development (Hayakawa, 1991), and stimulate plasmatocyte spreading and aggregation (Clark *et al.*, 1997). The 23-residue peptides are produced from larger inactive precursors in plasma. The precursor proteins are rapidly converted by a specific proteolysis to active molecules upon bleeding and perhaps after parasitization or infection.

We have found that paralytic peptide from *Manduca sexta* is also very active as a plasmatocyte attachment and spreading factor and that its injection suppresses bleeding. Like other members of this family (Hayakawa *et al.*, 1995; Hayakawa and Ohnishi, 1998; Clark *et al.*, 1998), *M. sexta* PP is synthesized as the carboxyl-terminal sequence of a larger precursor protein. The inactive precursor is present in larval plasma and is rapidly processed by an endogenous proteinase activity.

Phylogenetic Perspectives on the Vertebrate Immune System.
Edited by Beck *et al.*, Kluwer Academic/Plenum Publishers, 2001.

RESULTS AND DISCUSSION

We synthesized *Manduca sexta* paralytic peptide 1 (PP1) and found that within minutes after injecting the peptide into the hemocoel, it became very difficult to obtain even small amounts of hemolymph from the larvae. This result suggested that PP1 may be an important factor in hemostasis in hemolymph. Injection of PP1 also led plasmatocyte depletion in *M. sexta* and stimulated spreading and aggregation behavior of *M. sexta* plasmatocytes *in vitro.* Stimulation of plasmatocyte aggregation and adherence to the body wall may explain a decrease observed in the number of circulating plasmatocytes after injection of PP1. Such aggregates might rapidly form plugs in wounds to prevent bleeding.

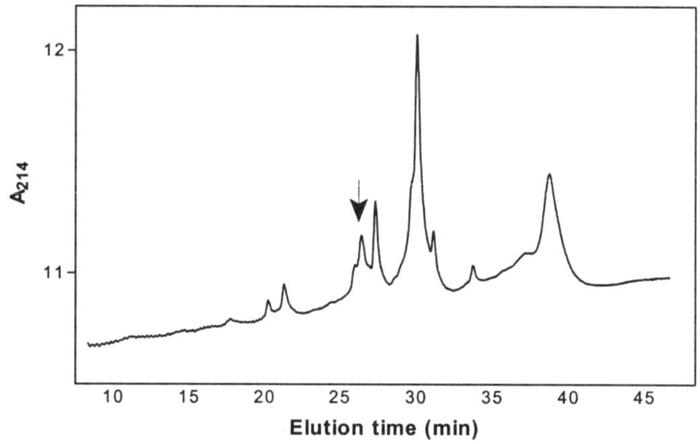

Figure 1. Purification of PP2 Released from its Recombinant Precursor *M. sexta* proPP2. Recombinant proPP2 was digested with Factor Xa at room temperature for 6 hours. After 95°C heating for 5 min to stop the reaction, the supernatant was filtered through centricon 10 device to remove proteins larger than 10 kDa. The flow through fraction was separated by reversed phase HPLC. A linear gradient of 15-35% acetonitrile in 0.1% trifluoroacetic acid was applied for 40 min. The arrow indicates the PP2 peak.

We cloned a cDNA for a *Manduca* paralytic peptide precursor, using polymerase chain reactions and cDNA library screening. The active 23-residue PP2 peptide encoded by this clone is at the carboxyl-terminal end of a precursor protein predicted to be 107 amino acid residues long after cleavage of a secretion signal peptide. A single proPP2 mRNA was present in fat body but not in hemocytes. The level of this mRNA was not affected by injection of bacteria into larvae. We produced recombinant proPP2 in *Escherichia coli* and purified the protein through Ni^{2+} affinity chromatograph, SP sepharose chromatograph and reversed phase HPLC, and used this protein to produce an antiserum. The antiserum detected proPP2 in plasma and was used to observe rapid proteolytic processing of proPP2 in plasma. After collection of plasma, the proPP band rapidly disappeared. But when we added *p*-aminobenzamidine, an inhibitor of serine proteinases with trypsin-like specificity, to the fresh plasma, the degradation of the proPP band was significantly delayed. This result suggested that after hemolymph is collected (or after wounding), an endogenous hemolymph proteinase processes proPP, resulting in production of active PP that can stimulate plasmatocyte adherence and spreading. The recombinant proPP2 band disappeared with the same kinetics as the natural proPP2, and its processing was similarly inhibited by *p*-aminobenzamidine. Thus, recombinant proPP2 promises to be useful as a substrate for assays to detect the proPP-processing proteinase in future work to purify and characterize this enzyme.

To test whether PP2 released from its precursor has the same biological activity as PP1, recombinant *M.sexta* proPP2 was digested with proteinase factor Xa. The released PP2 was purified by reversed phase HPLC (Fig.1). The peptide was confirmed to be PP2 by mass and N-terminal sequence analysis. We injected this purified PP2 into larvae and found that PP2 caused paralysis and decreased bleeding. In an *in vitro* assay, PP2 induced plasmatocyte spreading in the same manner as synthetic PP1.

REFERENCES

Clark, K.D., Pech, L.L., and Strand, M.R. (1997) Isolation and identification of a plasmatocyte-spreading peptide from the hemolymph of the lepidopteran insect *Pseudoplusia includens. J. Biol. Chem.* 272, 23440-23447.

Clark, K.D., Witherell, A., and Strand, M.R. (1998) Plasmatocyte spreading peptide is encoded by an mRNA differentially expressed in tissues of the moth *Pseudoplusia includens. Bioch. Biochem. Res. Commun.* 250, 479-485.

Hayakawa, Y. (1991) Structure of a growth-blocking peptide present in parasitized insect hemolymph. *J.Biol.Chem.* 266, 7982-7982.

Hayakawa, Y. and Ohnishi, A. (1998) Cell growth activity of growth-blocking peptide. *Biochem. Biophys. Res. Commun.* 250, 194-199.

Hayakawa, Y., Ohnishi, A., Yamanaka, A., Izumi, S., and Tomino, S. (1995) Molecular cloning and characterization of cDNA for insect biogenic peptide, growth blocking peptide. *FEBS Lett.* 376, 185-189.

Skinner, W.S., Dennis, P.A, Li, J.P., Summerfelt, R.M., Carney, R.L., and Quistad, G.B. (1991) Isolation and identification of paralytic peptides from hemolymph of the lepidopteran insects *Manduca sexta, Spodoptera exigua, and Heliothis virescens. J. Biol. Chem.* 266, 12873-12877.

INDEX OF CONTRIBUTORS

Ahmed, H., 275-287
Adelman, M.K., 13-30
Anderson, R.S., 131-139
Arisaka, F., 185-190
Armstrong, P.B., 141-160

Bartl, S., 1-12
Bayne, C.J., 209-222
Beauregard, K.A., 55-62, 63-69
Beck, G., ix-x, 41-54, 55-62, 63-69
Beisel, H.-G., 195-202
Bird, S., 89-94
Bode, W., 195-202

Chase, M.R., 349-362
Chisholm, J.R.S., 95-112
Chiu, H., 161-167
Cho, M.Y., 299-302, 303-307
Cohen, N., 237-249
Cooper , E.L., 169-183
Cossarizza, A., 169-183
Cueto, M., 77-87

Gokudan, S., 195-202
Govind, S., 161-167
Green, P., 229-236

Hong, S., 89-94
Howard, R.W., 265-273
Huber, R., 195-202

Inamori, K., 185-190
Iwaki, D., 185-190
Iwanaga, S., 185-190, 195-202

Jaqua, P.M., 13-30
Jiang, H., 313-317, 319-328, 373-375

Kanost, M.R., 191-194, 309-312, 313-317, 319-328, 373-375
Kawabata, S., 185-190, 195-202
Kawahara, T., 195-202
Kauschke, E., 169-183
Kimbrell, D.A., 251-263
Koori, K., 195-202
Koppenheffer, T.L., 203-207

Laing, K.J., 89-94
Lee, B.L., 299-302, 303-307
Lee, H.S., 299-302, 303-307
Lee, I.H., 71-76
Lehrer, R.I., 71-76
Lin, W., 55-62, 63-69
Loker, E.S., 209-222
Loo, P., 77-87
Lowenberger, C.A., 113-129

Ma, C., 309-312, 319-328
Mahajan, D., 229-236
Marchalonis, J.J., 13-30
Ménoret, A., 237-249
Menzel, L., 71-76
Mizunoe, Y., 195-202
Muta, T., 195-202

Nagira, T., 185-190
Nair, S., 229-236
Nappi, A.J., 329-348
Natori , S., 223-228
Newton, R., 229-236

O'Leary, N., 275-287

Pancer, Z., 31-40
Pearce, S., 229-236
Peppard, J.V., 77-87

Peters, R., 229-236
Pomponi, S., 77-87

Quesenberry, M.S., 275-287
Quigley, J.P., 141-160

Raftos, D., 229-236
Robbins, J., 229-236
Robert, J., 237-249
Rodriguez, A., 251-263

Saito, T., 185-190
Sarazin, P.M., 13-30
Schluter, S.F., 13-30
Secombes, C.J., 89-94
Seki, N., 195-202
Sills, M.A., 77-87
Smith, L. C., 363-372
Smith, V.J., 95-112
Sorrentino, R.P., 161-167
Srivastava, P.K., 237-249
Stanley, D.W., 265-273
Sugumaran, M., 289-298, 349-362

Truong, N.T., 41-54, 55-62, 63-69
Tsuda, R., 195-202

Vass, E., 329-348
Vasta, G.R., 275-287

Wai, S.N., 195-202
Wang, Y., 313-317, 319-328, 373-375
Waring, A., 71-76
Wennogle, L., 77-87
Wright, A., 77-87
Wu, C.-Y., 251-263

Yu, X.-Q., 191-194, 319-328

Zeitler, B.J., 13-30
Zhang, H., 55-62, 63-69
Zhao, C., 71-76
Zhou, Z., 251-263
Zhu, Y., 319-328
Zou, J., 89-94

INDEX

N-acetylgalactosamine binding lectin, 192
Actin, 47, 364
Acute phase proteins, 16, 101, 105
Aedes aegypti, 117–124
Aedes albopictus, 119–121
Agglutinin, 170, 180
Alkaline protease, 142
Ampicillin, 47
Anaphylotoxin, 233
Anopheles gambiae, 117–121, 291, 314, 350
Antibacterial proteins, 269, 289
Antigen binding cleft, (ABC), 1, 2, 4, 6, 9
Antimicorbial petides, 55, 56, 59, 62, 71, 102, 105, 106, 109, 113–115, 155, 170, 177, 185, 191, 196, 252, 329, 330
Antimicrobial proteins, 95, 97, 98, 102, 105, 107, 191, 275, 321, 322
Apoptosis, 171, 181, 279
Aprotinin 57, 58
Arachidonic acid, 265, 266, 268, 337
Armigeres subalbatus, 121, 291, 350
Aspartyl protease, 143, 145
Attacin, 41, 63, 122, 337

Bactenecin, 105
N-Benzoyl arginine *p*-nitroanilide, (BAPNA) 148–150
Big Bang 13, 14, 18, 26, 246
Blaberus discoidalis, 193, 281, 291, 350
Bombyx mori, 119, 121, 171, 192, 193, 269, 281, 290–292, 309, 315, 339, 350
Botulinus toxin, 144, 146

Caenorhabditis elegans, 14, 32, 370
Calmodulin, 331, 332
Cathepsin, 143
CD4, 5, 6, 9
CD5, 32, 34
CD6, 32, 34
CD14, 226
CD14 receptor, 101
Calliphora erythrocephala, 290, 350
Callyspongia, 77, 78, 80, 83–86
Carcinin, 106
Carcinus maenas, 87–100, 103–108
Cardiotrophin-1 (CT-1) 79
Caspase-1, 90
Cecropin, 41, 63, 114, 118–123, 269, 337
Cell adhesion molecule (CAM), 34
Chaperone, 15, 237
Chemokine, 92, 93
Chondrichthyans, 7
Chymotrypsin, 148, 326
Ciliary neurotropic factor (CNTF), 78
Clavanin, 71–75
Clavaspirin, 75, 76
Clavelina picta, 276–279, 282, 284, 285
Clustal W 350, 351, 367
Cobra venom factor (CoV), 203–207
Coelomic cytosolic factor (CCF) , 171–173
Coelomic fluid, 32, 170, 173–178
Coelomocytes, 32, 34–39, 56, 364, 365
Collagen, 201
Collagenous lectin, 229
Collectin, 192, 229, 231, 235
Combinatorial immune response, 13–16, 18
Complement, 16, 229, 235, 282, 332, 338
Complement activation, 229, 276, 277, 280–284, 293
Complement cascade, 34, 230, 232
Complement component, 232, 233, 338, 366–369
  C1q, 231, C2 367, C3, 229, 230, 233–235, 280, 281, 283, 284, 365, 366, 368–370, C4, 281, 368
Complement control protein (CCP) 366
Complement function, 370
Complement-like protein, 233

Complement pathway, 319
Complement proteins, 32, 294, 340
Complement system, 229, 232, 341, 363
Conglutinin, 231
C3 convertase, 365, 368
C5 convertase, 341
Crangon crangon, 98
C-reactive protein, 16, 21, 101, 105, 282
C-type lectin, 34, 192, 193, 215, 224, 225, 276–280, 282, 309, 322, 324, 325
Cucumari frondosa, 55–57, 62
Cucumaria japonicus, 278
Culex pipiens, 117
CXC-R2- receptor, 92
Cyclooxygenase, 266–268, 340
Cyclooxygenase inhibitor, 270, 271
Cysteine protease, 142–146, 155
Cytochrome c oxidases, 16
Cytochrome P450, 331, 332
Cytochrome P450 epoxygenase, 265, 267
Cytochrome P452, 16
Cytokine, 16, 42, 44, 45, 46, 56, 63, 77, 79, 82–85, 89, 320, 331, 332, 338, 341
Cytokine receptor, 79
Cytotoxic quinones, 330, 350

Defensin, 114–117, 121, 122, 124
Deuterostomes, 14, 15, 19, 20, 32, 230, 370
DHICA oxidase, 334
Didemnum teratanum, 276, 277
Digenean, 209, 212
5,6-Dihydroxyindole (DHI), 334–336
5,6-Dihydroxyindole–2-carboxylic acid (DHICA), 334–336
Dinitrophenyl hapten, 21, 23
Diptericin, 45, 337
Dopachrome isomerase, 294, 295
Dopachrome tautomerase, 334, 335
Dorothy, 251–259
Drosophila, 14, 50, 51, 115, 119, 121, 124, 161–163, 166, 210, 225, 251, 252, 261, 315, 320–322, 333, 335, 337, 339
Drosophila melanogaster, 119–121, 161, 162, 223, 225, 278, 290, 291, 335, 336, 349–352, 355, 357, 359, 360
Drosocin, 122, 337
Drosomycin, 115, 122, 337

Echinoderm, 55,56, 363, 364
EDTA, 46, 72, 369
EGTA, 369, 370
Eicosanoid, 265–271, 338
Eisenia fetida, 170–178
Eiseniapore, 178–190
Elastase, 143, 148
ELISA, 23, 24, 81, 188, 189
Encapsulation related factor (ERP), 303–306
Endocytosis, 34
Endotoxin, 42
Epitope promiscuity, 20
EST, 350, 351
Exon shuffling, 18
Exotoxin, 146, 155
Extracellular protease, 145

Ferritin, 104, 105, 145
Fetidin, 174, 175
Fibrinogen, 200, 201, 209
Fibrinogen related proteins (FREPs), 209, 210, 218–220
Fibronectin, 14, 79

β-galactosidase, 252, 253, 258–260
Gal binding lectin, 277–279
Galectin, 276, 277, 279, 284
Galleria mellonella, 45, 269, 290, 350
Ganaspis xanthopoda, 161–166
Genscan, 351
β-1,3-glucan, 171, 191, 192, 196, 213, 215, 281, 282, 294, 309, 323, 324, 331
Glucanase, 171
β-1,3-glucanase, 310, 324
β-1,3-glucan binding lectin, 282
β-1,3-glucan binding protein, 281, 309–311
β-1,3-glucan recognizing protein (GRP), 324, 325, 339
Glucose oxidase, 330
Gp-130, 78, 79, 85, 86
Granule exocytosis, 171
Granulocyte colony stimulating factor (GCSR), 79
Granulocytin, 226, 278

Haptoglobin, 81
Heat shock proteins (HSPs), 237–239, 243, 245
  HSPs 70, 239
  gp 96, 237, 239–245
Heliomicin, 337
Hemagglutination, 186, 200, 224
Hemagglutinin, 199, 200, 337
Hemocytin, 338
Hemolin, 322, 338, 339, 341
Hemolysin, 176
Hemomucin, 338
Holotrichia diomphalia, 290, 291, 299, 350
Homarus americanus, 97, 99, 276
Hyalophora cecropia, 119,–121, 290, 350
Hydrogen peroxide, 131, 132, 137, 330–334, 338, 339
Hydroxyeicosatetraenoic acid, 265, 267, 271
Hydroxyl radical, 131, 330

Hypervariable region (HVR), 2, 4, 6, 7, 9
Hyphantria cunea, 192, 291, 350
Hypochlorous acid (HOCl) 131, 330, 333

Immulectin, 192–194
Immunoglobulin (Ig) 2, 5, 14–16, 18–20, 63, 210, 218, 220, 230, 279, 285, 282, 341, 342, 370, 371
 IgA, 145
 IgG, 21, 305, 330
 IgM, 20–27
 Ig domain, 86
 Ig like domain 78, 79
 immunoglobulin superfamily, 210, 218
Innate immunity, 16, 147, 275, 277, 282, 329, 330
Integrin, 215
Interferon (INF), 42, 43, 64, 341
Interleukin, 42, 77, 238, 241, 245, 341
Interleukin-1 (IL-1), 41–51, 56,63–68, 81, 89–91, 101, 294, 332, 341
Interleukin-1 receptor, 321
Interleulin-2 (IL-2), 44, 45, 89, 341
Interleukin-4 (IL-4), 332
Interleukin–6 (IL-6), 46, 77, 78, 80–85, 341
Interleukin–6 inhibitor, 80
Interleukin-6 receptor (IL-6R), 78, 79, 85, 86
Interleukin-8 (IL-8), 17, 92
Interleukin-10 (IL-10), 332
Interleukin-11 (IL-11), 79
Interleukin-15, (IL-15), 45

Kinases, 14, 15, 48, 78, 213

Lactoferrin, 101
Laminarin, 173, 192, 281, 282
Lectins, 15, 185, 192, 196, 199, 200, 209, 219, 223–227, 231, 275, 277, 280, 282, 332, 338, 339, 340, 341
Leptin receptor, 79
Leptopilina heterotoma, 162–165
Leukemia inhibitory factor (LIF), 78, 85
Leukotactin, 92
Leukotrienes, 265–277
Limulin, 341
Limulus hemagglutinin, 22
Limulus polyphemus, 101, 147–154, 171, 276
Lipophorin, 338, 339
Lipopolysaccharide (LPS) 32, 42, 45, 64, 90, 91, 98, 100, 102, 107, 172, 173, 185–188, 191–194, 196, 197, 226, 276, 282, 309, 323–325, 331, 332, 336, 337, 364
Lipopolysaccharide binding lectin (LBL), 192, 193, 277, 325
Lipopolysaccharide binding protein (LBP), 101, 102, 192, 193
Lipoprotein related protein (LRP), 150–155
Lipoxygenase, 265–268, 271, 340
Locusta migratoria, 290, 292, 350
Lumbricin, 177
Lumbricus rubellus, 177
Lumbricus terrestris, 174
Luminol augmented chemiluminescence (LuCL), 132, 133, 135, 136
Lymphocyte activating factor (LAF), 42, 43
Lymphotactin, 92
Lysenin, 170, 175–178
Lysozyme, 41, 63, 99, 100, 102–104, 155, 269
Lytechnius pictus, 364

α2-macroglobulin, 16, 142, 147–155, 281, 340, 366, 368, 369
α2-macroglobulin receptor, 150, 151, 153–155
Magainin, 122, 123
Major histocompatibility complex (MHC), 1, 2, 4–9, 14–16, 18, 213, 237–247, 341
Manduca sexta, 46, 51, 63, 64, 66–68, 119–121, 191–193, 266, 282, 290–295, 309, 310, 313, 314, 319–321, 323–326, 332, 373–375
Mannose binding lectin (MBL), 276–279, 281–285, 324
Mannose binding protein (MBP), 192, 229, 230, 232
MBL associated serine protease (MASP), 281
McClade, 351
Melanin, 289, 292, 301, 313, 323, 334, 336, 340, 349, 350
Melanization, 161, 252, 280, 282, 289, 290, 293, 301, 303, 314, 323, 330, 333, 334, 350
Melanogenic complex, 294
Melanotic capsule, 162, 163, 165
Melanotic encapsulation, 113, 121, 333
Melanotic encapsule, 350
Melanotic tumor, 162
Metallo proteases, 143
Metallothioneins, 100, 104
Metchnikowin, 122, 337
β2-microglobulin, 7, 341
Monocyte chemotactic proteins (MCP), 92
Mosquitoes, 113–115, 122, 123
Musca domestica, 290, 350
Myeloperoxidase, 330, 331, 333

NADPH oxidase, 131, 132, 137, 330, 333
Nereis diversicolor, 173
NF-κB, 16, 321, 332
Nitric oxide (NO), 330, 334
Nitric oxide synthase (NOS), 331, 332, 336

Oncostatin M, 78, 85
Ovalbumin, 24
Oxidative stress, 16

Pacifastacus leniusculus, 99, 104, 281
PAHs, 134
p-aminobenzamidine, 374, 375
papain, 148
Pap D domain, 15
Paralytic peptide, 323, 324, 374, 375
Penaeidin, 106
Pentraxin, 16, 277
Peptidoglycan, 100, 191, 281, 309, 339
Peptidoglycan binding protein (PGBP), 339
Perforin, 170, 171, 174
Periplaneta americana, 193
Peroxidase, 175, 335, 336, 338, 340
Peroxidasin, 339
Peroxynitrite anion, 331, 334
Phagocytic cells, 15, 16, 42, 363
Phagocytosis, 41, 45, 55, 63, 101, 191, 280, 284, 288, 329, 339–341, 363, 370
Phenoloxidase (PO), 98, 155,192, 194, 196, 289–295, 300–302, 313, 320, 323, 334, 335, 349–352
Phenoloxidase inhibitor, 295
Phorbol myristyl acetate (PMA), 132
Phormia terranovae, 115
Phosphatase, 14
Phospholipase A2, 267
Phosphorylcholine, 21, 22, 25, 26, 101
Phylogenetic analysis, 7, 8, 117, 121, 367
Phylogenetic tree, 7–9, 18, 49, 79, 93, 367
Phylogeny, 89, 210
Plaminogen, 143, 144, 146
Plaminogen activator, 143, 146
Plasmin, 148
Plasmodium, 122, 123, 144, 145, 314
Projectin, 14
Prophenoloxidase (PPO, ProPO), 172, 173, 192, 269, 282, 290, 292, 295, 296, 299–302, 309, 313, 314, 320, 323, 324, 335, 340, 349–360
Prophenoloxidase activating enzyme (PPAE, PAE), 291, 315, 335
Prophenoloxidase activating factor, (PPAF), 299, 315
Prophenoloxidase activating protease (PAP), 314–316, 320, 324–326
Prophenoloxidase activation, 281, 290–293, 314, 316, 323, 326
Prophenoloxidase cascade, 280, 282, 289, 290, 295, 296, 299, 326, 339, 340
Prostaglandin H synthase, 340
Prostaglandins (PG), 265–271
Prostyelin, 74
Protease, 142–146, 149–152, 154, 171, 173, 290–292, 313–315, 319–323, 325, 339
Protease inhibitors, 141, 146, 147, 174, 175, 185, 290, 292, 326
Protein kinase, 14
Proteosomes, 352
Protostomes, 14, 15, 19, 20

Quinone isomerase, 290, 294, 295, 349
Quinone methide isomerase, 290, 294

Reactive intermediates of nitrogen (RNI), 331, 333, 334, 337
Reactive intermediates of oxygen (ROI), 330, 333, 336–338
Reactive nitrogen species (RNS), 131
Reactive oxygen species (ROS), 131–137
Receptor associated protein (RAP), 153
Recombinase, 14, 19
Recombinase activator genes (RAGs), 14, 16, 18, 19

S-type lectin, 276
Sarcophaga bullata, 290–293, 350, 351
Sarcophaga peregrina, 115, 119, 120, 223–227, 278
Sarcotoxin, 337
Scavenger receptor cysteine-rich repeats (SRCR), 32–39
Schistosoma mansoni, 212, 216
Sclerotization, 290, 293, 313, 334, 335, 349
Scorpion toxins, 115
Scyllin, 107
Sea cucumber, 56–59, 62
Sea urchin, 31, 32, 37–39, 364, 366, 367
Selectin, 220, 283
Serine esterase, 171
Serine protease, 146, 170, 185, 229, 232, 291, 292, 315, 319–321, 326, 324, 332, 366, 369, 374
Serine protease inhibitor, 233–235, 292, 325, 326
Serpin, 292, 315, 316, 325, 326
Shiva, 122, 123
Singlet oxygen, 131
Skeletal muscle c-protein, 14
Smooth muscle light chain kinase, 14
Soybean trypsin inhibitor, 149, 150, 175
Spätzle, 321, 322
Stichopus japonicus, 278
Strongylocentrotus purpuratus, 31, 32, 36, 58
Styela clava, 71, 75, 120, 121
Styela plicata, 230–234
Styelin, 71, 74
Subtilisin, 148
Superoxide, 45, 131, 132, 134, 135, 137, 279, 330, 333
Superoxide dismutase, 19, 330, 333
Synaptobrevin, 146

Tachylectin, 186–189, 195–200
Tachypleus tridentatus, 185, 195, 196
T-cell receptor, 1, 14–19, 21, 23, 24
Teichoic acid, 21
Tenebrio molitor, 290, 291, 299, 300, 301, 303, 306
Tetanus toxin, 144, 146
Thanatin, 337
Themolysin, 148
Thiolester, 150–152, 368, 369
Thyroglobulin, 21, 23, 25
Titin, 14
Toll, 16, 124, 321, 322
Toll-like receptor, 226
Toll receptor, 101
Transferrin, 101, 104, 115, 119, 121, 122, 145
Transforming growth factor (TGF), 89, 91, 92, 332
Tributyl tin, 133–135
Trichoplusia ni, 310
Trypsin, 81, 82, 144, 148–151, 153, 175, 322
Tumor necrosis factor (TNF), 42, 45, 64, 226, 332, 341
Twitchin, 14
Tyrosinase, 294, 334–336
Tyrosinase related protein 1 (TRP-1, DHICA oxidase), 334
Tyrosinase related protein 2 (TRP-2, dopachrome tautomerase), 334, 335
Tyrosine phospharylation, 78

UDP-glucoronosyl transferase, 254, 256, 257, 260, 261
UDP-glycosyl transferase, 251, 254, 256, 260

Vitamin D, 17
Vitamin D receptor (VDR) 17, 19

Xanthine oxidase, 330
Xenopus, 89, 90, 92, 240–245, 280
Xebra fish, 6, 7

Zinc finger motif, 17
Zymosan, 136, 191, 233, 234